THE NAPHTHYRIDINES

This Is the Sixty-Third Volume in the Series
THE CHEMISTRY OF HETEROCYCLIC COMPOUNDS

THE CHEMISTRY OF HETEROCYCLIC COMPOUNDS

A SERIES OF MONOGRAPHS

EDWARD C. TAYLOR and PETER WIPF, *Editors*

ARNOLD WEISSBERGER, *Founding Editor*

THE NAPHTHYRIDINES

D. J. Brown
Research School of Chemistry
Australian National University
Canberra

AN INTERSCIENCE PUBLICATION
JOHN WILEY & SONS, INC.

Copyright © 2008 by John Wiley & Sons, Inc. All rights reserved.

Published by John Wiley & Sons, Inc., Hoboken, New Jersey.
Published simultaneously in Canada.

No part of this publication may be reproduced, stored in a retrieval system, or transmitted in any form or by any means, electronic, mechanical, photocopying, recording, scanning, or otherwise, except as permitted under Section 107 or 108 of the 1976 United States Copyright Act, without either the prior written permission of the Publisher, or authorization through payment of the appropriate per-copy fee to the Copyright Clearance Center, Inc., 222 Rosewood Drive, Danvers, MA 01923, (978) 750-8400, fax (978) 750-4470, or on the web at www.copyright.com. Requests to the Publisher for permission should be addressed to the Permissions Department, John Wiley & Sons, Inc., 111 River Street, Hoboken, NJ 07030, (201) 748-6011, fax (201) 748-6008, or online at http://www.wiley.com/go/permission.

Limit of Liability/Disclaimer of Warranty: While the publisher and author have used their best efforts in preparing this book, they make no representations or warranties with respect to the accuracy or completeness of the contents of this book and specifically disclaim any implied warranties of merchantability or fitness for a particular purpose. No warranty may be created or extended by sales representatives or written sales materials. The advice and strategies contained herein may not be suitable for your situation. You should consult with a professional where appropriate. Neither the publisher nor author shall be liable for any loss of profit or any other commercial damages, including but not limited to special, incidental, consequential, or other damages.

For general information on our other products and services or for technical support, please contact our Customer Care Department within the United States at (800) 762-2974, outside the United States at (317) 572-3993 or fax (317) 572-4002.

Wiley also publishes its books in a variety of electronic formats. Some content that appears in print may not be available in electronic formats. For more information about Wiley products, visit our web site at www.wiley.com.

Wiley Bicentennical Logo: Richard J. Pacifico

Library of Congress Cataloging Number: 96-6182
Includes index.
ISBN: 978-0-471-75159-5

Classification Number: QD401.F96

Printed in the United States of America

10 9 8 7 6 5 4 3 2 1

*To the memory of
Marian Wozniak, a prolific contributor
to naphthyridine research*

The Chemistry of Heterocyclic Compounds
Introduction to the Series

The chemistry of heterocyclic compounds is one of the most complex and intriguing branches of organic chemistry, of equal interest for its theoretical implications, for the diversity of its synthetic procedures, and for the physiological and industrial significance of heterocycles.

The Chemistry of Heterocyclic Compounds has been published since 1950 under the initial editorship of Arnold Weissberger, and later, until his death in 1984, under the joint editorship of Arnold Weissberger and Edward C. Taylor. In 1997, Peter Wipf joined Prof. Taylor as editor. This series attempts to make the extraordinarily complex and diverse field of heterocyclic chemistry as organized and readily accessible as possible. Each volume has traditionally dealt with syntheses, reactions, properties, structure, physical chemistry, and utility of compounds belonging to a specific ring system or class (e.g., pyridines, thiophenes, pyrimidines, three-membered ring systems). This series has become the basic reference collection for information on heterocyclic compounds.

Many broader aspects of heterocyclic chemistry are recognized as disciplines of general significance that impinge on almost all aspects of modern organic chemistry, medicinal chemistry, and biochemistry, and for this reason we initiated about 1971 a parallel series entitled *General Heterocyclic Chemistry*, which treated such topics as nuclear magnetic resonance, mass spectra, and photochemistry of heterocyclic compounds, the utility of heterocycles in organic synthesis, and the synthesis of heterocycles by means of 1,3-dipolar cycloaddition reactions. These volumes were intended to be of interest to all organic, medicinal, and biochemically oriented chemists, as well as to those whose particular concern is heterocyclic chemistry. It has, however, become increasingly clear that the above distinction between the two series was unnecessary and somewhat confusing, and we have therefore elected to discontinue *General Heterocyclic Chemistry* and to publish all forthcoming volumes in this general area in *The Chemistry of Heterocyclic Compounds* series.

Dr. Des J. Brown is again to be applauded and profoundly thanked for still another fine contribution to the literature of heterocyclic chemistry. This volume on The *Naphthyridines* is the first book devoted in its entirety to the six pyridopyridine ring systems, and it covers the last 80 years of the literature until 2007. It must be noted with admiration that many of the books in this series that have come to be regarded as classics in heterocyclic chemistry (*The Pyrimidines, The Pyrimidines*

Supplement I, The Pyrimidines Supplement II, Pteridines, Quinazolines Supplement I, The Quinoxalines Supplement II, The Pyrazines Supplement I, and Cinnolines and Phthalazines) are also from the pen of Dr. Brown.

Department of Chemistry
Princeton University
Princeton, New Jersey

EDWARD C. TAYLOR

Department of Chemistry
University of Pittsburgh
Pittsburgh, Pennsylvania

PETER WIPF

Preface

This is the first book devoted entirely to the chemistry of the six naphthyridine systems.

Because of structural uncertainties and confusing nomenclature, much of the early literature is, frankly, of little more than historical interest. However, after 1930, when most authors adopted the present *Chemical Abstracts* naming (see "Note on Nomenclature" that follows), a reliable body of literature gradually accumulated on all six systems, albeit at very different rates. For example, data on the 1,8-naphthyridines greatly outnumbered those on the other systems, probably as a result of the discovery of significant antimicrobial properties associated with nalidixic acid as well as the intrinsic suitability of the 1,8-system for metal complexation.

In this book, the respective volumes of reported data have led to treatment of the first four of the other naphthyridines in seven chapters each, whereas each-naphthyridine system is covered in a single rather long chapter. The appendix tables aim to list all simple naphthyridines, along with an indication of their physical properties, that have been reported before 2006.

I am greatly indebted to the Dean of the Research School of Chemistry, Professor Denis Evans, for the provision of postretirement facilities within the School; to the Librarian, Joan Smith, for her kindly assistance in all library matters; and to my wife, Jan, for her patient encouragement and practical help during the years of writing.

Research School of Chemistry DES J. BROWN
Australian National University
Canberra

Note on Chemical Nomenclature

The six possible pyridopyridine systems have been known almost universally since circa 1930 as 1,5- (**1**), 1,6- (**2**), 1,7- (**3**), 1,8- (**4**), 2,6- (**5**), and 2,7-naphthyridine (**6**), However, the occasional use of an appropriate pyridopyridine or diazanaphthalene may still be found in some publications, Historically, the word "naphthyridin(e)" was coined by Arnold Reissert in 1893 specifically for the 1,8-naphthyridine system (**4**) and was so used for some years, especially in the German literature. Other terms, such as "isonaphthyridine" [for 1,5-naphthyridine (**1**)], "benzodiazines" (very misleading), "pyridinopyridines", "2,5-naphthyridine" [for 1,6-naphthyridine (**2**)], and "copyrin(e)"[13,40] or copurine[39] [for 2,7-naphthyridine (**6**)] have appeared in the literature.

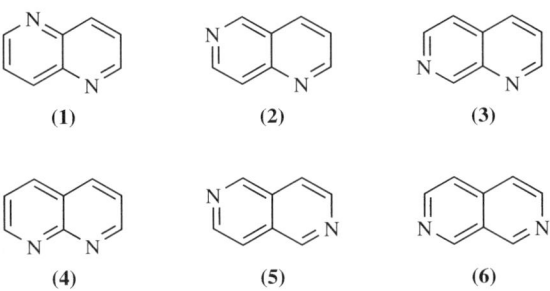

In this book, all such systems are designated as the appropriate naphthyridines; other ring systems are also named and numbered according to recommendations of the Chemical Abstracts Service [*Ring Systems Handbook* (eds. anonymous, American Chemical Society, Columbus, Ohio, 1998 edition and supplements)]; and general chemical nomenclature follows current IUPAC rules [*Nomenclature of Organic Chemistry, Sections A–E, H* (J. Rigaudy and S. P. Klesney, eds., Pergamon, Oxford, 1979)], with one important exception: in order to keep "naphthyridine" as the principal part of each name, those groups that would normally qualify as principal suffixes but are not attached directly to the nucleus are rendered as prefixes. Thus, 3-carboxymethyl-1,5-naphthyridin-2(1*H*)-one would be used instead of 2-(2-oxo-1,2-dihydro-1,5-naphthyridin-3-yl)acetic acid; secondary, tertiary, or quaternary amino groups are also rendered as prefixes.

Contents

CHAPTER 1 PRIMARY SYNTHESES OF 1,5-NAPHTHYRIDINES   1

1.1 From a Single Aliphatic Substrate   1
1.2 From a Single Pyridine Substrate   2
    1.2.1 By Formation of the N1,C2-Bond   2
    1.2.2 By Formation of the C3,C4-Bond   3
    1.2.3 By Formation of the C4,C4a-Bond   4
1.3 From a Pyridine Substrate with One Synthon   5
    1.3.1 Where the Synthon Supplies One Atom   5
    1.3.2 Where the Synthon Supplies Two Atoms   6
    1.3.3 Where the Synthon Supplies Three Atoms   7
1.4 From a Pyridine Substrate and Two Synthons   9
1.5 From Other Heterocyclic Substrates   10

CHAPTER 2 1,5-NAPHTHYRIDINE, ALKYL-1,5-NAPHTHYRIDINES, AND ARYL-1,5-NAPHTHYRIDINES   13

2.1 1,5-Naphthyridine   13
    2.1.1 Preparation of 1,5-Naphthyridine   13
    2.1.2 Properties of 1,5-Naphthyridine   14
    2.1.3 Reactions of 1,5-Naphthyridine   16
2.2 Alkyl- and Aryl-1,5-Naphthyridines   19
    2.2.1 Preparation of Alkyl- and Aryl-1,5-Naphthyridines   20
    2.2.2 Reactions of Alkyl- and Aryl-1,5-Naphthyridines   22

CHAPTER 3 HALOGENO-1,5-NAPHTHYRIDINES   25

3.1 Preparation of Halogeno-1,5-Naphthyridines   25
    3.1.1 By Direct Halogenation   25
    3.1.2 By Halogenolysis of 1,5-Naphthyridinones or the Like   26
    3.1.3 By the Meissenheimer Reaction on 1,5-Naphthyridine *N*-Oxides   29
    3.1.4 By Miscellaneous Procedures   31
3.2 Reactions of Halogeno-1,5-Naphthyridines   31
    3.2.1 Alcoholysis or Phenolysis of Halogeno-1,5-Naphthyridines   31
    3.2.2 Aminolysis of Halogeno-1,5-Naphthyridines   34
    3.2.3 Other Reactions of Halogeno-1,5-Naphthyridines   38

CHAPTER 4 OXY-1,5-NAPHTHYRIDINES   43

4.1 Tautomeric 1,5-Naphthyridinones and Extranuclear
    Hydroxy-1,5-Naphthyridines   43
    4.1.1 Preparation of Tautomeric 1,5-Naphthyridinones and the Like 44
    4.1.2 Reactions of Tautomeric 1,5-Naphthyridinones and the Like   46
4.2 Alkoxy- and Aryloxy-1,5-Naphthyridines   48
4.3 Nontautomeric 1,5-Naphthyridinones   49
4.4 1,5-Naphthyridine *N*-Oxides   50

CHAPTER 5 THIO-1,5-NAPHTHYRIDINES   53

CHAPTER 6 NITRO-, AMINO-, AND RELATED 1,5-NAPHTHYRIDINES   55

6.1 Nitro-1,5-Naphthyridines   55
    6.1.1 Preparation of Nitro-1,5-Naphthyridines   55
    6.1.2 Reactions of Nitro-1,5-Naphthyridines   56
6.2 Amino- and (Substituted-Amino)-1,5-Naphthyridines   57
    6.2.1 Preparation of Amino-1,5-Naphthyridines   58
    6.2.2 Reactions of Amino-1,5-Naphthyridines   58

CHAPTER 7 1,5-NAPHTHYRIDINECARBOXYLIC ACIDS AND RELATED DERIVATIVES   61

7.1 1,5-Naphthyridinecarboxylic Acids   61
7.2 1,5-Naphthyridinecarboxylic Esters   64
7.3 1,5-Naphthyridinecarboxamides, Carbonitriles, Carbaldehydes,
    and Ketones   65

CHAPTER 8 PRIMARY SYNTHESES OF 1,6-NAPHTHYRIDINES   67

8.1 By Condensation of Two or More Aliphatic Substrates/Synthons   67
8.2 From a Single Pyridine Substrate   69
8.3 From a Pyridine Substrate with One Synthon   75
    8.3.1 Where the Synthon Supplies One Ring Atom   76
    8.3.2 Where the Synthon Supplies Two Ring Atoms   78
    8.3.3 Where the Synthon Supplies Three or More Ring Atoms   81
8.4 From a Pyridine Substrate with Two or More Synthons   83
8.5 From Other Heterocyclic Systems   84

CHAPTER 9 1,6-NAPHTHYRIDINE, ALKYL-1,6-NAPHTHYRIDINES, AND ARYL-1,6-NAPHTHYRIDINES   91

9.1 1,6-Naphthyridine   91
    9.1.1 Preparation of 1,6-Naphthyridine   91
    9.1.2 Properties of 1,6-Naphthyridine   93
    9.1.3 Reactions of 1,6-Naphthyridines   94

9.2	Alkyl- and Aryl-1,6-Naphthyridines	97
	9.2.1 Preparation of Alkyl- and Aryl-1,6-Naphthyridines	97
	9.2.2 Reactions of Alkyl- and Aryl-1,6-Naphthyridines	99

CHAPTER 10 HALOGENO-1,6-NAPHTHYRIDINES   103

10.1	Preparation of Halogeno-1,6-Naphthyridines	103
	10.1.1 By Direct Halogenation	103
	10.1.2 By Halogenolysis of 1,6-Naphthyridinones or the Like	104
	10.1.3 By Other Methods	107
10.2	Reactions of Halogeno-1,6-Naphthyridines	108
	10.2.1 Alcoholysis or Phenolysis of Halogeno-1,6-Naphthyridines	108
	10.2.2 Aminolysis of Halogeno-1,6-Naphthyridines	110
	10.2.3 Dehalogenation of Halogeno-1,6-Naphthyridines	111
	10.2.4 Other Reactions of Halogeno-1,6-Naphthyridines	113

CHAPTER 11 OXY-1,6-NAPHTHYRIDINES   115

11.1	Tautomeric/Nontautomeric 1,6-Naphthyridinones and Extranuclear Hydroxy-1,6-Naphthyridines	115
	11.1.1 Preparation of 1,6-Naphthyridinones and the Like	116
	11.1.2 Reactions of 1,6-Naphthyridinones and the Like	118
11.2	Alkoxy- and Acyloxy-1,6-Naphthyridines	120
11.3	1,6-Naphthyridine *N*-Oxides	121

CHAPTER 12 THIO-1,6-NAPHTHYRIDINES   125

CHAPTER 13 NITRO-, AMINO-, AND RELATED 1,6-NAPHTHYRIDINES   127

13.1	Nitro-1,6-Naphthyridines	127
	13.1.1 Preparation of Nitro-1,6-Naphthyridines	127
	13.1.2 Reactions of Nitro-1,6-Naphthyridines	128
13.2	Amino- and (Substituted-Amino)-1,6-Naphthyridines	129
	13.2.1 Preparation of Amino-1,6-Naphthyridines	130
	13.2.2 Reactions of Amino-1,6-Naphthyridines	131

CHAPTER 14 1,6-NAPHTHYRIDINECARBOXYLIC ACIDS AND RELATED DERIVATIVES   135

14.1	1,6-Naphthyridinecarboxylic Acids	135
	14.1.1 Preparation of 1,6-Naphthyridinecarboxylic Acids	135
	14.1.2 Reactions of 1,6-Naphthyridinecarboxylic Acids	137
14.2	1,6-Naphthyridinecarboxylic Esters	139
14.3	1,6-Naphthyridinecarboxamides	139
14.4	1,6-Naphthyridinecarbonitriles, Carbaldehydes, and Ketones	141

CHAPTER 15   PRIMARY SYNTHESES OF 1,7-NAPHTHYRIDINES   143

15.1   From a Single Aliphatic Substrate   143
15.2   From a Single Pyridine Substrate   143
15.3   From a Pyridine Substrate with a Synthon   147
15.4   From Other Heterocyclic Systems   150

CHAPTER 16   1,7-NAPHTHYRIDINE, ALKYL-1,7-NAPHTHYRIDINES, AND ARYL-1,7-NAPHTHYRIDINES   153

16.1   1,7-Naphthyridine   153
   16.1.1   Preparation of 1,7-Naphthyridine   153
   16.1.2   Properties of 1,7-Naphthyridine   154
   16.1.3   Reactions of 1,7-Naphthyridine   155
16.2   Alkyl- and Aryl-1,7-Naphthyridines   157
   16.2.1   Preparation of Alkyl- and Aryl-1,7-Naphthyridines   157
   16.2.2   Reactions of Alkyl- and Aryl-1,7-Naphthyridines   158

CHAPTER 17   HALOGENO-1,7-NAPHTHYRIDINES   161

17.1   Preparation of Halogeno-1,7-Naphthyridines   161
17.2   Reactions of Halogeno-1,7-Naphthyridines   163

CHAPTER 18   OXY-1,7-NAPHTHYRIDINES   167

18.1   1,7-Naphthyridinones   167
   18.1.1   Preparation of 1,7-Naphthyridinones   167
   18.1.2   Reactions of 1,7-Naphthyridinones   168
18.2   Alkoxy- and Acyloxy-1,7-Naphthyridines   170
18.3   1,7-Naphthyridine N-Oxides   171

CHAPTER 19   THIO-1,7-NAPHTHYRIDINES   173

CHAPTER 20   NITRO-, AMINO-, AND RELATED 1,7-NAPHTHYRIDINES   175

20.1   Nitro-1,7-Naphthyridines   175
20.2   Amino-1,7-Naphthyridines   176

CHAPTER 21   1,7-NAPHTHYRIDINECARBOXYLIC ACIDS AND RELATED DERIVATIVES   179

21.1   1,7-Naphthyridinecarboxylic Acids   179
   21.1.1   Preparation of 1,7-Naphthyridinecarboxylic Acids   179
   21.1.2   Reactions of 1,7-Naphthyridinecarboxylic Acids   180
21.2   1,7-Naphthyridinecarboxylic Esters   182
21.3   1,7-Naphthyridinecarboxamides   182

| 21.4 | 1,7-Naphthyridinecarbonitriles | 182 |
| 21.5 | 1,7-Naphthyridinecarbaldehydes and Ketones | 182 |

CHAPTER 22 PRIMARY SYNTHESES OF 1,8-NAPHTHYRIDINES   183

22.1	From an Aliphatic Substrate	183
22.2	From a Single Pyridine Substrate	184
22.3	From a Pyridine Substrate and Synthon(s)	187
22.4	From Other Heterocyclic Substrates	192

CHAPTER 23 1,8-NAPHTHYRIDINE, ALKYL-1,8-NAPHTHYRIDINES, AND ARYL-1,8-NAPHTHYRIDINES   197

23.1	1,8-Naphthyridine and Hydro Derivatives		197
	23.1.1	Preparation of 1,8-Naphthyridine	197
	23.1.2	Properties of 1,8-Naphthyridine	198
	23.1.3	1,8-Naphthyridine Complexes	199
	23.1.4	Reactions of 1,8-Naphthyridine	200
23.2	Alkyl- and Aryl-1,8-Naphthyridines		203
	23.2.1	Preparation of Alkyl- and Aryl-1,8-Naphthyridines	203
	23.2.2	Reactions of Alkyl- and Aryl-1,8-Naphthyridines	205

CHAPTER 24 HALOGENO-1,8-NAPHTHYRIDINES   209

| 24.1 | Preparation of Halogeno-1,8-Naphthyridines | 209 |
| 24.2 | Reactions of Halogeno-1,8-Naphthyridines | 214 |

CHAPTER 25 OXY-1,8-NAPHTHYRIDINES   221

25.1	1,8-Naphthyridinones and the Like		221
	25.1.1	Preparation of 1,8-Naphthyridinones and the Like	221
	25.1.2	Reactions of 1,8-Naphthyridinones and the Like	224
25.2	Alkoxy- and Aryloxy-1,8-Naphthyridines		227
25.3	1,8-Naphthyridine N-Oxides		228

CHAPTER 26 THIO-1,8-NAPHTHYRIDINES   231

26.1	1,8-Naphthyridinethiones	231
26.2	Alkylthio- and Arylthio-1,8-Naphthyridines	232
26.3	1,8-Naphthyridine Sulfoxides and Sulfones	232
26.4	1,8-Naphthyridinesulfonic Acids and the Like	233

CHAPTER 27 NITRO-, AMINO-, AND RELATED 1,8-NAPHTHYRIDINES   235

27.1	Nitro-1,8-Naphthyridines		235
	27.1.1	Preparation of Nitro-1,8-Naphthyridines	235
	27.1.2	Reactions of Nitro-1,8-Naphthyridines	237

27.2	Nitroso-1,8-Naphthyridines	238
27.3	Amino-1,8-Naphthyridines	238
	27.3.1 Preparation of Amino-1,8-Naphthyridines	238
	27.3.2 Reactions of Amino-1,8-Naphthyridines	241

CHAPTER 28 1,8-NAPHTHYRIDINECARBOXYLIC ACIDS AND RELATED DERIVATIVES 247

28.1	1,8-Naphthyridinecarboxylic Acids	247
	28.1.1 Preparation of 1,8-Naphthyridinecarboxylic Acids	247
	28.1.2 Reactions of 1,8-Naphthyridinecarboxylic Acids	249
28.2	1,8-Naphthyridinecarbonyl Halides	251
28.3	1,8-Naphthyridinecarboxylic Esters	252
	28.3.1 Preparation of 1,8-Naphthyridinecarboxylic Esters	252
	28.3.2 Reactions of 1,8-Naphthyridinecarboxylic Esters	252
28.4	1,8-Naphthyridinecarboxamides	254
	28.4.1 Preparation of 1,8-Naphthyridinecarboxamides	254
	28.4.2 Reactions of 1,8-Naphthyridinecarboxamides	255
28.5	1,8-Naphthyridinecarbonitriles	257
28.6	1,8-Naphthyridinecarbaldehydes and Related Ketones	257
	28.6.1 Preparation of the Carbaldehydes and Ketones	257
	28.6.2 Reactions of the Carbaldehydes and Ketones	259

CHAPTER 29 THE 2,6-NAPHTHYRIDINES 261

29.1	Primary Syntheses of 2,6-Naphthyridines	261
	29.1.1 2,6-Naphthyridines by Cyclization of Pyridine Derivatives	261
	29.1.2 2,6-Naphthyridines by Cyclocondensation of a Pyridine Derivative with a Synthon	263
	29.1.3 2,6-Naphthyridines from Other Heterocyclic Substrates	264
29.2	2,6-Naphthyridine and Alkyl-2,6-Naphthyridines	265
	29.2.1 Preparation of 2,6-Naphthyridine	266
	29.2.2 Properties of 2,6-Naphthyridine	266
29.3	Halogeno-2,6-Naphthyridines	267
	29.3.1 Preparation of Halogeno-2,6-Naphthyridines	267
	29.3.2 Reactions of Halogeno-2,6-Naphthyridines	268
29.4	Oxy-2,6-Naphthyridines	269
29.5	Thio-2,6-Naphthyridines	270
29.6	Nitro-, Amino-, and Related 2,6-Naphthyridines	271
	29.6.1 Preparation of Amino-2,6-Naphthyridines	271
	29.6.2 Reactions of Amino-2,6-Naphthyridines	272
29.7	2,6-Naphthyridinecarboxylic Acids and Related Derivatives	272

CHAPTER 30 THE 2,7-NAPHTHYRIDINES 275

30.1	Primary Syntheses of 2,7-Naphthyridines	275
	30.1.1 2,7-Naphthyridines from Nonheterocyclic Precursors	275

Contents

	30.1.2	2,7-Naphthyridines by Cyclization of Pyridine Substrates	277
	30.1.3	2,7-Naphthyridines by Cyclocondensation of Pyridine Substrates with Synthons	278
	30.1.4	2,7-Naphthyridines from Other Heterocyclic Substrates	280
30.2	2,7-Naphthyridine and Alkyl-2,7-Naphthyridines		283
	30.2.1	Preparation of 2,7-Naphthyridine	283
	30.2.2	Properties of 2,7-Naphthyridine	283
	30.2.3	Reactions of 2,7-Naphthyridine	284
30.3	Halogeno-2,7-Naphthyridines		285
	30.3.1	Preparation of Halogeno-2,7-Naphthyridines	286
	30.3.2	Reactions of Halogeno-2,7-Naphthyridines	287
30.4	Oxy-2,7-Naphthyridines		289
30.5	Thio-2,7-Naphthyridines		290
30.6	Nitro-, Amino-, and Related 2,7-Naphthyridines		290
30.7	2,7-Naphthyridinecarboxylic Acids and Related Derivatives		291
	30.7.1	2,7-Naphthyridinecarboxylic Acids	291
	30.7.2	2,7-Naphthyridinecarboxylic Esters	292
	30.7.3	2,7-Naphthyridinecarboxamides and Carbonitriles	293
	30.7.4	2,7-Naphthyridinecarbaldehydes and Ketones	293

REFERENCES   295

APPENDIX   TABLES OF SIMPLE NAPHTHYRIDINES   337

Table A.1	Alphabetical List of Simple 1,5-Naphthyridines	339
Table A.2	Alphabetical List of Simple 1,6-Naphthyridines	348
Table A.3	Alphabetical List of Simple 1,7-Naphthyridines	361
Table A.4	Alphabetical List of Simple 1,8-Naphthyridines	366
Table A.5	Alphabetical List of Simple 2,6-Naphthyridines	409
Table A.6	Alphabetical List of Simple 2,7-Naphthyridines	411

INDEX   415

THE NAPHTHYRIDINES

This Is the Sixty-Third Volume in the Series
THE CHEMISTRY OF HETEROCYCLIC COMPOUNDS

CHAPTER 1

Primary Syntheses of 1,5-Naphthyridines

The primary synthesis of 1,5-naphthyridines may be accomplished by double cyclization of appropriate aliphatic substrates; by cyclization of appropriately substituted pyridines; by cyclocondensation of pyridine substrates with one or more aliphatic synthons; or from other heterocyclic substrates by degradation, rearrangement, or the like. Partially or fully reduced 1,5-naphthyridines are often made by somewhat similar procedures; such cases are usually illustrated toward the end of each subsection. Some reviews of naphthyridine chemistry contain material on the primary synthesis of 1,5-naphthyridines.[49–52,61,231,265,670,1260,1273,1430,1432]

1.1. FROM A SINGLE ALIPHATIC SUBSTRATE

This unlikely type of synthesis is represented by the reduction of 5,6-dimethyl-5,6-dinitrodecane-2,9-dione (**1**) to give a mixture from which 2,4a,6,8a-tetramethyl-3,4,4a,7,8,8a-hexahydro-1,5-naphthyridine 1,5-dioxide (**2**) was isolated via its picrate (Zn, NH_4Cl, H_2O, EtOH, 20°C, 6 h: ?%);[1022] the structural configuration of the foregoing compound and related products were studied later.[196]

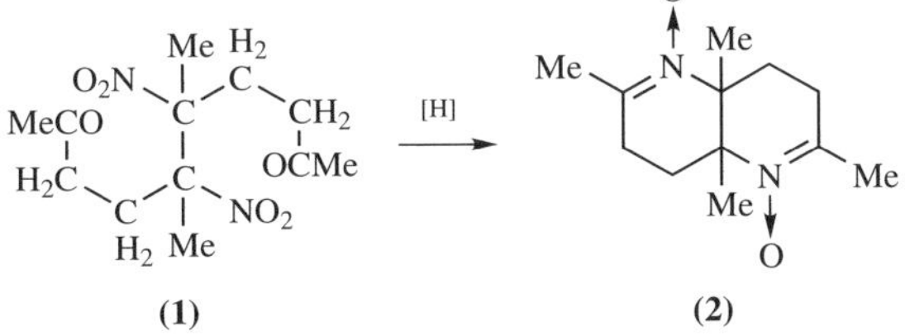

An essentially similar type of reaction has been used to prepare fused 1,5-naphthyridines such as 2,8-dimethyl-6,12-bis(p-tolylimino)-5,6,11,12-tetrahydrodibenzo[b, g][1,5]-naphthyridine (**3**).[688]

The Naphthyridines: The Chemistry of Heterocyclic Compounds, Volume 63, by D.J. Brown
Copyright © 2008 John Wiley & Sons, Inc.

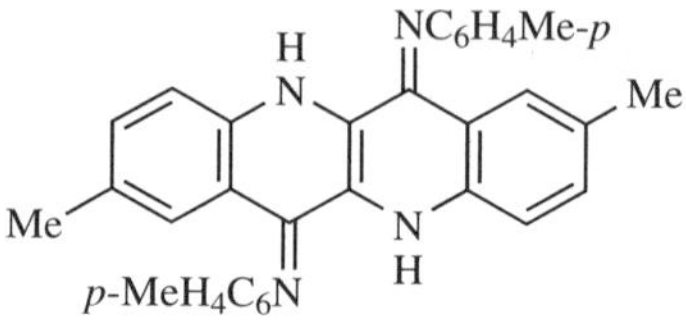

1.2. FROM A SINGLE PYRIDINE SUBSTRATE

Such syntheses may be subdivided according to which bond of the resulting 1,5-naphthyridine is formed in the process. Of the five possibilities, only three are represented in the literature.

1.2.1. By Formation of the N1,C2-Bond

Several preparations within this category are illustrated in the following examples.

2-(2-Ethoxycarbonylvinyl)-3-pyridinamine (**4**, R = H) gave 1,5-naphthyridin-2(1*H*)-one (**5**, R = H) (EtONa, EtOH, reflux, 3 h: 72%);[1044] the homologous substrate (**4**, R = Me) likewise gave 6,8-dimethyl-1,5-naphthyridin-2(1*H*)-one (**5**, R = Me) (reflux, 2 h: 71%);[1044] and the analogous carboxy substrate, 2-(2-carboxyvinyl)-3-pyridinamine (**6**), also afforded 1,5-naphthyridine-2(1*H*)-one (**5**, R = H) using the Posner[1428] technique (H_2NOH, MeONa, MeOH, reflux, 12 h: 31%).[1000]

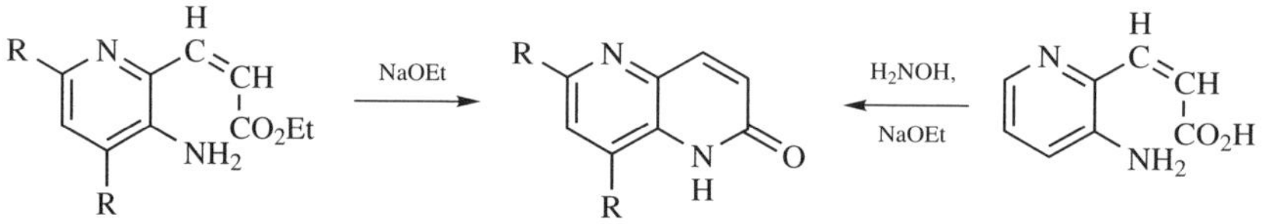

2-(3-Cyano-1-isopropylprop-1-enyl)-6-methoxy-3-nitropyridine (**7**, R = Pr^i, X = CN) gave 6-methoxy-4-isopropyl-1,5-naphthyridine-2-carbonitrile 1-oxide (**8**, R = Pr^i, X = CN) (Me_3SiCl, Et_3N, Me_2NCHO, 20°C, 6 h: 48%);[588] somewhat similarly, 2-(3-ethoxycarbonylprop-1-enyl)-6-methoxy-3-nitropyridine (**7**, R = H, X = CO_2Et) gave ethyl 6-methoxy-1,5-naphthyridine-2-carboxylate 1-oxide (**8**, R = H, X = CO_2Et) [Me_3SiCl, diazabicycloundecene (DBU), Me_2NCHO, 20°C, 1 h: 85%].[588]

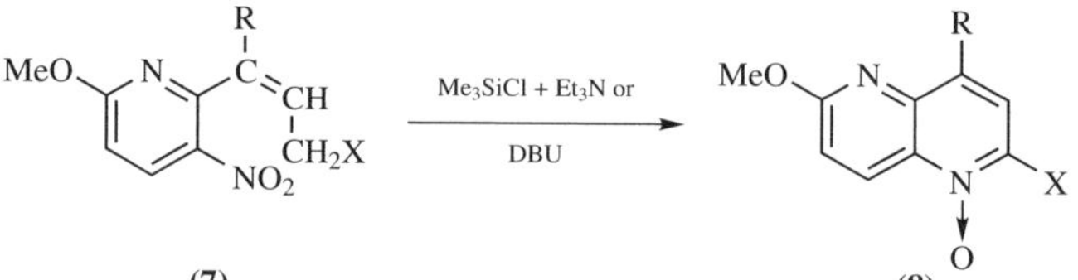

2-Ethoxalylmethyl-6-methoxy-3-nitropyridine (**9**) underwent reductive cyclization to 3-hydroxy-6-methoxy-3,4-dihydro-1,5-naphthyridin-2(1*H*)-one (**10**) (PtO$_2$, H$_2$, 3 atm, EtOH, 20°C, 2 h: 69%) that was easily aromatized to give 6-methoxy-1,5-naphthyridine-2,3(1*H*, 5*H*)-dione (**11**) [TsCl, pyridine, 150°C (?), 4 h: 83%].[234]

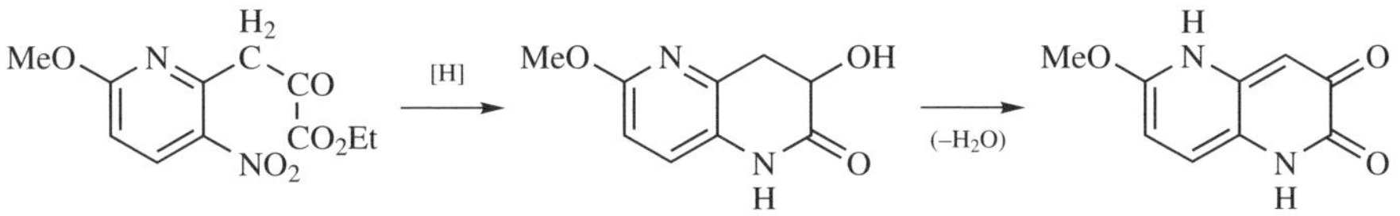

2-(1-Hydroxy-3-phenylallyl)-6-methoxy-3-methylaminopyridine (**12**, R = Me) underwent thermolytic cyclization to give 2-methoxy-5-methyl-6-phenyl-5,6-dihydro-1,5-naphthyridine (**13**, R = Me) (C$_6$H$_4$Cl$_2$, 180°C, 4 h: 31%). However, when the allylamino substrate (**12**, R = CH$_2$CH:CH$_2$) was so treated, the initial product, 1-allylamino-6-methoxy-2-phenyl-1,5-naphthyridine (**13**, R = CH$_2$CH:CH$_2$) (180°C, 20 min: 65%), underwent subsequent aromatization to 2-methoxy-6-phenyl-1,5-naphthyridine (**14**), presumably by loss of propane [180°C, 20 h: 43%; or F$_3$CCO$_2$H, (Ph$_3$P)$_4$RhH, EtOH, reflux, 4 h: 54%].[619]

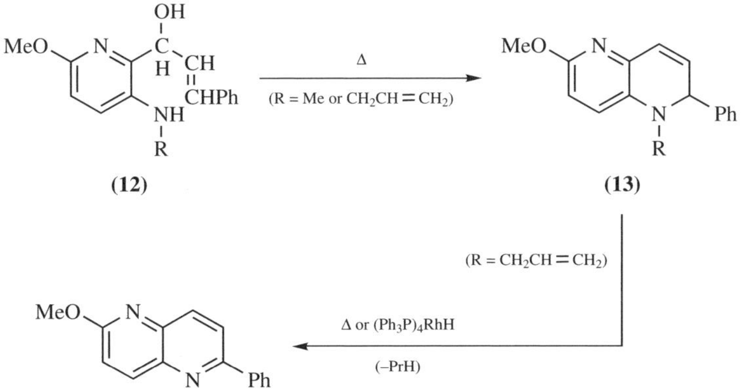

1.2.2. By Formation of the C3,C4-Bond

Such a synthesis is evident in the Dieckmann cyclization[981] of ethyl 3-[2-(ethoxycarbonyl)acetamido]-2-pyridinecarboxylate (**15**) (prepared *in situ*) to give

4-hydroxy-1, 8-naphthyridin-2(1H)-one (**16**) (EtONa, EtOH, reflux, 5 h; resulting solid, NaOH, H$_2$O, reflux until gas-ceased: 69%).[1023]

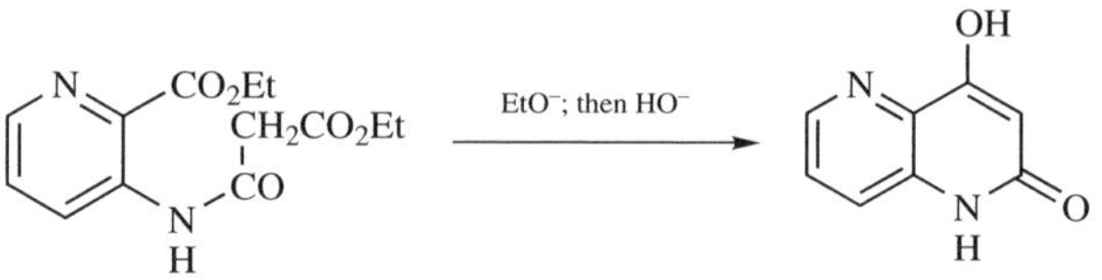

1.2.3. By Formation of the C4,C4a-Bond

Although it involves the formation of a C,C-bond, this synthesis has been widely used, as illustrated in the following typical examples.

3-[(2,2-Diethoxycarbonylvinyl)amino]-5,6-dimethylpyridine (**17**) underwent regioselective thermal cyclization to give ethyl 6,7-dimethyl-4-oxo-1,4-dihydro-1,5-naphthyridine-3-carboxylate (**18**) (Dowtherm A, reflux, 15 min: 89%; analogs likewise).[1398]

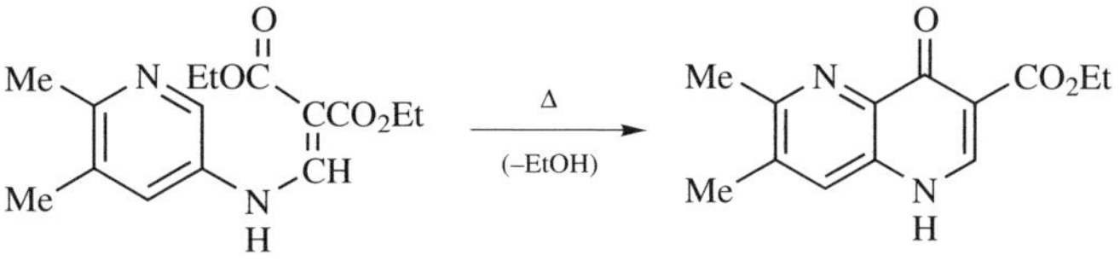

3-[(2,2-Diethoxycarbonylvinyl)amino]-4-methoxypyridine (**19**) gave ethyl 8-methoxy-4-oxo-1,4-dihydro-1,5-naphthyridine-3-carboxylate (**20**) (Ph$_2$O, reflux, 25 min: 72%);[301] a similar substrate likewise gave ethyl 6-butoxy-4-oxo-1,4-dihydro-1,5-naphthyridine-3-carboxylate (Ph$_2$O, reflux, 15 min: ~75%).[1036]

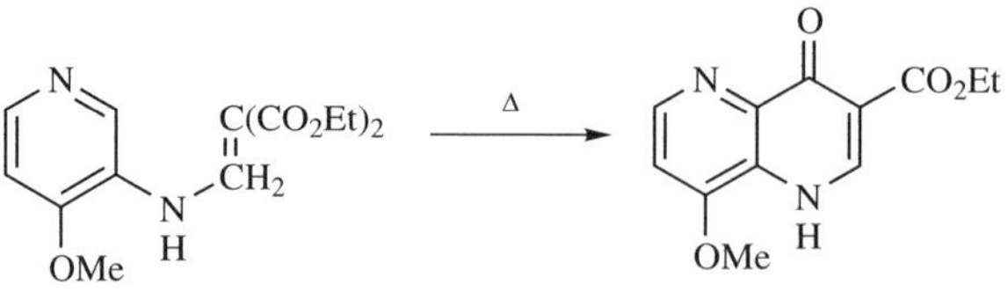

In contrast, regioselectively was less marked in the thermal cyclization of 3-[(2-ethoxycarbonyl)amino]pyridine (**21**), which gave a 4:1 mixture of

2-methyl-1,5-naphthyridin-4(1*H*)-one (**22**) and 2-methyl-1,7-naphthyridin-4(1*H*)-one (**23**) (Ph$_2$O, reflux, 1 h: 75% of mixture).[828]

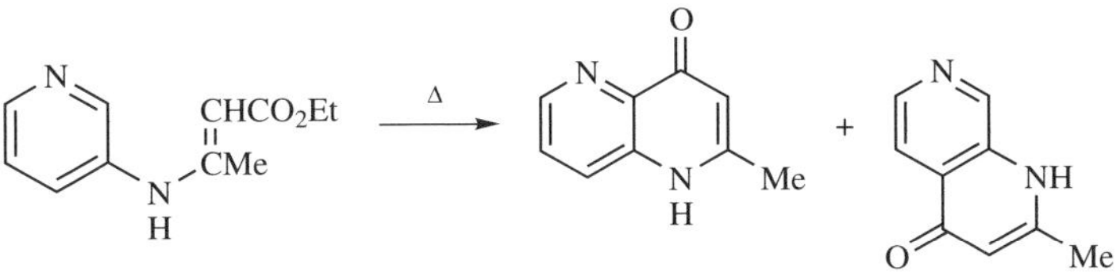

3-[(3-Hydroxy-2-methyl-3-phenylpropyl)amino]pyridine (**24**) underwent dehydrative cyclization to give 3-methyl-4-phenyl-1,2,3,4-tetrahydro-1,5-naphthyridine (**25**) (70% H$_2$SO$_4$, 0°C → 25°C, ~15 h: 65%) that was easily dehydrogenated to give 3-methyl-4-phenyl-1,5-naphthyridine (**26**) (neat substrate, Pd/C, 200°C, 3 h: 56%).[879] Also other examples.[48,485,865,967,1232,1245]

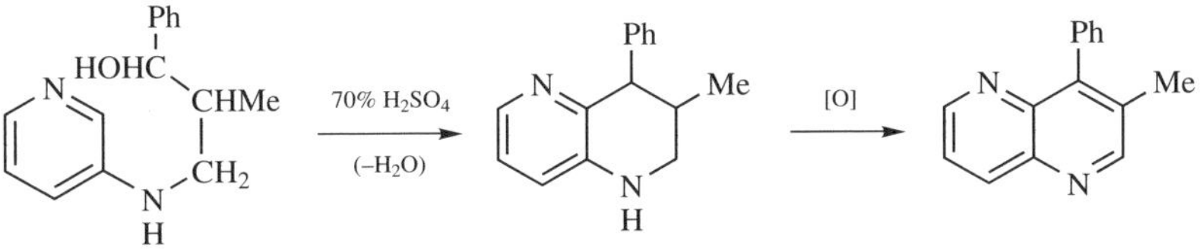

1.3. FROM A PYRIDINE SUBSTRATE WITH ONE SYNTHON

Most primary syntheses of 1,5-naphthyridines fall into this basket. Procedures are divided initially into categories in which the synthon provides one, two, or three atoms to the final 1,5-naphthyridine; when necessary, such categories are further subdivided according to which specific atoms are provided by the synthon.

1.3.1. Where the Synthon Supplies One Atom

There appears to be no example of such a synthesis in the parent 1,5-naphthyridine series, but benzo-1,5-naphthyridines have been so made. Thus ethyl 3-dimethylamino-methyleneamino-6,7-dimethoxyquinoline (**27**) gave 7,8-dimethoxy-4-oxo-1,4-dihydrobenzo[*b*]-1,5-naphthyridine-3-carbonitrile (**28**)[NaCH$_2$CN (made *in situ*), THF, −78°C; substrates dropwise, −78°C, 3 h; then 20°C, 12 h: 58%].[208]

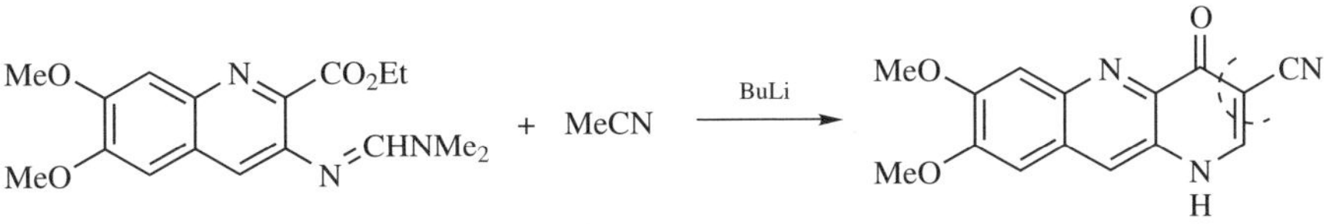

1.3.2. Where the Synthon Supplies Two Atoms

Nearly all the reported syntheses in this category have used the synthon to supply C2 + C3 of the resulting 1,5-naphthyridine, but C3 + C4 may be so supplied, as illustrated at the end of the following list of typical examples.

Ethyl 3-aminopicolinate (**29**) with diethyl malonate gave 4-hydroxy-1,5-naphthyridin-2(1H)-one (**30**) [neat reactants, 120°C, 5 h; solid, EtONa, EtOH, reflux, 5 h; solid, 2.5M NaOH, reflux until gas ceased: 96%].[1151]

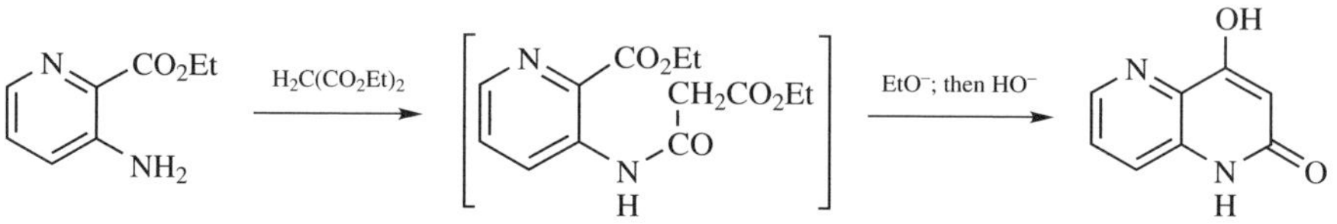

Somewhat similarly, 3-aminopicolinic acid with ethyl acetoacetate gave 2-methyl-1,5-naphthyridin-4(1H)-one (neat reactants, reflux, 4 h: 13%).[1000]

3-Amino-2-pyridinecarbonitrile (**31**) with diethyl malonate gave 4-amino-1,5-naphthyridin-2(1H)-one (**32**) (neat reactants, EtONa, reflux, 7.5 h: 36%).[725]

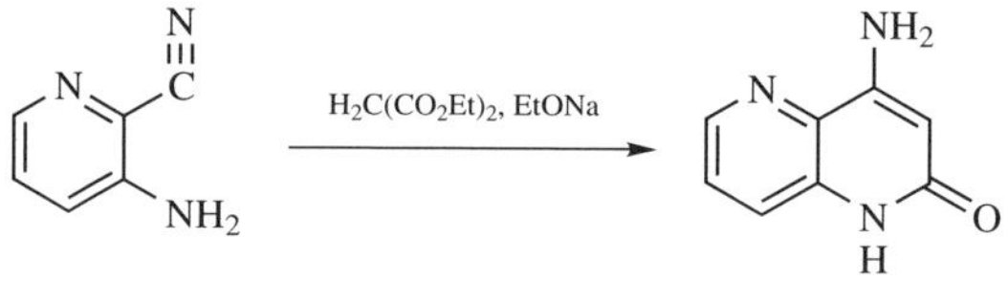

2-(α-Hydroxybenzyl)-6-methoxy-3-methylaminopyridine (**33**) with allyltrimethylsilane gave 2-methoxy-5-methyl-8-phenyl-6-(trimethylsilylmethyl)-5,6,7,8-tetrahydro-1,5-naphthyridine (**34**) (reactants, BF$_3$,Et$_2$O. 60–80°C, 4.5 h: 50%); analogs likewise.[619]

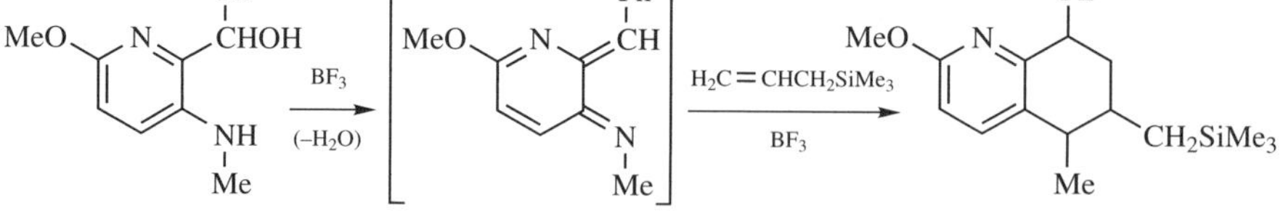

3-Benzylideneamino-6-methoxypyridine (**35**) underwent partially regioselective cyclocondensation with phenylacetylene in the presence of oxidizing agents to afford a separable mixture of 2-methoxy-6,8-diphenyl-1,5-naphthyridine (**36**) and 6-methoxy-2,4-diphenyl-1,7-naphthyridine (**37**) (reactants, FeCl$_3$, tetrachlorobenzoquinone, MeCN, reflux, ~45 min; 50% and 7%, respectively, after separation).[455] Also other examples.[312,1302]

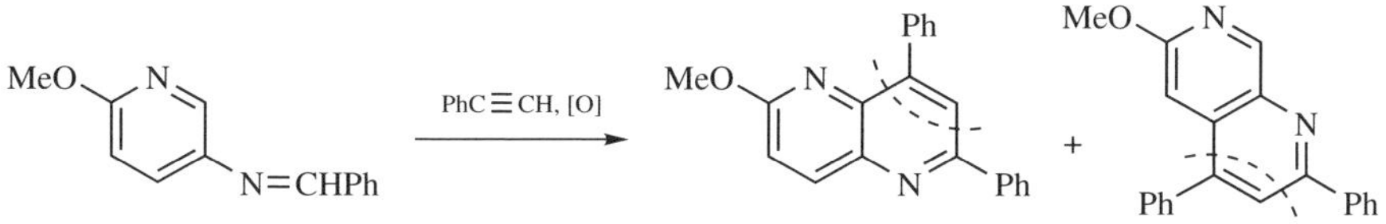

(35) → (36) + (37)
PhC≡CH, [O]

1.3.3. Where the Synthon Supplies Three Atoms

Of the two possibilities in this category (to supply N1 + C2 + C3 or C2 + C3 + C4), only the latter has been employed, often by submission of a 3-pyridinamine to a Skraup-like synthesis. Typical examples follows.

3-Pyridinamine (**38**) and acrolein (**39**) (formed from glycerol under Skraup-like conditions) gave 1,5-naphthyridine (**40**) [$O_2NC_6H_4SO_3Na$-m, H_2SO_4, H_3BO_3, $FeSO_4$, 0°C; HOHC(CH_2OH)$_2$↓, substrate ↓, H_2O ↓; 135°C, 4 h: 90%;[873] HOCH(CH_2OH)$_2$, $O_2NC_6H_4SO_3Na$-m, ~70% H_2SO_4, 135°C, 4 h: 50%;[45] earlier versions, using As_2O_5 as the oxidizing agent, gave <30% yields.[14,32,113,155,104]

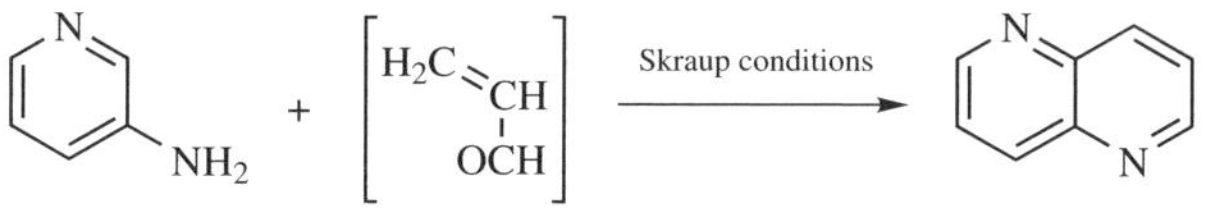

(38) + (39) → (40)
Skraup conditions

6-Methyl-3-pyridinamine (**41**, R = Me) gave 2-methyl-1,5-naphthyridine (**42**, R = Me) [96% H_2SO_4, $O_2NC_6H_4SO_3Na$-m, H_3BO_3, $FeSO_4$, $7H_2O$, HOHC(CH_2OH)$_2$↓, substrate↓, H_2O↓, then 135°C, 18 h: 55%];[268] 2,5-pyridinediamine (**41**; R = NH_2) gave 1,5-naphthyridine-2-amine (**42**, R = NH_2) [70% H_2SO_4, $O_2NC_6H_4SO_3Na$-m, HOCH(CH_2OH)$_2$, 135°C, 5 h: 88%].[811,1371]

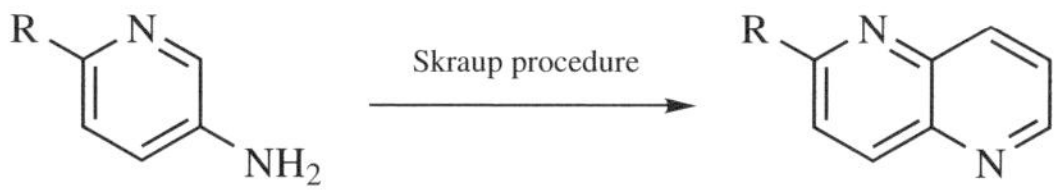

(41) → (42)
Skraup procedure

Appropriate pyridine substrates similarly gave 1,5-naphthyridine-2(1H)-one (**43**) [HOHC(CH_2OH)$_2$, As_2O_5, 165°C, ~90 min: 57%;[1037] likewise but 160°C, 45 min: 15%];[1051] 1,5-naphthyridin-4(1H)-one (**44**)(likewise, 160°C, ~2.5 h:

37%);[1047] and 1,5-naphthyridine 1-oxide (**45**) [$O_2NC_6H_4SO_3H$, H_2SO_4, $FeSO_4$, $7H_2O$, H_3BO_3, 0°C; HOCH(CH$_2$OH)$_2$↓, substrate ↓, H$_2$O ↓; then 130°C, 5 h: 19%].[1033]

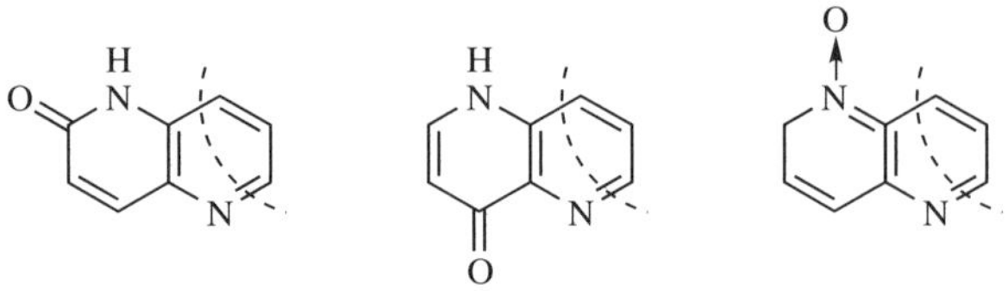

(**43**)   (**44**)   (**45**)

3-Pyridinamine (**47**) with crotonaldehyde gave 2-methyl-1,5-naphthyridine (**46**)(substrate, $O_2NC_6H_4SO_3H$, H_2SO_4, H_2O, 125°C; MeCH=CHCHO in dropwise, 1 h: then 150°C, 12 h: ~10%), with methacrylaldehyde gave 3-methyl-1,5-naphthyridine (**48**) (substrate, $FeSO_4.7H_2O$, As_2O_5, 96% H_2SO_4, 120°C; H_2CCMeCHO in dropwise during 7 h; the 170°C, 15 h: 30%); or with methyl vinyl ketone gave 4-methyl-1,5-naphthyridine (**49**)[as for isomer (**46**) but MeCOCH=CH$_2$; 11%].[155]

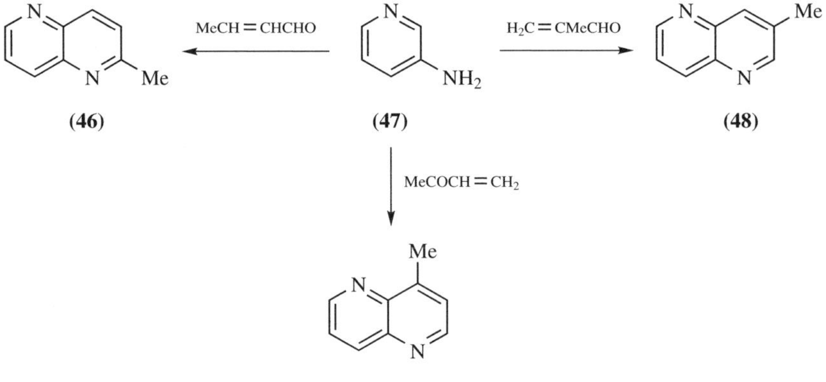

(**46**)   (**47**)   (**48**)

(**49**)

6-Methyl-3-pyridinamine (**50**) with crotonaldehyde gave 2,6-dimethyl-1,5-naphthyridine (**51**) [substrate, $O_2NC_6H_4SO_3H$, H_2SO_4 (?), H_2O, 125°C; MeCH=CHCHO in slowly; then 150°C, 8 h: ~10%];[1169] analogs somewhat similarly.[269]

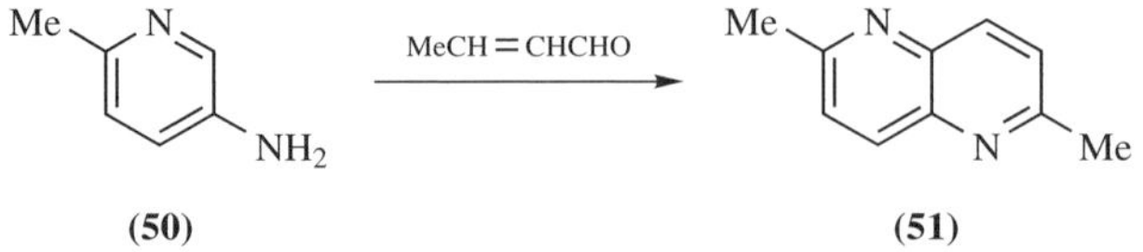

(**50**)   (**51**)

2-Bromo-6-ethoxy-3-pyridinamine (**52**) reacted with methyl 3-methoxyacrylate (2 equiv) to give 6-ethoxy-4-methoxy-1-(1-methoxy-2-methoxycarbonylethyl)-1,5-naphthyridin-2(1*H*)-one (**53**) [Pd(OAc)$_2$, P(C$_6$H$_4$Me-*p*)$_3$, Et$_3$N, N$_2$, 100°C bath, 5 days: 28%].[1015]

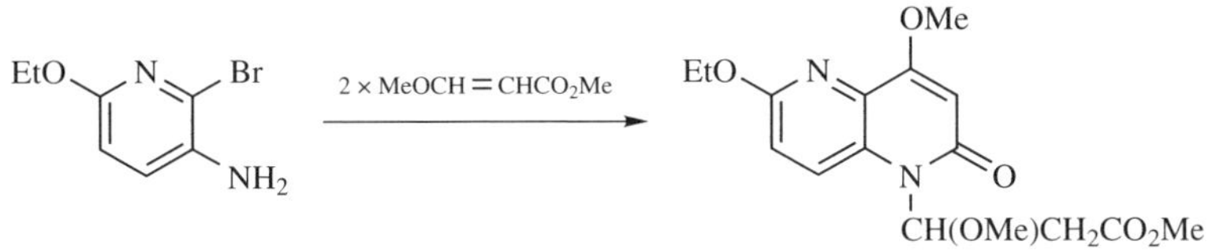

3-Pyridinamine (**54**) with diethyl ethoxymethylenemalonate gave ethyl 4-oxo-1,4-dihydro-1,5-naphthyridine-3-carboxylate (**55**) (reactants, Dowtherm A, 150°C, reflux, 1 h: 80%);[101] somewhat similarly, 5-amino-2(1*H*)-pyridinone gave ethyl 4,6-dioxo-1,4,5,6-tetrahydro-1,5-naphthyridine-3-carboxylate (**56**) [EtOCH=C(CO$_2$Et)$_2$, Ph$_2$O, reflux, 1 h: 20%].[233]

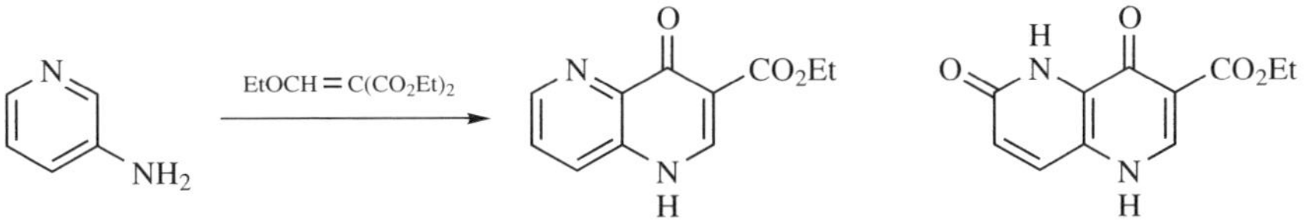

3-Aminopicolinic acid (**57**) with ethyl acetoacetate gave 2-methyl-1,5-naphthyridin-4(1*H*)-one (**58**) (neat reactants, reflux, 4 h: 13%).[828,1000]

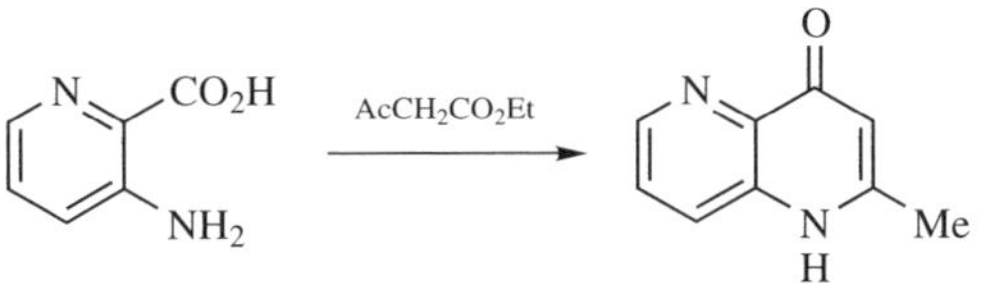

2-Methoxy-5-nitropyridine (**59**) with 1-phenyl-3-phenylsulfonylpropane (**60**) gave 2-methyl-6-phenyl-8-phenylsulfonyl-1,5-naphthyridine (**61**) (reactants, MeCN, Bu^{t}Me$_3$SiCl, DBU, 20°C, 3 days: 54%),[521,554] Also other examples,[181,273,316,760,827,1272,1377]

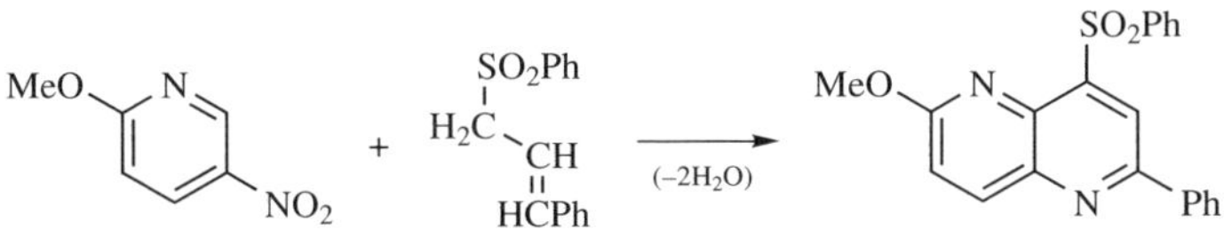

1.4. FROM A PYRIDINE SUBSTRATE AND TWO SYNTHONS

This type of synthesis is represented only by several procedures akin to the Doebner–Miller quinoline synthesis, in which both synthons are the same.

4-Methyl-3-pyridinamine (**62**) and an excess of acetaldehyde gave 2,8-dimethyl-1,5-naphthyridine (**63**) (substrate, 10M HCl, 5°C; MeCHO in dropwise; 0°C, 1 h; then reflux, 1 h; 16%; note the necessity for oxidation, either aerial or by the excess of MeCHO?),[268]

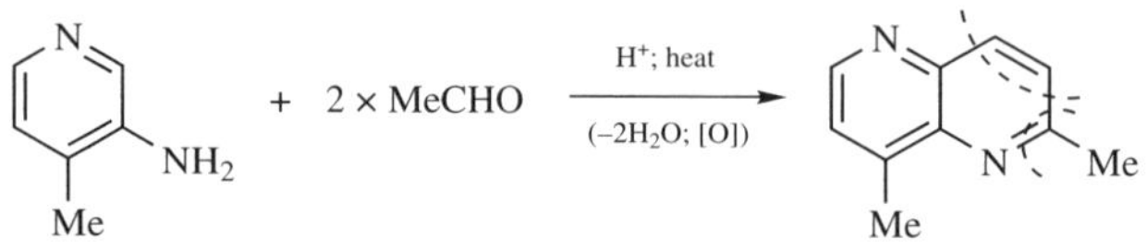

(**62**)   +   2 × MeCHO   →[H+; heat] (−2H$_2$O; [O])   (**63**)

Similar treatment of 6-methoxy-3-pyridinamine (**64**) involved an additional hydrolysis of the methoxy group to afford 6-methyl-1,5-naphthyridin-2(1H)-one (**65**)(36%).[268]

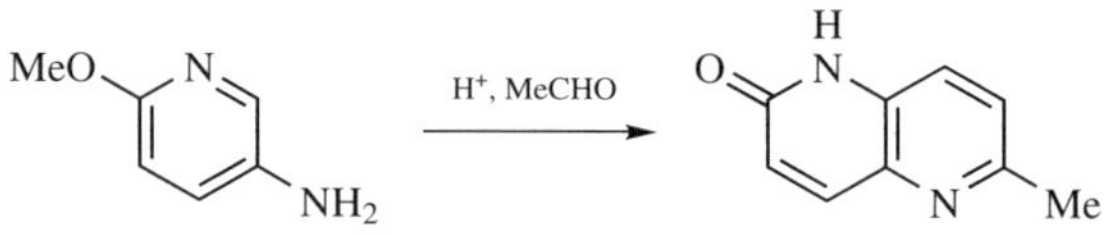

(**64**)   →[H+, MeCHO]   (**65**)

Also other examples.[94]

1.5. FROM OTHER HETEROCYCLIC SUBSTRATES

This general approach to 1,5-naphthyridines has seldom been used, but classified examples follow.

From Benzo[b]-1,5-naphthyridines

7,9-Dihydroxybenzo[b]-1,5-naphthyridin-10(5H)-one (**66**), easily made by fusion of 3-aminopicolinic acid with phloroglucinol, underwent oxidation to 4-oxo-1,4-dihydro-1,5-naphthyridine-2,3-dicarboxylic acid (**67**) [HNO$_3$ (d. 1.5), 20°C, then 95°C, 40 min: ?%, after separation from a byproduct].[17]

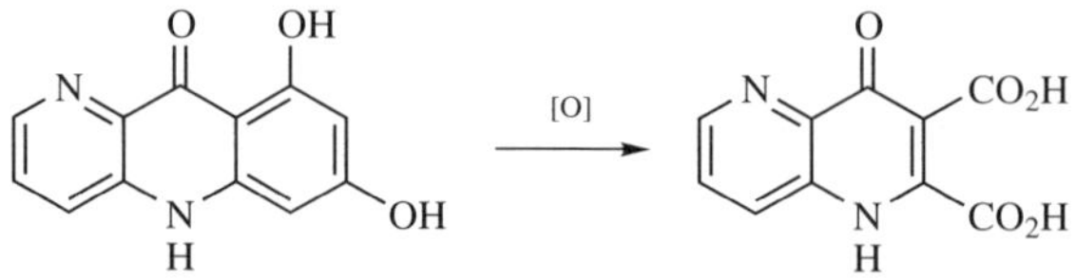

(**66**)   →[[O]]   (**67**)

From 1,3-Dioxolanes

4,5-Diamino-2,7-bis(1,3-dioxolan-2-yl)-4,5-dimethyloctane (**68**) [a bis(cycloacetal) of the corresponding dialdehyde] gave 2,4a,6,8a-tetramethyl-3,4,4a,7,8,8a-hexahydro-1,5-naphthyridine (**69**) (1M H_2SO_4: no details).[196]

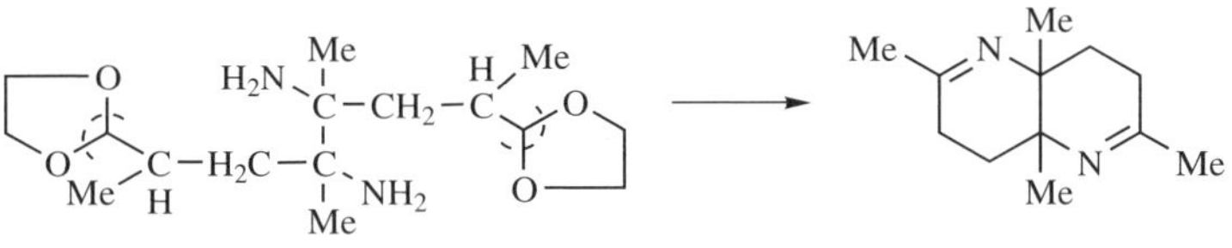

From 1,2,4-Triazines

5,6-Diphenyl-3-[3-(2-phenylimidazol-1-yl)propyl]-1,2,4-triazine (**70**) underwent thermal intramolecular addition (with loss of nitrogen) to give the tricyclic intermediate (**71**) and thence (by loss of benzonitrile) 2,3-diphenyl-5,6,7,8-tetrahydro-1,5-naphthyridine (**72**) [substrate, antioxidant (2,6-di-*tert*-butyl-4-methylphenol), 1,3,5-$Pr^i_3C_6H_3$, reflux, 3 h: 92%] that could be aromatized to 2,3-diphenyl-1,5-naphthyridine (**73**) (1,3,5-$Pr^i_3C_6H_3$, reflux, air, 24 h: 91%); the latter product (**73**) was also made directly from the triazine (**70**) (neat substrate, Se, 330°C, 10 h: 85%); analogs likewise.[137,522]

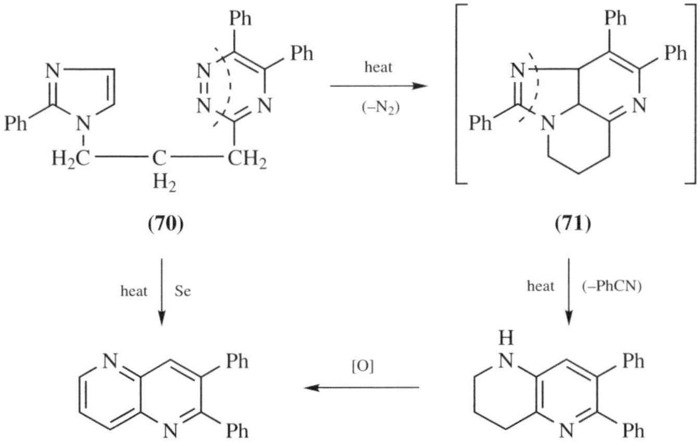

CHAPTER 2

1,5-Naphthyridine, Alkyl-1, 5-naphthyridines, and Aryl-1, 5-naphthyridines

This chapter covers information on the preparation, physical properties, and reactions of 1,5-naphyhyridine and its *C*-alkyl, *C*-aryl, *N*-alkyl, and *N*-aryl derivatives as well as their respective ring-reduced analogs. In addition, it includes methods for introducing alkyl or aryl groups (substituted or otherwise) into 1,5-naphthyridines already bearing substituents and the reactions specific to the alkyl or aryl groups in such compounds. For simplicity, the term *alkyl-1,5-naphthyridine* in this chapter is intended to include alkyl-, alkenyl-, alkynyl-, aralkyl-, and cycloalkyl-1,5-naphthyridines; likewise, *aryl-1,5-naphthyridine* includes both aryl- and heteroaryl-1,5-naphthyridines.

The general chemistry of 1,5-naphthyridines, usually in association with those of other naphthyridines and often quite brief, has been reviewed from time to time.[49–52,55–58,61,231,265,670,1260,1357,1430,1432]

2.1. 1,5-NAPHTHYRIDINE

2.1.1. Preparation of 1,5-Naphthyridine

1,5-Naphthyridine is usually made from 2-pyridinamine by a Skraup-like reaction (see Section 1.3.3), but it has also been prepared by removal of unwanted substituents as illustrated in the following examples.

1,5-Naphthyridine 1-oxide (**1**) underwent deoxygenation by phosphorus trichloride to give 1,5-naphthyridine (**2**) (substrate, CHCl$_3$, PCl$_3$↓, 0°C→reflux, 1 h: 60%).[1033]

2-Chloro-1,5-naphthyridine (**3**) gave 1,5-naphthyridine (**2**) (H$_2$, Pd/CaCO$_3$, KOH, MeOH: ?%).[87]

The Naphthyridines: The Chemistry of Heterocyclic Compounds, Volume 63, by D.J. Brown
Copyright © 2008 John Wiley & Sons, Inc.

Both 2-*p*-toluenesulfonylhydrazino- (**4**, R = H) and 2,6-bis(*p*-toluenesulfonyl-hydrazino)-1,5-naphthyridine (**4**, R=NHNHTs) afforded 1,5-naphthyridine (**2**) (KOH, H$_2$O, 95°C, until N$_2$↑ ceased: ?%).[1047]

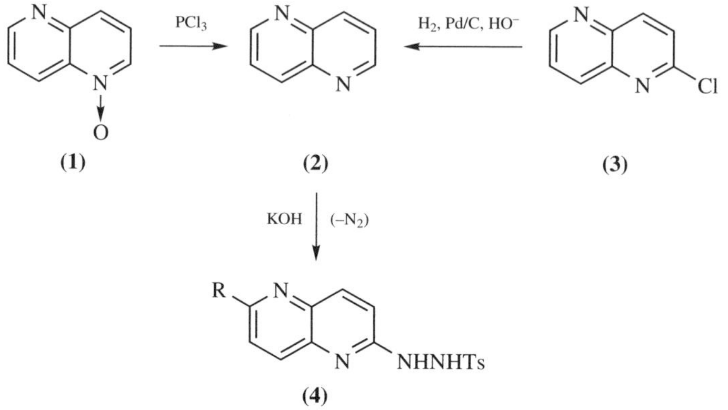

1,5-Naphthyridin-2 (1*H*)-one gave 1,5-naphthyridine (Zn: ~2%).[17]

Two parent hydro-1,5-naphthyridines have been made by reductive methods as exemplified here.

1,5-Naphthyridine (**5**) gave 1,2,3,4-tetrahydro-1,5-naphthyridine (**6**) (H$_2$, Pd/C, EtOH, 20°C, 2 h: 90%,[47,155] cf. H$_2$, PtO, AcOH: ?%)[87] and thence *trans*-decahydro-1,5-naphthyridine (**7**) (H$_2$, PtO, EtOH, until uptake ceased: 93%;[47] or Na, C$_5$H$_{12}$OH, reflux: 91%).[237,cf. 557]

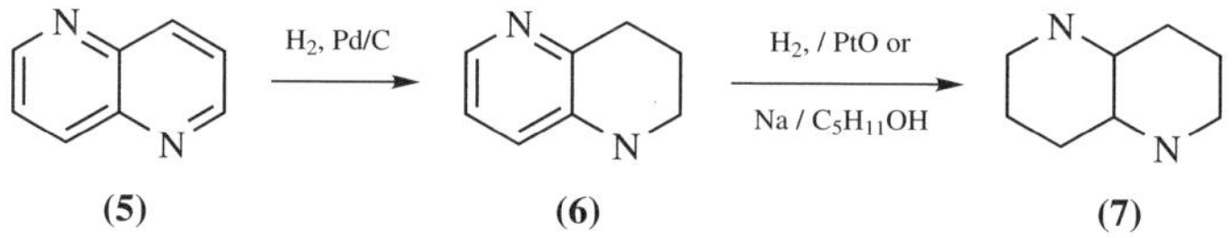

2.1.2. Properties of 1,5-Naphthyridine

Reported physical data for 1,5-naphthyridine may be found by reference to its entry in the Appendix Table A1 toward the end of this book. More notable studies of such data are briefly listed here in alphabetical order.

Complexes. The iodine complexes of 1,5-naphthyridine and other heterocycles have been used under anhydrous conditions to estimate the pK_a values for such heterocycles.[885] The structures of 1 : 1 complexes of 1,5-naphthyridine with oxalic acid,[1024] fumaric acid,[1024] or *meso*-1,2-diphenyl-1,2-ethane-diol[1021] have been studied. Complexes of 1,5-naphthyridine with Co(II), Ni(II), Cu(II), Zn(II), and Ag(I) salts have been prepared;[705] the mono- and di(tungstenpentacarbonyl) complexes have been made for comparison with such complexes of pteridine and related heterocycles;[888] the characteristic crystals of the Pt and Au complexes may prove useful for the detection of such noble metals in low concentration;[1366] and the 1:1 complex with decamethylytterbocene has an unusual structure.[321]

Crystal Structure. The X-ray crystal structures for anhydrous and hydrated 1,5-naphthyridine have been measured using various refinements.[193,703,766,798] The bond lengths so obtained appear to confirm those predicted by calculation.[1241]

Electron Density. The π-electron densities for 1,5-naphthyridine and related systems have been calculated[1126] and their correlation with ease of electrophilic[173] and nucleophilic reactions[998] discussed.

Electron Spin Resonance. The ESR spectra for anions from 1,5-naphthyridine and related heterocycles have been measured and interpreted;[973,1079,1083] the dicationic radical from 1,5-naphthyridine has also been studied.[1057]

Infrared/Raman Spectra. The IR and Raman spectra for 1,5-naphthyridine have been measured in the solid state and in water, respectively, for comparison with those of other azanaphthalenes;[44] they have also been recorded in mineral oil and in the vapor state for comparison with those of naphthylene.[1321]

Ionization Phenomena. The first basic pK_a of 1,5-naphthyridine is reported to be 2.90–2.91 (by potentiometry[45,1027,1214,1303]), 2.93 [by a ^{13}CNMR (i.e., nuclear magnetic resonance using carbon-13 isotope) technique[1192]], 3.05 (by spectrometry[331]), or 3.2 (by partition coefficient[155]); the second is −1.10 (by spectrometry?).[1303] The pK_a values for 1,5-naphthyridine and related compounds have been correlated with fundamental energy calculations[806,813] and with ease of quaternization.[1306]

Mass Spectra. The MS of 1,5- and other naphthyridines have been compared and contrasted.[975]

Nuclear Magnetic Resonance Spectra. The ^{1}HNMR chemical shifts and coupling constants for 1,5-naphthyridine and related molecules have been measured[42,1140] and the effect of added tris(dipivaloylmethanato)europium[13] or dissolution in *p*-butyl-*N*-(*p*-methoxybenzylidene)aniline[995] reported. A correlation of chemical shifts with electron densities in 1,5-naphthyridine has been attempted.[969] The ^{14}NNQR (i.e., nuclear quadrupole resonance using nitrogen-14 isotope) spectra of 1,5- and isomeric naphthyridines have been studied in some detail,[365,882] and aspects of their ^{13}CNMR spectra have been reported.[1174,1182]

Phosphorescence Spectra. The phosphorescence spectra for 1,5-naphthyridine and structurally related compounds have been measured in heptane[1344] or in (solid) 1,2,4,5-tetramethylbenzene at very low temperatures.[970,1114,1193]

Polarography. Several aspects of the polarographic reduction[833,848,1002,1231,1329,1342] and electrochemical oxidation[1092] of 1,5- and isomeric naphthyridines have been reported in some depth.

Resonance Energies. Resonance energies and the like have been calculated by various procedures for 1,5- and other naphthyridines.[209,676,812,840]

Ultraviolet and Visible Spectra. As well as those obtained from solution, such spectra of 1,5-naphthyridine have been measured in the vapor state[830,1119] and the solid state.[1094] Interpretation of the various bands has been proposed with assistance from magnetic circular dichroism data[1153] and other sources.[1016,1113,1336] Several types of calculation have been used to rationalize the position and intensity of each band.[807,820,910,1086,1173,1312,1426]

2.1.3. Reactions of 1,5-Naphthyridine

Reported reactions of 1,5-naphthyridine or the corresponding hydronaphthyridines are illustrated by the following classified examples.

C-Alkylation

1,5-Naphthyridine (**8**) gave 4,8-dimethyl-1,5-naphthyridine (**9**) (excess Me_2SO, NaH, N_2, 70°C; substrate↓, 70°C, 4 h: 19%).[880]

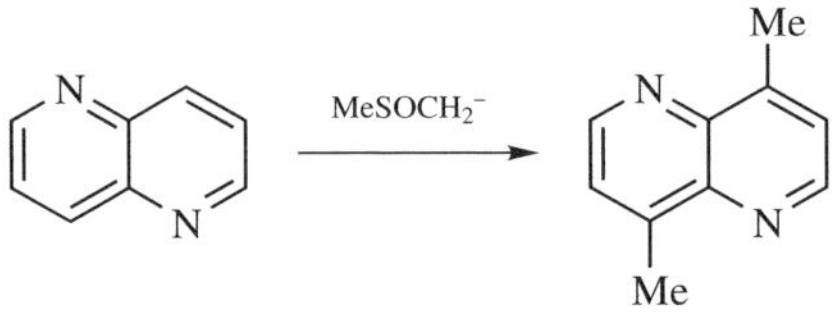

For an example of heteroarylation, see "Reissert Reactions" below.

N-Alkylation or Quaternization

Decahydro-1,5-naphthyridine underwent *N*-alkylation by 2-[4-(*N*-acetylanilino)-buta-1,3-dienyl]-1,3,3-trimethyl-3*H*-indolium perchlorate to give 1,5-bis[4-(1,3,3-trimethyl-3*H*-indolio)buta-1,3-dienyl] decahydro-1,5-naphthyridine bisperchlorate (**10**) with loss of acetanilide (substrate, EtOH, Me_2NCHO, 40°C; synthon↓ dropwise; 6 h: 29%);[1203] also methylation to 1,5-dimethyldecahydro-1,5-naphthyridine (no details).[1089]

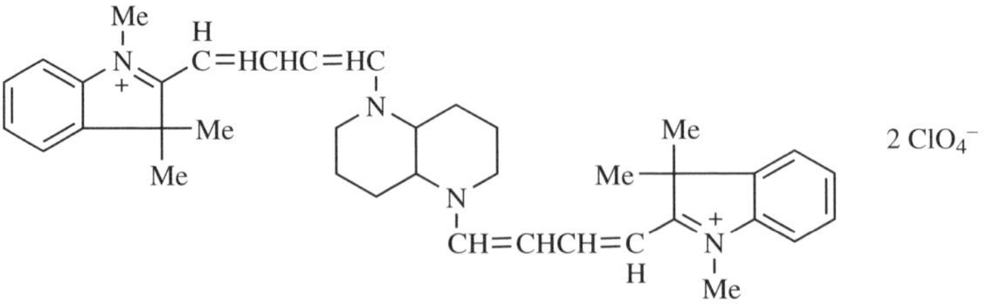

1,5-Naphthyridine (**12**) gave its 1-monomethiodide (**11**) (MeI, MeOH, reflux, 12 h: 85%;[155] kinetics;[961] other data[993]), the bisdimethosulfate (**13**) (neat Me_2SO_4, reflux, 1 h: 33%);[906] or the bis(methofluorosulfonate) [neat $MeFSO_3$ (exothermic), then 95°C 1 h; more synthon↓, 95°C, 30 min: 85%].[1061]

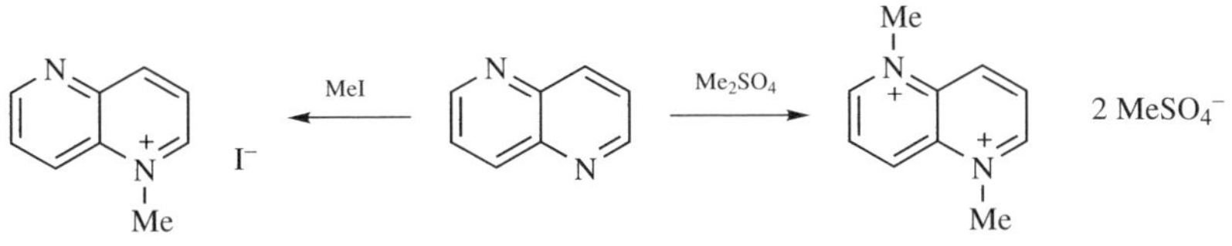

Amination

Note: 1,5-Naphthyridine can undergo semidirect *C*- or *N*-amination, but in the latter case the product must be a salt or a zwitterion.

1,5-Naphthyridine (**14**) with potassium amide in liquid ammonia gave a solution of the adduct (**15**) that underwent oxidation to afford 1,5-naphthyridin-2-amine (**16**) [substrate, KNH_2 (made *in situ*), liquid NH_3, 10 min; then $KMnO_4$↓, 10 min: 36%].[425,cf. 173,302,643,692,1047,1353,1371]

1,5-Naphthyridine (**14**) with *O*-mesitylenesulfonylhydroxylamine ($MSONH_2$) gave 1-amino-1,5-naphthyridin-1-ium mesitylene sulfonyloxide (**17**) (substrate, MeCN, 5°C; $MSONH_2$↓ dropwise; 20°C, 10 min: 81%) and thence, by treatment with benzoyl chloride, 1,5-naphthyridinium-1-benzimidate (**18**) (neat BzCl, 90°C, 2.5 h: 60%).[1003,cf. 1004]

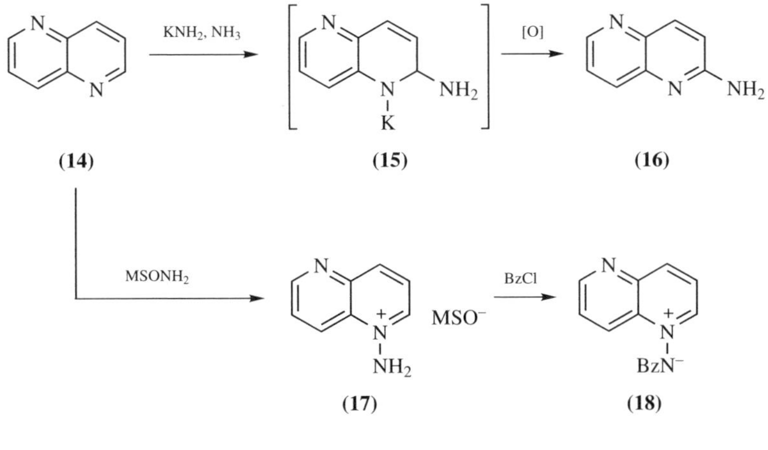

Cycloaddition

1,5-Naphthyridine (**19**) with chloromethyl phenyl sulfone gave 6,7-bisphenylsulfonyl-6,6a,6b,7-tetrahydro-1*H*-azirino [1,2-*a*] cyclopropa[*c*]-1,5-naphthyridine

(20) (KOH, Me$_2$SO, 20°C; reactants↓ during 5 min; 20°C, tlc monitored: 40%); analogs likewise.[293,803]

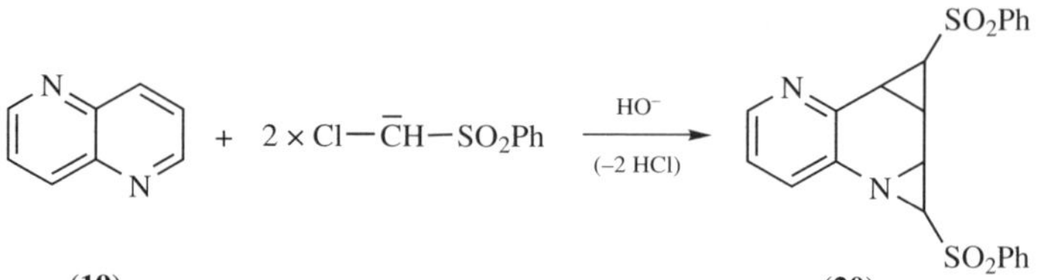

Deuteration

1,5-Naphthyridine gave mainly 2,6-d_2-1,5-naphthyridine (21) (substrate, D$_2$O, 220°C, sealed: 88%)[302] or mainly perdeutero-1,5-naphthyridine (22) (Pt/asbestos, D$_2$O, 170°C, N$_2$, sealed, 24 h; then repeated: 96%).[629]

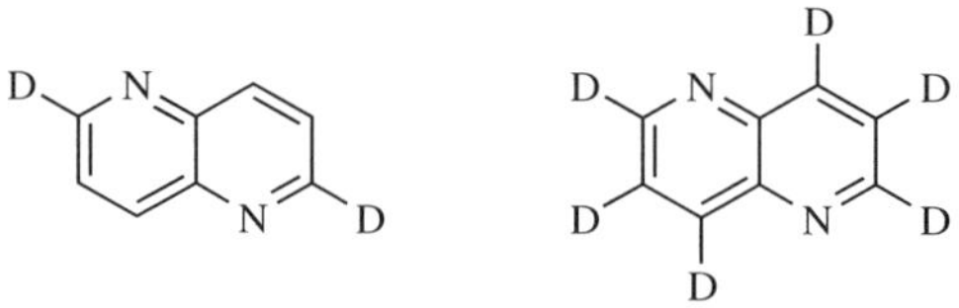

Halogenation

Vapor-phase bromination of 1,5-naphthyridine gave a (separable?) mixture of 2-bromo-, 2,6-dibromo-, 2,3,6-tribromo-, and 2,4,6-tribromo-1,5-naphthyridine (500°C, N$_2$; see original for details);[259] in contrast, bromination in fuming sulfuric acid gave 2-bromo- and an unidentified dibromo-1,5-naphthyridine (Br$_2$, H$_2$SO$_4$-SO$_3$, 135°C, sealed, 45 h, see original for details).[149] For another bromination procedure, see Section 3.1.1.

N-Oxidation

1,5-Naphthyridine gave 1,5-naphthyridine 1-oxide (23) and/or 1,5-naphthyridine 1,5-dioxide (24) according to the ratio of oxidant to substrate (MeCO$_3$H, 55°C, 20 h: ~20% monoxide or ~80% dioxide, after purification;[1047] 30% H$_2$O$_2$, Na$_2$WO$_4$·2H$_2$O, 55°C, 2 h: mono- and dioxide, 10% and 67% after separation;[232] or 30% H$_2$O$_2$, AcOH, 60°C, 2 h: dioxide, 98%).[971,cf. 262]

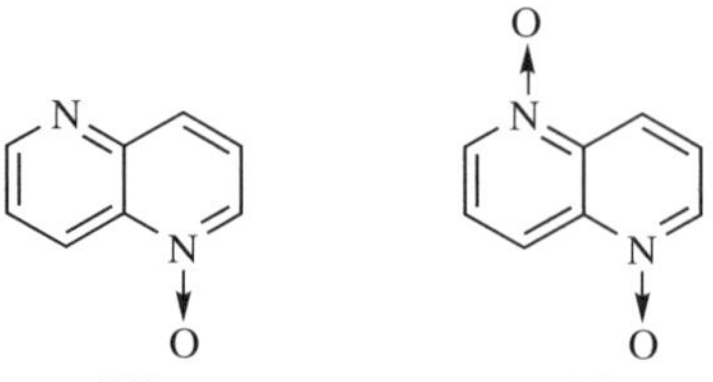

Reduction

Note: The preparative nuclear reduction of 1,5-naphthyridine has been covered in Section 2.1.1.

A spectral study of the electrochemical reduction of 1,5-naphthyridine has been reported.[829]

Reissert Reactions

1,5-Naphthyridine underwent a Reissert reaction with indole and benzoyl chloride to give 1-benzoyl-2-(indol-3-yl)-1,2-dihydro-1,5-naphthyridine (**25**) (reactants, PhMe, reflux, 4 h: 68%); some 2,6-bisheteroarylation occurred when Me$_2$NCHO was used as solvent.[791]

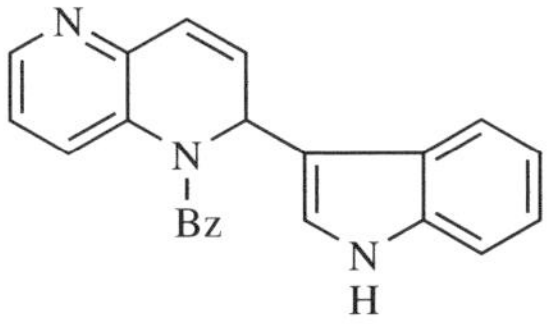

(**25**)

***N*-Trimethylsilylation**

1,5-Naphthyridine suffered reductive trimethylsilylation by chlorotrimethylsilane in the presence of potassium metal to give 1-trimethylsilyl-1,4-dihydro-1,5-naphthyridine (**26**) (reactants, THF, 20°C, until K gone: 19%).[889]

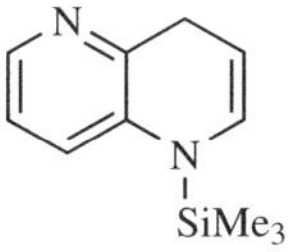

(**26**)

2.2. ALKYL- AND ARYL-1,5-NAPHTHYRIDINES

Although alkyl and aryl groups attached to a heterocyclic nucleus can undergo a wide variety of reactions and do have significant steric and electronic effects on adjacent substituents, little such data on 1,5-naphthyridines can be found in the literature.

The polarography of mono- and dimethiodides of 1,5-naphthyridine has been examined,[997,1303] the influence of *N*-methylation on the ^{13}NNMR spectra of 1,5- and other naphthyridines has been reported,[1319] and the NMR spectra of three monomethyl-1,5-naphthyridines have been compared with those of isomeric naphthyridines.[964]

2.2.1. Preparation of Alkyl- and Aryl-1,5-naphthyridines

Most known *C*-alkyl-1,5-naphthyridines have been made by primary syntheses (see Chapter 1) and most *N*-alkyl-1,5-naphthyridinium salts by quaternization. Other reported approaches are illustrated in Section 2.1.3 and by the following examples.

By *C*-Alkylation

Note: Regular and Mannich-type reactions have been used. The alkylation of unsubstituted 1,5-naphthyridine has been exemplified in Section 2.1.3.

3-Nitro-1,5-naphthyridine (**27**, R = H) gave 3-nitro-4-phenylsulfonylmethyl-1, 5-naphthyridine (**28**, R = H) (NaOH, Me$_2$SO; reactants↓; 20°C, 2 h: 87%); 2-substituted substrates reacted similarly to give, for example, 2-chloro-3-nitro-4-phenylsulfonylmethyl-1,5-naphthyridine (**28**, R=Cl) (73%).[854]

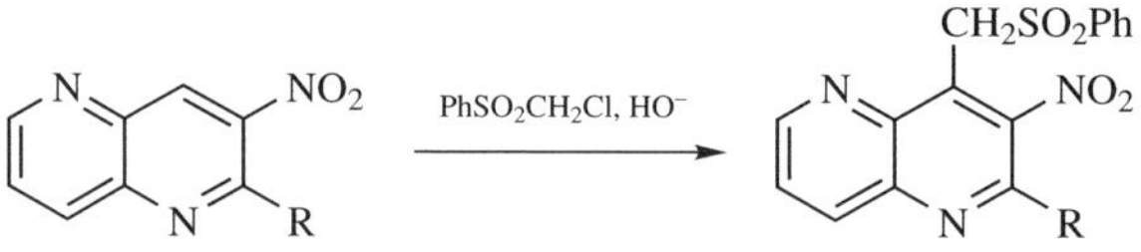

1,5-Naphthyridin-3(5*H*)-one (**29**) underwent Mannich alkylation to give 4-dimethylaminomethyl-1,5-naphthyridin-3(5*H*)-one (**30**) (Me$_2$NH, CH$_2$O, EtOH, H$_2$O, 25°C, 4 h: 73%; analogs likewise).[932]

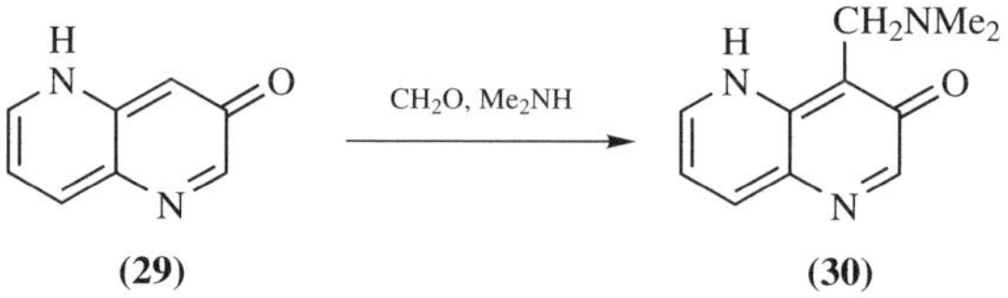

3-Bromo-8-(*p*-hydroxyanilino)-1,5-naphthyridine underwent extranuclear Mannich alkylation to give 3-bromo-8-[4-hydroxy-3,5-bis(4-methylpiperazin-1-yl)anilino]-1,5-naphthyridine (**31**) (substrate, CH$_2$O, 1-methylpiperazine, EtOH, reflux, 30 h: ~50%); analogs likewise.[1379]

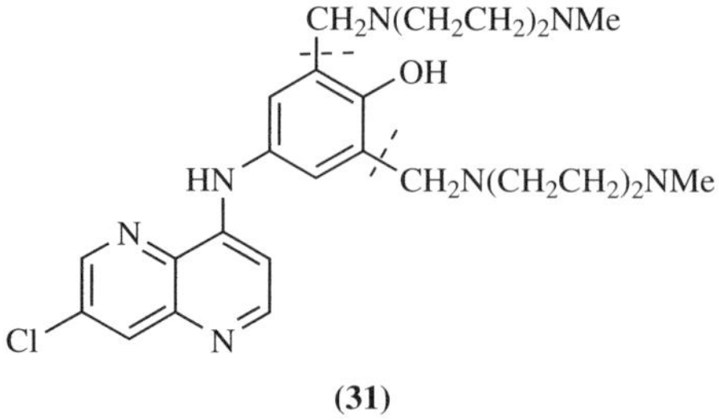

By Interconversion

2-Methyl-1,5-naphthyridine gave 2-styryl-1,5-naphthyridine (**32**) (PhCHO, AcOH, Ac$_2$O, reflux, 48 h: 72%).[155]

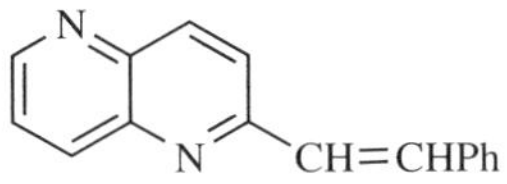

(**32**)

By Quaternization

Note: Quaternization of unsubstituted 1,5-naphthyridine has been exemplified in Section 2.1.3; other quaternizations are illustrated here.

2,6-Dimethyl-1,5-naphthyridine (**34**) gave 1,2,6-trimethyl-1,5-naphthyridinium iodide (**33**) (MeI, MeCN, reflux, 5 h: 82%) or 1,2,5,6-tetramethyl-1, 5-naphthyridinedium bisperchlorate (**35**) (neat Me$_2$SO$_4$, 125°C, 1 h: solid, HClO$_4$: 95%).[1169]

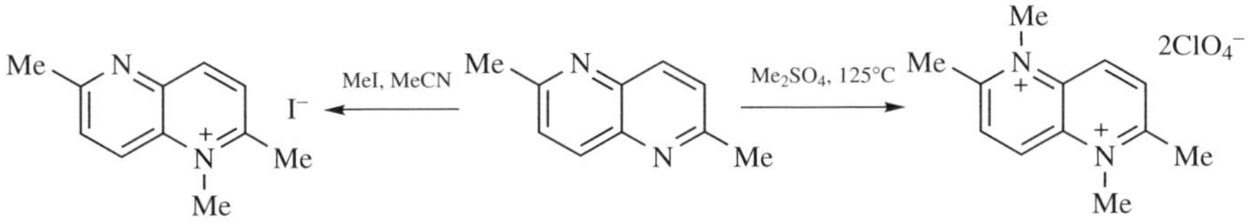

(**33**) (**34**) (**35**)

1,6-Dimethyl-1,5-naphthyridin-2(1*H*)-one gave 1,2,5-trimethyl-6-oxo-5,6-dihydro-1,5-naphthyridin-l-ium iodide (**36**) (MeI, EtOH, reflux, 16 h: 18%).[1051]

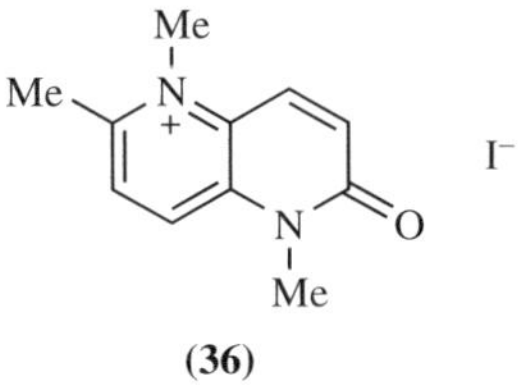

(**36**)

By Rearrangement

4,8-Dimethoxy-1,5-naphthyridine (**37**) rearranged into 1,5-dimethyl-1, 5-naphthyridine-4,8(1*H*,5*H*)-dione (**38**) (MeI, 225°C, sealed, 2.5 h: 78%) and thence, by prolonged similar treatment, into 3,7-dimethyl-1,5-naphthyridine-4,8 (1*H*,5*H*)-dione (**39**) (12 h: 55%); the initial substrate (**37**) gave the final product (**39**) in a one-pot reaction under similar conditions (12 h: 52%).[978,cf. 301]

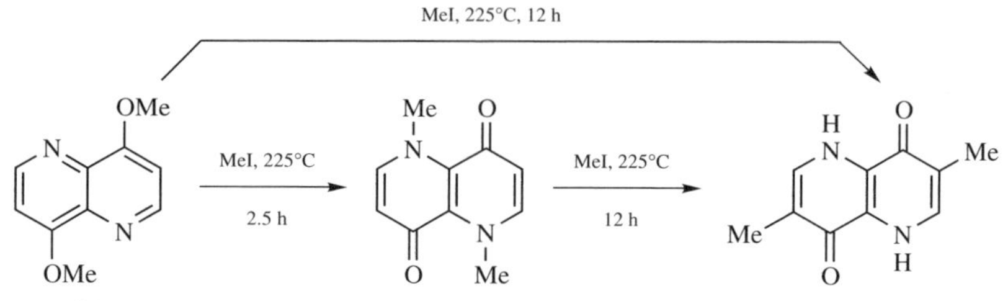

22 1,5-Naphthyridine, Alkyl-1, 5-naphthyridines, and Aryl-1,5-naphthyridines

2.2.2. Reactions of Alkyl- and Aryl-1,5-naphthyridines

Only reactions specific to the alkyl or aryl groups in such 1,5-naphthyridines are illustrated in the following examples.

Acylation

2-Methyl-1,5-naphthyridine (**40**) with ethyl picolinate gave 2-(picolinoylmethyl)-1,5-naphthyridine (**41**) [reactants, THF, N_2, $-78°C$; $(Me_3Si)_2NK$/PhMe↓ 10 min; then $-78°C$, 1 h: >95%]; several analogs likewise.[268]

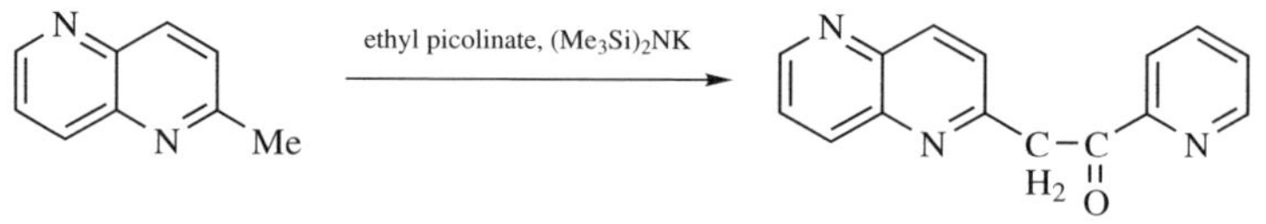

Halogenation

2-Methyl-1,5-naphthyridine 1,5-dioxide (**42**, R=H) gave 2-bromomethyl-1,5-naphthyridine 1,5-dioxide (**42**, R=Br) (dioxane dibromide: 81%; for further details, see original).[290] See also Section 3.1.1.

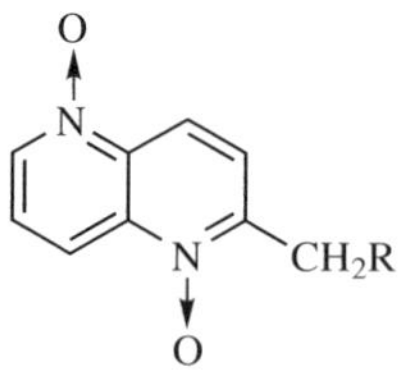

Imination

1,5-Naphthyridine methiodide (**43**) underwent oxidative imination to 1-methyl-1,5-naphthyridin-2 (1*H*)-imine hydriodide (**44**), accompanied by partial hydrolysis to 1,5-naphthyridin-2(1*H*)-one hydriodide (**45**) during workup (liquid NH_3, $KMnO_4$, 20 min: 3% and 11%, respectively, after separation).[1093,cf. 460]

(**43**) (**44**) (**45**)

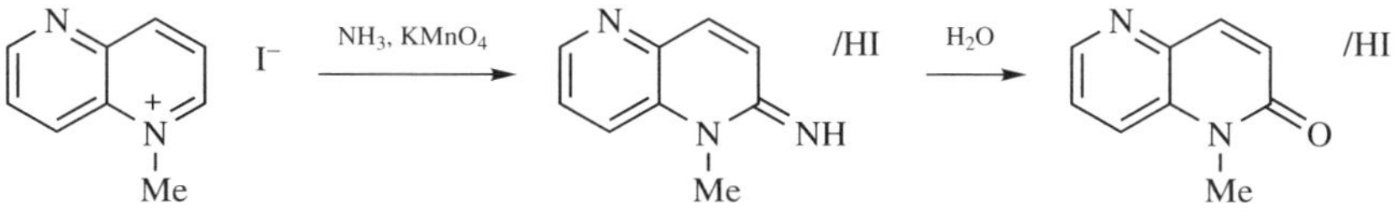

Oxidative Reactions

3-Ethyl-1,5-naphthyridine (**46**) gave 1,5-naphthyridine-3-carboxylic acid (**47**) (substrate, H_2O, 70°C; $KMnO_4$↓ during 1 h: then 70°C, 30 min: ~35%, allowing for some recovered substrate).[155]

(**46**) (**47**)

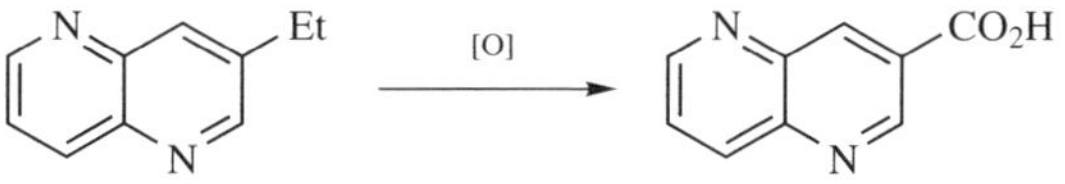

2-Methyl-1,5-naphthyridine (**49**) with permanganate gave 3-acetamido-2-pyridinecarboxylic acid (**48**) (substrate, H_2O, 70°C; $KMnO_4$↓ slowly, 70°C, 15 min: 40%), with selenium dioxide gave 1,5-naphthyridine-2-carboxylic acid (**50**) (SeO_2, pyridine, reflux, 6 h: 16%), or with peroxide/tungstate gave 2-methyl-1,5-naphthyridine 1,5-dioxide (**52**) (H_2O_2, Na_2WO_4, 40°C, 75 h: 71%) and thence 1,5-naphthyridine-2-carbaldehyde 1,5-dioxide (**51**) (SeO_2 Me_2NCHO, 20°C, 7 days: 68%);[797] in contrast, 2-styryl-1,5-naphthyridine (**53**) with permanganate gave 1,5-naphthyridine-2-carboxylic acid (**50**)) $KMnO_4$, AcMe, 25°C, 15 min: 62%).[797]

(**48**) (**49**) (**50**) (**51**) (**52**) (**53**)

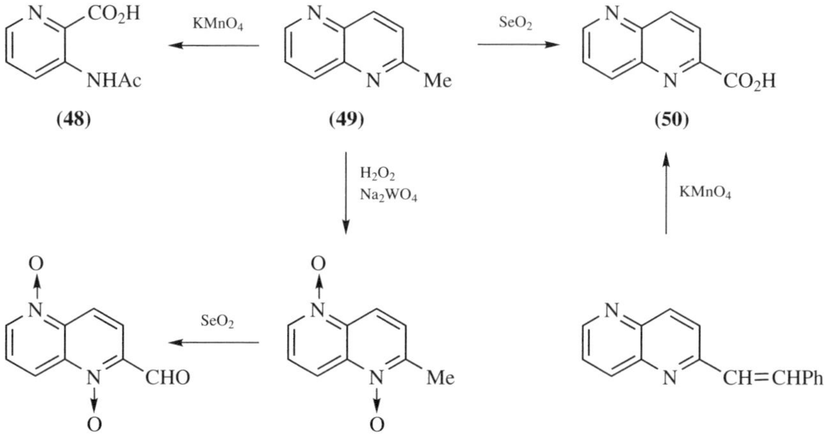

1-Allyl-6-methoxy-2-phenyl-1,2-dihydro-1,5-naphthyridine (**54**) underwent nuclear dehydrogenation (by loss of propene?) to give 2-methoxy-6-phenyl-1,5-naphthyridine (**55**) [(Ph$_3$P)$_4$RhH, EtOH, F$_3$CCO$_2$H, reflux, 4 h: 54%: or neat substrate, 180°C, 20 h: 43%]: analogs likewise.[619]

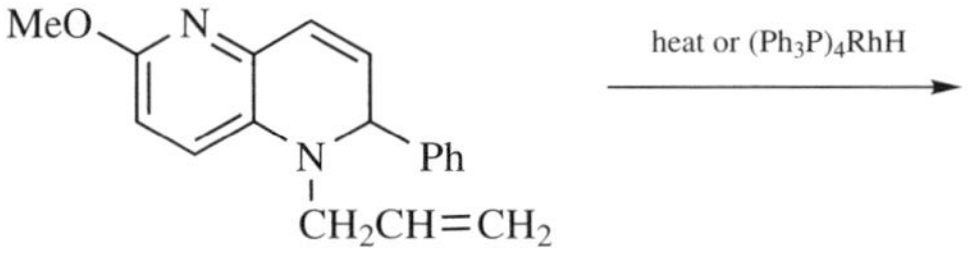

4-Methyl-1,5-naphthyridine 1-oxide (**56**, R=Me) gave 1,5-naphthyridine 4-carbaldehyde 1-oxide (**56**, R=CHO) (SeO$_2$, AcOEt, reflux, 3.5 h: 66&).[814]

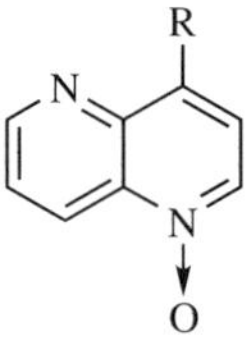

CHAPTER 3

Halogeno-1,5-naphthyridines

A halogeno substituent in the 2-, 4-, 6-, or 8-position of 1,5-naphthyridine will be activated by either an *ortho-* or a *para*-ring nitrogen, so its reactivity will resemble *o-* or *p*-chloronitrobenzene unless it is affected electronically or structurally by other substituents. In contrast, a halogeno substituent in the 3- or 7-position will be only marginally more active than a halogenonaphthalene. An extranuclear halogeno substituent will approximate in activity that of benzyl chloride unless affected by an adjacent group on the side chain.

Halogeno-1,5-naphthyridines have proved very useful as intermediates for the synthesis of many other 1,5-naphthyridine derivatives; in this respect, the more easily available chloro and bromo derivatives have been used almost exclusively.

Halogeno-1,5-naphthyridines have also shown useful bioactivity in their own right, for example, as the herbicide 2-chloro-7-[p-(2-carboxyethoxy)phenoxy]-1,5-naphthyridine.[470] The mass spectra of a variety of halogeno-1,5-naphthyridines have been reported.[546] Antimalarial activity has been detected.[1161, 1165]

3.1. PREPARATION OF HALOGENO-1,5-NAPHTHYRIDINES

Some halogeno-1,5-naphthyridines have been made by *primary syntheses* (see Chapter 1), but several other preparative routes have been more widely used, as indicated in the following subsections.

3.1.1. By Direct Halogenation

This route has occasionally been followed to make both nuclear and extranuclear halogeno-1,5-naphthyridines. Some examples follow.

1,5-Naphthyridin-2(1*H*)-one (**1**, R = H) gave 3-bromo-1,5-naphthyridin-2(1*H*)-one (**1**, R = Br) (substrate, H_2O, "warm"; $Br_2\downarrow$ dropwise: 71%); 1,5-

---

The Naphthyridines: The Chemistry of Heterocyclic Compounds, Volume 63, by D.J. Brown
Copyright © 2008 John Wiley & Sons, Inc.

naphthyridin-4(1H)-one (**2**, R = H) likewise gave 3-bromo-1,5-naphthyridin-4(1H)-one (**2**, R = Br) (65%).[1037]

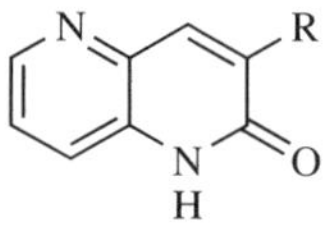

(1)

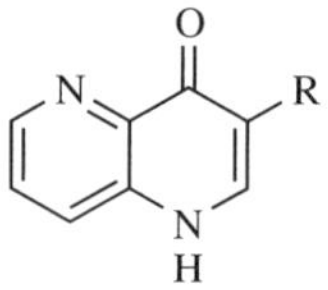

(2)

1,5-Naphthyridine (**3**, Q = R = H) gave a separable mixture of 3-bromo- (**3**, Q = Br, R = H) and 3,7-dibromo-1,5-naphthyridine (**3**, Q = R = Br) (substrate, Br$_2$, CCl$_4$, reflux, 1 h: then pyridine↓, reflux, 12 h: 27% and 10%, respectively);[173] for other procedures affording different brominated products, see Section 2.1.3.

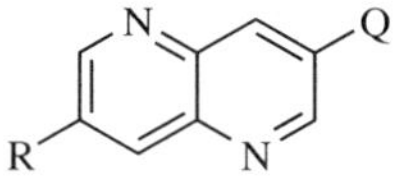

(3)

4-Diethoxycarbonylmethyl-1,5-naphthyridine 1-oxide (**4**, R = H) gave 4-(α-bromo-α,α-diethoxycarbonylmethyl)-1,5-naphthyridine 1-oxide (**4**, R = Br) (Br$_2$, CHCl$_3$, reflux, 4 h: 80%).[814]

For another example, see Section 2.2.2

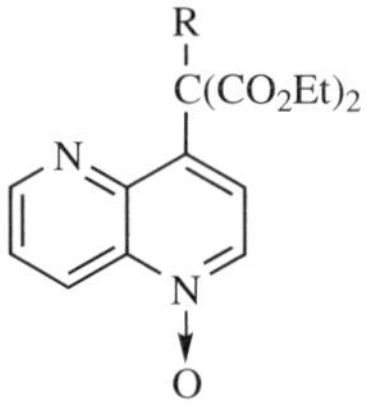

(4)

3.1.2. By Halogenolysis of 1,5-Naphthyridinones or the Like

The conversion of a tautomeric 1,5-naphthyridinone into the corresponding chloro- or bromo-1,5-naphthyridine is usually done by heating for several hours with neat phosphoryl chloride or phosphoryl bromide, respectively. A mixture of phosphoryl chloride and phosphorus pentachloride has been used for a few difficult cases. In addition, exceptional reactions have been reported in which a nontautomeric N-alkyl-1,5-naphthyridinone or a hydroxymethyl-1,5-naphthyridine was used successfully as substrate. The following examples illustrate the foregoing possibilities.

Using Neat Phosphoryl Chloride

1,5-Naphthyridin-4(1H)-one (5) gave 4-chloro-1,5-naphthyridine (6) (POCl$_3$, 100°C: ~60%);[101] somewhat similarly, 1,5-naphthyridin-2(1H)-one gave 2-chloro-1,5-naphthyridine (POCl$_3$, reflux, 30 min: >95%).[87]

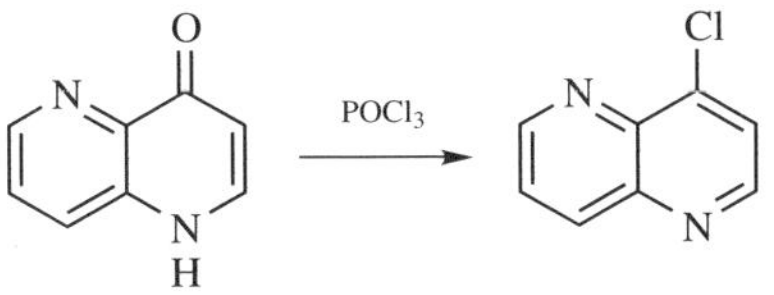

6-Methyl-1,5-naphthyridin-2(1H)-one gave 2-chloro-6-methyl-1,5-naphthyridine (7) (POCl$_3$, reflux, 45 min: ~85%;[1051] or likewise, 90 min: 65%);[268] 6-methoxy-1,5-naphthyridin-2(1H)-one gave 2-chloro-6-methoxy-1,5-naphthyridine (8) (POCl$_3$, reflux, 12 h: 70%);[234] and 8-butoxy-1,5-naphthyridin-2(1H)-one gave 4-butoxy-6-chloro-1,5-naphthyridine (9) (POCl$_3$, 95°C, ~83%).[1036]

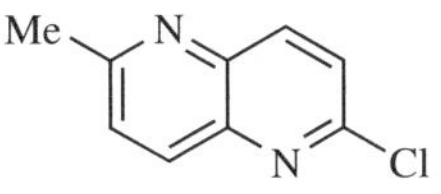

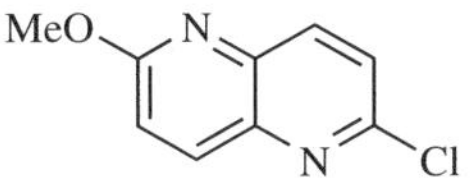

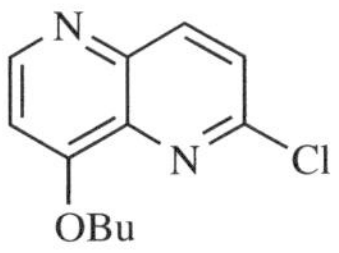

4-Hydroxy-1,5-naphthyridin-2(1H)-one (10) gave 2,4-dichloro-1,5-naphthyridine (11) (POCl$_3$, reflux, 6 h: 70–80%);[1023,1151] 1,5-naphthyridine-4,8 (1H,5H)-dione gave 4,8-dichloro-1,5-naphthyridine (12) (POCl$_3$, 180°C, sealed, 6 h: 82%).[301]

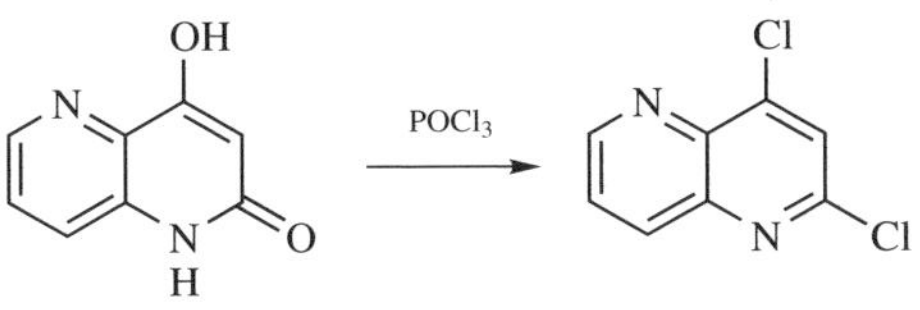

Ethyl 4-oxo-1,4-dihydro-1,5-naphthyridine-3-carboxylate (13, R = H) gave ethyl 4-chloro-1,5-naphthyridine-3-carboxylate (14, R = H) (POCl$_3$: for details, see original);[433] ethyl 6-p-chlorophenoxy-4-oxo-1,4-dihydro-1,5-naphthyridine-3-carboxylate (13, R = OC$_6$H$_4$Cl-p) gave ethyl 4-chloro-6-

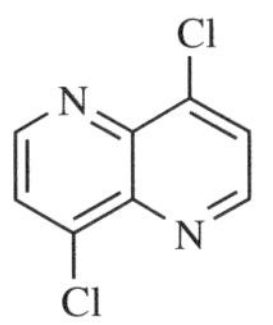

p-chlorophenoxy-1,5-naphthyridine-3-carboxylate (**14**, R = OC$_6$H$_4$Cl-p) (POCl$_3$, reflux, 1 h: 96%).[967]

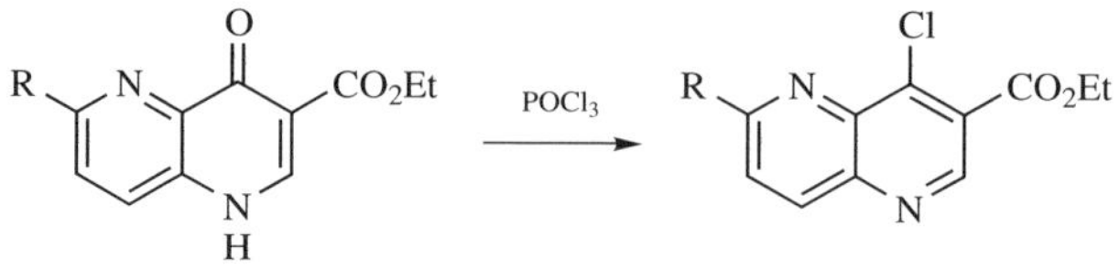

7-Bromo-1,5-naphthyridin-4(1H)-one gave 3-bromo-8-chloro-1,5-naphthyridine (**15**) (POCl$_3$, reflux, 10 h: ~85%);[1377] 3-nitro-1,5-naphthyridin-4 (1H)-one gave 4-chloro-3-nitro-1,5-naphthyridine (**16**) (POCl$_3$, reflux, 2 h: 78%);[48] and 3-nitro-1,5-naphthyridin-2(1H)-one gave 2-chloro-3-nitro-1,5-naphthyridine (**17**) (POCl$_3$, reflux, 15 h: 60%).[818]

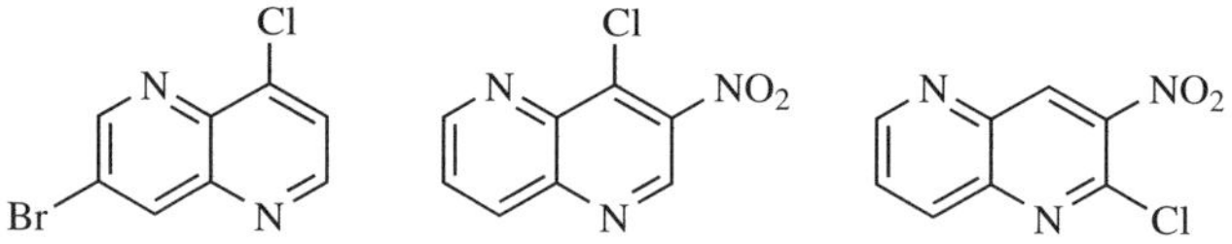

Using Phosphorus Pentachloride in Phosphoryl Chloride

Note: This mixture has been used for chlorolyses that were perceived as difficult, but phosphorus pentachloride can introduce complications.

3-Nitro-1,5-naphthyridin-2 (1H)-one (**18**) gave a 1 : 1 mixture of 2-chloro-3-nitro- (**19**, R = NO$_2$) and 2,3-dichloro-1,5-naphthyridine (**19**, R = Cl) (PCl$_5$, POCl$_3$, reflux, 2.5 h: ~40% of a mixture that afforded ~10% yield of each product after separation);[818] thus neat POCl$_3$ was better (see preceding example).

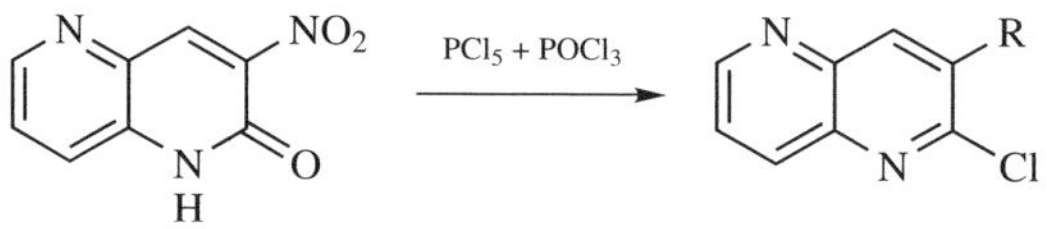

8-Nitro-1,5-naphthyridin-4(1H)-one gave 4-chloro-8-nitro-1,5-naphthyridine (**20**) (PCl$_5$, POCl$_3$, reflux, 75 min; 68%).[48]

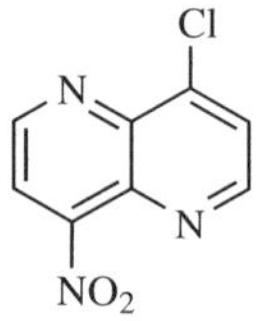

1,5-Naphthyridine-4,8 (1H, 5H)-dione gave 4, 8-dichloro-1,5-naphthyridine (**21**) (PCl$_5$, POCl$_3$, reflux, 1 h: 62%).[233] The yield was lower than that obtained by using neat phosphoryl chloride (see preceding list of examples), but a sealed-tube reaction was avoided by this procedure.

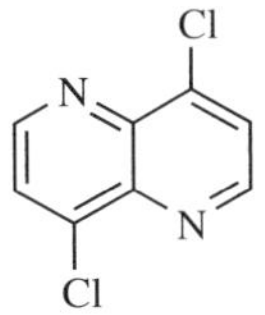

(**21**)

1,5-Naphthyridin-2 (1H)-one gave 2-chloro-1,5-naphthyridine (PCl$_5$, POCl$_3$, reflux, 4 h: 70%;[262] again a lower yield than when neat POCl$_3$ was used).

Using Phosphoryl Bromide

1,5-Naphthyridin-2(1H)-one (**22**) gave 2-bromo-1,5-naphthyridine (**23**) (neat POBr$_3$, 125°C, sealed, 3 h: 80%); similarly 1,5-naphthyridin-4 (1H)-one gave 4-bromo-1,4-naphthyridine (**24**) (130°C, 4 h: 60%).[811]

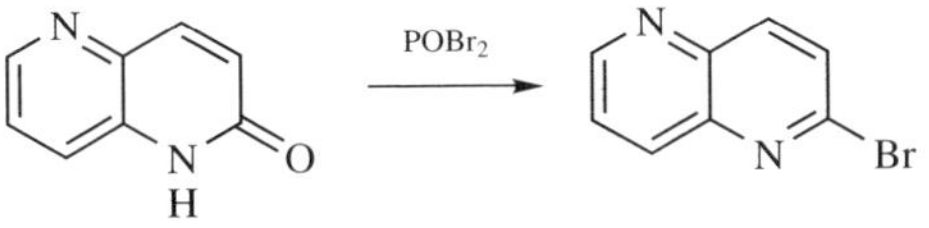

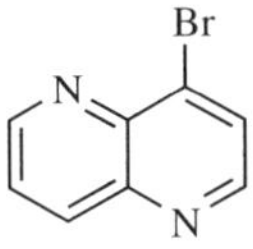

(**22**)   (**23**)   (**24**)

7-Bromo-1,5-naphthyridin-4(1H)-one (**25**, R = Br) gave 4,7-dibromo-1,5-naphthyridine (**26**, R = Br) (neat POBr$_3$, 140°C, 4 h: 70%);[1179] similarly, 7-trifluoromethyl-1,5-naphthyridin-4 (1H)-one (**25**, R = CF$_3$) gave 4-bromo-7-trifluoromethyl-1,5-naphthyridine (**26**, R = CF$_3$) (~80%).[1179]

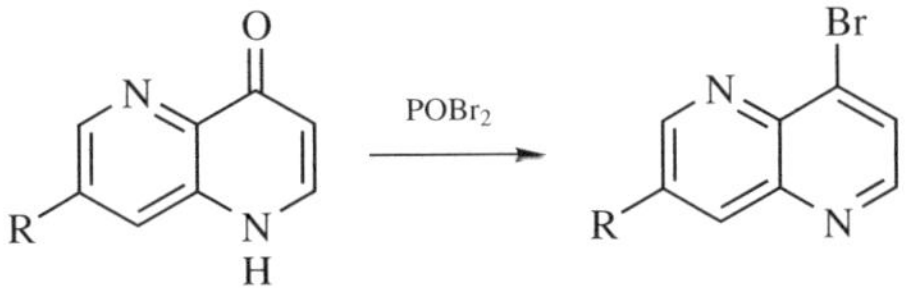

(**25**)   (**26**)

Also other examples.[575,822]

3.1.3. By the Meissenheimer Reaction on 1,5-Naphthyridine N-Oxides

The conversions of 1,5-naphthyridine N-oxides into C-halogeno-1,5-naphthyridines by the Meissenheimer reaction (POCl$_3$ or POBr$_3$) or a related reaction (Br$_2$ in Ac$_2$O) both suffer from the fact that two or more products are usually formed in each case. This introduces a sometimes difficult separation procedure, so that such an approach to halogeno-1,5-naphthyridines is, as a rule, more of interest than utility. The following typical examples illustrate these points.

Submission of 1,5-naphthyridine 1-oxide (**27**) to Meissenheimer conditions has been reported to afford only 2-chloro-1,5-naphthyridine (**28**) ($POCl_3$, PCl_5, reflux, 20 min: ~75%);[1047] a difficult-to-separate 2 : 3 mixture of 2- (**28**) and 4-chloro-1,5-naphthyridine (**30**) ($POCl_3$, PCl_5, reflux, 20 min: low yields after separation);[262] or a chromatographically separable mixture of 2- (**28**), 3- (**29**), and 4-chloro-1,5-naphthyridine (**30**) ($POCl_3$, reflux, 20 min: 34%, 3%, and 43%, respectively, after separation).[232,cf. 225] Meissenheimer brominarion of the same substrate gave broadly similar results.[439,448]

(**27**) (**28**) (**29**) (**30**)

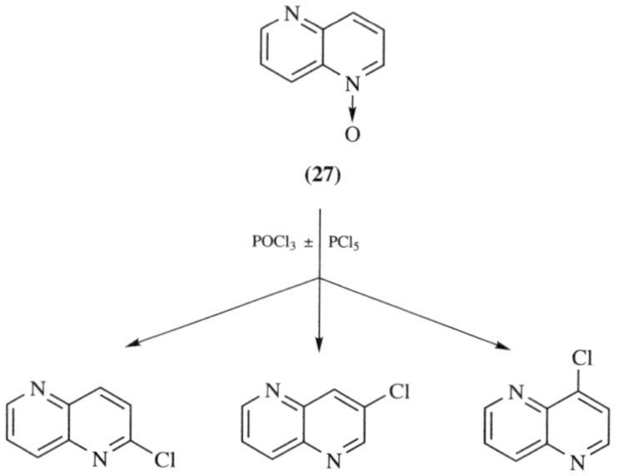

Similar treatment of 1,5-naphthyridine 1,5-dioxide (**31**) has been reported to give only 2,6-dichloro-1,5-naphthyridine (**33**) ($POCl_3$, reflux, 20 min: 37 or 54%)[971,1047] or a mixture of 2,4- (**32**), 2,6- (**33**), 2,7- (**34**), 2,8- (**35**), and 3,8-dichloro-1,5-naphthyridine (**36**) from which all except the 2,7-isomer were isolate eventually in a pure state.[233]

(**31**) (**32**) (**33**) (**34**) (**35**) (**36**)

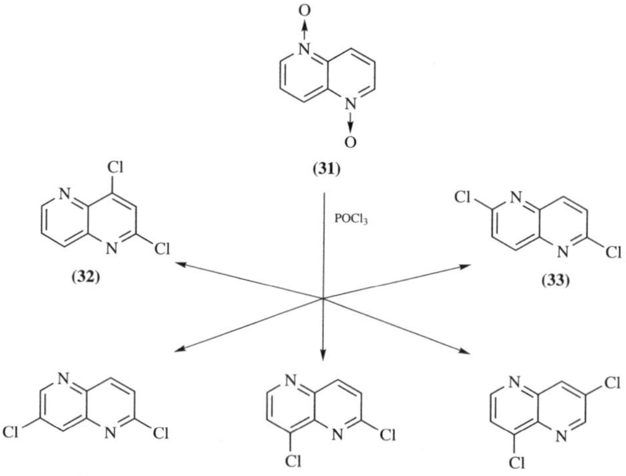

Treatment of 1,5-naphthyridine 1-oxide with bromine in acetic anhydride gave a complicated mixture in which 2-bromo- and 2,7-dibromo-1,5-naphthyridine, 3-bromo- and 3,6-dibromo-1,5-naphthyridine 1-oxide, and 3-bromo-1, 5-naphthyridin-2 (1H)-one were identified.[317]

3.1.4. By Miscellaneous Procedures

Minor routes to halogeno-1,5-naphthyridines are illustrated by the following examples.

By Transhalogenation

2-Chloro- gave 2-iodo-1,5-naphthyridine (NaI, 50% HT, AcEt, 95°C, 8 h: 47%); likewise, 4-chloro- gave 4-iodo-1,5-naphthyridine (68%).[692]

From Nontautomeric Naphthyridinones

1,5-Dimethyl-1,5-naphthyridine-2,6 (1H, 5H)-dione (**37**) gave 2,6-dichloro-1, 5-naphthyridine (**38**) (PCl$_5$, POCL$_3$, reflux, 9 h: 14%; mechanism unclear).[233]

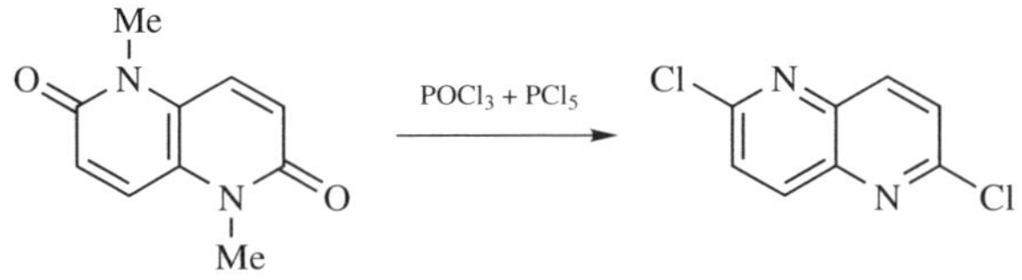

3.2. REACTIONS OF HALOGENO-1,5-NAPHTHYRIDINES

The utility of halogeno-1,5-naphthyridines as intermediates is indicated by the variety of their reactions. In terms of nucleophilic displacement, the positional order of reactivity is 2(6)-halogeno > 4(8)-halogeno ≫ 3(7)-halogeno-1,5-naphthyridines, but individual reactivities may be affected significantly by the presence of electron-releasing or electron-withdrawing substituents. This situation is especially evident in Sections 3.2.1 and 3.2.2.

3.2.1. Alcoholysis or Phenolysis of Halogeno-1,5-naphthyridines

Although alkoxy ion is a more powerful nucleophile than aryloxy ion, both types react vigorously with activated halogeno-1,5-naphthyridines and even (albeit sluggishly) with 3(7)-halogeno-1,5-naphthyridines. In the absence of kinetic data, the following preparative examples illustrate the practicalities of such displacement reactions.

2(6)-Halogeno-1,5-naphthyridines as Substrates

2-Chloro-1,5-naphthyridine (**39**, R = H) gave 2-methoxy-1,5-naphthyridine (**40**, R = H) (MeONa, MeOH, reflux, 4 h: 38%);[234] 2-chloro-6-methyl- (**39**, R = Me) gave 2-methoxy-6-methyl- (**40**, R = Me) (likewise, 7 h: 75%), 2-butoxy-6-methyl- (BuONa, BuOH: no further details), or 6-methyl-2-phenoxy-1,5-naphthyridine (substrate, PhOH, 180°C, NH$_3$↓, 7 h: ?%).[1051]

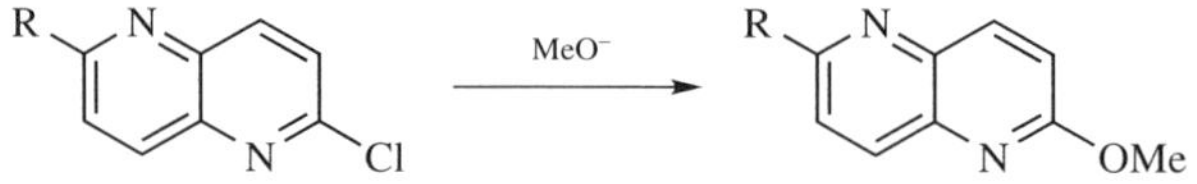

(**39**) → MeO⁻ → (**40**)

2-Chloro-3-nitro-1,5-naphthyridine (**41**, R = Cl) gave 2-ethoxy-3-nitro-1,5-naphthyridine (**41**, R = OEt) (KOH, EtOH, 20°C, 12 h: 90%; note mild conditions due to the activating effect of the nitro group).[827]

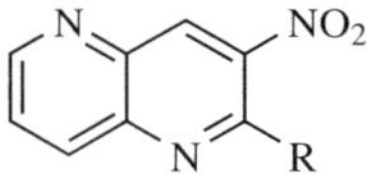

(**41**)

2-Iodo-1,5-naphthyridine with sodium nitrite in dimethylformamide gave bis(1,5-naphthyridin-2-yl) ether (140°C, 3 h: 44%; mechanism unclear).[692] Also other examples.[271]

4(8)-Halogeno-1,5-naphthyridines as Substrates

4-Chloro-1,5-naphthyridine (**42**) gave 4-methoxy-1,5-naphthyridine (**43**, R = Me) (MeONa, MeOH, reflux, 4 h: 75%)[814] somewhat similarly: 48%)[202] or 4-phenoxy-1,5-naphthyridine (**43**, R = Ph) [substrate, PhOH, 95°C, 10 min; MeCHNH$_2$(CH$_2$)$_3$NEt$_2$↓, 95°C, 3 h: 48%; note lack of aminolysis].[101]

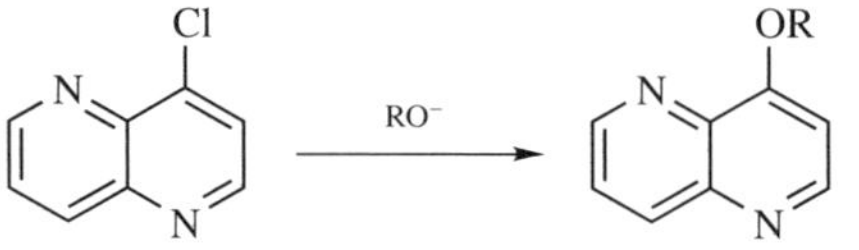

(**42**) → RO⁻ → (**43**)

2-Butoxy-8-chloro- (**44**, R = Cl) gave 2-butoxy-8-phenoxy-1,5-naphthyridine (**44**, R = OPh) (KOH, PhOH, 160°C, 3 h: 84%).[1036]

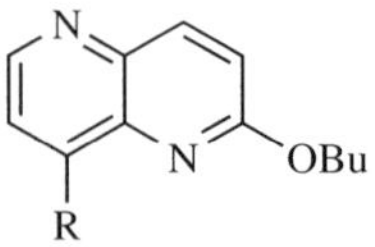

(**44**)

3(7)-Halogeno-1,5-naphthyridines as Substrates

Ethyl 7-chloro-1-ethyl-4-oxo-1,4-dihydro-1,5-naphthyridine-3-carboxylate (**45**) gave 1-ethyl-7-methoxy-4-oxo-1,4-dihydro-1,5-naphthyridine-3-carboxylic acid (**46**) (KOH, MeOH, dibenzo-1,8-crown-6; MeCN↓, reflux, 7 h: 67%; it would seem that the crown ether was required to assist alcoholysis of the unactivated chloro substituent; note incidental saponification of the ester group).[1398]

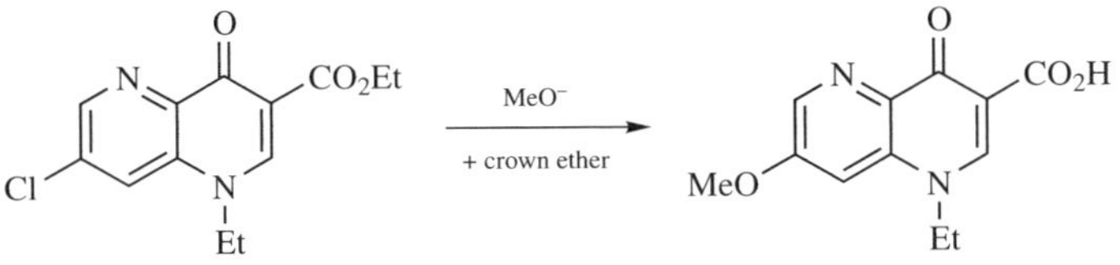

Dihalogeno-1,5-naphthyridines as Substrates

Note: It is clearly possible to achieve monoalcoholysis of such substrates, especially when one halogeno substituent is in a less activated position.

2,4-Dichloro-1,5-naphthyridine (**48**) gave 2-methoxy-4-chloro-1,5-naphthyridine (**47**) (limited MeONa, MeOH, reflux, 1 h: ~65%).[1151] or 2,4-diphenoxy-1,5-naphthyridine (**49**) (excess PhOH, reflux, 6 h: 89%).[1023]

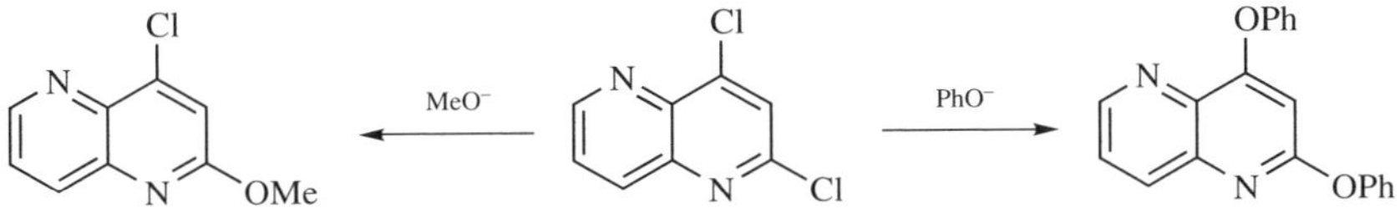

4,8-Dichloro- (**50**, R = Cl) gave 4,8-dimethoxy-1,5-naphthyridine (**50**, R = OMe) (MeONa, MeOH, reflux, 4 days: 87%).[301]

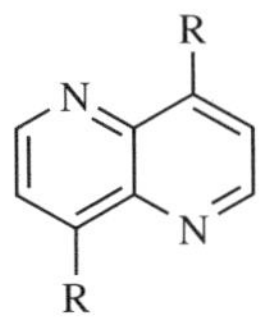

2,3-Dibromo- (**51**, R = Br) gave 3-bromo-2-ethoxy-1,5-naphthyridine (**52**, R = OEt) (EtONa, EtOH, reflux, 2.5 h: ~50%);[827] and 3-bromo-8-chloro-

(52, R = Cl) gave 3-bromo-8-methoxy-1,5-naphthyridine (52, R = OMe) (MeONa, MeOH, reflux; ~65%).[1377]

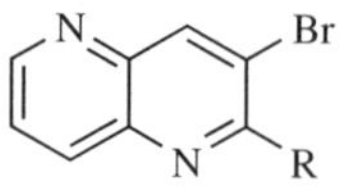

(51)

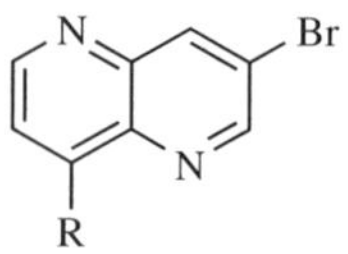

(52)

Also macrocyclic products from 2,6-dichloro-1,5-naphthyridine and triethylene glycol or the like.[389]

3.2.2. Aminolysis of Halogeno-1,5-naphthyridines

The ease of aminolysis of halogeno-1,5-naphthyridines is governed by several factors already outlined for nucleophilic displacement in general (Section 3.2) and alcoholysis (Section 3.2.1). The examples that follow are divided according to the number and position(s) of the halogeno substituent(s).

2(6)-Halogeno-1,5-naphthyridines as Substrates

2-Chloro-3-nitro- (53, R = Cl) gave 2-methylamino-3-nitro-1,5-naphthyridine (53, R = NHMe) (MeNH$_2$, MeOH, 20°C, 1 h: 58%).[854]

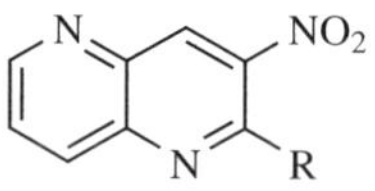

(53)

2-Chloro-6-methyl- (54, R = Cl) gave 2-anilino-6-methyl- (54, R = NHPh) (neat PhNH$_2$, reflux, 2 h: >95%) or 2-methyl-6-piperidino-1,5-naphthyridine [54, R = N(CH$_2$)$_5$] [neat HN(CH$_2$)$_5$, 180°C, sealed, 7 h: 30% (as picrate)].[1051]

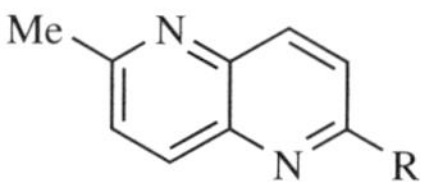

(54)

Also other examples.[137,271,818,837,1036]

3(7)-Halogeno-1,5-naphthyridines as Substrates

Note: Because they are essentially unactivated, these substrates would be expected to require vigorous conditions for aminolysis.

3-Bromo-1,5-naphthyridine (**55**, R = Br) gave 1,5-naphthyridin-3-amine (**55**, R = NH$_2$) (NH$_4$OH, CuSO$_4$ 170°C, sealed, 40 h: 75%).[811]

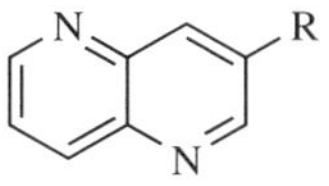

(**55**)

7-Bromo-1-ethyl-4-oxo-1,4-dihydro-1,5-naphthyridine-3-carboxylic acid (**56**, R = Br) gave 7-dimethylamino-1-ethyl-4-oxo-1,4-dihydro- (**56**, R = NMe$_2$) (Me$_2$NH, Me$_2$NCHO, EtOH, CuSO$_4$, 110°C, sealed, 5 h: 20%) or 1-ethyl-4-oxo-7-(pyrrolidin-1-yl)-1,4-dihydro-1,5-naphthyridine-3-carboxylic acid [**52**, R = N(CH$_2$)$_4$] [HN (CH$_2$)$_4$, Me$_2$NCHO, 125°C (reflux), 6 h: 44%].[1398]

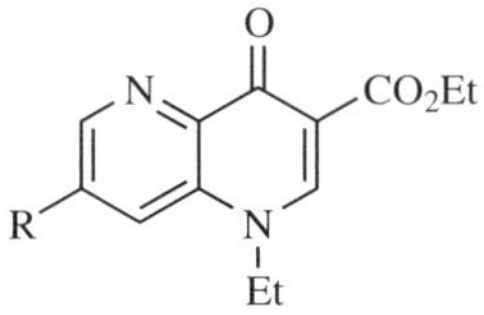

(**56**)

4(8)-Halogeno-1,5-naphthyridines as Substrates

4-Chloro-1,5-naphthyridine (**57**) gave 1,5-naphthyridin-4-amine (**58**, R = H) (substrate, PhOH, AcNH$_2$, 170°C, NH$_3$↓, 6 h: 86%),[164,cf.1371] 4-(5-diethylaminopentylamino)-1,5-naphthyridine [**58**, R = (CH$_2$)$_5$NEt$_2$] [neat H$_2$N(CH$_2$)$_5$NEt$_2$, 90°C, 36 h: 85%],[101] or 4-hydrazino-1,5-naphthyridine (**58**, R = NH$_2$) (H$_2$NNH$_2$·H$_2$O, EtOH, reflux, 4.5 h: 51%).[814]

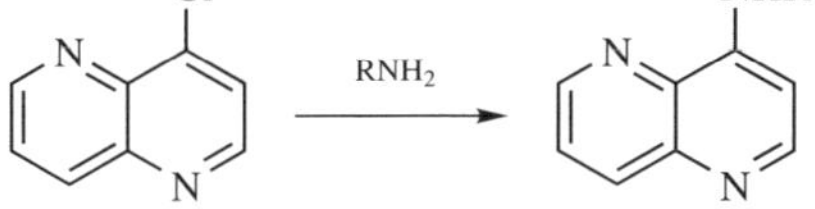

(**57**) → (**58**)

4-Chloro-1,5-naphthyridine 1-oxide (**59**, R = Cl) gave 4-isopropylamino- (**59**, R = NHPr^i) (Pr^{i}NH$_2$, Pr^{i}OH, 110°C, sealed, 7 h: 51% as hydrochloride) or 4-hydrazino-1,5-naphthyridine 1-oxide (**59**, R = NHNH$_2$) (H$_2$NNH$_2$.H$_2$O, EtOH, 50°C, 90 min: 62%);[814] 4-chloro-3-nitro-1,5-naphthyridine (**60**, R = Cl) gave 3-nitro-4-tosylhydrazino-1,5-naphthyridine (**60**, R = NHNHTs) (TsNHNH$_2$, CHCl$_3$, 20°C, 24 h: 52%).[48] Note the apparent facilitation of aminolysis by the type of nucleophile or by activation of the halogeno leaving group by the *N*-oxide entity.

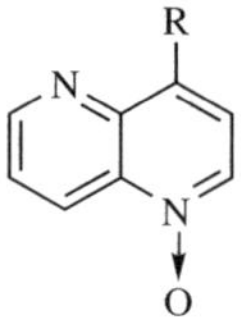

(59)

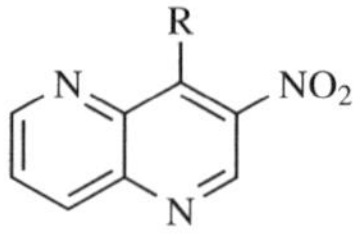

(60)

4-Chloro-2-methoxy-1,5-naphthyridine gave 4-[3-(diethylaminomethyl)-4-hydroxyanilino]-2-methoxy-1,5-naphthyridine (**61**) [4-amino-3-(diethylaminomethyl)phenol. 2HCl, EtOH, H$_2$O, reflux, 4 h: ~50%];[1151] also analogous aminolyses.[432]

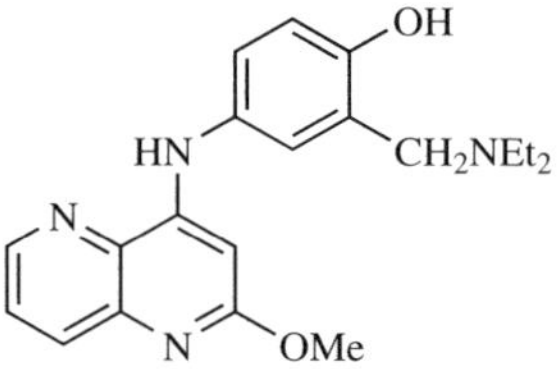

(61)

Also other examples.[433,811,837,967,1443]

2,3(6,7)-Dihalogeno-1,5-naphthyridines as Substrates

2,3-Dibromo-1,5-naphthyridine gave 3-bromo-1,5-naphthyridin-2-amine (**62**) (NH$_3$, EtOH, 160°C, sealed, 48 h: ?%).[822]

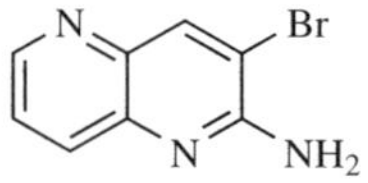

(62)

2,4(6,8)-Dihalogeno-1,5-naphthyridines as Substrates

2,4-Dichloro- (**63**, Q = R = Cl) gave 2,4-bis(benzylamino)- (**63**, Q = R = NHCH$_2$Ph) (neat PhCH$_2$NH$_2$, reflux, 4 h: 88%), 2,4-dianilino- (**63**, Q = R = NHPh) (neat PhNH$_2$, 180°C, 5 h: 86%), or 4-chloro-2-hydrazino-1,5-naphthyridine (**63**, Q = NHNH$_2$, R = Cl) (H$_2$NNH$_2$·H$_2$O, dioxane, reflux, 16 h: 68%);[1023] the same substrate (**63**, Q = R = Cl) gave 4-chloro-1,5-naphthyridin-2-amine (**63**, Q = NH$_2$, R = Cl) (NH$_3$, EtOH, 170°C, sealed, 20 h: 47%).[1023]

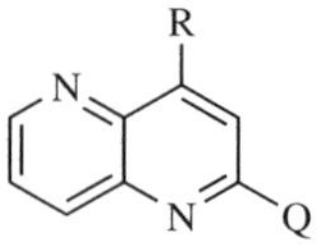

(63)

2,6-Dihalogeno-1,5-naphthyridines as Substrates

2,6-Dichloro-1,5-naphthyridines gave 2-chloro-6-hydrazino-1,5-naphthyridine (**64**) (H$_2$NNH$_2$,H$_2$O, EtOH, 100°C, sealed?, 1 h: 70%); since an excess of hydrazine was used and both chloro substituents are equally activated, the entry of one hydrazino group must have deactivated the second chloro substituent sufficiently to resist aminolysis under the conditions used);[234] the same substrate gave 2,6-dianilino-1,5-naphthyridine (**65**) (neat PhNH$_2$, reflux, 5 min; then 95°C, 1 h: ?%);[1047] and 2,6-dibromo-1,5-naphthyridine gave 6-bromo-1,5-naphthyridin-2-amine (**66**) (NH$_4$OH, 160°C, sealed, 6 h: 60%).[246]

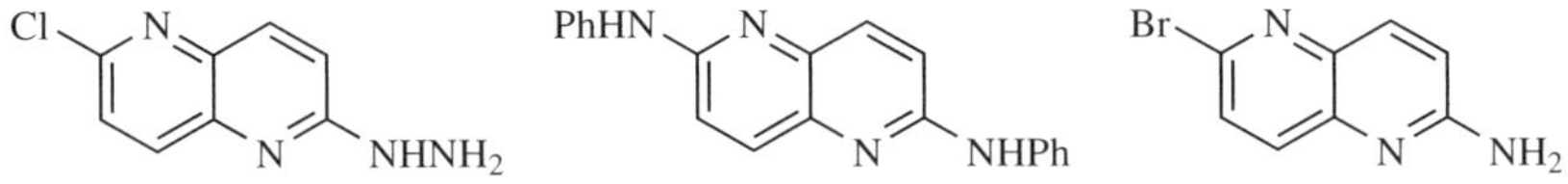

(64)   (65)   (66)

3,4(7,8)-Dihalogeno-1,5-naphthyridines as Substrates

3,4-Dibromo- (**67**, R = Br) gave 3-bromo-1,5-naphthyridin-4-amine (**67**, R = NH$_2$) (NH$_3$, solvent?, 160°C, ? h: ?%).[374,cf. 149]

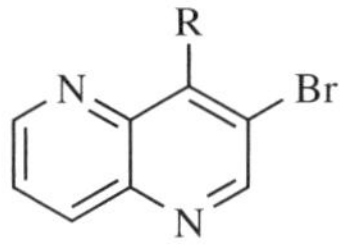

(67)

3,8(4,7)-Dihalogeno-1,5-naphthyridines as Substrates

3-Bromo-8-chloro-1,5-naphthyridine (**69**) with an appropriate amine gave 3-bromo-8-(2-diethylaminoethyl)amino-1,5-naphthyridine (**68**) (Et$_2$NCH$_2$CH$_2$NH$_2$, C$_7$H$_{16}$, 160°C, sealed, 20 h: 90%),[1377] 3-bromo-8-[3,5-bis(dipropylaminomethyl)-2-hydroxyanilino]-1,5-naphthyridine (**70**) [substrate, 2-amino-4,6-bis(dipropylaminomethyl)phenol, EtOH, H$_2$O, HCl to pH 2.5, reflux, 8.5 h: 76%],[1152] or a great variety of analogous products, most of which showed significant antimalarial activities.[480,1108,1152,1184,1194,1254,1377]

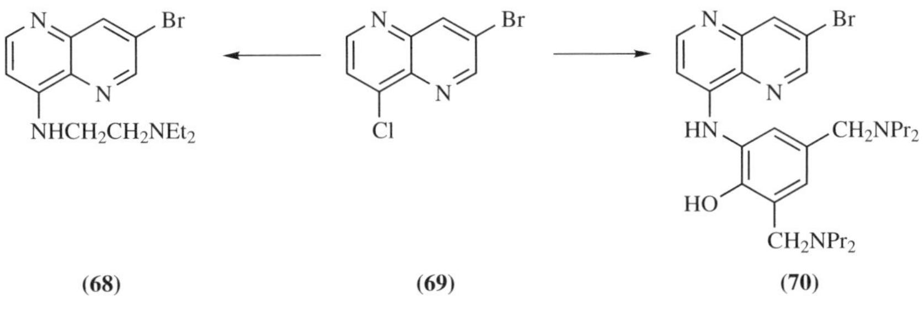

(68)   (69)   (70)

4,8-Dihalogeno-1,5-naphthyridines as Substrates

4,8-Dichloro-1,5-naphthyridine gave 1,5-naphthyridine-4,8-diamine (**71**) (PhOH, warm, $NH_3\downarrow$, 10 min; then substrate$\downarrow$, $NH_3\downarrow$, 175°C, 10 h: 56%).[301]

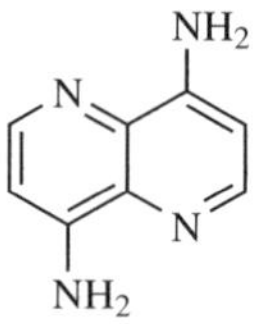

(**71**)

3.2.3. Other Reactions of Halogeno-1,5-naphthyridines

Each of the other reported reactions of halogeno-1,5-naphthyridines is illustrated by the following classified examples.

Alkanelysis

3,8-Dibromo-1,5-naphthyridine with *N*-(hex-5-enyl)phthalimide gave 3-bromo-8-(6-phthalimidohex-1-enyl)-1,5-naphthyridine (**72**) [reactants, $Pd(OAc)_2$, $P(C_6H_4Me\text{-}o)_3$, NaI, Bu_3N, Me_2NCHO, 100°C, N_2, 17 h: ~25%].[1179]

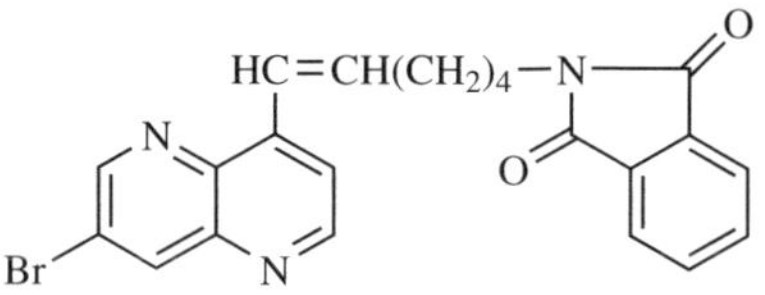

(**72**)

4-Chloro-1,5-naphthyridine-1-oxide with diethyl α-sodiomalonate gave 4-diethoxycarbonylmethyl-1,5-naphthyridine 1-oxide (**73**) [$NaCH(CO_2Et)_2$ (made *in situ*), PhMe; substrtae$\downarrow$ 105°C, 3 h: 57%].[814]

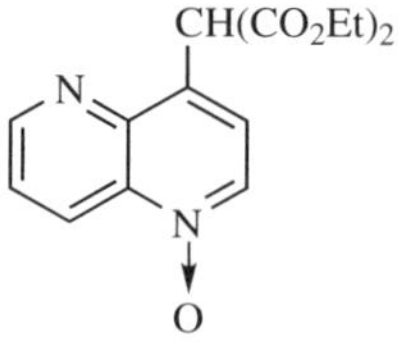

(**73**)

Alkane- or Arenethiolysis

7-Bromo-1-ethyl-4-oxo-1,4-dihydro-1,5-naphthyridine-3-carboxylic acid gave 1-ethyl-7-methylthio-4-oxo-1,4-dihydro-1,5-naphthyridine-3-carboxylic acid

(**74**, R = Me) [MeSNa (made *in situ*), Me$_2$NCHO, 15°C; substrate↓ slowly, 20°C, 12 h, then 70°C, 2.5 h: 77%], or 1-ethyl-7-isopropylthio-4-oxo-1,4-dihydro-1,5-naphthyridine-3-carboxylic acid (**74**, R = Pr^i) (substrate, Pr^{i}SH, K$_2$CO$_3$, Me$_2$NCHO, 75°C, 20 h: 51%).[1398]

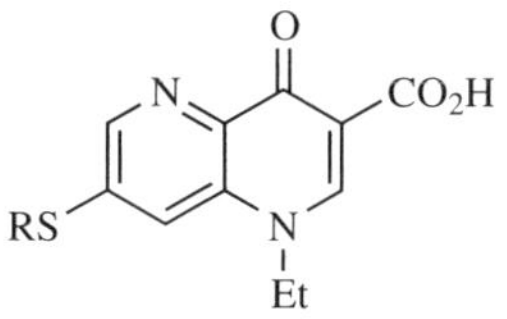

(**74**)

4-Chloro-2-ethyl-1,5-naphthyridine with 4′-mercaptobiphenyl-2-carbonitrile (deprotected *in situ*) gave 4-(2′-cyanobiphenyl-4-ylthio)-2-ethyl-1,5-naphthyridine (**75**) (reactants, Et$_3$N, MeOH, reflux, N$_2$, 1 h: 38%; see original for more synthon details).[614]

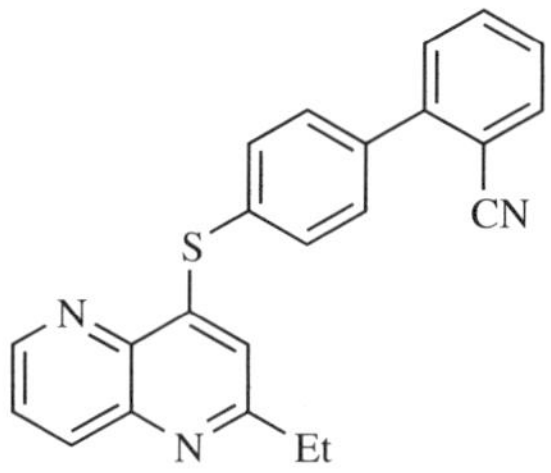

(**75**)

4-Chloro-1,5-naphthyridine gave bis(1,5-naphthyridin-4-yl) sulfide (**76**) [substrate, S=C (NH$_2$)$_2$, EtOH, reflux, 10 min: ∼90%].[814]

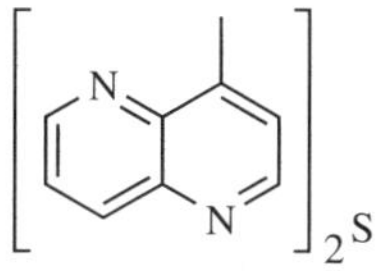

(**76**)

Dehalogenation

Note: Because hydrogenolysis of halogeno-1,5-naphthyridines frequently induces at least some additional nuclear reduction, dehalogenation has usually been done indirectly by hydrazinolysis and subsequent oxidation (see Section 2.1.1).

Ethyl 4-chloro-6-*p*-fluorophenoxy-1,5-naphthyridine-3-carboxylate (**77**) gave ethyl 1-*p*-fluorophenoxy-1,4-dihydro-1,5-naphthyridine-3-carboxylate (**78**) (substrate, Pd/C, Et$_3$N, dioxane, H$_2$, 20°C; 51%).[967]

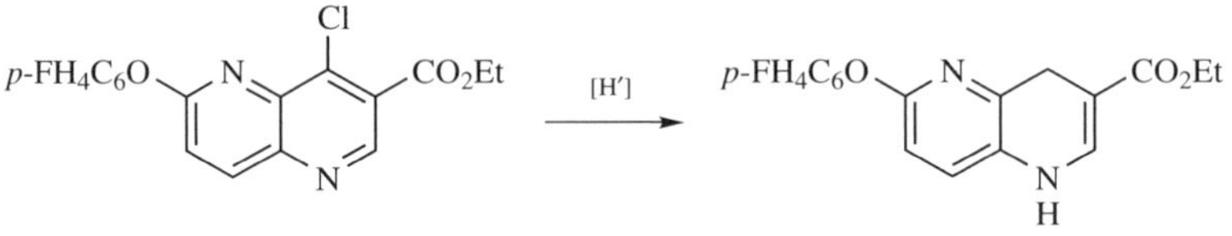

2,6-Dichloro- (**79**, R = Cl) gave 2-chloro-6-hydrazino- (**79**, R = NHNH$_2$) (70%; see Section 3.2.2) and thence 2-chloro-1,5-naphthyridine (**79**, R = H) (substrate, AcOH, H$_2$O; CuSO$_4$ in H$_2$O↓ dropwise; 95°C, ~1 h: 50%).[234]

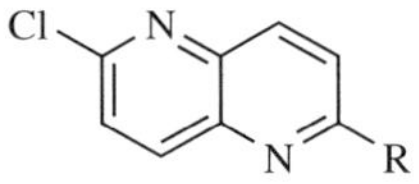

Also other examples.[575,827,1023,1118,1151]

Hydrolysis

4-Chloro-1,5-naphthyridine 1-oxide (**80**) gave 1-hydroxy-1,5-naphthyridin-4(1*H*)-one (**81**) (0.1M NaOH, reflux, 3 h: 76%);[814] 4-(α-bromo-α,α-diethoxycarbonylmethyl)-1,5-naphthyridine 1-oxide (**82**, R = Br) gave 4-(α,α-diethoxycarbonyl-α-hydroxymethyl)-1,5-naphthyridine 1-oxide (**82**, R = OH) (H$_2$O, reflux, 4 h: 37%).[814]

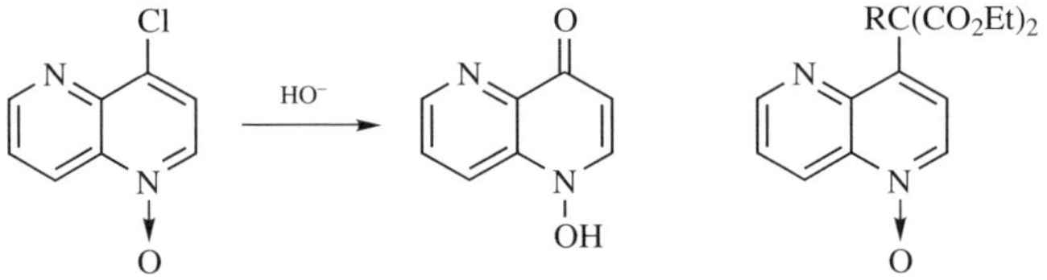

2,4-Dichloro-1,5-naphthyridine (**83**) gave selectively 4-chloro-1,5-naphthyridin-2(1*H*)-one (**84**) 5M HCl, reflux, 3 h: 66%).[1023,1151] In contrast, 2,6-dichloro-1,5-naphthyridine gave 1,5-naphthyridine-2,6(1*H*, 5*H*)-dione (10% Na$_2$CO$_3$, reflux, 2 h: ?%).[1647]

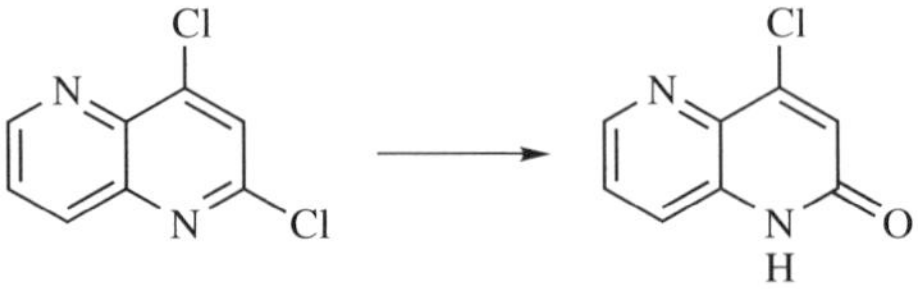

Hydrolysis may also be done indirectly: for example, 2-bromomethyl-(**85**, R = Br) to 2-acetoxymethyl- (**85**, R = OAc) to 2-hydroxymethyl-1, 5-naphthyridine (**85**, R = OH) (66% overall; for details, see original).[290]

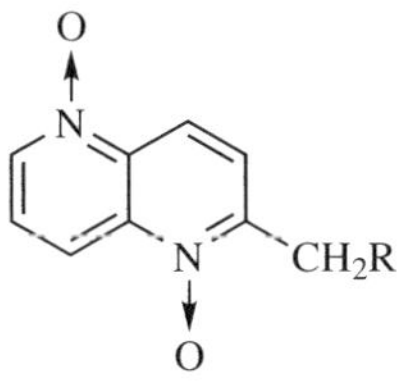

(**85**)

Thiolysis

4,8-Dichloro-1,5-naphthyridine (**86**) gave 1,5-naphthyridine-4,8(1H, 5H)-dithione (**87**) (KHS, EtOH, H_2O, reflux, $H_2S\downarrow$, 12 h: 96%; considerable detail in original).[301]

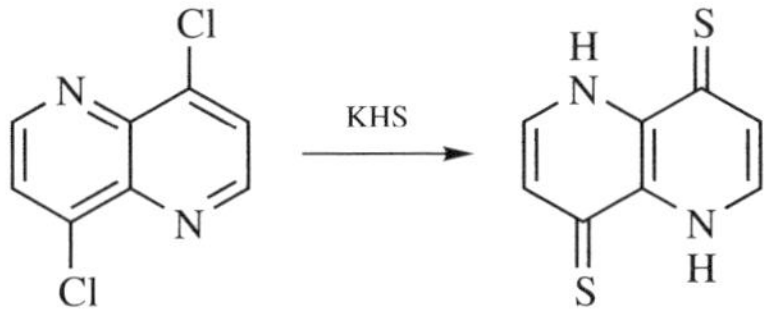

4-Chloro-1,5-naphthyridine (**88**) with thiourea gave 1,5-naphthyridine-4 (1H)-thione (**90**) via the thiouronium chloride (**89**) [reactants, EtOH, 20°C, 19 h: 80% (intermediate); that solid, KOH, H_2O, 20°C, 30 min: 81%].[814]

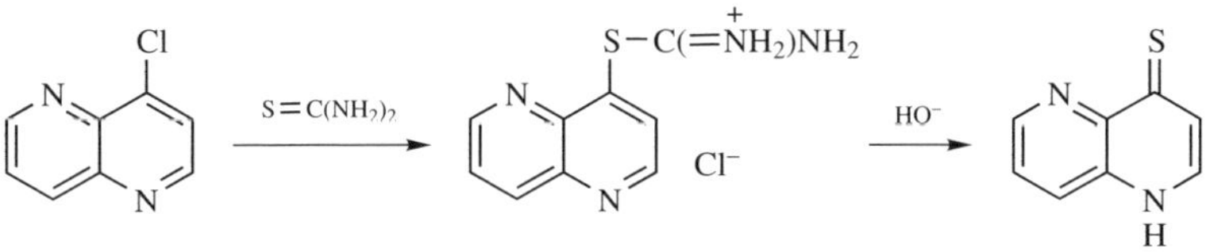

Also other examples.[271]

CHAPTER 4

Oxy-1,5-Naphthyridines

The term *oxy-1,5-naphthyridine* includes the tautomeric and nontautomeric 1,5-naphthyridinones, extranuclear hydroxy-1,5-naphthyridines, alkoxy-and aryloxy-1,5-naphthyridines, and 1,5-naphthyridine *N*-oxides.

As well as their fundamental chemical importance, oxy-1,5-naphthyridines have been implicated as antimicrobials,[107,110,323,572,1417] in which category cephalosporin derivatives such as Apalcillin (PC-904) (**1**) showed wide-spectrum activities.[324,327,337,351,354,361,420,491] The antibiotic naphthyridinomycin (a fused 1,5-naphthyridine derivative isolated from a culture of *Streptomyces lusitanus*[111]) also showed wide-spectrum activities,[111,446,517] as did analogs.[908]

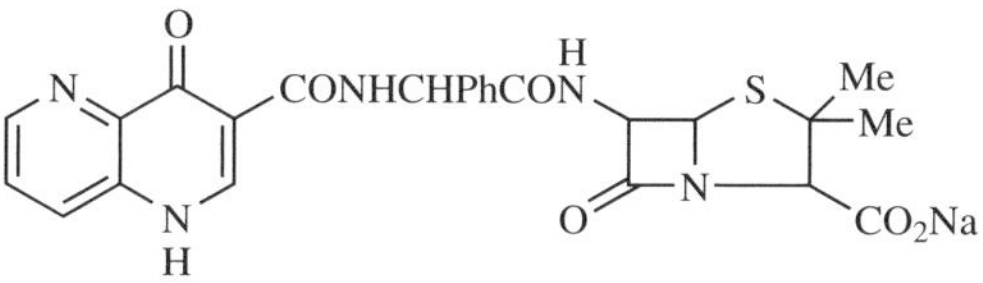

(**1**)

4.1. TAUTOMERIC 1,5-NAPHTHYRIDINONES AND EXTRANUCLEAR HYDROXY-1,5-NAPHTHYRIDINES

From an overwhelming mass of data on π-deficient nitrogenous heterocycles, it is usually considered axiomatic that (wherever possible) hydroxy-1,5-naphthyridines will exist predominantly as their respective 1,5-naphthyridinone tautomers; for example, 1,5-naphthyridin-2-ol (**2**) will exist as 1,5-naphthyridin-2(1*H*)-one (**3**). This postulate has been strengthened by ionization constant, infrared spectral, and ultraviolet spectral measurements on simple 1,5-naphthyridinones;[887,1026,1027,1035,1040] theoretical calculations have been less meaningful.[1295]

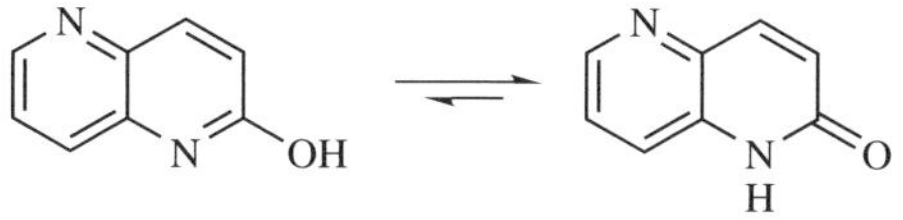

(**2**) ⇌ (**3**)

The Naphthyridines: The Chemistry of Heterocyclic Compounds, Volume 63, by D.J. Brown
Copyright © 2008 John Wiley & Sons, Inc.

The mass spectra of 1,5-naphthyridin-2(1*H*)-one, 1,5-naphthyridin-4(1*H*)-one, and 4-hydroxy-1,5-naphthyridin-2(1*H*)-one have been compared with those of isomeric naphthyridinones.[1253]

Various aspects of the Cd, Co, Cu, Fe, Mg, Mn, Ni, Pb, and Zn complexes from simple 1,5-naphthyridinones have been studied in some detail.[202,281,1040,1292]

4.1.1. Preparation of Tautomeric 1,5-Naphthyridinones and the Like

Most such naphthyridinones have been made by *primary synthesis* (see Chapter 1) and some by *hydrolysis of halogeno-1,5-naphthyridines* (see Section 3.2.3). Other preparative routes are illustrated by the following examples.

By Direct Oxylation

3-Methyl-1,5-naphthyridine (**4**) gave 3-methyl-1,5-naphthyridin-2(1*H*)-one (**5**) (substrate, 3M H_2SO_4, 100°C, $Na_2Cr_2O_7/H_2O$↓ during 2 h; then 100°C, 12 h: ~25%, allowing for recovered substrate).[155]

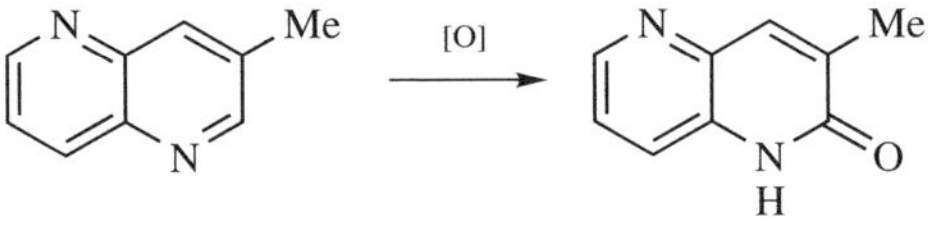

From 1,5-Naphthyridinamines

1,5-Naphthyridin-2-amine 1,5-dioxide (**6**) underwent alkaline hydrolysis to give 1-hydroxy-1,5-naphthyridin-2(1*H*)-one 5-oxide (**7**) (2M NaOH, reflux, 45 min: 54%).[787]

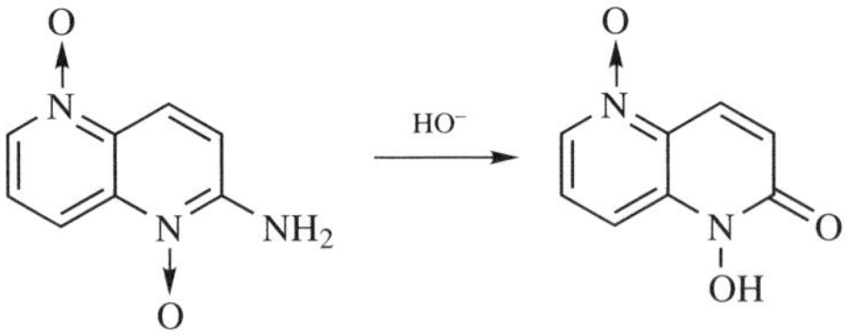

1,5-Naphthyridin-2-amine (**8**, R = H) gave 1,5-naphthyridin-2(1*H*)-one (**9**, R = H) (substrate, HCl, H_2O, $NaNO_2/H_2O$↓, 20→100°C: ?%);[1047] 4-chloro-1,5-naphthyridin-2-amine (**8**, R = Cl) gave 4-chloro-1,5-naphthyridin-2(1*H*)-one (**9**, R = Cl) (substrate, H_2O, H_2SO_4, 0°C; $NaNO_2$↓ slowly; then 100°C, 30 min: ?%).[1023]

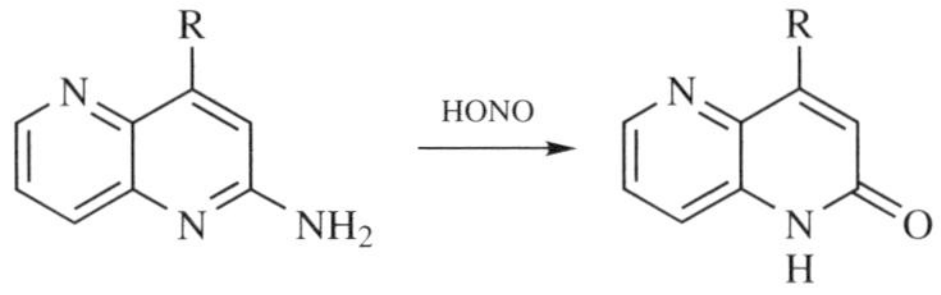

Tautomeric 1,5-Naphthyridinones and Extranuclear Hydroxy-1,5-Naphthyridines 45

(8) → (9) [HONO]

Also other examples.[1443]

From Alkoxy-1,5-naphtyridines

6-Methoxy-1,5-naphthyridin-2(1H)-one (10) gave 1,5-naphthyridine-2,6(1H,5H)-dione (11) (48% HBr, reflux, 2 h: 83%); the 3,4-dihydro-substrate likewise gave 3,4-dihydro-1,5-naphthyridine-2,6(1H, 5H)-dione (38%).[234]

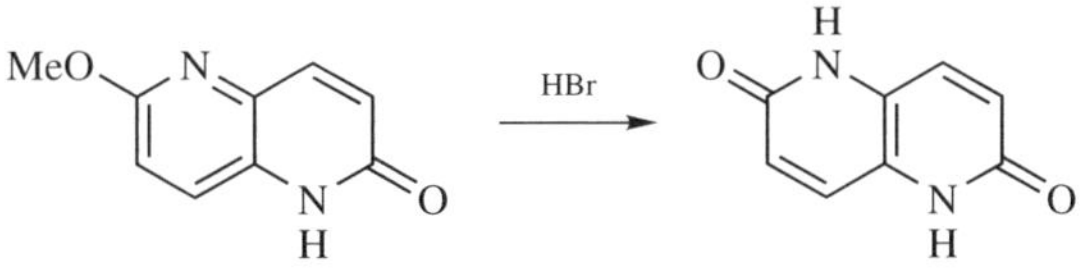

(10) → (11) [HBr]

4-Chloro-2-methoxy-1,5-naphthyridine (12) gave 4-chloro-1,5-naphthyridin-2(1H)-one (13) (5M HCl, dioxane, reflux, 1 h: ~50%; note survival of the chloro substituent under these mild conditions).[1151]

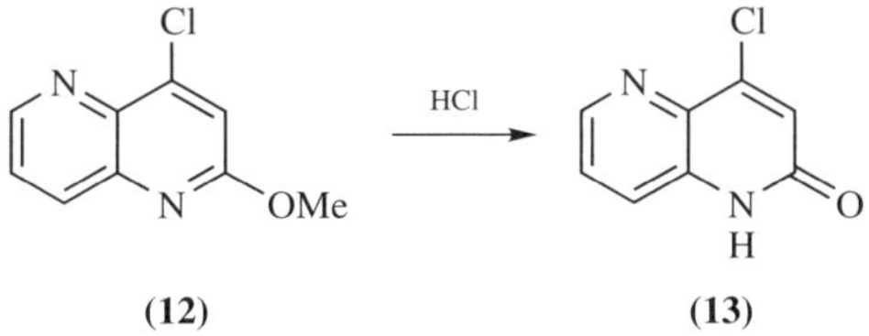

(12) → (13) [HCl]

By Reduction of Esters

Ethyl 6-p-chlorophenoxy-4-(3-diethylaminopropyl)amino-1,5-naphthyridine-3-carboxylate (9a, R = CO_2Et) gave 2-p-chlorophenoxy-8-(3-diethylaminopropyl)amino-7-hydroxymethyl-1,5-naphthyridine (9a, R = CH_2OH) [lAlH$_4$/ Et$_2$O, 10 min; substrate/Et$_2$O↓ dropwise; then 20°C, 1 h; 32%].[967]

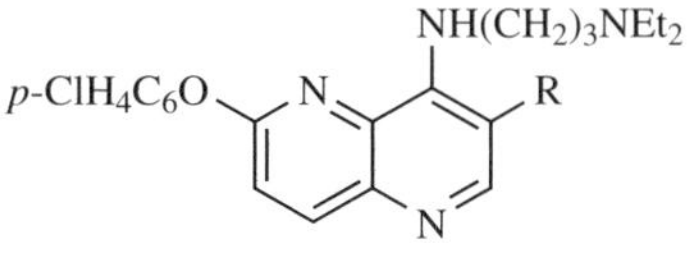

(9a)

4.1.2. Reactions of Tautomeric 1,5-Naphthyridinones and the Like

The *reductive deoxygenation* (Section 2.1.1) and *halogenolysis* (Section 3.1.2) of such naphthyridinones have been covered already. Their other reactions are illustrated in the following examples.

O- or N-Acylation

3-Benzyl-4-hydroxy-1,5-naphthyridin-2(1H)-one (**14**) (or tautomer) gave a single *o*- or *N*-monoacetyl derivative of indeterminate structure (no details).[760]

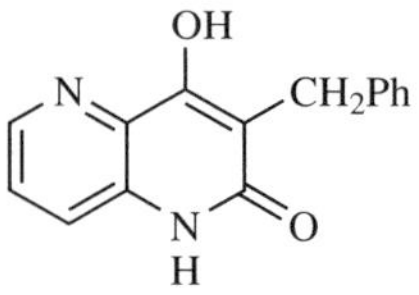

(**14**)

5-Acetyl-l-benzyl-3,4,4a,5,6,7-hexahydro-1,5-naphthyridine-2,7(1H)-dione (**15**) (a byproduct from a primary synthesis) appears to have undergone partial dehydrogenation and $N \rightarrow O$-transacylation to afford 7-acetoxy-l-benzyl-3,4-dihydro-1,5-naphthyridin-2(1H)-one (**16**) (Pd/C, xylene, 130°C, 30 h: 25%).[1272]

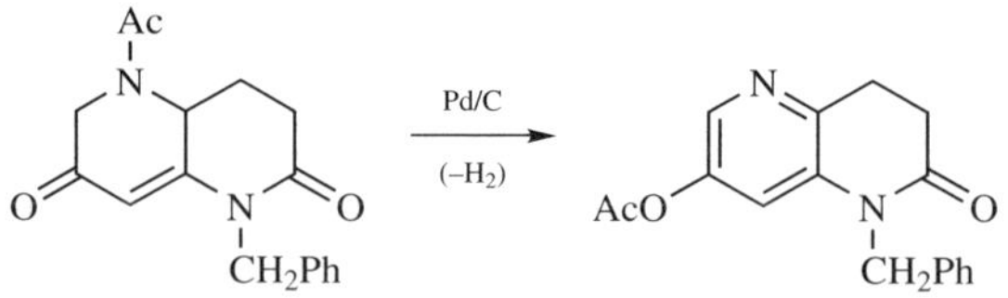

(**15**) → (**16**)

O- or N-Alkylation

Note: The alkylation of tautomeric 1,5-naphthyridinones usually gives *N*-alkyl derivatives, but *O*-alkylation has been achieved by treatment of the substrate silver salt with an alkyl halide; the other likely route to *O*-alkylation, treatment with an diazoalkane, has not been used in this series.

The silver salt (**17**) of 3-nitro-1,5-naphthyridin-2(1H)-one gave 2-ethoxy-3-nitro-1,5-naphthyridine (**18**) (neat EtI, reflux, 2 h: 20%).[837]

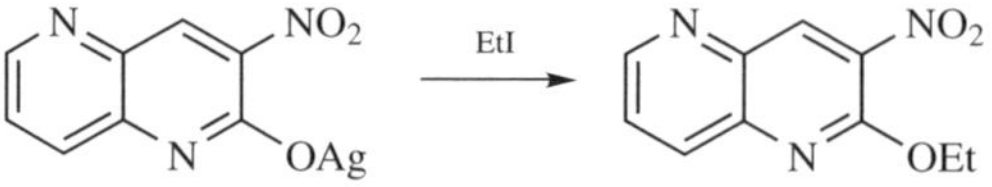

(**17**) → (**18**)

1-Hydroxy-1,5-naphthyridin-2(1*H*)-one 5-oxide (**19**, R = H) gave 1-methoxy-1,5-naphthyridin-2(1*H*)-one 5-oxide (**19**, R = Me) (MeI, NaOH, H$_2$O, MeOH, 20°C, 28 h: 78%).[787]

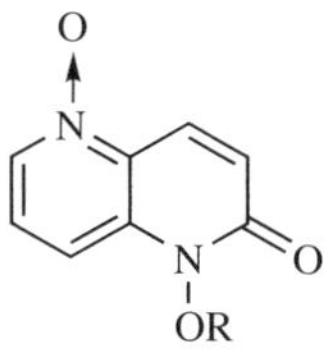

(**19**)

1,5-Naphthyridin-4(1*H*)-one (**20**, R = H) gave 1-ethyl-1,5-naphthyridin-4(1*H*)-one (**20**, R = Et) substrate, NaH, Me$_2$NCHO, 25°C, 4 h; EtI↓, 95%C, 12 h: 42%).[814]

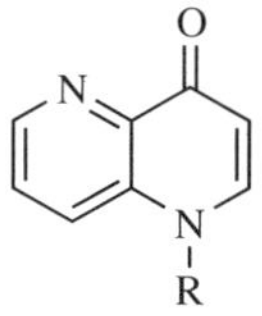

(**20**)

4-Oxo-1,4-dihydro-1,5-naphthyridine-3-carboxylic acid (**21**, R = H) gave 1-ethyl-4-oxo-1,4-dihydro-1,5-naphthyridine-3-carboxylic acid (**21**, R = Et) (EtI, KOH, H$_2$O, EtOH, reflux, 3.5 h: 50%);[911] the analogous substrate, ethyl 7-methyl-4-oxo-1,4-dihydro-1,5-naphthyridine-3-carboxylate (**22**, R = H), gave ethyl 1-ethyl-7-methyl-4-oxo-1,4-dihydro-1,5-naphthyridine-3-carboxylate (**22**, R = Et) (substrate, NaH, Me$_2$NCHO, 20°C, 20 min; then EtI↓, 50°C, 5 h, then 20°C, 12 h: 50%).[1398]

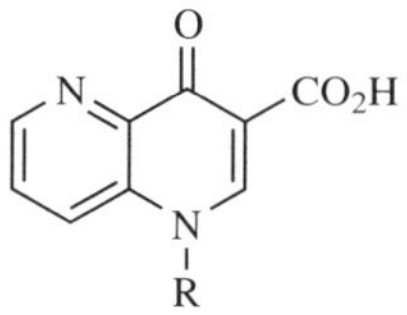

(**21**) 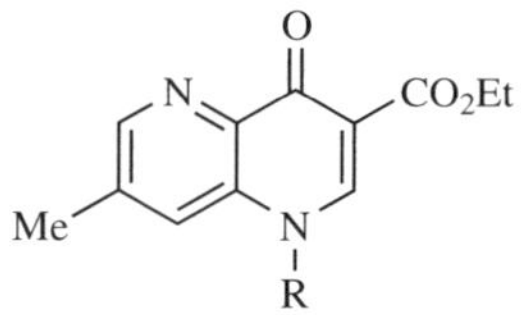   (**22**)

Also other examples.[1051]

Aminolysis

4-Hydroxy-1,5-naphthyridin-2(1*H*)-one (**23**, R=OH) with aniline hydrochloride in aniline gave 4-anilino-1,5-naphthyridin-2(1*H*)-one (**23**, R = NHPh) (neat reactants, reflux, 12 h: 88%).[1023]

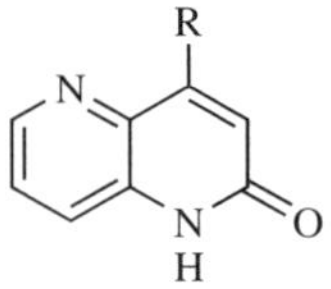

(23)

Deoxygenation

3-Hydroxy-6-methoxy-3,4-dihydro-1,5-naphthyridin-2(1H)-one (24) underwent dehydration to afford 6-methoxy-1,5-naphthyridin-2(1H)-one (25) (TsCl, pyridine, 150°C, sealed?, 4 h: 83%).[234]

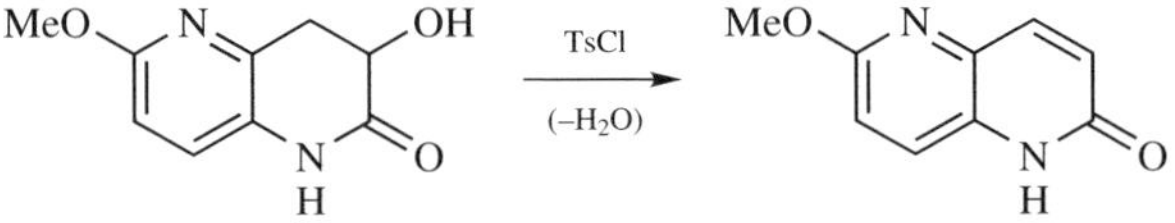

See also Section 2.1.1 for an example of reductive deoxygenation.

4.2. ALKOXY- AND ARYLOXY-1,5-NAPHTHYRIDINES

Not many such 1,5-naphthyridines have been reported, but the X-ray structure of 4,8-dimethoxy-1,5-naphthyridine has been determined.[777] Their preparation and reactions are illustrated by examples in the following lists.

Preparation of Alkoxy-1,5-naphthyridines

Note: All known alkoxy-1,5-naphthyridines have been made by *primary synthesis* (see Chapter 1), by *alcobolysis of halogeno-1,5-naphthyridines* (see Section 3.2.1), or by *alkylation of 1,5-naphthyridinones* (see Section 4.1.2).

Reactions of Alkoxy-1,5-naphthyridines

Note: For their apparent *rearrangement into C-alkyl-1,5-naphthyridinones*, see Section 2.2.1; for their *hydrolysis*, see Section 4.1.1; and for other reactions, see the following examples.

4-Phenoxy-1,5-naphthyridine (26) underwent *aminolysis* to afford 4-(4-diethylamino-1-methylbutyl) amino-1,5-naphthyridine (27) [neat H$_2$NCHMe(CH$_2$)$_3$NEt$_2$, reflux, 3 h: 90%].[101]

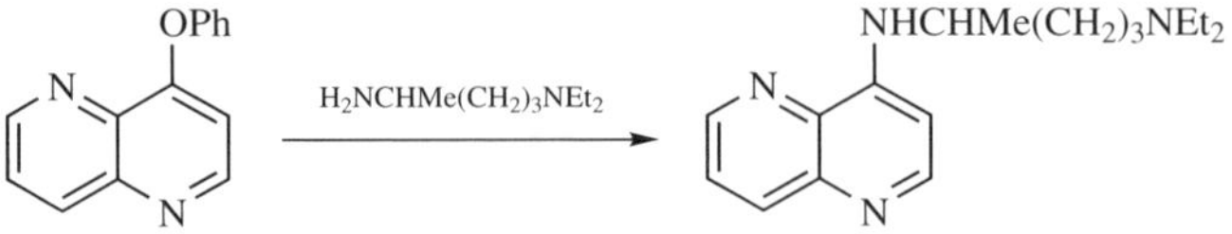

Ethyl 4,8-dimethoxy-1,5-naphthyridine-3-carboxylate (**28**, R = CO$_2$Et) underwent *thermal rearrangement* to ethyl 1,5-dimethyl-4,8-dioxo-1,4,5,8-tetrahydro-1,5-naphthyridine-3-carboxylate (**29**, R = CO$_2$Et) (Ph$_2$O, reflux, 6.5 h: ∼5% after a difficult separation from other products);[301] somewhat similarly, 4,8-dimethoxy-1,5-naphthyridine (**28**, R = H) gave 1,5-naphthyridine-4,8(1*H*,5*H*)-dione (**29**, R = H) (neat substrate, 225°C, 10 h: 62%).[301]

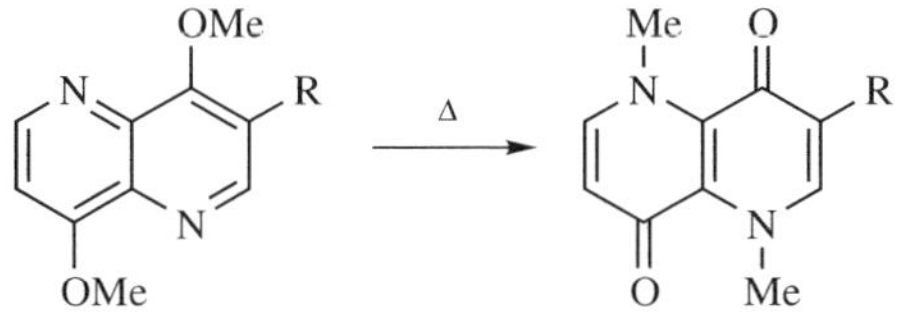

4.3. NONTAUTOMERIC 1,5-NAPHTHYRIDINONES

Reports on the chemistry of these naphthyridinones are scarce. Available information is summarized here.

Preparation of Nontautomeric 1,5-Naphthyridinones

Note: Most of these naphthyridinones have been made by either *primary synthesis* (see Chapter 1) or *alkylation of tautomeric 1,5-naphthyridinones* (see Section 4.1.2). A third route, involving *direct oxylation*, is illustrated here.

1,5-Naphthyridine methiodide (**30**) gave 1-methyl-1,5-naphthyridin-2(1*H*)-one (**31**) [substrate, H$_2$O, <0°C; NaOH/H$_2$O↓ during 5 min; then K$_3$Fe(CN)$_6$↓ during 30 min; then <0°C, 90 min; then 20°C, 5 h: 56%]; this was converted into its 5-methiodide (**32**) (MeI, PhH, reflux, 2 days: ?%) which underwent oxidation (as above) to afford 1,5-dimethyl-1,5-naphthyridine-2,6(1*H*,5*H*)-dione (**33**) (∼70%); analogs likewise.[155]

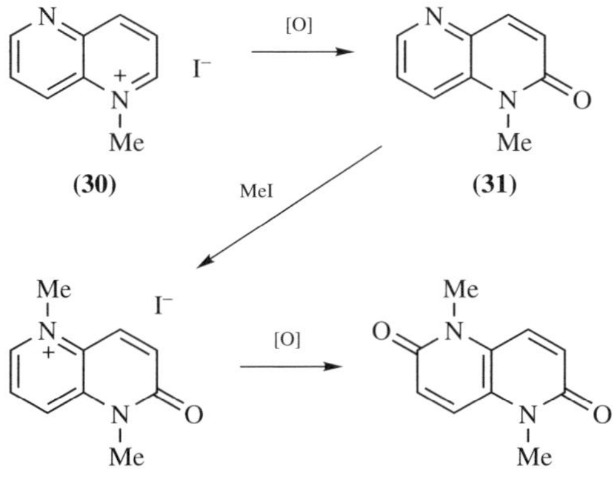

Reactions of Nontautomeric 1,5-Naphthyridinones

Note: The *rearrangement* of such a fixed naphthyridinone to a tautomeric *C*-alkylnaphthyridinone has been described in Section 2.2.1. The *halogenolysis* of a fixed naphthyridinone with concomitant *N*-dealkylation has been exemplified in Section 3.1.4.

4.4. 1,5-NAPHTHYRIDINE *N*-OXIDES

The available data on these oxides are summarized in the following paragraphs.

Preparation of 1,5-Naphthyridine *N*-Oxides

Note: There is at least one *primary synthesis* for these oxides (see Section 1.1). All other known compounds in this class have been made by *direct oxidation*, as illustrated in Sections 2.1.3 and 2.2.2, as well as by the following examples.

1,5-Naphthyridin-2-amine gave 1,5-naphthyridin-2-amine 1,5-dioxide (**34**, R = H) (30% H_2O_2, Na_2WO_4, $2H_2O$, 20°C, 5 days: 68%);[787] 2-acetamido-1,5-naphthyridine gave a separable mixture of the same product (**34**, R = H) and 2-acetamido-1,5-naphthyridine 1,5-dioxide (**34**, R = Ac) (likewise but 72 h: 54% and 25%, respectively);[787] and 4-acetamido-1,5-naphthyridine gave its 1,5-dioxide (**35**) (30% H_2O_2, $Na_2WO_4.2H_2O$, 40°C, 12 h: 37%).[814]

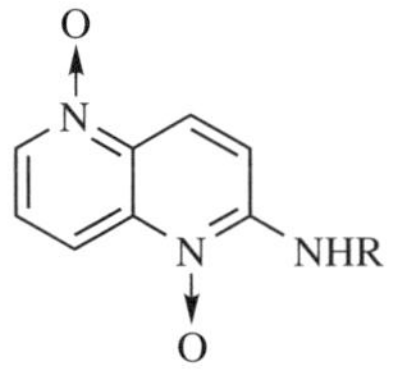

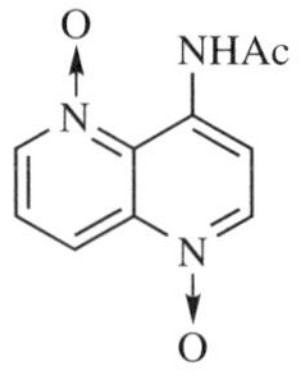

3-Bromo-1,5-naphthyridine gave its 1,5-dioxide (**36**) ($PhCO_3H$, $CHCl_3$, 20°C, 3 days: 11%).[811]

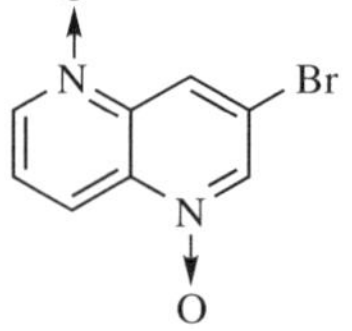

3-Bromo-8-(6-phthalimidohex-1-enyl)-1,5-naphthyridine (**37**) gave its 5-oxide (**38**) [*m*-$ClC_6H_4CO_3H$ (1 equiv), $CHCl_3$, 65°C, 3 h: ~75%] or 3-bromo-8-

[3-(4-phthalimidobutyl)oxiran-2-yl]-1,5-naphthyridine 5-oxide (**39**) [*m*-ClC$_6$H$_4$CO$_3$H (2 equiv), CHCl$_3$, reflux, 6 h: ~80%]; analogs likewise.[1179]

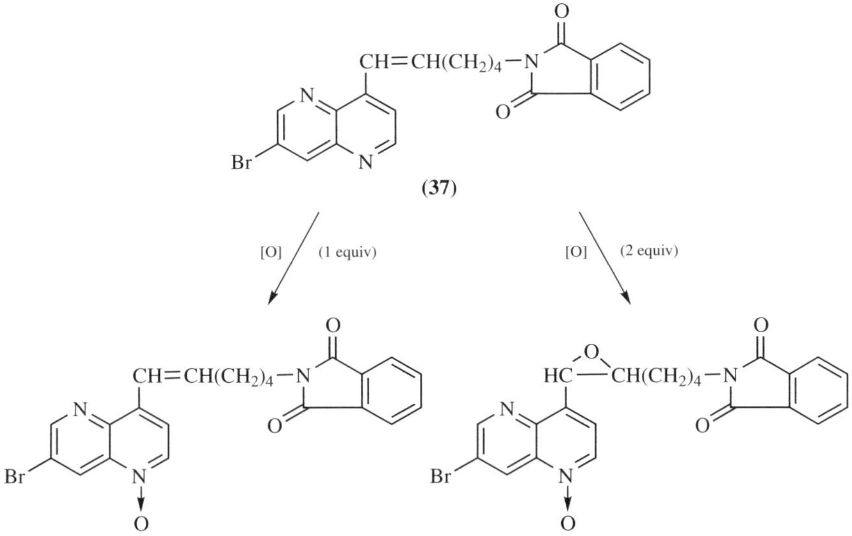

Also other examples.[271,797,1051,1212]

Reactions of 1,5-Naphthyridine *N*-Oxides

Note: The conversion of 1,5-naphthyridine *N*-oxides to *C*-halogeno-1,5-naphthyridines (*Meissenheimer reaction*) has been discussed in Section 3.1.3. Other reported reactions are illustrated here.

3-Bromo-8-[α-hydroxy-α-(piperinin-2-yl)methyl]-1,5-naphthyridine was prepared by *deoxygenation* of its 5-oxide (**40**) (substrate, Na$_2$S$_2$O$_4$, H$_2$O, MeOH, 60°C, 45 min: ~10%);[1179] for another example of deoxygenation by phosphorus trichloride, see Section 4.1.1.

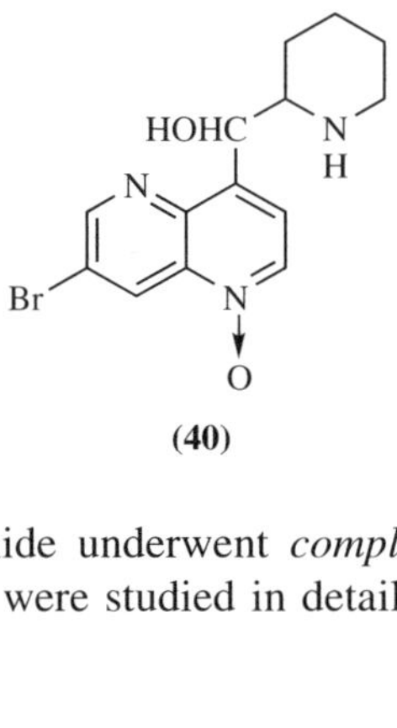

1,5-Naphthyridine 1,5-dioxide underwent *complexation* with Cu(II) salts; the highly colored products were studied in detail.[349]

CHAPTER 5

Thio-1,5-naphthyridines

This chapter summarizes the sparse data on 1,5-naphthyridines with substituents that are joined directly or indirectly to the nucleus through a sulfur atom. Included are any tautomeric or nontautomeric 1,5-naphthyridinethiones, extranuclear mercapto-1,5-naphthyridines, alkylthio-1,5-naphthyridines, bis(1,5-naphthyridinyl) sulfides or disulfides, 1,5-naphthyridine sulfoxides or sulfones, and 1,5-naphthyridinesulfonic acids or their derivatives. However, several categories have no known representatives.

Where possible, examples of preparative routes and reactions are given (or cross-referenced) in the following classified lists.

1,5-Naphthyridinethiones—Preparation

Note: 1,5-Naphthyridinethiones have been made by *thiolysis of halogeno-1,5-naphthyridines* (see Section 3.2.3).

1,5-Naphthyridinethiones—Reactions

1,5-Naphthyridine-4 (1H)-thione (**2**, R = H) underwent *S-alkylation* to give 4-methylthio-1,5-naphthyridine (**1**) (MeI, IM NaOH, 20°C, 90 min: 97%);[814] 1-hydroxy-1,5-haphthyridine-4(1H)-thione (**2**, R = OH) gave 4-benzylthio-1,5-naphthyridine 1-oxide (**3**) (PhCH$_2$Cl, H$_2$O, HO$^-$?, 50°C, 1 h: > 54%).[814]

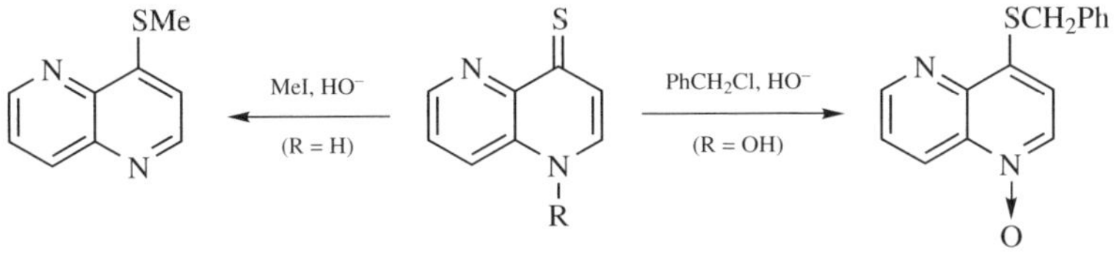

The Naphthyridines: The Chemistry of Heterocyclic Compounds, Volume 63, by D.J. Brown
Copyright © 2008 John Wiley & Sons, Inc.

1,5-Naphthyridine-4, 8 (1H,5H)-dithione with transition metal salts gave organometallic coordination polymers [e.g., substrate, Ni(NO$_3$)$_2$, Me$_2$NCHO, warm, 3 h: >90%].[1392]

Alkylthio-1,5-naphthyridines—Preparation

Note: These alkylthionaphthyridines have been made by *alkanethiolysis of halogeno-1,5-naphthyridines* (see Section 3.2.3) or by *S-alkylation of 1,5-naphthyridinethiones* (see immediately above).

Alkylthio-1,5-naphthyridines—Reactions

4-Benzylthio-1,5-naphthyridine 1-oxide underwent *oxidation* to give 4-benzylsulfonyl-1,5-naphthyridine 1-oxide (**4**) (30% H$_2$O$_2$, AcOH, 20°C, 2 days: 63%).[814]

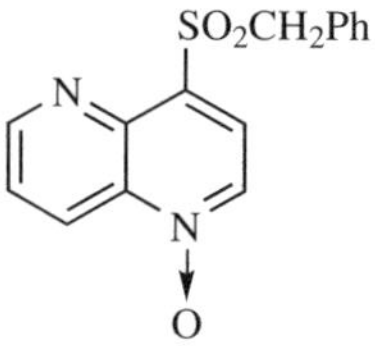

(4)

Bis(1,5-naphthyridinyl) Sulfides—Preparation

Note: One such sulfide has been made *from a halogeno-1,5-naphthyridine* with thiourea (see Section 3.2.3).

1,5-Naphthyridine Sulfones

Note: One such sulfone has been prepared by *primary synthesis* (see Section 1.3.3); another by *oxidation of an alkylthio-1,5-naphthyridine* (see a preceding subsection).

CHAPTER 6

Nitro-, Amino-, and Related 1,5-Naphthyridines

This chapter covers 1,5-naphthyridines bearing nitrogenous substituents that are joined to the nucleus through their nitrogen atoms.

6.1. NITRO-1,5-NAPHTHYRIDINES

The main use of nitro-1,5-naphthyridines has been as precursors for 1,5-naphthyridinamines. However, the presence of a powerfully electron-withdrawing nitro group is often of use to activate adjacent leaving groups such as halogeno toward nucleophilic replacement reactions.

The mass spectral fragmentation patterns of a variety of nitro-1,5-naphthyridine derivatives have been reported.[434,492]

6.1.1. Preparation of Nitro-1,5-naphthyridines

Some such nitro derivatives have been prepared by *primary synthesis* (see Chapter 1). Other preparative routes are illustrated in the following examples.

By Nitration

1,5-Naphthyridin-2(1*H*)-one (**1**, R = H) gave its 3-nitro derivative (**1**, R = NO$_2$) [HNO$_3$ (d. 1.5), H$_2$SO$_4$ (20% SO$_3$), 95°C, 4 h: ∼ 70%].[1037, cf. 1443]

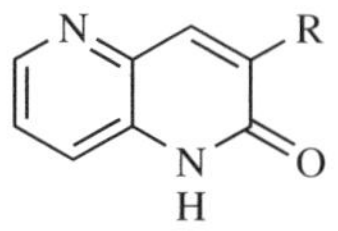

(**1**)

The Naphthyridines: The Chemistry of Heterocyclic Compounds, Volume 63, by D.J. Brown
Copyright © 2008 John Wiley & Sons, Inc.

1,5-Naphthyridin-4(1*H*)-one (**2**, R = H) gave its 3-nitro derivative (**2**, R = NO$_2$) [HNO$_3$ (d. 1.46), reflux, 2 h: 55–70%;[48,722] HNO$_3$ (d. 1.5), H$_2$SO$_4$ (20% SO$_3$), 95°C, 4 h: ?%].[1037]

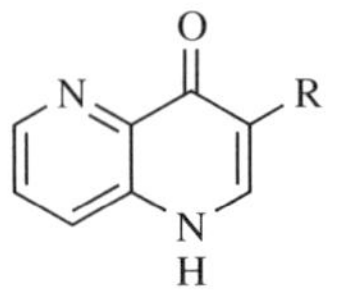

(**2**)

4-Hydroxy-1,5-naphthyridin-2(1*H*)-one (**3**, R = H) gave its 3-nitro derivative (**3**, R = NO$_2$) [HNO$_3$ (d. 1.42), AcOH, 95°C, ~7 min: 30%].[312]

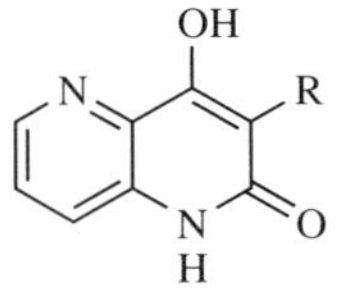

(**3**)

1,5-Naphthyridin-2-amine appeared to give 2-nitroamino-1,5-naphthyridine [HNO$_3$ (d. 1.5), H$_2$SO$_4$, 0°C, 4 h: 70%].[811]

By Oxidation of Dimethylsulfimido-1,5-naphthyridines

Note: This procedure offers an indirect route from 1,5-naphthyridinamines to nitro-1,5-naphthyridines.

1,5-Naphthyridin-2-amine (**4**) gave 2-dimethylsulfimido-1,5-naphthyridine (**5**) [Me$_2$SO, CH$_2$Cl$_2$, −78°C; (F$_3$CSO$_2$)$_2$O↓ dropwise, N$_2$; substrate/CH$_2$Cl$_2$↓, then −78°C, 2 h, −55°C, 1 h: 68%], which underwent oxidation to give 2-nitro-1,5-naphthyridine (**6**) (*m*-ClC$_6$H$_4$CO$_3$H, CH$_2$Cl$_2$, −5°C; substrate/ CH$_2$Cl$_2$↓ dropwise, <0°C; then 0°C, 40 min; then O$_3$↓, 0°C, 30 min: 27%).[692]

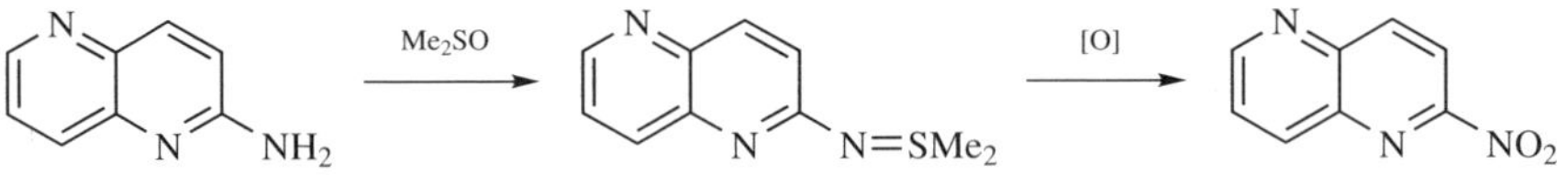

(**4**)   (**5**)   (**6**)

6.1.2. Reactions of Nitro-1,5-naphthyridines

The *halogenolysis of a nitro-1,5-naphthyridine* (using phosphorus pentachloride) has been recorded in Section 3.1.2. The *reduction of nitro-1,5-naphthyridines* to 1,5-naphthyridinamines is illustrated in the following examples.

3-Nitro- (**7**, R = NO$_2$) gave 3-amino-1,5-naphthyridin-2(1*H*)-one **7**, R = NH$_2$) (Raney Ni, NaOH, H$_2$O, EtOH, H$_2$, ?°C, ? h; 83%),[722]

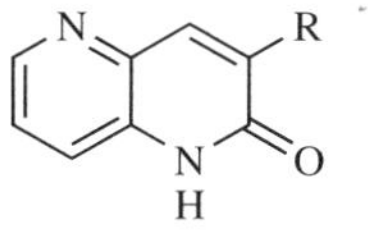

(**7**)

2-Ethoxy-3-nitro-1,5-naphthyridine (**8**, R = NO$_2$) gave 2-ethoxy-1,5-naphthyridin-3-amine (**8**, R = NH$_2$) [Pd/C, KOH, H$_2$O, EtOH, H$_2$ (~3 atm), ?°C: ~50%].[827]

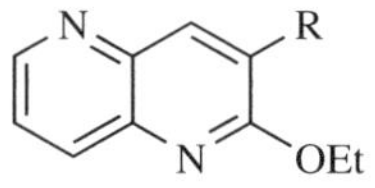

(**8**)

6.2. AMINO- AND (SUBSTITUTED-AMINO) -1,5-NAPHTHYRIDINES

The physical and biological properties of these important derivatives have been widely investigated, usually for comparison with those of isomeric or related systems.

There is infrared spectral evidence that 1,5-naphthyridin-2-amine may exist (at least in the solid state) as its 1,5-naphthyridin-2(1*H*)-imine tautomer (**9**),[148] but confirmation of this unexpected postulate by other means has not been forthcoming. The ionization constants of 1,5-naphthyridinamines have been compared with those of comparable 1,6-naphthyridinamines;[331] the mass spectra of several 1,5-naphthyridinamines have been compared with those of a variety of other heterocycles;[434, 492,1227] and correlation of amino proton chemical shifts (in 1,5-naphthyridin-2-amine and related amines) with Hammett constants and electron densities has been attempted.[174] Some highly substituted amino-1,5-naphthyridines proved disappointing in antimalarial screens,[333] but others showed great promise (see toward the end of Section 3.2.7 for references).

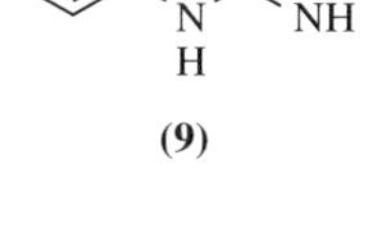

(**9**)

6.2.1. Preparation of Amino-1,5-naphthyridines

Many amino-1,5-naphthyridines have been made by *primary synthesis* (see Chapter 1), by *aminolysis of halogeno-1,5-naphthyridines* (see Section 3.2.2), or by *aminolysis of 1,5-naphthyridinones* (see Section 4.1.2). Others have been made by *direct amination*, as illustrated in the following examples.

> *Note*: A definitive procedure for the amination of 1,5-naphthyridine has been given in Section 2.1.3, along with references to its vexed history.
>
> 3-Nitro-1,5-naphthyridine (**10**, R = H) gave 3-nitro-1,5-naphthyridin-4-amine (**10**, R = NH$_2$) (substrate, liquid NH$_3$↓; KMnO$_4$;↓ portionwise; then 15 min: 74%); 3-nitro-1,5-naphthyridin-2-amine (**11**, R = H) gave 3-nitro-1,5-naphthyridine-2,4-diamine (**11**, R = NH$_2$) (likewise: 33%); and 3-nitro-1,5-naphthyridin-2(1H)-one (**12**, R = H) gave 4-amino-3-nitro-1,5-naphthyridin-2(1H)-one (**12**, R = NH$_2$) (likewise: 51%).[818]

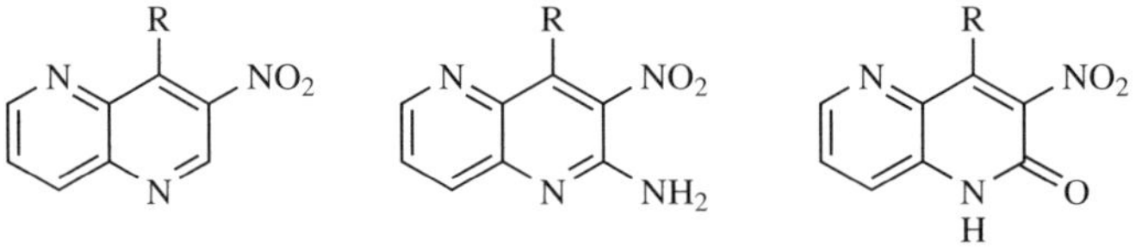

(10)   **(11)**   **(12)**

1-Amino-1,5-naphthyridin-1-ium mesitylenesulfonyloxide (**13**) (see Section 2.1.3) with base gave, not the free ylide (**14**), but its nonzwitterionic dimer (**15**) (K$_2$CO$_3$, H$_2$O, 20°C, 30 min: 63%).[1028]

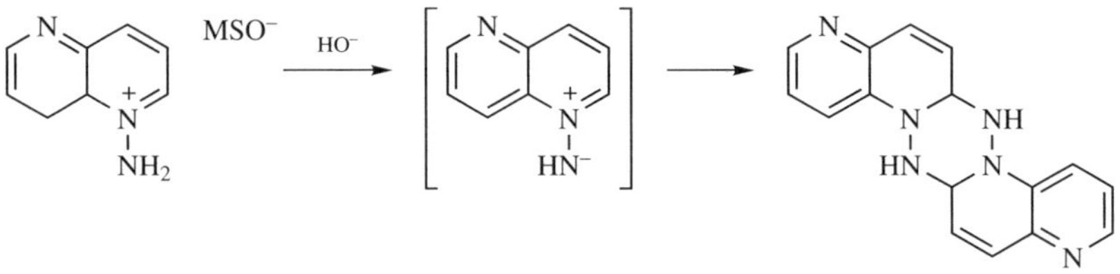

(13)   **(14)**   **(15)**

Also analogous methylaminations.[1443]

6.2.2. Reactions of Amino-1,5-naphthyridines

The *hydrolysis* of amino-1,5-naphthyridines has been covered in Section 4.1.1. Other reported reactions are illustrated by the following examples.

Acylation

1,5-Naphthyridin-4-amine (**16**, R = H) gave 4-acetamido-1,5-naphthyridine (**16**, R = Ac) (Ac$_2$O, PhH, reflux, 4 h: 72%);[814] 2-acetamido-1,5-naphthyridine (**17**) was made by a similar acetylation.[811,1443]

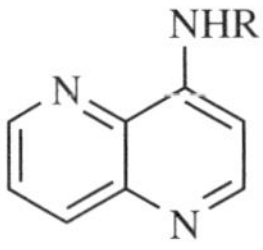

(**16**) 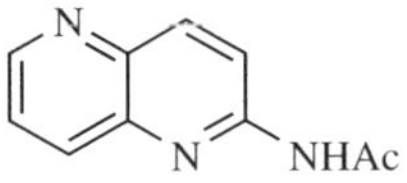   (**17**)

Cyclizations

1,5-Naphthyridin-4-amine (**18**) with diethyl α-(ethoxymethylene)malonate gave 4-(2,2-diethoxycarbonylvinyl)amino-1,5-naphthyridine (**19**) (neat reactants, 145°C, 3 h: 88%), which underwent thermal cyclization to afford ethyl 7-oxo-7,10-dihydropyrido[3,2-c]-1,5-naphthyridine-8-carboxylate (**20**) (Dowtherm A, reflux, 1 h: 89%).[164]

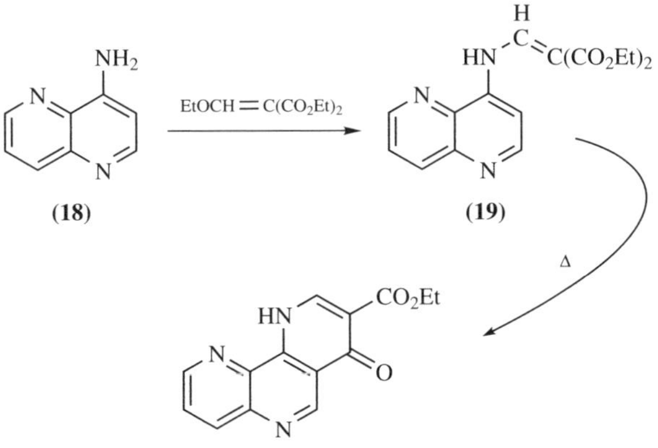

Also other examples.[643]

Ring Contractions

3-Amino-1,5-naphthyridin-4(1H)-one (**21**) underwent diazotization to 3-diazonio-1,5-naphthyridin-4(1H)-one chloride (**22**) (substrate, Me$_2$NCHO, 0°C, Me$_2$CHCH$_2$CH$_2$ONO↓, HCl/EtOH↓: ~75%), which on irradiation gave 1H-pyrrolo[3,2-b]pyridine-3-carboxylic acid hydrochloride (**23**) [substrate, AcOH, H$_2$O, 0°C, hν (arc lamp), ? h: ?%].[722]

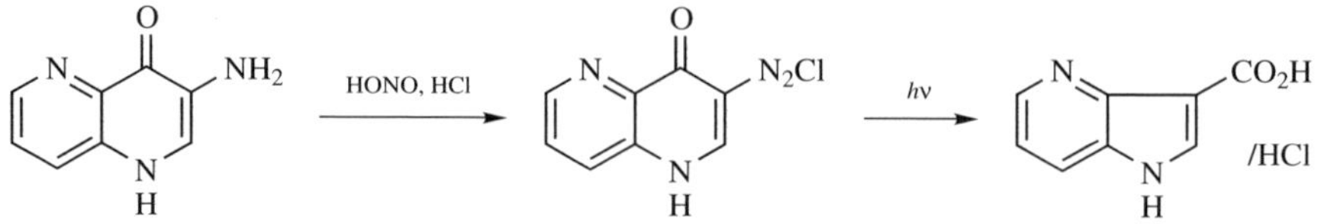

Charge Transfer Complexation

1,5-Naphthyridin-2-amine formed a well-defined complex with tetracyanoquinodimethane (**24**) that had interesting electrical properties.

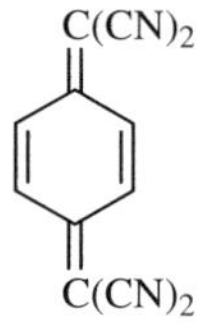

CHAPTER 7

1,5-Naphthyridinecarboxylic Acids and Related Derivatives

This chapter summarizes the sparse existing data on those 1,5-naphthyridines that bear functional groups that are joined to the nucleus through their carbon atoms: carboxylic acids, esters, amides, nitriles, aldehydes, and ketones.

The marked antimalarial activity of some derivatives of oxodihydro-1, 5-naphthyridinecarboxylic acids were noted at the beginning of Chapter 4.

7.1   1,5-NAPHTHYRIDINECARBOXYLIC ACIDS

The reported chemistry of such carboxylic acids is summarized by the following list of examples.

1,5-Naphthyridinecarboxylic Acids—Preparation

Note: 1,5-Naphthyridinecarboxylic acids have been made by *primary synthesis* (see Chapter 1) and by *oxidation of alkyl-1,5-naphthyridines* (see Section 2.2.2). The remaining routes involving *oxidation of aldehydes* or *hydrolysis of esters* are exemplified here.

1,5-Naphthyridine-2-carbaldehyde 1,5-dioxide (**1**) underwent *oxidation* to 1, 5-naphthyridine-2-carboxylic acid 1,5-dioxide (**2**) (30% H_2O_2, AcOH, 25°C, 3 days: 61%).[797,cf. 831]

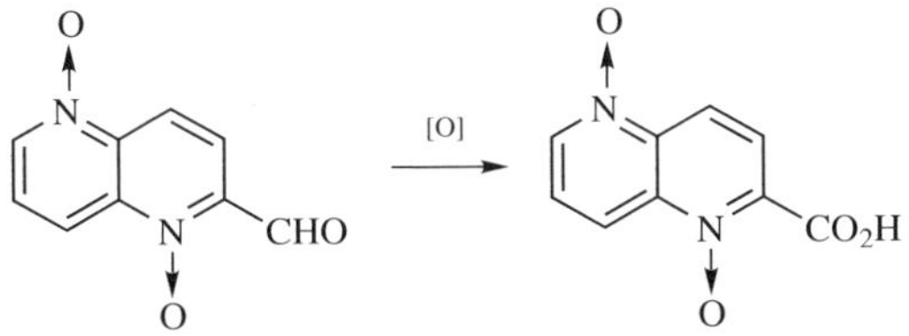

The Naphthyridines: The Chemistry of Heterocyclic Compounds, Volume 63, by D.J. Brown
Copyright © 2008 John Wiley & Sons, Inc.

Ethyl 4-oxo-1,4-dihydro-1,5-naphthyridine-3-carboxylate (**3**, R = H) underwent alkaline *hydrolysis* to 4-oxo-1,4-dihydro-1,5-naphthyridine-3-carboxylic acid (**4**, R = H) (M NaOH, reflux, 6 h: ∼ 75%);[101,1245] ethyl 6-ethoxy-4-oxo-1,4-dihydro-1,5-naphthyridine-3-carboxylate (**3**, R = OEt) gave 6-ethoxy-4-oxo-1,4-dihydro-1,5-naphthyridine-3-carboxylic acid (**4**, R = OEt) (NaOH, H_2O, EtOH, 40°C-reflux, 2 h: 94%; see original for large-scale details);[1015] and ethyl 6-butoxy-4-oxo-1,4-dihydro-1,5-naphthyridine-3-carboxylate (**3**, R = OBu) gave 6-butoxy-4-oxo-1,4-dihydro-1,5-naphthyridine-3-carboxylic acid (**4**, R = OBu) (M NaOH, 95°C, 4 h: ∼ 70%).[1036]

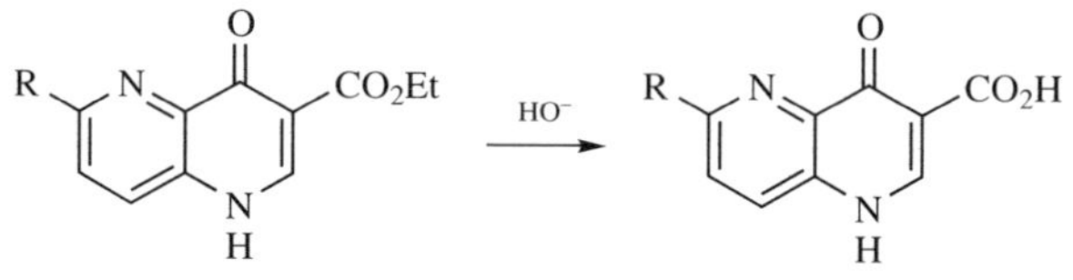

Ethyl 7-bromo-4-oxo-1,4-dihydro-1,5-naphthyridine-3-carboxylate (**5**, Q = H, R = Br) gave 7-bromo-4-oxo-1,4-dihydro-1,5-naphthyridine-3-carboxylic acid (**6**, Q = H, R = Br) (2.5M NaOH, reflux, 1 h: ∼ 85%);[1377] ethyl 1-ethyl-7-methyl-4-oxo-1,4-dihydro-1,5-naphthyridine-3-carboxylate (**5**, Q = Et, R = Me) gave 1-ethyl-7-methyl-4-oxo-1,4-dihydro-1,5-naphthyridine-3-carboxylic acid (**6**, Q = Et, R = Me) (0.7M KOM, reflux, 2.5 h: 56% analogs somewhat similarly).[1398]

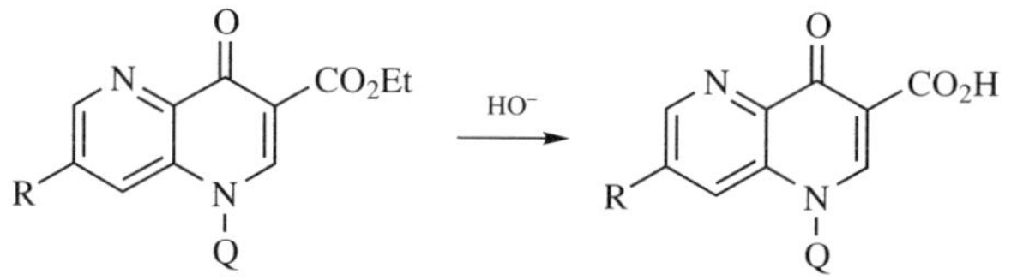

Ethyl 4,8-dioxo-1,4,5,8-tetrahydro-1,5-naphthyridine-3-carboxylate (**7**, R = Et) gave 4,8-dioxo-1,4,5,8-tetrahydro-1,5-naphthyridine-3-carboxylic acid (**7**, R = H) (M NaOH, reflux, 4 h: 77%;[233] 48% HBr, reflux, 24 h: 88%).[301]

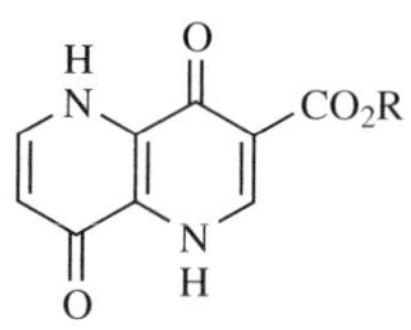

4-Diethoxycarbonylmethyl-1-5,-naphthyridine 1-oxide (**8**) underwent rapid hydrolysis and monodecarboxylation to give 4-carboxymethyl-1,5-naphthyridine 1-oxide (**9**) (M NaOH, reflux, 3 min: 987%).[814]

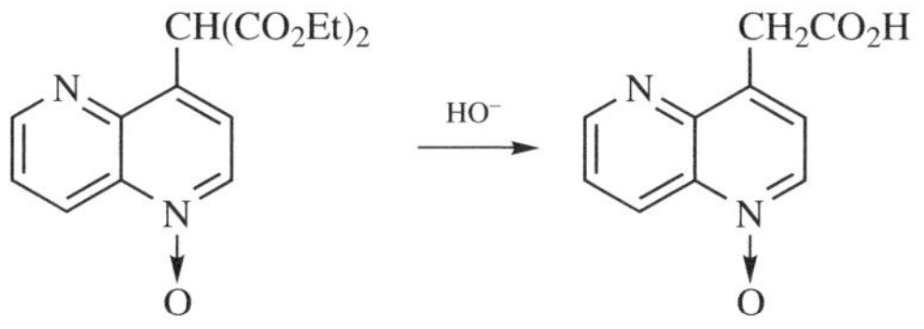

(**8**) → (**9**)

Also other examples.[575]

1,5-Naphthyridinecarboxylic Acids—Reactions

Note: The *decarboxylation, esterification*, and *amide formation* of such carboxylic acids are represented in these examples.

4-Oxo-1,4-dihydro-1,5-naphthyridine-3-carboxylic acid (**10**, R = H) underwent *decarboxylation* to 1,5-naphthyridin-4(1H)-one (**11**, R = H) (substrate, "mineral oil," 325°C, 30 min: ∼ 85%;[101] neat substrate, 315°C, ? h: ∼ 25%);[1245, of. 17] 6-butoxy-4-oxo-1,4-dihydro-1,5-naphthyridine-3-carboxylic acid (**10**, R = 6–OBu) gave 6-butoxy-1,5-naphthyridin-4(1H)-one (**11**, R = 6–OBu) (Ph$_2$O, reflux, 15 min: ∼ 85%);[1036] and 7-bromo-4-oxo-1,4-dihydro-1,5-naphthyridine-3-carboxylic acid (**10**, R = 7–Br) gave 7-bromo-4-oxo-1,4-dihydro-1,5-naphthyridin-4(1H)-one (**11**, R = 7–Br) (quinoline, reflux; substrate during 10 min; reflux, 1 h: 85%).[1377]

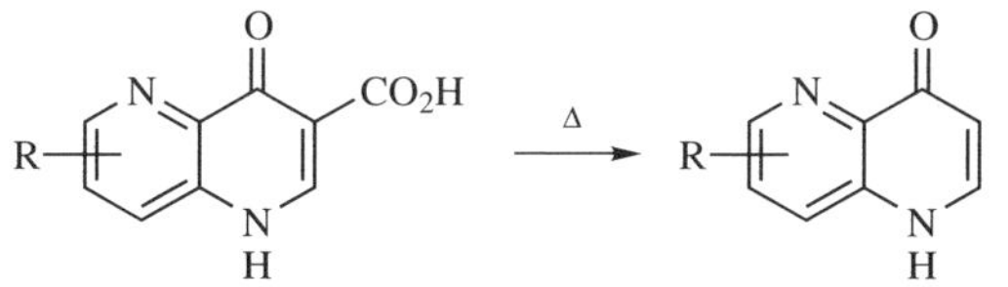

(**10**) → (**11**)

4,8-Dioxo-1,4,5,8-tetrahydro-1,8-naphthyridine-3-carboxylic acid (**12**, R = CO$_2$H) gave 1,5-naphthyridine-4,8(1H,5H)-dione (**12**, R = H) (neat substrate, ∼ 275°C: 62%;[233] quinoline, reflux, 10 h: 88%).[301]

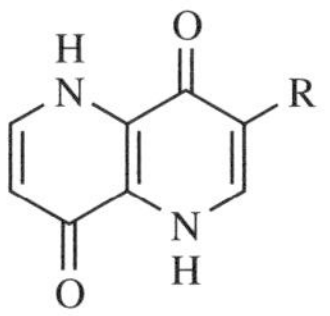

(**12**)

4-Carboxymethyl-1,5-naphthyridine 1-oxide (**13**, R = H) underwent *esterification* to 4-ethoxycarbonylmethyl-1,5-naphthyridine 1-oxide (13, R = Et) (substrate, SOCl$_2$, EtOH, reflux until clear: 88%)[814] also analogs (see original for details).[330]

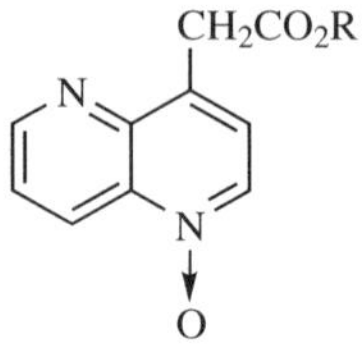

(13) [structure: 1,5-naphthyridine with CH₂CO₂R at 4-position, N-oxide]

4-Oxo-1,4-dihydro-1,5-naphthyridine-3-carboxylic acid (**14**) and *N*-hydroxysuccinimide (**15**) gave the ester-like product, 3-(succinimidooxycarbonyl)-1,5-naphthyridin-4(1*H*)-one (**16**) (reactants, pyridine, Me₂NCHO, < 5°C; SOCl₂↓ dropwise, < 5°C; then 25°C, 5 h: ~80%).[1232]

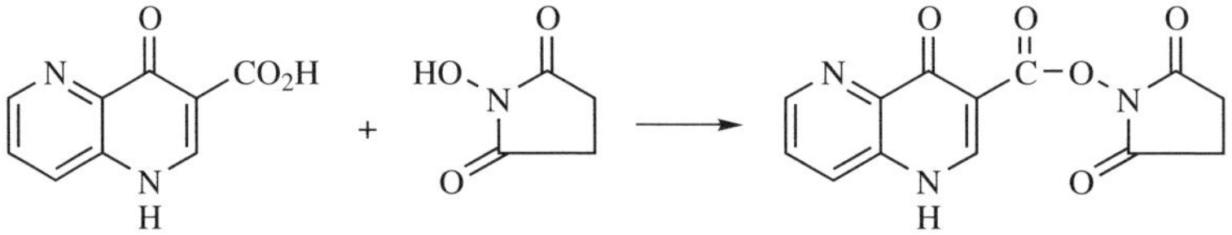

(**14**) + (**15**) → (**16**)

1,5-Naphthyridine-3-carboxylic acid (**17**, R=OH) underwent indirect *amide formation* to give 1,5-naphthyridine-3-carboxamide (**17**, R=NH₂) (SOCl₂, heat; acyl chloride, NH₄OH: ?%; no details given);[155] in a more direct way, 6-ethoxy-4-oxo-1,4-dihydro-1,5-naphthyridine-3-carboxylic acid (**18**, R=OH) gave *N*-benzyl-6-ethoxy-4-oxo-1,4-dihydro-1,5-naphthyridine-3-carboxamide (**18**, R=NHCH₂Ph) (substrate, Me₂NCHO, 1,1′-carbonyldiamidazole, 90°C, 2 h; then PhCH₂NH₂↓, 35°C, 2 h: 97%).[1015]

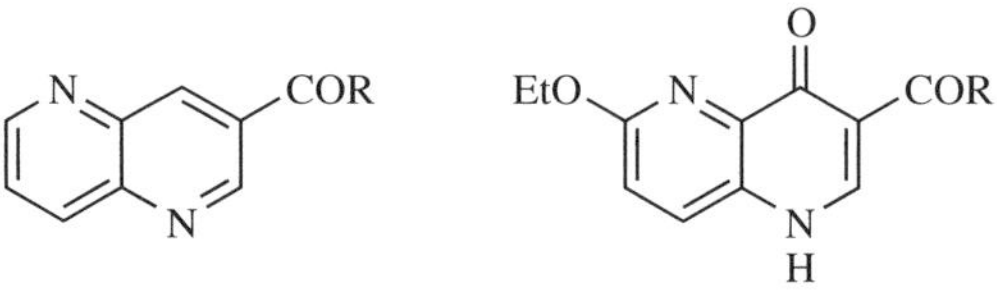

(**17**)   (**18**)

7.2. 1,5-NAPHTHYRIDINECARBOXYLIC ESTERS

The reported chemistry of such esters is summarized briefly in the following paragraphs.

1,5-Naphthyridinecarboxylic Esters—Preparation

Note: These esters have been made by *primary synthesis* (see Chapter 1), by the *Reissert reaction* (see Section 2.1.3), by *esterification of 1,5-naphthyridinecarboxylic acids* (see Section 7.1), and by various *passenger introductions* (see several chapters). Other possible routes have not been used.

1,5-Naphthyridinecarboxylic Esters—Reactions

Note: The *reduction of these esters to hydroxymethyl-1,5-naphthridines* (Section 4.1.1) and their *hydrolysis to carboxylic acids* (Section 7.1) have been exemplified already; the *nuclear dehydrogenation* of a dihydro ester is seen in the following example.

Ethyl 6-*p*-fluorophenoxy-1,4-dihydro- (**19**) gave ethyl 6-*p*-fluorophenoxy-1,5-naphthyridine-3-carboxylate (**20**) (substrate, ACMe; KMnO$_4$/ACMe↓ dropwise: 97%).[967]

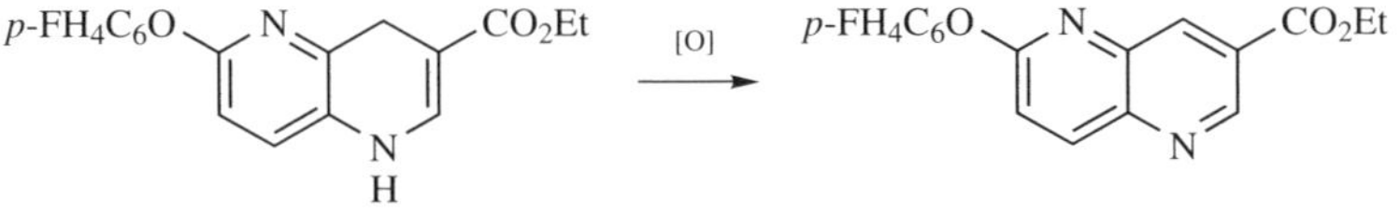

7.3. 1,5-NAPHTHYRIDINECARBOXAMIDES, CARBONITRILES, CARBALDEHYDES, AND KETONES

The scarce information about these derivatives is a summarized in the following paragraphs.

1,5-Naphthyridinecarboxamides

These amides have been made by *primary synthesis* (see Chapter 1) or by direct or indirect *aminolysis of corresponding carboxylic acids* (see Section 7.1). No reactions appear to have been described.

1,5-Naphthyridinecarbonitriles

Such nitriles have been made by *primary synthesis.* (see Chapter 1) but apparently not by other means; no reactions have been reported.

1,5-Naphthyridinecarbaldehydes

These aldehydes have been made by *oxidation of alkyl-1,5-naphthyridines* (see Section 2.2.2).

The equilibrium between 1,5-naphthyridine-4-carbaldehyde and its hydrate (**21**) has been studied.[953]

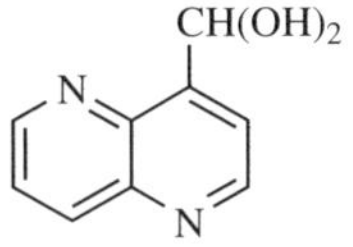

1,5-Naphthyridine-2-carbaldehyde 1,5-dioxide has been converted into its phenylhydrazone (**22**, R=MHPh), semicarbazone (**22**, R=NHCONH$_2$), and thiosemicarbazone (**22**, R=NHCSNH$_2$).[797]

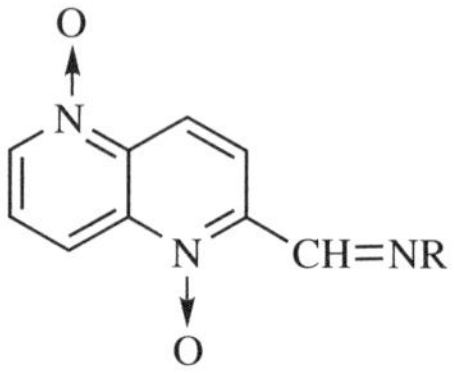

(**22**)

1,5-Naphthyridine Ketones

Such ketones have been made by direct *C-acylation of alkyl-1,5-naphthyridines* (see Section 2.2.2).

The ketone, 2-picolinoylmethyl-1,5-naphthyridine (**23**), underwent α-bromination and subsequent cyclization with thiourea to give 2-[2-amino-4-(pyridin-2-yl)thiazol-5-yl]-1,5-naphthyridine (**24**) [substrate, Br$_2$, dioxane, 20°C, 1 h; solid, EtOH, (H$_2$N)$_2$CS↓, reflux, 4 h: 16%].[268]

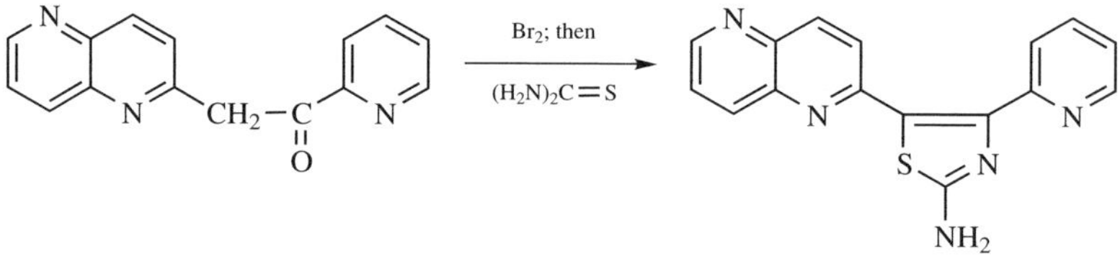

(**23**) (**24**)

CHAPTER 8

Primary Syntheses of 1,6-Naphthyridines

The primary synthesis of 1,6-naphthyridines has been accomplished by condensation of two or more aliphatic substrates; by cyclization of a single pyridine substrate; by condensation of a pyridine substrate with an aliphatic synthon that provides one, two, three, or even four ring atoms; by condensation of a pyridine substrate with two or more synthons; or from other heterocyclic substrates by degradation, rearrangement, or other elaborative processes.

Some existing reviews of naphthyridine chemistry contain at least some information on the primary synthesis of 1,6-naphthyridine derivatives.[49–52,55,57,58,61,265,328,407,1260,1273,1357,1430,1432]

8.1. BY CONDENSATION OF TWO OR MORE ALIPHATIC SUBSTRATES/SYNTHONS

Although some such syntheses are quite useful, most appear to have been observed in the course of other work and their mechanisms remain unproven. Accordingly, in presenting the following examples, no attempt will be made here to correlate substrate/synthon atoms with ring atoms in the final product.

1-Benzoyl-2-dimethylaminoethylene (**1**) with malononitrile (**2**) (as its dimer) gave 5,7-dioxo-4-phenyl-1,5,6,7-tetrahydro-1,6-naphthyridine-8-carboxamide (**3**) 10M HCl, 100°C, 5 min: 71%; homologs likewise).[217]

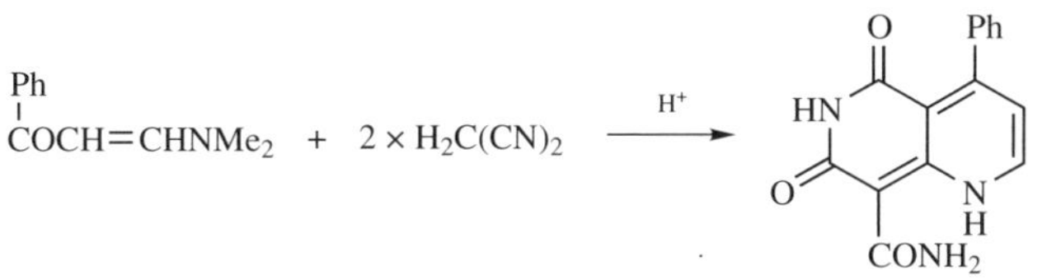

The Naphthyridines: The Chemistry of Heterocyclic Compounds, Volume 63, by D.J. Brown
Copyright © 2008 John Wiley & Sons, Inc.

Ethyl 2-(phenylhydrazino)acetoacetate (**4**) with malononitrile (**5**) (as its dimer) gave 7-hydroxy-2,5-dioxo-3-phenylhydrazino-1,2,5,6-tetrahydro-1,6-naphthyridine-8-carbonitrile (**6**) (reactants, AcONH$_4$, 160°C, 1 h: 88%).[1360]

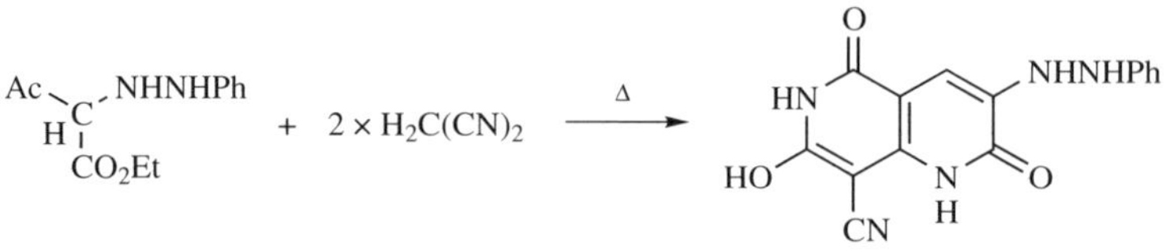

Ethyl *N*-benzylacetimidate (**7**) with ketene dimer (**8**) gave 3-acetyl-1,6-dibenzyl-4,7-dimethyl-1, 6-naphthyridine-2,5 (1*H*,6*H*) dione (**9**) [substrate (**7**), AcOH, 20°C; synthon (**8**)↓ dropwise; 20°C. 12 h: 51%].[1014]

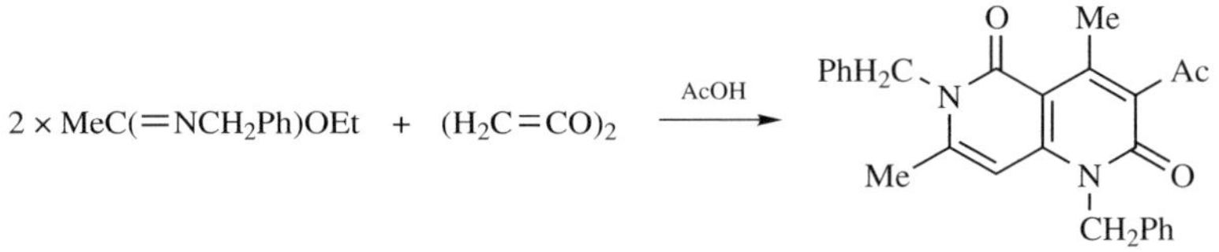

4-Methylpent-3-en-2-one (**11**) with malononitrile (**10**) in the presence of pyrrolidine gave 5-amino-2, 4, 4-trimethyl-7-(pyrrolidin-1-yl)-1,4-dihydro-1,6-naphthyridine-8-carbonitrile (**12**) [reactants, EtOH, 20°C, HN(CH$_2$)$_4$↓ dropwise; reflux. 7 h: 32%].[170]

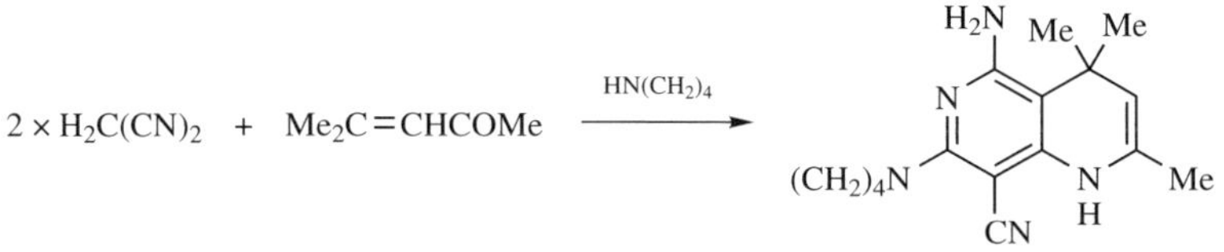

4-Ethoxy-4-methylaminobut-3-en-2-one (**13**) with acetyl chloride (**14**) gave 3-acetyl-1,4,6,7-tetramethyl-1,6-naphthyridine-2,5 (1*H*,6*H*)-dione (**14**) (reflux, 6 h: 46%); similar procedures furnished several homologs, of which 3-benzoyl-7-methyl-4-phenyl-1,5-naphthyridine-2,5(1*H*,6*H*)-dione was confirmed in structure by X-ray analysis.[636]

Also other examples.[72,481,566,774,784,920,950,1286,1448]

8.2. FROM A SINGLE PYRIDINE SUBSTRATE

An appropriately substituted pyridine may be converted into a 1,6-naphthyridine by cyclization to complete any 1 of the 10 bonds in the product. Nine such possibilities have been employed in the literature, and the examples that follow are classified according to which bond is formed in the synthesis. Fused 1,6-naphthyridines have been made similarly.[19,21]

By Completion of the 1,2-Bond

3-(Ethoxycarbonylvinyl)-4-pyridinamine (**15**) gave 1,6-naphthyridin-2(1H)-one (**16**) (EtONa, EtOH, reflux, 2 h: 67%).[1044]

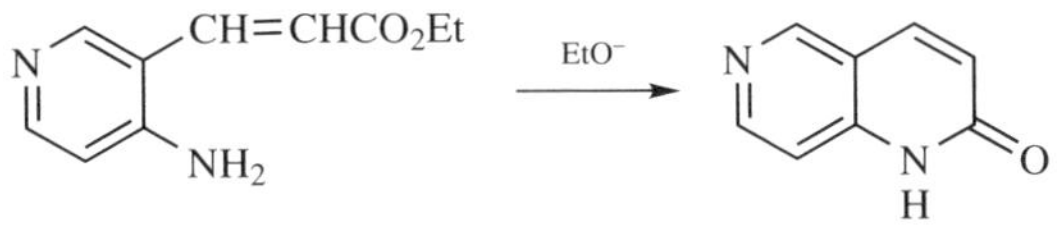

4-*tert*-Butyramido-3-(3,3-diethoxy-2-hydroxy-2-phenylpropyl amino)pyridine (**17**) underwent cyclization to 2,3-dihydroxy-3-phenyl-1,2,3,4-tetranydro-1,6-naphthyridine-5-carbaldehyde (**18**) (M H$_2$SO$_4$, 5°C, 2 h; then 100°C, 2 h: 65%), which underwent aromatization by loss of water to afford 3-phenyl-1,6-naphthyridine-5-carbaldehyde (**19**) [diazabicycloundecene (DBU), dioxane, reflux, 6 h: 30%].[903]

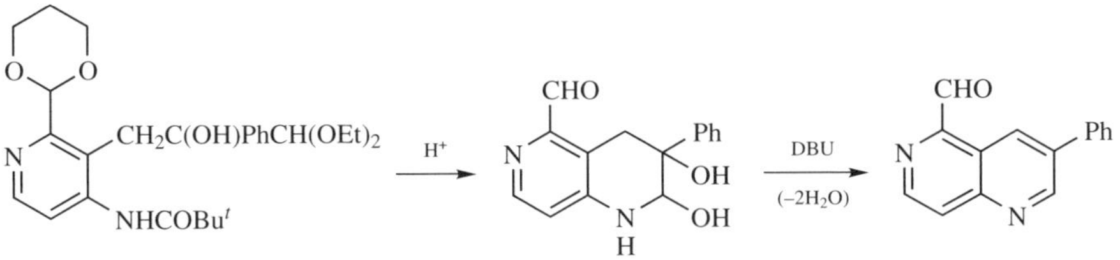

Ethyl 3-(2-ethoxycarbonylethyl)-4-phenlimino-1-piperidinecarboxylate (**20**) underwent reductive cyclization to ethyl 2-oxo-1-phenyl-1,2,3,4,5,6,7,8-octahydro-1,6-naphthyridine-1-carboxylate (**21**) (NaBH$_4$, dioxane, 20°C, 12 h: 27%) or its decahydro analog (substrate, THF, HCl gas↓, NaBH$_3$CN/MeOH↓ slowly, 20°C, 3.5 h: 22%).[1104]

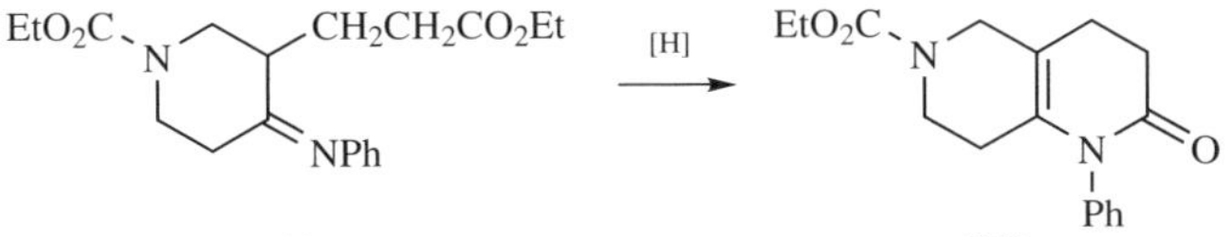

Also other examples.[477]

By Completion of the 1,8a-Bond

2,4-Dichloro-5-(3-cyclopropylamino-2-ethoxycarbonylacryloyl)-3-methylpyridine (**22**, R = Me) gave ethyl 7-chloro-1-cyclopropyl-8-methyl-4-oxo-1,4-dihydro-1,6-naphthyridine-3-carboxylate (**23**, R = Me) (substrate, THF, 0°C; NaH↓ slowly, 0°C, 1 h, N_2: 61%);[648] the 3-fluoro substrate (**22**, R=F) likewise gave ethyl 7-chloro-1-cyclopropyl-8-fluoro-4-oxo-1,4-dihydro-↑1,6-naphthyridine-3-carboxylate (**23**, R = F) (2 h: 96%).[1175]

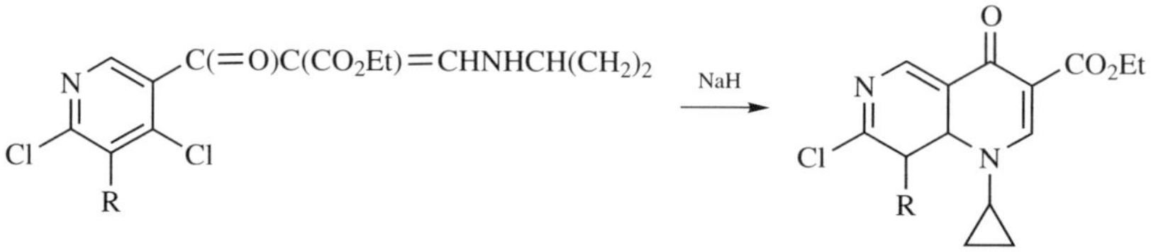

(22)   (23)

3-(2-Cyanoethyl)-1,3,6-trimethyl-4-piperidinone (**24**) gave 4a,5,7-trimethyldecahydro-1,6-naphthyridine (**25**) (Raney Ni, MeOH, H_2, 100 atm, 100°C: 83%).[106]

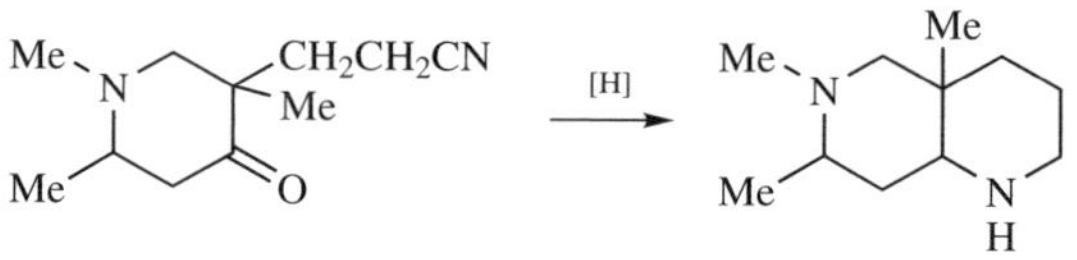

(24)   (25)

By Completion of the 3,4-Bond

Methyl 6-bromo-4-[N-(2-ethoxycarbonylethyl)-N-ethylamino]-3-pyridinecarboxylate (**26**) gave ethyl 7-bromo-1-ethyl-4-oxo-1,2,3,4-tetrahydro-1,6-naphthyridine-3-carboxylate (**27**) (substrate, Bu^{t}OH, NaH, 20°C, 30 min: 85%); an analog likewise.[1408]

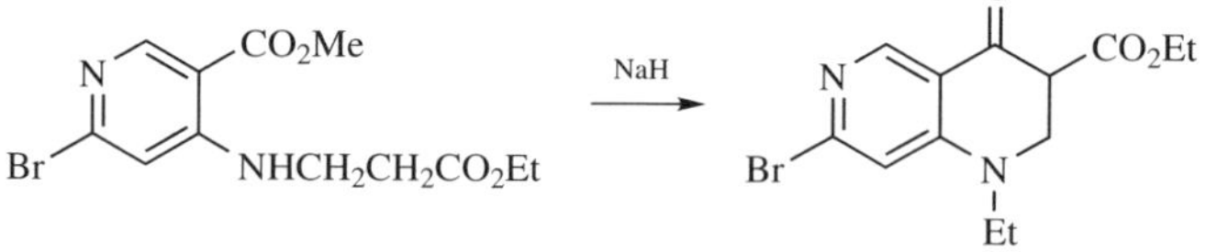

(26)   (27)

By Completion of the 4,4a-Bond

4-(2,2-Diethoxycarbonylvinylamino) pyridine (**28**) gave ethyl 4-oxo-1,4-dihydro-1,6-naphthyeidine-3-carboxylate (**29**) (Dowtherm, reflux, 15 min: 82%;[113] Ph_2O, 245°C, N_2, ? h: 53%);[48] the analogous ethy 7-chloro- (49%)[1284] and ethyl 5,7-dimethyl-4-oxo-1,4-dihydro-1,6-naphthyridine-3-carboxylate (80%)[255] were made somewhat similarly.

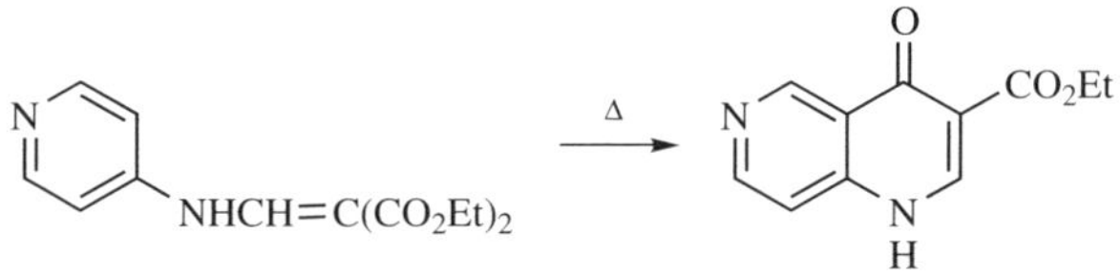

(28) → (29)

4-(3,3-Dimethoxypropylamino)-3-nitro-6-(2,2,2-trifluoroethoxy)pyridine (30) gave 8-nitro-5-(trifluoromethoxy)-1,2-dihydro-1,6-naphthyridine (31) (85% H$_3$PO$_4$, 100°C, 2 h: 776%), which was aromatized in unstated yield (tetrachloro-1,4-benzoquinone, CHCl$_3$, reflux, 2.5 h).[1116]

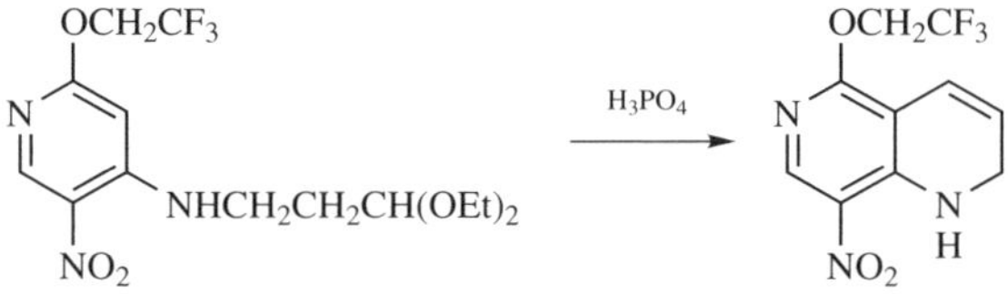

(30) → (31)

4-Methacrylamidopyridine (32) underwent photocyclization to give 3-methyl-3,4-dihydro-1,6-naphthyridin-2(1H)-one (33) (PhH, AcOH, hv, 3 h: 72%) and thence 3-methyl-1,6-naphthyridin-2(1H)-one (SeO$_2$, AcOH, reflux, 24 h: 36%).[865]

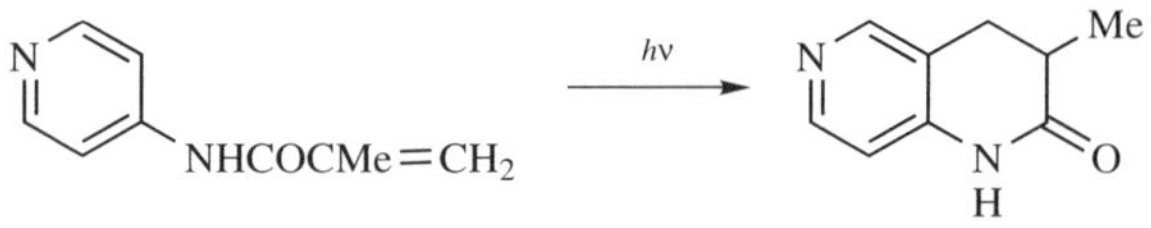

(32) → (33)

Also other examples,[863] including fused analogs.[146]

By Completion of the 4a,5-Bond

2-{C-Cyano-C-[N-(dimethylaminomethylene) carbamoyl]methylene}-1-methyl-piperidine (34) gave 1-methyl-7-oxo-1,2,3,4,6,7-hexahydro-1,6-naphthyridine-8-carbonitrile (35) (neat reactants, ~140°C, 20 min: >95%).[845]

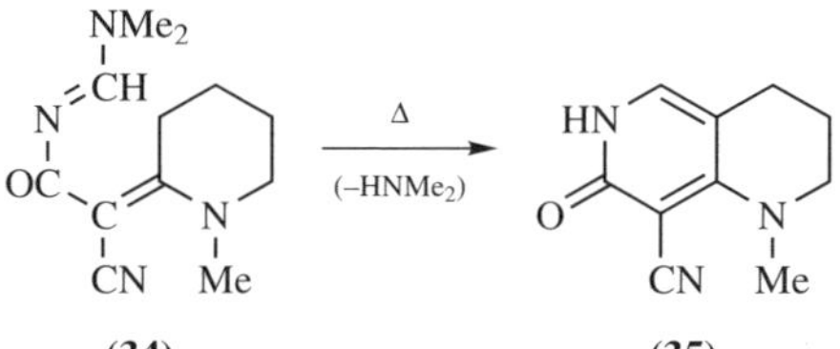

(34) → (35)

Also other examples.[1078]

By Completion of the 5,6-Bond

Note: This type of synthesis has been used extensively but abounds in pitfalls for the unwary. Some typical factual examples are given here.

3-Acetyl-2-dicyanomethylene-4,6-dimethyl-1-phenyl-1, 2-dihydropyridine (**36**) gave 2,4,5-trimethyl-7-oxo-1-phenyl-1,7-dihydro-1,6-naphthyridine-8-carbonitrile (**37**) (H_3PO_4, 130°C, 40 min: 73%).[742]

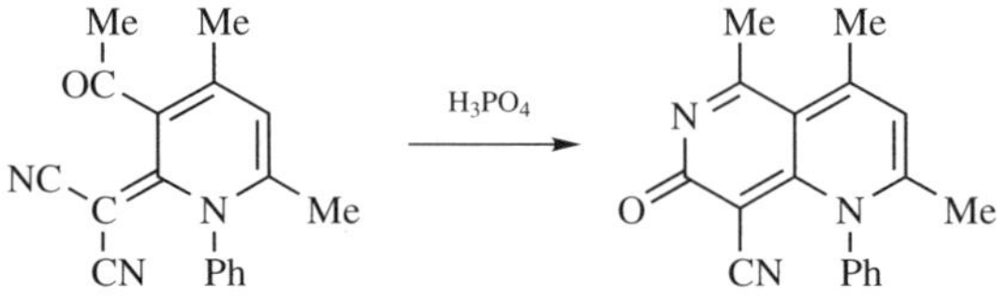

Methyl 2-[2-(hydroxyimino)-2-methoxycarbonylmethyl]-3-pyridinecarboxylate (**38**) underwent reductive cyclization to methyl 8-hydroxy-5-oxo-5,6-dihydro-1,6-naphthyridine-7-carboxylate (**39**) (Pd/C, HCl, H_2O, H_2, ~30 min: 88%).[6]

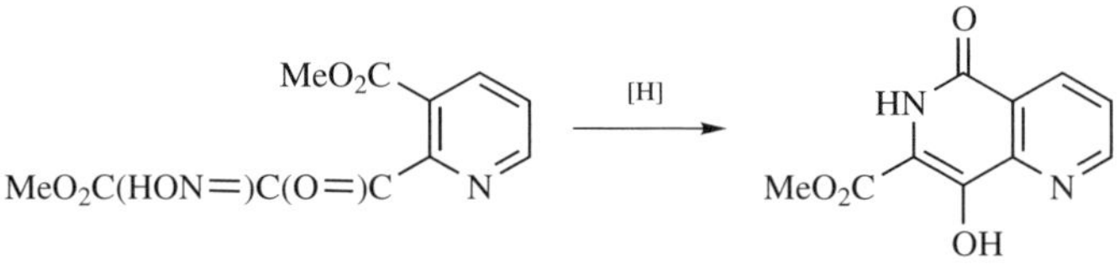

2-{[α-(Hydroxymethylphenethylamino]ethyl}-3-pyridinecarbonitrile (**40**) gave 5-[α-(hydroxymethyl)phenethylamino]-7,8-dihydro-1,6-naphthyridin-5(6*H*)-one (**41**) (H_2O, EtOB, reflux, 48 h: >56%),[443] further characterized as its 1-methiodide.[443,539,1111]

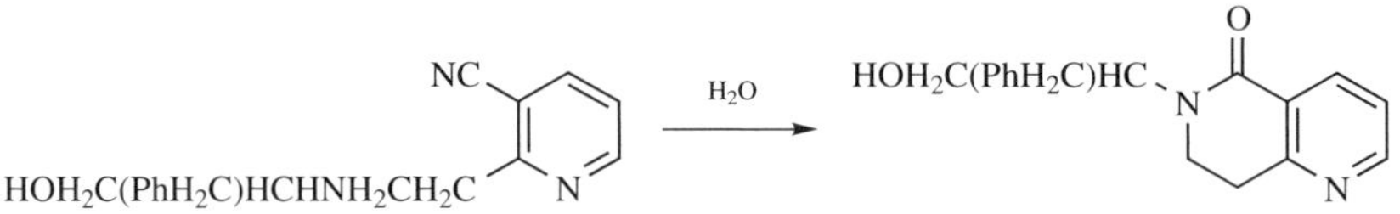

Ethyl 5-cyano-2-(α-cyanoethoxycarbonylmethyl)-4-oxo-1,4-dihydro-3-pyridinecarboxylate (**42**) gave ethyl 3-cyano-7-hydroxy-4,5-dioxo-1,4,5,6-tetrahydro-1,6-naphthyridine-8-carboxylate (**43**) (Et_3N, EtOH, reflux, 3 h: 65%).[698]

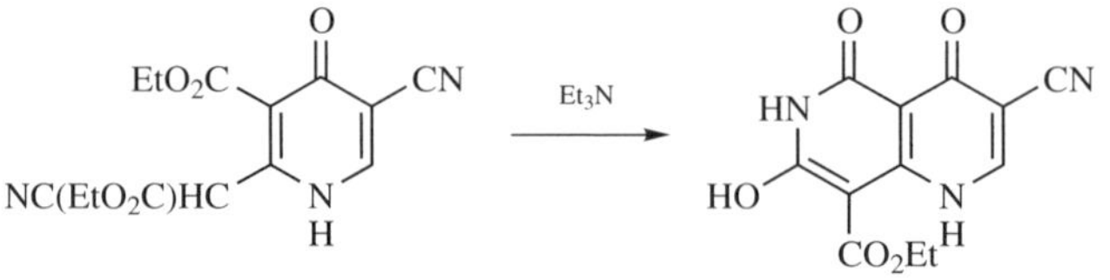

2-(Dicyanomethyl)-4,6-dimethyl-3-pyridinecarbonitrile (**44**) gave 5-amino-7-methoxy-2,4-dimethyl-1,6-naphthyridine-8-carbonitrile (**45**) (MeONa, MeOH, reflux, 12 h: 16%).[994,cf. 895,913,930]

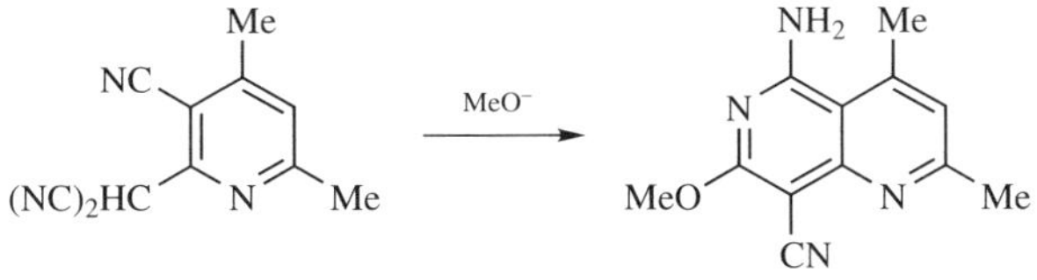

Also other examples.[392,673,708,751,951,1215,1322]

By Completion of the 6,7-Bond

Note: Cyclizations of this type may be achieved in many ways, some of the more interesting of which are exemplified here.

3-*tert*-Butyliminomethyl-2-phenylethynylpyridine (**46**) gave 8-iodo-7-phenyl-1,6-naphthyridine (**47**) (I$_2$ or ICl, NaMeCO$_3$, MeCN; substrate/MeCN↓ dropwise, A, 20°C, 30 min: 90%); 3-*tert*-butyliminomethyl-2-(hex-1-ynyl)pyridine gave 7-hexyl-8-phenylseleno-1,6-naphthyridine (somewhat similarly but using PhSeCl in place of I$_2$ or ICl: 61%).[119]

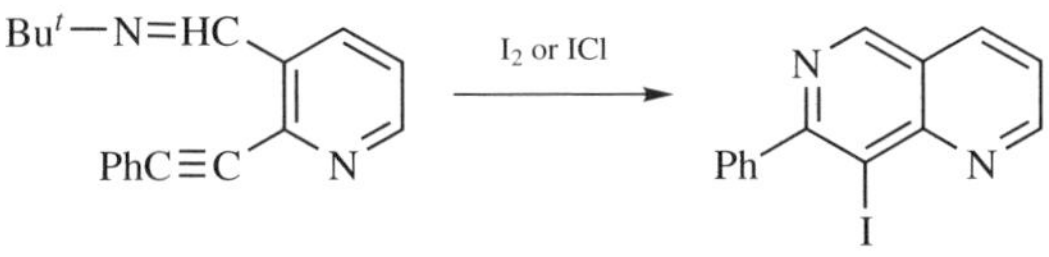

3-*N*-Hydroxyiminomethyl-2-phenylethynylpyridine (**48**) gave 7-phenyl-1,6-naphthyridine 6-oxide (**49**) (K$_2$CO$_3$, EtOH, reflux, 15 h: 67%; analogs likewise).[621]

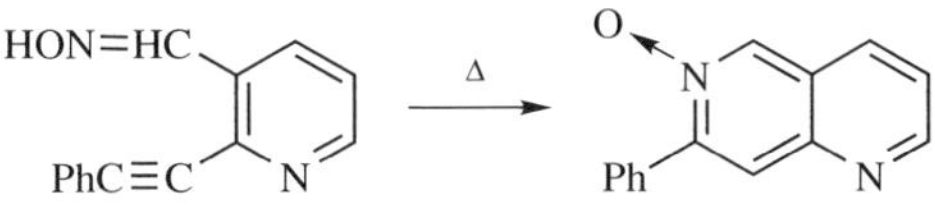

3-*N*-Benzylaminomethyl-2-(2-chloropropyl)pyridine (**50**) gave 6-benzyl-7-methyl-5,6,7,8-tetrahydro-1,6-naphthyridine (**51**) (Et$_3$N, EtOH, reflux, 1 h: 91%).[1058]

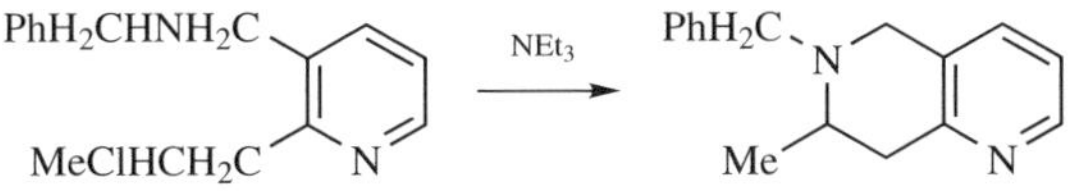

4,6-Dimethyl-2-phenylethynyl-3-pyridinecarboxamide (**52**) gave 2,4-dimethyl-7-phenyl-1,6-naphthyridin-5(6*H*)-one (**53**) (EtONa, EtOH, reflux, 3 h: 88%).[1055]

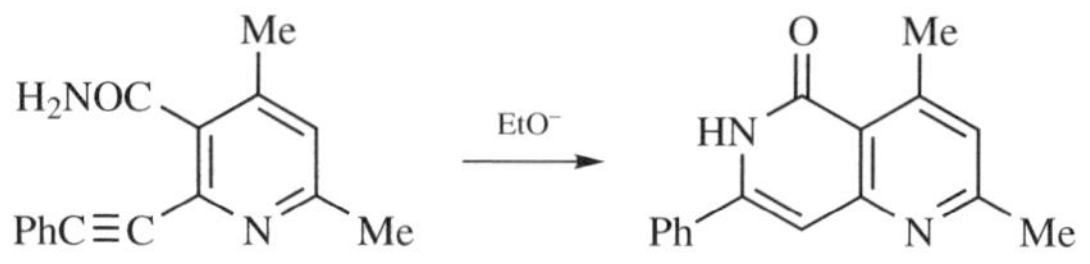

2-(2-Hydroxyethyl)-3-pyridinecarboxamide (**54**) underwent oxidative cyclization, probably via the aldehyde (**55**), to give 1,6-naphthyridin-5(6*H*)-one (**56**) (substrate, AcOH, 45°C; CrO$_3$ in 90% AcOH↓ during 2 h; 45°C, 3 h: 48%; homologs likewise).[1285]

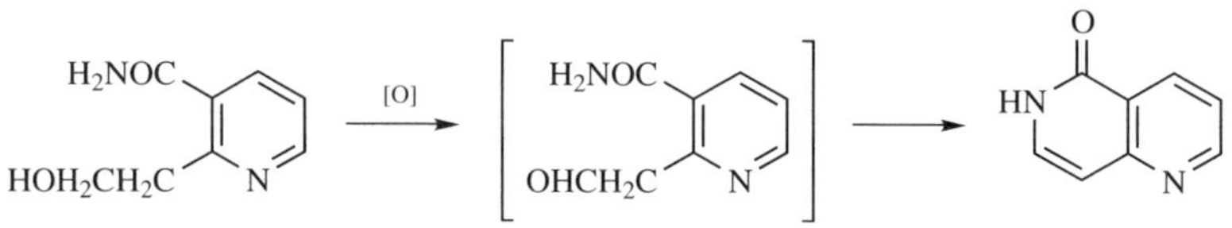

6-Phenyl-2-styryl-3-pyriidnecarboxanilide (**57**) gave 2,6,7-triphenyl-7,8-dihydro-1,6-naphthyridin-5(6*H*)-one (**58**) (P$_2$O$_5$, H3PO$_4$, 135°C, 4 h: 66%);[992] analogs somewhat similarly.[793,972,1025]

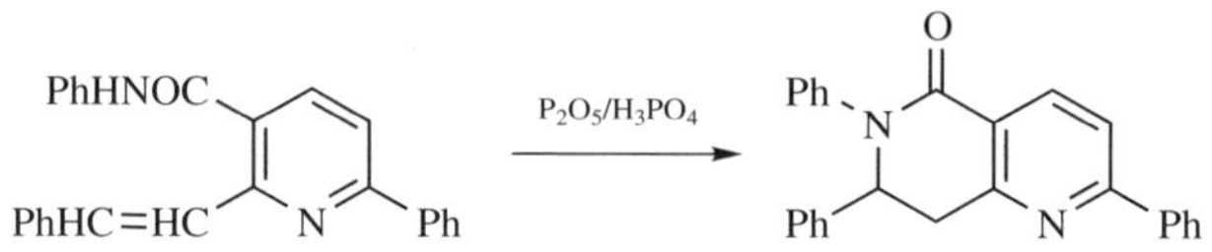

2-(2-Dimethylaminovinyl)-5-methyl-6-oxo-1,4,5,6-tetrahydro-3-pyridinecarbonitrile (**59**) gave 5-bromo-3-methyl-3,4-dihydro-1,6-naphthyridin-2(1*H*)one (**60**) (substrate, CHCl$_3$, HBr gas↓, 5°C, 30 min after dissolution: 95%).[716,cf. 339]

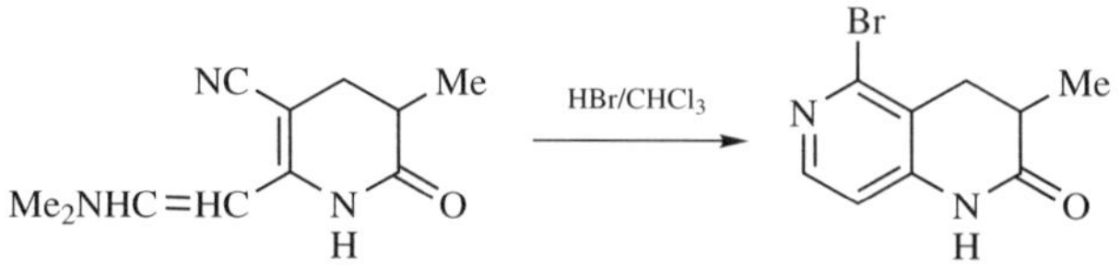

2,4-Diamino-6-ethoxycarbonylmethyl-3,5-pyridinedicarbonitrile (**61**) gave 2,4-diamino-5,7-dioxo-1,5,6,7-tetrahydro-1,6-naphthyridine-3-carbonitrile (**62**) (HCl, AcOH, reflux, 2 h: 65%).[1096]

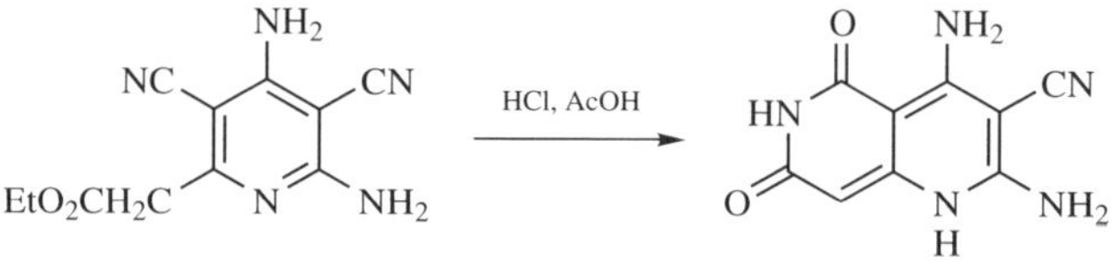

(**61**)   (**62**)

Also other examples.[719,779,936,1070,1270]

By Completion of the 7,8-Bond

Methyl 3-[*N*-(ethoxycarbonylmethyl)-*N*-methylcarbamoyl]-2-pyridinecarboxylate (**63**) gave methyl 8-hydroxy-6-methyl-5-oxo-5,6-dihydro-1,6-naphthyridine-7-carboxylate (**64**) (MeONa, MeOH, reflux, 15 min: 50%; note transesterification); analogs likewise.[654]

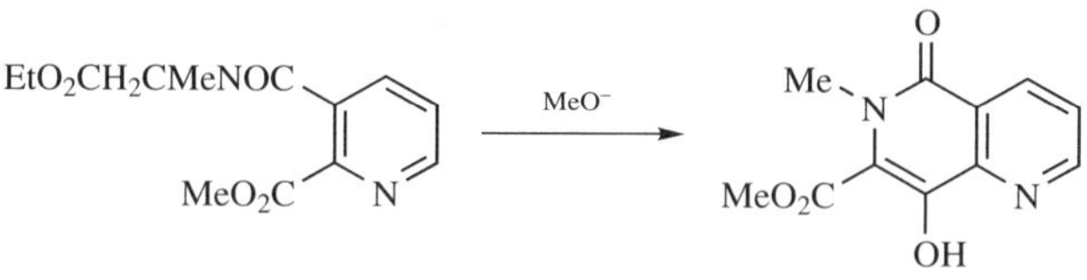

(**63**)   (**64**)

Also other examples.[141]

By Completion of the 8,8a-Bond

3-[1-(*N*-Prop-1-enylamino)ethyl]pyridine (**65**) gave a separable mixture of 5,8-dimethyl-7,8-dihydro-1,6-naphthyridine (**66**) and an isomeric 2,7-naphthyridine (substrate, 600°C, He: 80% net; see original for considerable detail).[400]

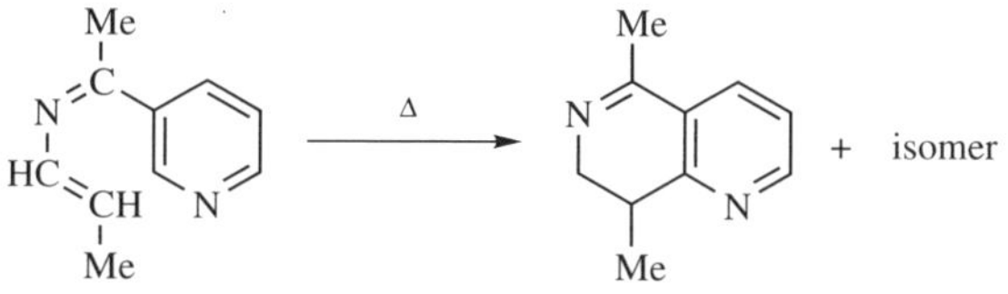

(**65**)   (**66**) + isomer

8.3. FROM A PYRIDINE SUBSTRATE WITH ONE SYNTHON

A pyridine substrtae may be condensed with one synthon to furnish 1,6-naphthyridines; the synthon can provide one, two, three, or four of the ring atoms in the product.

8.3.1. Where the Synthon Supplies One Ring Atom

Of the 10 possibilities, only the N1, N6, or C7 atoms of 1,6-naphthyridines have been provided by the synthon in such primary syntheses. Classified examples follow.

Provision of N1 by the Synthon

3-(2,2-Dicyanovinyl)-1-*p*-ethoxycarbonylphenyl-4-(pyrrolidin-1-yl)-1,2,5,6-tetrahydropyridine (**67**) with ammonia gave 2-amino-6-*p*-ethoxycarbonylphenyl-5,6,7,8-tetrahydro-1,6-naphthyridine-3-carbonitrile (**68**) (NH$_3$/MeOH, 30°C, 12 h: 96%); analogs likewise.[378]

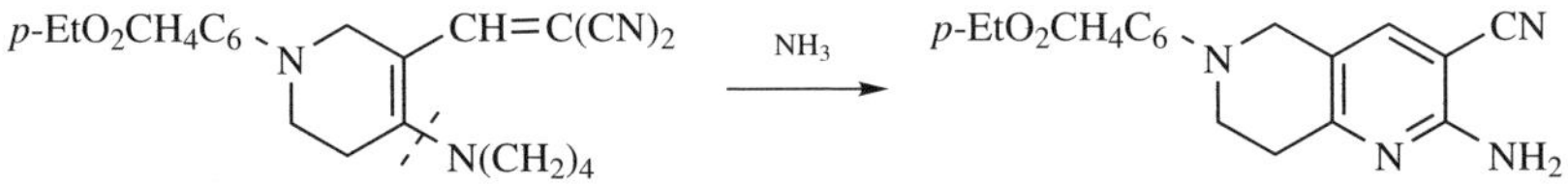

3-(2-Benzoylethyl)-2,2,6,6-tetramethyl-4-piperidinone (**69**) with ammonium ion gave 5,5,7,7-tetramethyl-2-phenyl-5,6,7,8-tetrahydro-1,6-naphthyridine (**70**) (substrate. HCl, NH$_4$Cl, EtOH, reflux, 4 h: 77%); analogs likewise.[933,1351]

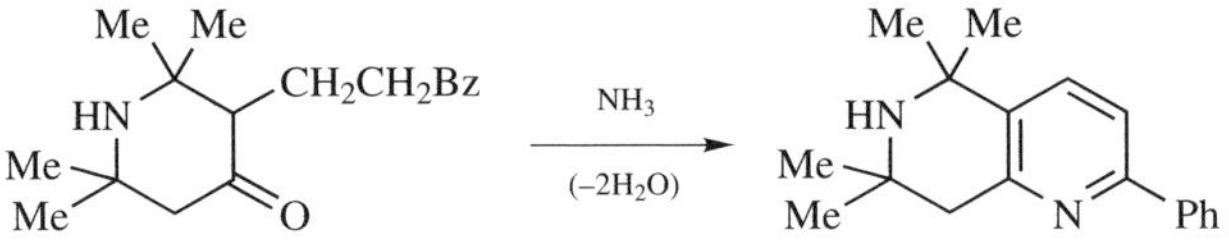

5-[2-(Carboxymethyl)heptyl]-1-veratryl-2,4-piperidinedione (**71**) with veratrylamine gave 4-hexyl-1,6-diveratryl-3,4,4a,5-tetrahydro-1,6-naphthyridine-2,7-dione (**72**) (PhH, reflux, water removal, 12 h: 60%).[549]

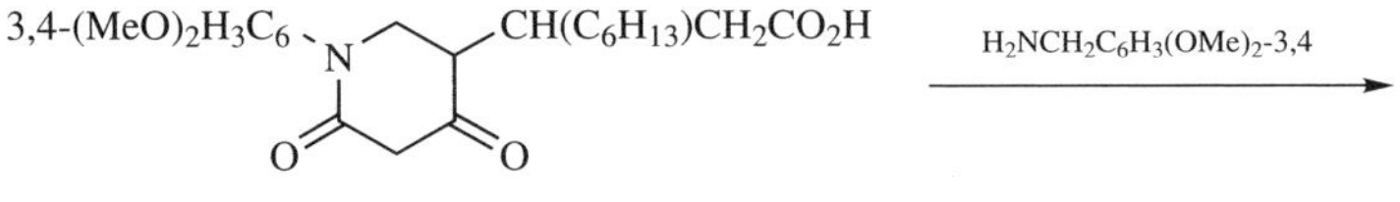

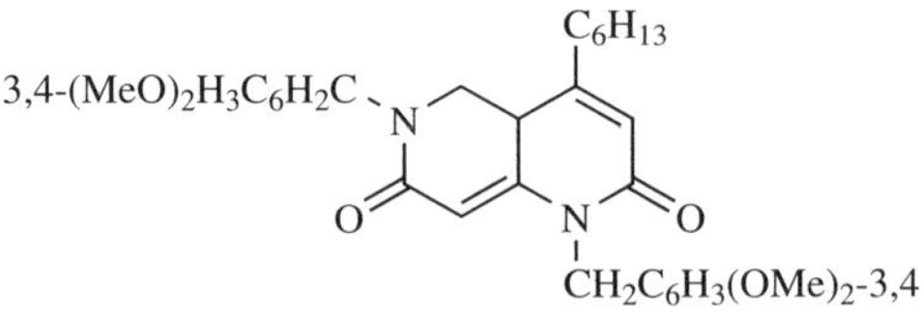

Also other examples.[127,371]

Provision of N6 by the Synthon

5-Acetyl-3-bromo-6-(2-dimethylaminovinyl)-2(1H)-pyridinone (**73**) with ammonium acetate gave 3-bromo-5-methyl-1,6-naphthyridin-2(1H)-one (**74**) (AcONH$_4$, Me$_2$NCHO, 95°C, 5 h: 54%);[928] analogs likewise.[928,1166]

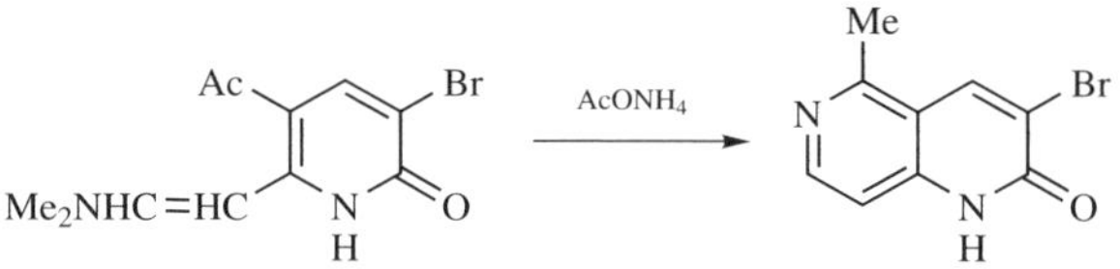

2-Acetonyl-3-pyridinecarboxylic acid (**75**) gave 7-methyl-1,6-naphthyridin-5(6H)-one (**76**) (NH$_4$OH, reflux, 18 h: 80%; analogs likewise).[1072]

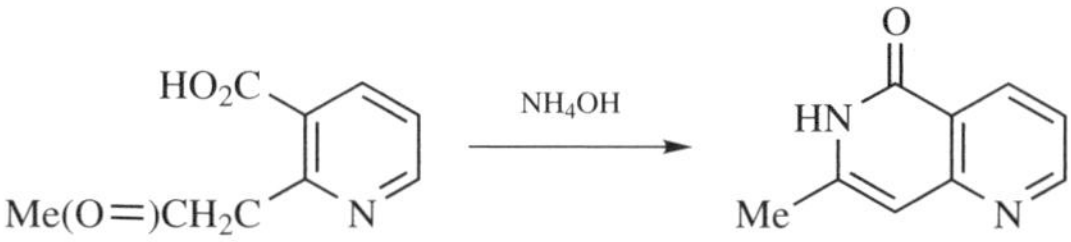

6-(2-Dimethylaminovinyl)-3-propionyl-2(1H)-pyridinone (**77**) with hydroxylamine hydrochloride gave 5-ethyl-1,6-naphthyridin-2 (1H)-one 5-oxide (**78**) (reactants, H$_2$O, 20°C, 12 h: 89%; analogs likewise).[490]

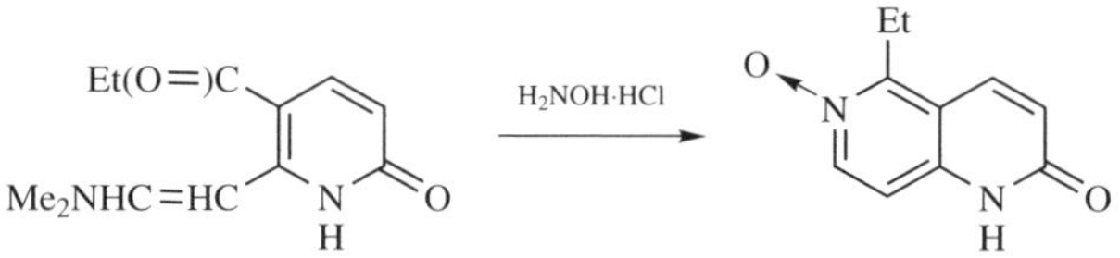

Also other examples.[4,754]

Provision of C7 by the Synthon

2-Methyl-6-oxo-1-p-tolyl-1,6-dihydro-3,5-pyridinedicarbonitrile (**79**) with carbon disulfide followed by methyl iodide gave 5,7-bismethylthio-2-oxo-1-p-tolyl-1,2-dihydro-1,6-naphthyridine-3-carbonitrile (**81**), presumably via the intermediate (**80**) (NaH, PhH, Me$_2$NCHO, CS$_2$↓, reflux, 2 h; MeI↓, 20°C; then reflux, 30 min: 28%).[507]

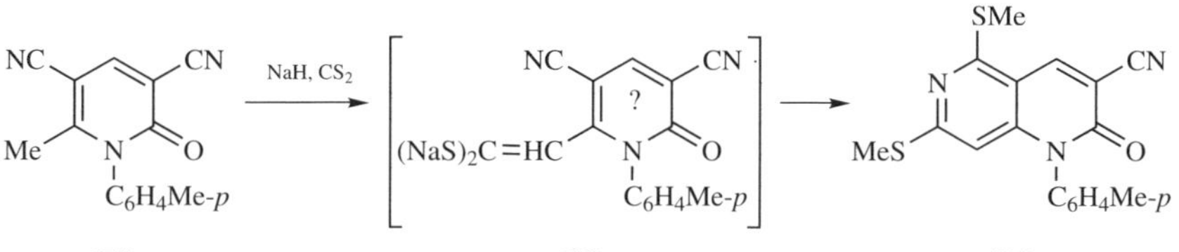

Also other examples.[1447]

8.3.2. Where the Synthon Supplies Two Ring Atoms

Of the many possibilities, only C2 + C3, N6 + C7, and C7 + C8 have been supplied by the synthon in such primary syntheses. Moreover, most of the published procedures involve the provision of C2 + C3, as illustrated by the typical examples that follow. Such condensations may also be used to make fused 1,6-naphthyridines.[24,27,29,41]

Provision of C2 + C3 by the Synthon

2-Amino-3-pyridinecarbaldehyde (**82**) with ethyl cyanoacetate gave 2-oxo-1,2-dihydro-1,6-naphthyridine-3-carbonitrile (**83**) (reactants, trace piperidine, EtOH, reflux, 1 h: 92%);[247] also analogs similarly.[924]

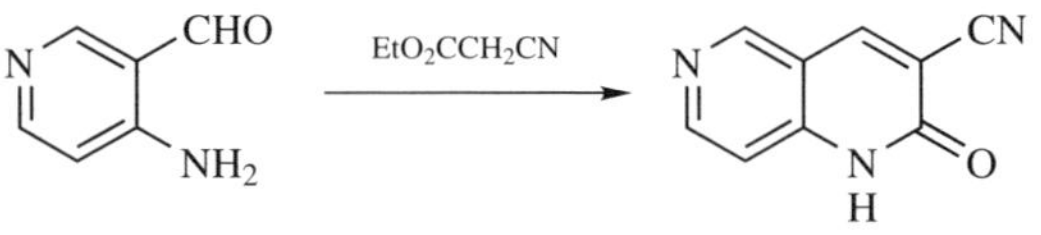

4,6-Diamino-3-pyridinecarbaldehyde (**84**) with 2-(2,6-dichlorophenyl)acetonitrile gave 3-(2,6-dichlorophenyl)-1,6-naphthyridine-2,7-diamine (**85**) (EtOCH$_2$CH$_2$ONa, EtOCH$_2$CH$_2$OH, reflux, 30 min: 82%);[239] also analogous condensations.[641,683,974]

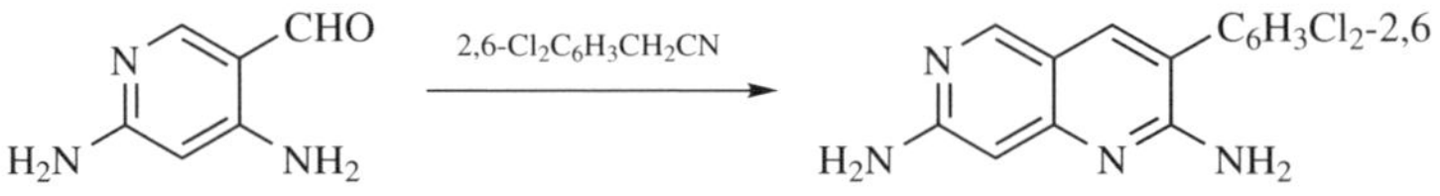

4-Amino-3-pyridinecarbaldehyde (**86**) with limited 2,3-butanedione gave 2,2′-bi-1,6-naphthyridine (**87**) (substrate, NaOH, H$_2$O, EtOH, reflux; synthon/EtOH↓ dropwise during 3 h; reflux, 8 h: 81%).[620]

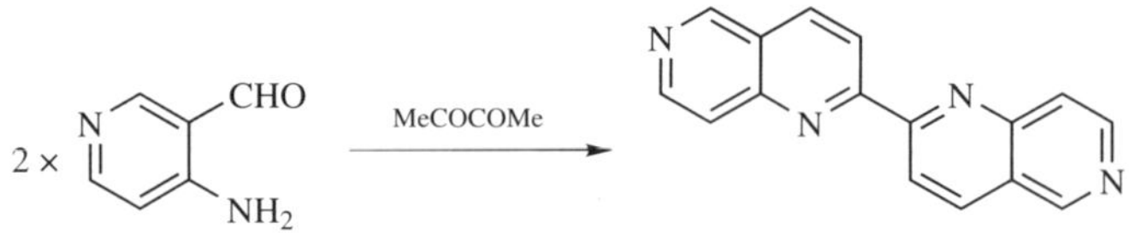

4-*tert*-Butyramido-3-ethoxalylpyridine (**88**) with methyl phenyl ketone gave 1-phenyl-1,6-naphthyridine-4-carboxylic acid (**89**) (substrate, KOH, EtOH,

H$_2$O, reflux 2 h: synthon↓, reflux 24 h: 89%; note initial saponification);[1219] also analogous condensations.[244,1219]

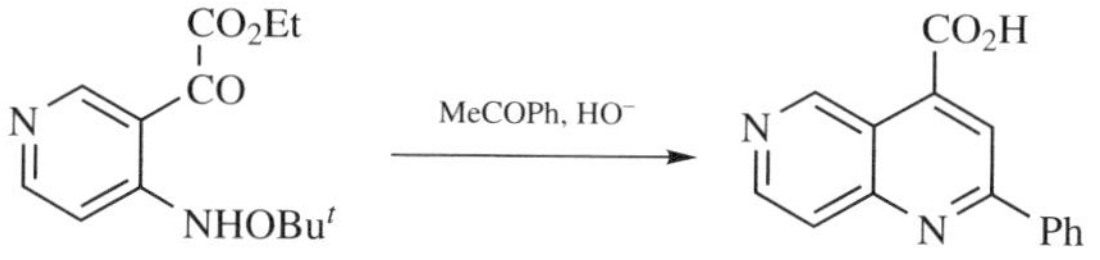

(88) → (89) [MeCOPh, HO⁻]

Ethyl 4-amino-3-pyridinecarboxylate (90) in ethyl acetate gave 4-hydroxy-1,6-naphthyridin-2(1H)-one (91) (NaH, "mineral oil," 20°C; AcOEt↓; then 100°C, 2 h: ?%);[312] analogs somewhat similarly.[312,996]

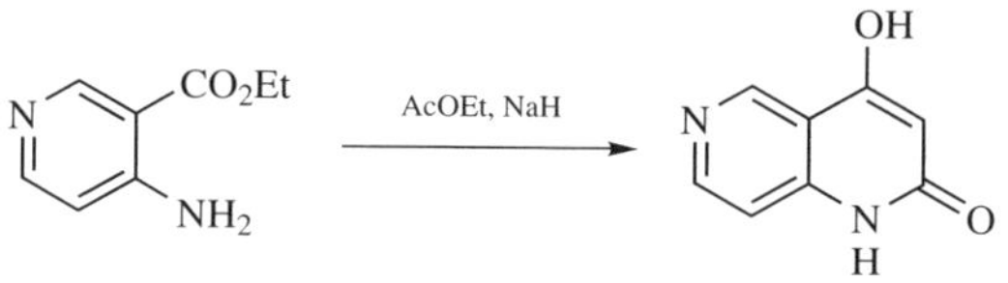

(90) → (91) [AcOEt, NaH]

Also other examples.[391,640,725,1068]

Provision of N6 + C7 by the Synthon

2-Methyl-3-pyridinecarboxylic acid (92) with propionitrile gave 7-ethyl-1,6-naphthyridin-5(6H)-one (93) [Et$_2$NNa (made *in situ*), THF, A: synthon/THF↓ dropwise, −78°C, then 20°C, 24 h: 35%]; 7-propyl homolog (50%) similarly.[131]

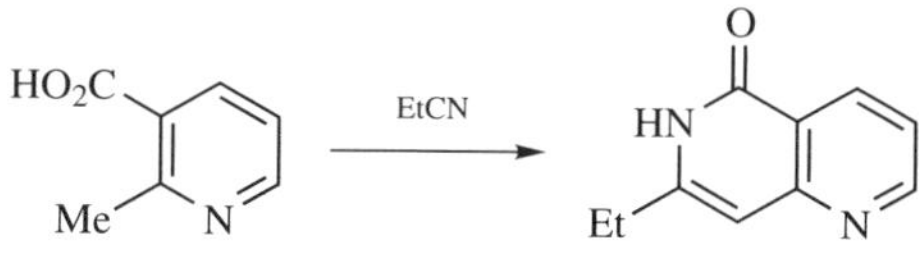

(92) → (93) [EtCN]

2,4-Diamino-6-ethoxycarbonylmethyl-3,5-pyridinedicarbonitrile (94) with trichloracetonitrile was first reported to afford ethyl 2,4,5-tritriamino-3-cyano-7-oxo-6,7-dihydro-1,6-naphthyridine-8-carboxylate (95) (reactants, Et$_3$N, Me$_2$NCHO, 20°C, 12 h: 69%)[1070] but later was reported to afford ethyl 2,4-diamino-3-cyano-5-oxo-7-trichloromethyl-5,6-dihydro-1,6-naphthyridine-8-carboxylate (96) (reactants, Et$_3$N, dioxane, reflux, 3 h: 70%);[1096,cf. 495] the reason for such a difference is not evident.

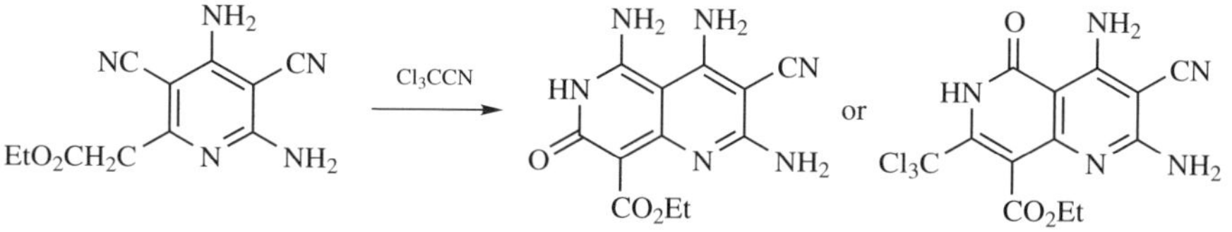

(94) → (95) or (96) [Cl$_3$CCN]

Diethyl 2,6-dimethyl-3,5-pyridinedicarboxylate (**97**) with 1,3,5-triazine gave ethyl 2-methyl-5-oxo-4,5-dihydro-1,6-naphthyridine-3-carboxylate (**98**) (reactants, Et_3N, EtOH, reflux, 2 h: 80%; structure confirmed by X-ray analysis);[1147] also other examples.[754,1105]

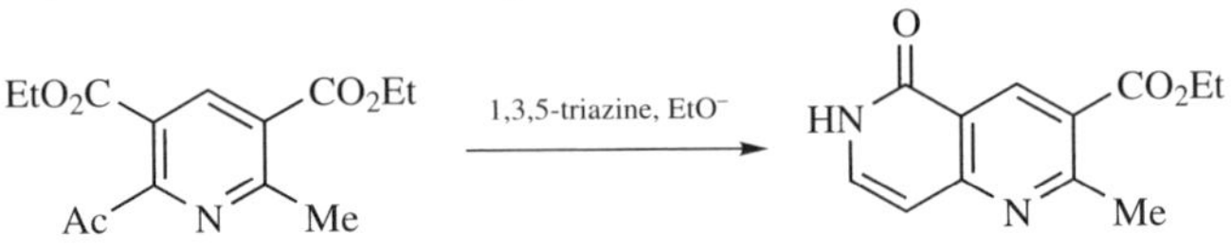

(**97**) (**98**)

2-Ethoxycarbonylmethyl-3-pyridinecarboxylic acid (**99**) with a Vilsmeier reagent (made *in situ*) gave 6-methyl-5-oxo-5,6-dihydro-1,6-naphthyridine-8-carboxylic acid (**100**) (substrate, Me_2NCHO, $POCl_3$, 0°C, 1 h, then 95°C, 6 h: 69%; note the hydrolysis of the ester grouping).[876]

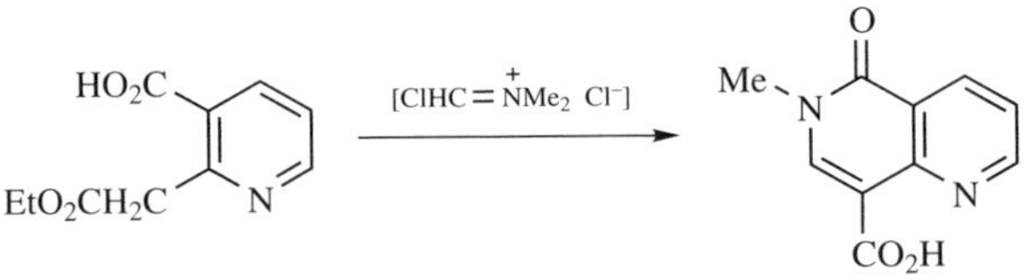

(**99**) (**100**)

Also other examples.[586]

Provision of C7 + C8 by the Synthon

2-Bromo-3-*tert*-butyliminomethylpyridine (**101**) with hex-1-yne gave the crude intermediate (**102**) [reactants, $PdCl_2(PPh_3)_2$, CuI, Et_3N, 55°C, 2 h: solid], which underwent cyclization to 7-butyl-1,6-naphthyridine (**103**) (CuI, Me_2NCHO, 100°C, 15 h: 72% overall).[544]

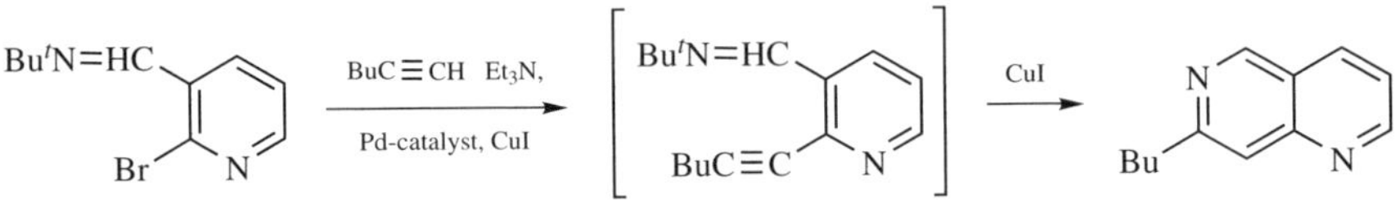

(**101**) (**102**) (**103**)

N-Benzyl-2-chloro-3-pyridinecarboxamide (**104**) with malononitrile gave 7-amino-6-benzyl-5-oxo-5,6-dihydro-1,6-naphthyridine-8-carbonitrile (**105**) (reactants, K_2CO_3, Me_2NCHO, reflux, 6 h: 73%); analogs likewise.[1220]

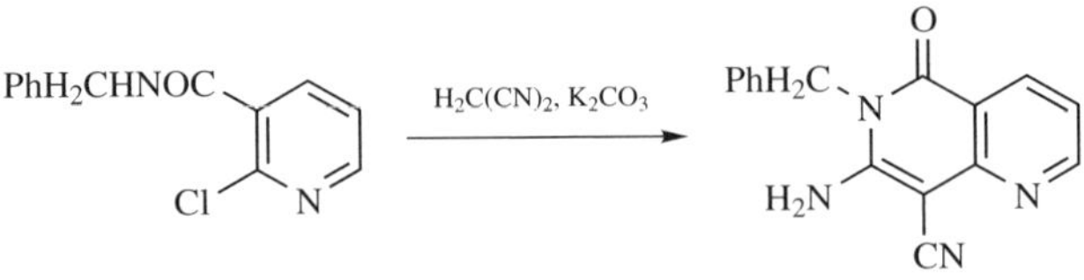

(**104**) (**105**)

Also other examples.[770,884]

8.3.3. Where the Synthon supplies Three or More Ring Atoms

Of the possibilities, only N1 + C2 + C3, C2 + C3 + C4, N6 + C7 + C8, N1 + C2 + C3 + C4, and C5 + N6 + C7 + C8 have been supplied by the synthon in this category of primary synthesis. Such cyclocondensations are illustrated here by typical classified examples. Fused 1,6-naphthyridines may also be made by analogous procedures.[169]

Provision of N1 + C2 + C3 by the Synthon

1-Methyl-4-oxo-3-piperidinecarbaldehyde (**106**), as its sodio derivative, with 2-cyanoacetamide, gave 6-methyl-2-oxo-1,2,5,6,7,8-hexahydro-1,6-naphthyridine-3-carbonitrile (**107**) (reactants, PhH, AcOH, reflux, water removal, 6 h: 73%).[180]

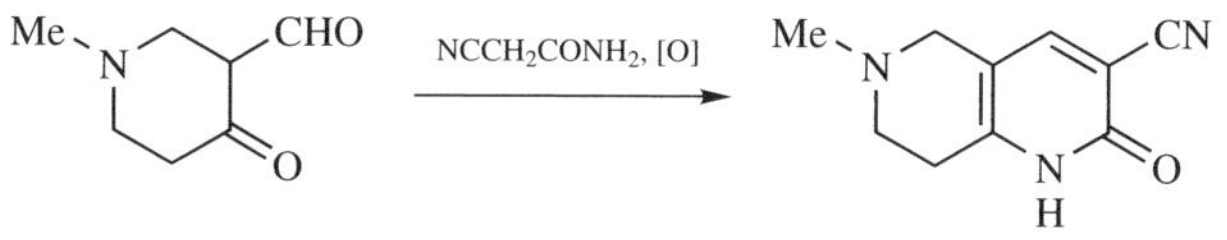

3,5-Dibenzylidene-1-isopropyl-4-piperidinone (**108**) (or tautomer) with 2-cyano(thioacetamide) gave 8-benzylidene-6-isopropyl-4-phenyl-2-thioxo-1,2,5,6,7,8-hexahydro-1,6-naphthyridine-3-carbonitrile (**109**) (or tautomer) (reactants, MeONa, MeOH, 50°C, 8 h: 55%).[946]

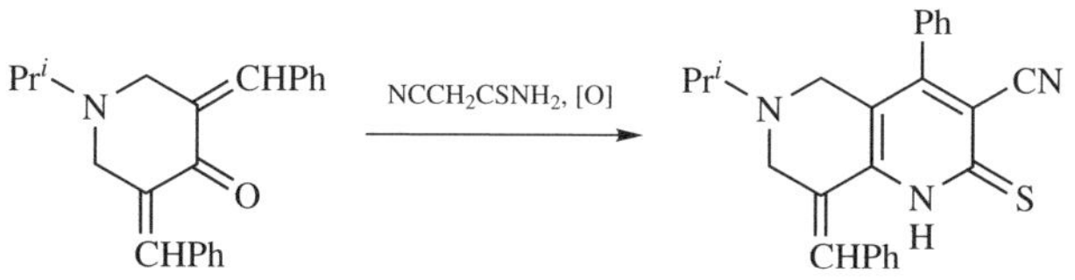

Also other examples.[952]

Provision of C2 + C3 + C4 by the Synthon

4-Pyridinamine (**110**) underwent modified Skraup reactions to afford 1,6-naphthyridine (**111**) (m-$O_2NC_6H_4SO_3H$, H_2SO_4, 5°C $HOCH_2CHOHCH_2OH\downarrow$, substrate$\downarrow$, $H_2O\downarrow$, 20°C, 10 min, 130°C, 5 h: 40%;[968] H_2SO_4, SO_3, $PhNO_2$, H_3BO_3, $FeSO_4$, 5°C, $HOCH_2CHOHCH_2OH\downarrow$, substrate$\downarrow$, $H_2O\downarrow$, 130°C, 5 h: 70%).[873,cf. 880]

3-Iodo-4-pyridinamine (**112**) with allyl alcohol also gave 1,6-naphthyridine (**111**) [reactants, $PdCl_2$, $NaHCO_3$, Pd(*p*-$MeC_6H_4)_3$, $OP(ONMe_2)_3$, 140°C. 4 h: 51%].[165]

(110) **(111)** **(112)**

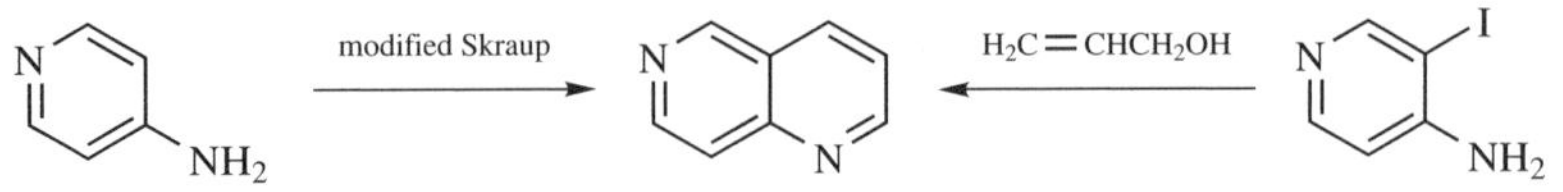

4-Amino-6-hydroxy-2(1*H*)-pyridinone (**113**) with 1-benoyl-2-dimethylaminoethylene gave 4-phenyl-1,6-naphthyridine-5,7(1*H*,6*H*)-dione (**114**) (reactants, AcOH, reflux, 5 min: 95%);[757] analogs likewise.[469,757,1310]

(113) **(114)**

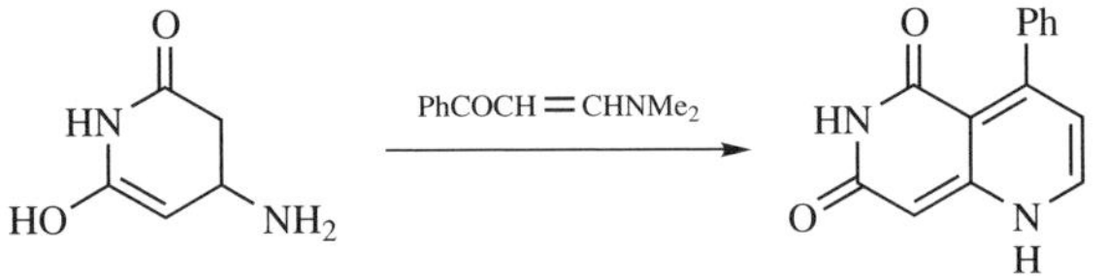

4-Benzylamino-6,6-dimethyl-3,6-dihydro-2(1*H*)-pyridinethione (**115**) with bis-(2,4,6-trichlorophenyl) α-ethylmalonate gave 1-benzyl-3-ethyl-4-hydroxy-7,7-dimethyl-5-thioxo-5,6,7,8-tetrahydro-1,6-naphthyridin-2(1*H*)-one (**116**)-(reactants, PhBr, reflux, 2.5 h: 30%);[679] analogs somewhat similarly.[672,679,691]

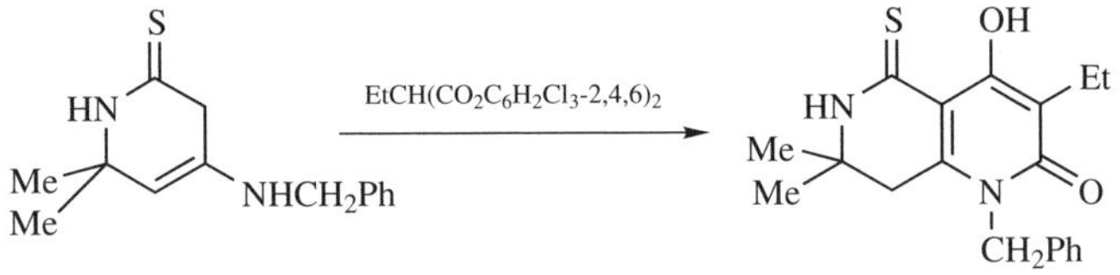

(115) **(116)**

Also other examples.[135,269,464,545,964,1244]

Provision of N6 + C7 + C8 by the Synthon

2-Chloro-3-pyridinecarbonyl chloride (**117**) with 1-benzoyl-3-phenyl-2-propylaminopropane gave 8-benzoyl-7-benzyl-6-propyl-1,6-naphthyridin-5(6*H*)-one (**118**) (reactants, Et_3N, PhMe, reflux, 3 h; diazabicycloundecene, reflux, 3 h: 52% after separation from other products; analogs likewise.[183]

(117) **(118)**

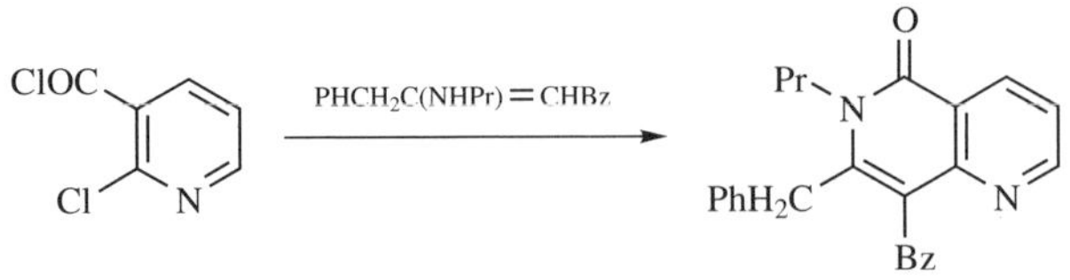

Also other examples.[1442]

Provision of N1 + C2 + C3 + C4 by the Synthon

1-Benzyl-4-piperidinone (**119**) with 3-amino-2-methylacrolein gave 6-benzyl-3-methyl-5,6,7,8-tetrahydro-1,6-naphthyridine (**120**) (reactants, BF$_3$, Et$_2$O, pyridine, xylene, reflux, water removal, 16 h: 23%).[277] in a somewhat similar way, 1-methyl-4-piperidinone with 3-aminoacrolein gave 6-methyl-5,6,7,8-tetrahydro-1,6-naphthyridine (reactants, AcONH$_4$, Et$_3$N, 120°C 2 h: 35% as its oxalate).[571]

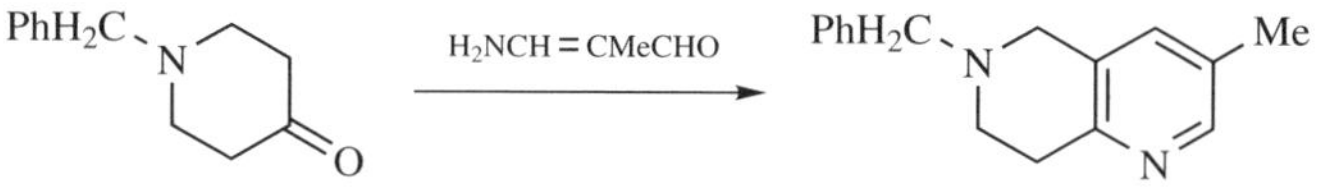

Provision of C5 + N6 + C7 + C8 by the Synthon

1-Methyl-3,5-dinitro-2(1H)-pyridinone (**122**) with with 1,3,3-trimethyl-4-piperidinone (**121**) gave 6,8,8-trimethyl-3-nitro-5,6,7,8-tetrahydro-1,6-naphthyridine (**123**) (reactants, NH$_4$OH, MeOH, reflux, 5 h: 57%; mechanism?);[1294] also analogous condensations.[204,1294]

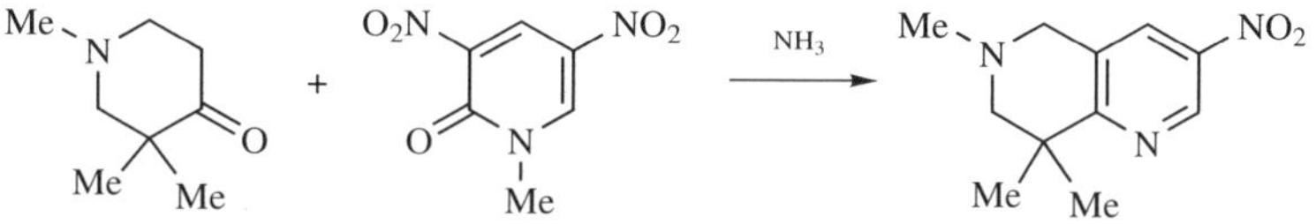

8.4. FROM A PYRIDINE SUBSTRATE WITH TWO OR MORE SYNTHONS

Only a few such syntheses have been reported, as illustrated in the following examples.

3-Bromo-6-methyl-5-propionyl-2(1H)-pyridinone (**124**) with *tert*-butoxybis(dimethylamino)methane followed by ammonium acetate gave 3-bromo-5-ethyl-1,6-naphthyridin-2(1H)-one (**125**) [substrate, dioxane, 20°C, Bu^{t}OCH(NMe$_2$)$_2$/dioxane↓, reflux, 5 h: solid; AcONH$_4$, Me$_2$NCHO, 20°C, foregoing solid↓, 95°C, 5 h: 86%]; analogs likewise.[590]

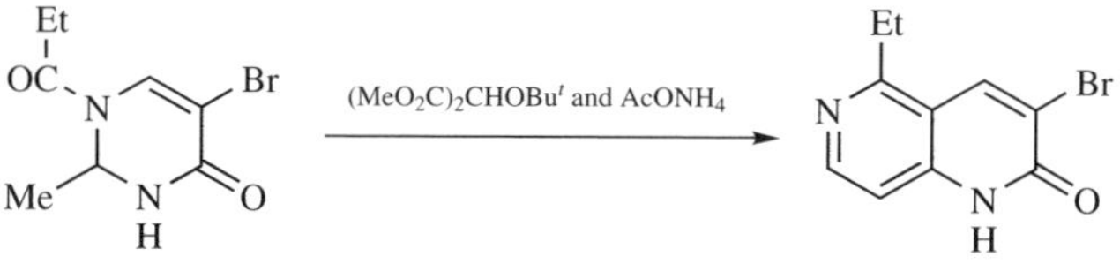

1-Methyl-4-piperidinone (**126**) with 2-cyano(thioacetamide) and 2-thiophene-carbaldehyde gave 6-methyl-4-(thien-2-yl)-2-thioxo-1,2,5,6,7,8-tetrahydro-1,6-naphthyridine-3-3-carbonitrile (**127**) (reactants, $Na_2S_2O_3$, Et_3N, EtOH, reflux, 30 min 64%); analogs likewise.[1429]

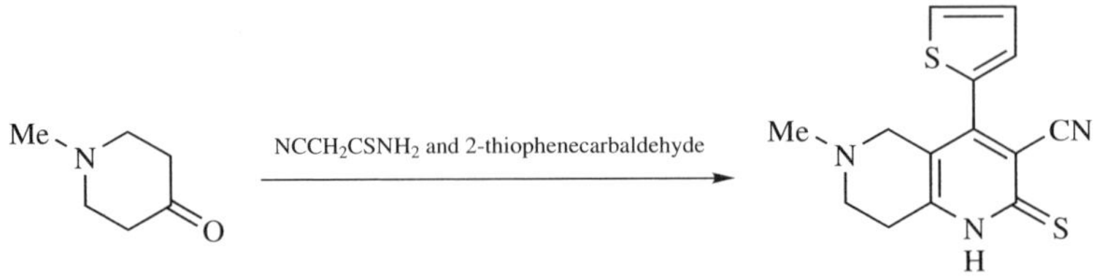

2,4-Piperidinedione with methyl 3-aminocrotonate and *p*-nitrobenzaldehyde (**128**) gave methyl 2-methyl-5-oxo-4-*p*-nitrophenyl-1,4,5,6,7,8-hexahydro-1,6-naphthyridine-3-carboxylate (**129**) (reactants, MeOH, reflux, ~7 h: 65%); analogs likewise.[869]

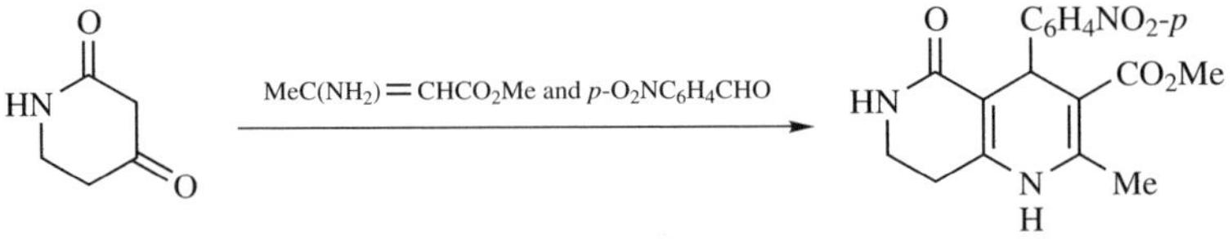

Dimethyl 1,2,6-trimethyl-4-phenylpiperidine-3,5-dicarboxylate (**130**) with paraformaldehyde and methylamine hydrochloride gave dimethyl 2,6,7-trimethyl-4-phenyl-1,4,4a,5,6,7-hexahydro-1,6-naphthyridine-2,4a-dicarboxylate (**131**) (reactants, $CHCl_3$, reflux, 26 h: 30%);[1210] also analogous condensations[532,1210]

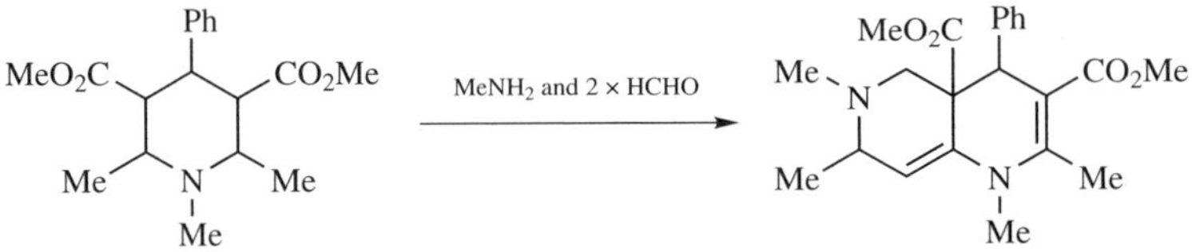

Also other examples.[413]

8.5. FROM OTHER HETEROCYCLIC SYSTEMS

At least 10 heterocyclic systems, other than pyridine, have been used as substrates for the synthesis of 1,6-naphthyridines (or occasionally their benzo derivatives[144]). However, only a few such procedures appear to have much practical potential. Some typical examples follow.

Benzo-1,4-thiazines as Substrates

4-Methylbenzo-1,4-thiazin-3(4H)-imine hydrochloride (**132**) with acetic anhydride gave the hexacyclic product (**133**) (neat reactants, reflux, 90 min: 57%), which underwent desulfurization by Raney nickel to afford 4,5-dimethyl-2,7-bis(N-methylanilino)-1,6-naphthyridine (**134**) (PhH, reflux, 12 h: ~20%).[1032]

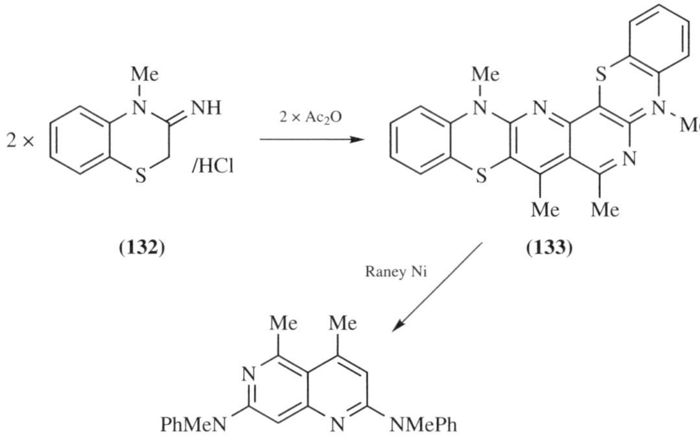

Cyclopenta[b] pyridines (pyrindines) as Substrates

An easily prepared mixture of 5H-cyclopenta[b]pyridine (**135**) and its 7aH-isomer underwent a one-pot ozonization and subsequent treatment with ammonium hydroxide to give a separable mixture of 1,6- (**136**) and 1,7-naphthyridine (**137**) (substrate, MeOH, −78°C, O₃↓ until blue; then N₂↓; NaHCO₃↓; then Me₂S↓, 20°C, 6 h; NH₄OH↓, 20°C, 6 h: 24% and 48%, respectively); homologs likewise.[943]

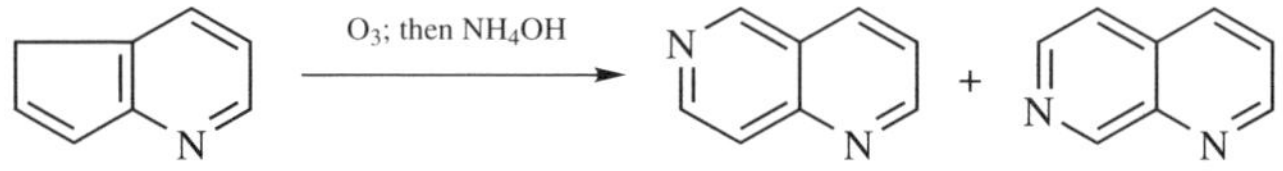

Pyrans as Substrates

2-Amino-6-phenyl-3,5-pyrandicarbonitrile (**138**) underwent a multistep reaction with formaldehyde and malononitrile to give 5-oxo-2-phenyl-5,6-dihydro-1,6-naphthyridine-3,8-dicarbonitrile (**139**) [substrate, HCHO, H₂C(CN)₂,

Et₃N, EtOH, 20°C, 2 h: 66%; see original for proposed steps];[1378] analogs somewhat similarly.[685]

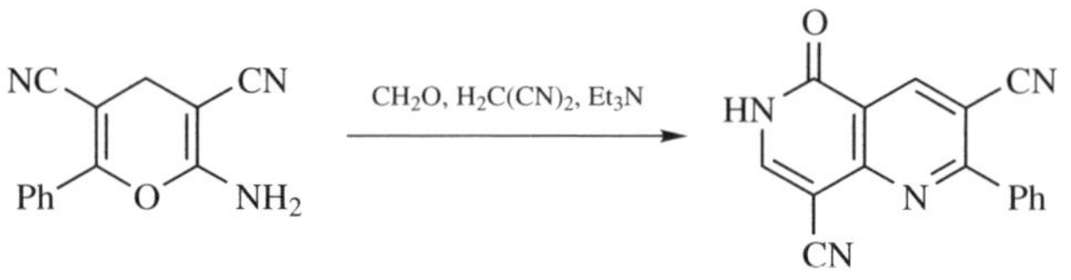

(138) → (139)

Pyrano[4,3-b] pyridines as Substrates

7-Phenyl-5H-pyrano [4,3-b] pyridin-5-one (140) with ethanolamine gave 6-(2-hydroxyethyl)-7-phenyl-1,6-naphthyridin-5(6H)-one (141) (reactants, MeOH, reflux, 3 h: 91%).[902]

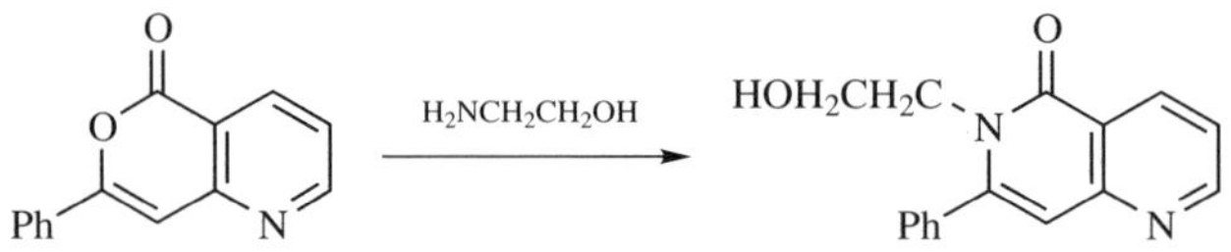

(140) → (141)

5-Oxo-7-phenyl-2-styryl-7,8-dihydro-5H-pyrano[4,3-b]pyridine-3-carboxylic acid (142) with a Vilsmeier reagent followed by methylamine gave 8-(α-hydroxybenzyl)-6-methyl-5-oxo-2-styryl-5,6-dihydro-1,6-naphthyridine-3-carboxylic acid (143) (Me₂NCHO, POCl₃↓ dropwise, 5°C, then substrate↓; 80°C, 12 h; POCl₃↓, H₂O↓, 5°C; MeNH₂·HCl↓, 30 min: 57%); analogs likewise.[1038]

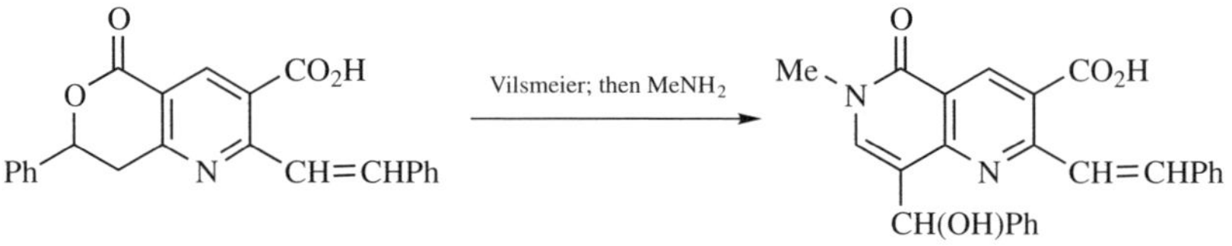

(142) → (143)

Also other examples.[383,867,987,1077]

Pyrido[1,2-c] pyrimidines as Substrates

5-Dimethylaminomethylene-3-oxo-5,6,7,8-tetrahydro-3H-pyrido[1,2-c] pyrimidinecarbonitrile (144) suffered ring fission and reclosure to give 7-oxo-1,2,3,4,6,7-hexahydro-1,6-naphthyridine-8-carbonitrile (145) (substrate, H₂O, reflux, 5 h: 79%);[823] analogs likewise.[805,839] (Note? Such transformations have been used to make fused derivatives of 1,6- and other naphthyridines.[161])

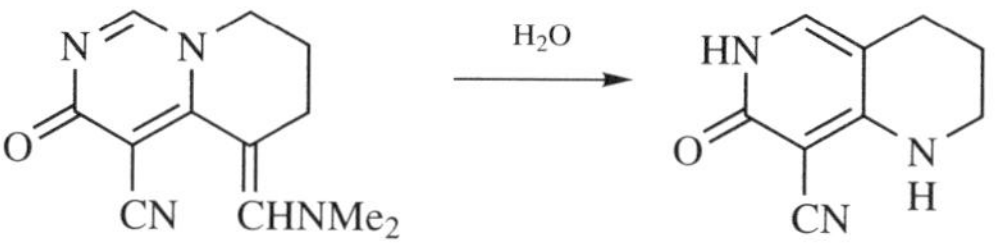

(144) → (145)

Pyrimidines as Substrates

5-Nitropyrimidine (147) underwent a Diels–Alder-type addition by the enamine, 1-acetyl-4-(pyrrolidin-1-yl)-1,2,3,6-tetrahydropyridine (146), and subsequent reactions to afford 6-methyl-3-nitro-5,6,7,8-tetrahydro-1,6-naphthyridine (148, R = Ac) (reactants, EtOH, reflux, 3 h: 48%) and thence deacetylation to 3-nitro-5,6,7,8-tetrahydro-1,6-naphthyridine (148, R = H) (6M HCl, reflux, 90 min: 89%) and oxidation to 3-nitro-1,6-naphthyridine (substrate, AcOH, EtOH, reflux, I$_2$/EtOH I$_2$EtOH↓ dropwise, reflux. 2 h: 64%).[656]

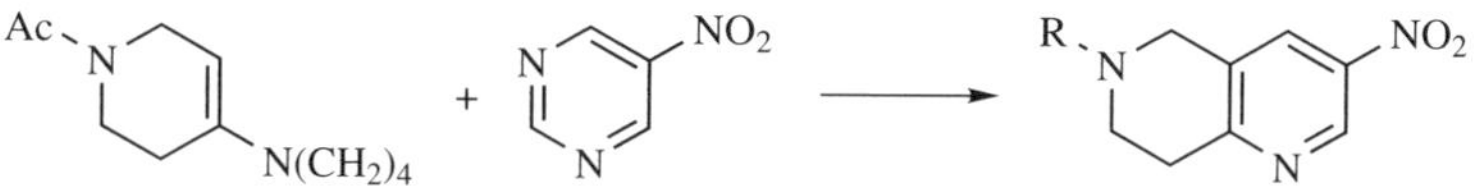

(146) + (147) → (148)

Pyrrolo[3,4-b]pyridines as Substrates

6-Ethoxycarbonylmethyl-6,7-dihydro-5H-pyrrolo[3,4-b]pyridine (149) isomerized to give a separable mixture of ethyl 8-hydroxy-5-oxo-5,6-dihydro-1,6-naphthyridine-7-carboxylate (150) and ethyl 5-hydroxy-8-oxo-7,8-dihydro-1,7-naphthyridine-6-carboxylate (151) (EtONa, EtOH, reflux, 30 min: 40% and 18%, respectively;[1221] when MeONa in MeOH was used, an additional transesterification occurred to afford the corresponding methyl esters in 50% and 20% yields, respectively).[1048,cf. 109]

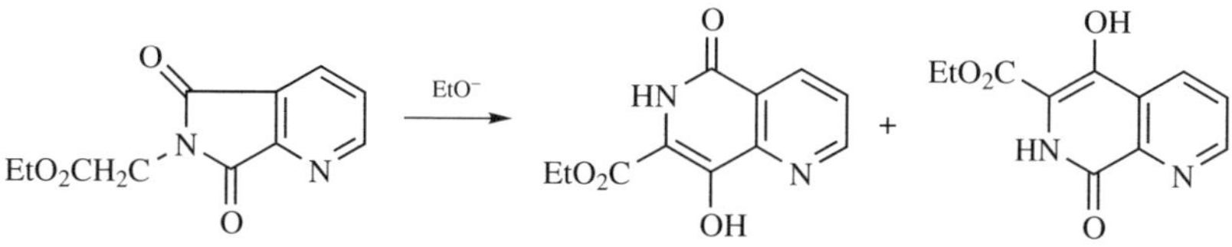

(149) → (150) + (151)

Also other examples.[1444]

Thiazolo[2,3-*f*] [1,6]-naphthyridin-4-iums as Substrates

8-Carboxy-6-hydroxythiazolo[2,3-*f*][1,6]-naphthyridin-4-ium-3-carboxylate (**152**) (berninamycinic acid) underwent exhaustive methylation to afford methyl 8-methoxy-6-(1-methoxycarbonyl-2-methylthiovinyl)-5-oxo-5,6-dihydro-1,6-naphthyridine-2-carboxylate (**153**) (MeI, Ag$_2$O: low yield)[293] or hydrogenolysis to furnish a separable mixture of 6-(1-carboxylatoethyl)-8-hydroxy-1,2,3,4-tetrahydro-1,6-naphthyridin-6-ium-2-carboxylic acid (**154**) and 8-hydroxy-1,2,3,4-tetrahydro-1,6-naphthyridine-2-carboxylic acid (**155**) (Raney Ni, NaOH, H$_2$O: unstated yields).[274]

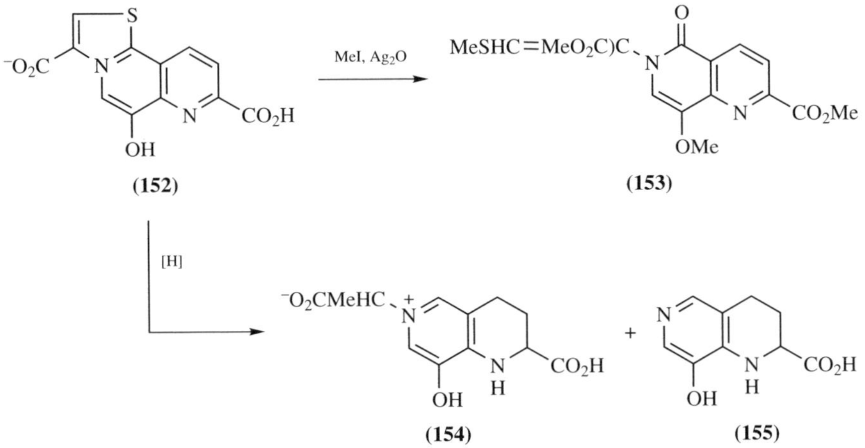

1,2,4-Triazines as Substrates

Triethyl 1,2,4-triazine-3,5,6-tricarboxylate (**156**) underwent a Diels–Alder reaction with 2-methoxy-1-methyl-1,4,5,6-tetrahydropyridine (**157**) and subsequent degradation to give triethyl 1-methyl-1,6-naphthyridine-5,7,8-tricarboxylate (**158**) (reactants, PhH, until N$_2$↓ ceased; then 20°C, 12 h: 83%).[795]

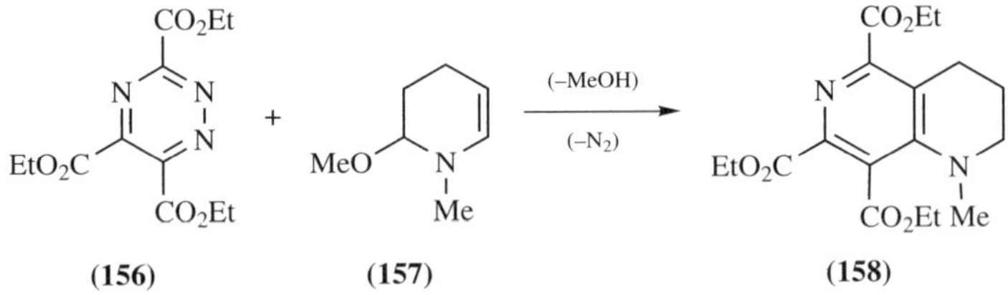

1,3,5-Triazines as Substrates

Note: Examples of such syntheses have been given in Section 8.3.2.

Benzo[*de*] [1,6] naphthyridines as Substrates

4-Acetyl-8,9-dimethoxy-5,6-dihydro-4*H*-benzo[*de*][1,6]naphthyridine (159) underwent ozonolysis to afford methyl 1-acetyl-4-methoxycarbonylmethylene-1,2,3,4-tetrahydro-1,6-naphthyridine-5-carboxylate (**160**) (or tautomer) (substrate, MeOH, $O_3\downarrow$, $-78°C$, 1 h: $Me_2S\downarrow$, $20°C$: ?%).[1143]

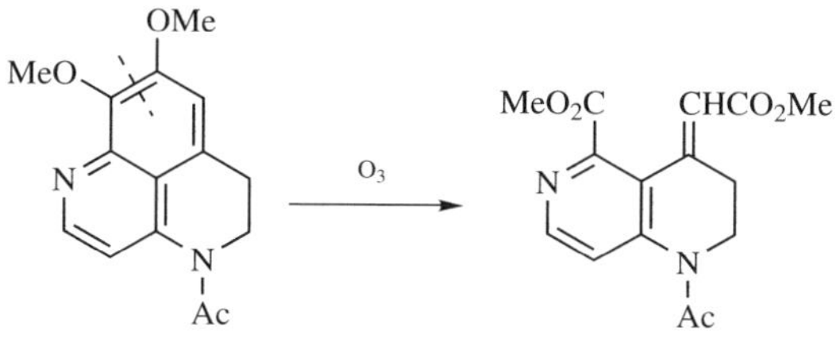

CHAPTER 9

1,6-Naphthyridine, Alkyl-1, 6-naphthyridines, and Aryl-1, 6-naphthyridines

The scope of this chapter resembles that on the corresponding 1,5-naphthyridines; see the introduction to Chapter 2.

The general chemistry of 1,6-naphthyridines has been reviewed frequently but usually quite briefly.[54–57,59–61,265,328,407,542,670,1260,1268,1273,1357,1430,1432] Derivatives show bioactivities.[907,908]

9.1. 1,6-NAPHTHYRIDINE

9.1.1. Preparation of 1,6-Naphthyridine

The parent 1,6-naphthyridine is probably best made by a modified Friedlander[1441] or another *primary syntheses* based on 4-pyridinamine[873,880,968] or on 3-iodo-4-pyridinamine[165] (see Section 8.3.3). However, other routes have also been used, as illustrated in the following examples.

4-Chloro-1,6-naphthyridine (**1**) underwent direct hydrogenolysis to afford 1,6-naphthyridine (**3**) (AcOH, Pd/C, EtOH, H_2, 20°C, 1 atm: 60%).[42]

The same substrate (**1**) also underwent indirect hydrogenolysis by hydrazinolysis to 4-hydrazino-1,6-naphthyridine (**2**) ($H_2NNH_2 \cdot H_2O$, EtOH, 20°C, 4 days: 90%) and subsequent oxidation to 1,6-naphthyridine (**3**) ($CuSO_4$, kieselguhr, H_2O, reflux, 5 min: 60%);[45] a somewhat similar procedure converted 5-chloro- into 5-hydrazino-1,6-naphthyridine and thence into 1,6-naphthyridine

---

The Naphthyridines: The Chemistry of Heterocyclic Compounds, Volume 63, by D.J. Brown
Copyright © 2008 John Wiley & Sons, Inc.

(~65% overall).[1285]

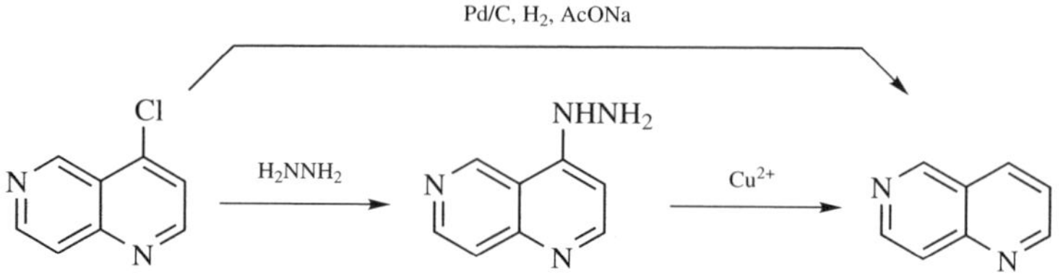

1,2,3,4-Tetrahydro-1,6-naphthyridine (**4**) underwent dehydrogenation to 1,6-naphthyridine (**5**) (neat substrate, Pt/C, 225°C, 2 h: ~10%).[134]

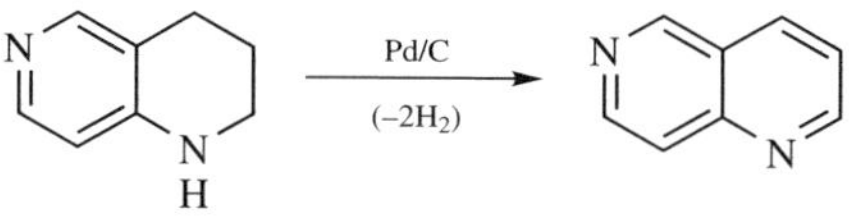

Also other routes.[215]

At least three hydro derivatives of the parent 1,6-naphthyridine have been prepared and characterized as illustrated here.

Submission of 1,6-naphthyridine to sodium/ethanol reduction followed by hydrogenation over platinum oxide afforded *trans*-decahydro-1,6-naphthyridine (**6**) in 65% yield (for details, see original).[47]

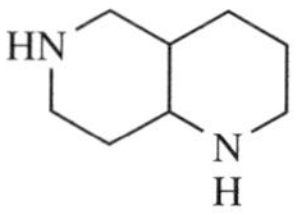

Less profound reduction of 1,6-naphthyridine afforded 1,2,3,4-tetrahydro-1,6-naphthyridine (**7**) (Pd/C, EtOH, H_2, 20°C, 1 atm, 6 h: 75%);[47] a somewhat similar procedure gave the same product (**7**) and a minor isomeric product, possibly 5,6,7,8-tetrahydro-1,6-naphthyridine.[1243] Ostensibly, the hydro derivative (**7**) has also been made by a primary synthesis involving reductive cyclization of 4-(2-carboxyethyl)-3-nitropyriidne 1-oxide,[134] but the melting points differ markedly.

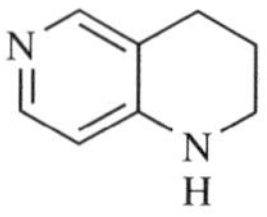

Gentle borohydride reduction of 1,6-naphthyridine afforded 1,2-dihydro-1,6-naphthyridine (**8**) (substrate, NaBH$_4$, MeOH, H$_2$O, HCl, 5→15°C, N$_2$, 1 h: 68%).[1275]

(**8**)

9.1.2. Properties of 1,6-Naphthyridine

The reported physical data for 1,6-naphthyridine may be gleaned from its entry in Appendix Table A2 toward the end of this book. Some wider studies of such phenomena are listed here.

Electron Density. Electron density calculations for 1,6-naphthyridine have been made for comparison with those of other nitrogenous heterocycles and for the rationalization of ionization phenomena and electronic spectra.[676,840,1126,1173]

Electron Spin Resonance. Radicals derived from 1,6-naphthyridine have been studied,[1329] especially with respect to their ESR spectra.[1083]

Infrared/Raman Spectra. Assignments for the major bands in IR/Raman spectra of 1,6-naphthyridine and related compounds have been reported.[44,1124,1251]

Ionization. Theoretical calculations for the protonation of 1,6-naphthyridine have been carried out for comparison with those of the other naphthyridines and related heterocycles.[806,813]

Mass Spectra. The main fragmentation patterns for 1,6-naphthyridine and other diazanaphthalenes have been reported.[975]

Nuclear Magnetic Resonance. The ^{1}HNMR spectra of 1,6-naphthyridine and related systems have been measured, especially in connection with the shielding effects of the ring nitrogen atoms,[42] long-range spin–spin coupling effects,[1140] the enhancing effects of lanthanide reagents on chemical shifts,[43] the site (s) of covalent hydration,[1170] and the shielding/deshielding effects of the substituents.[296] The ^{13}CNMR spectra of 1,6-naphthyridine have been compared with those of other naphthyridines[923] and used to measure the first and second pK_a values for protonation.[492] The ^{14}N and ^{15}NNMR spectra of 1,6-naphthyridine have been measured for comparison with those of related heterocycles and to reveal any mutual trans-ring shielding effects.[1189,1228]

Polarography. Some aspects of the polarographic reduction of 1,6-naphthyridine and related substrates have been examined.[829,1303]

Ultraviolet Spectra. The high-resolution photoelectronic spectra[1181] and mixed-crystal spectra[1188] for 1,6-naphthyridine and related systems have been reported; also several theoretical approaches to such spectra.[820,878,1312]

9.1.3. Reactions of 1,6-Naphthyridines

Reported reactions of 1,6-naphthyridine or its hydro derivatives are illustrated in the following classified examples.

C-Alkylation

Note: Although simple C-alkylation of 1,6-naphthyridine is rare, concomitant C-alkylation and N-acylations are well represented (see Reissert reactions below).

1,6-Naphthyridine (**9**, R = H) gave 4-methyl-1,6-nsphthyridine (**9**, R = Me) (substrate, NaH, Me$_2$SO, 70°C, 4 h: 4%).[880]

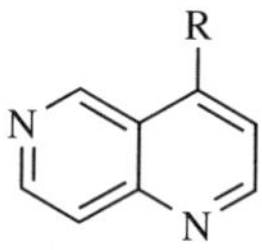

(**9**)

N-Alkylation and Quaternization

Decahydro-1,6-naphthyridine has been 1,6-dimethylated, as was its 1,5-naphthyridine isomer (see Section 2.1.3).[1203]

1,6-Naphthyridine gave only its 6-methiodide (**10**) (MeI, AcMe, reflux, 12 h: >85%;[1214] kinetics[961]) or its 1,6-bis(methofluorosulfonate) (**11**) (neat FSO$_3$Me, exothermic: 90%).[1061] Other such quaternary salts have been prepared[1002] and their physical chemistry examined.[993,1002]

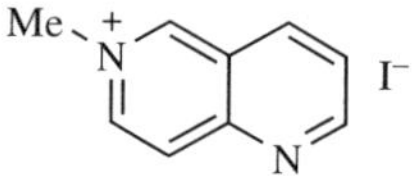

(**10**) 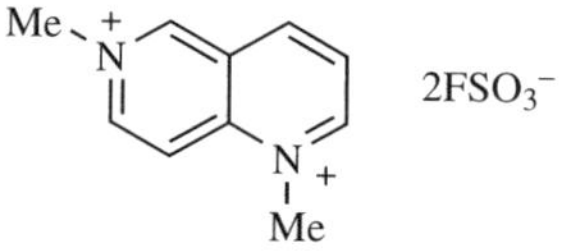 (**11**)

1,6-Naphthyridine with tetracyanoethylene oxide gave 1,6-naphthyridin-6-ium-6-dicyanomethanide (**12**) (substrate, THF, <5°C; synthon/THF↓ dropwise, <5°C, 24 h: 63%);[857] also analogous reactions.[101,857]

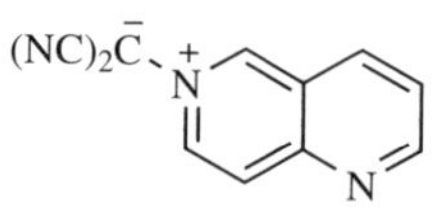

(**12**)

Amination

1,6-Naphthyridine (**13**) with potassium amide gave 1,6-naphthyridin-2-amine (**15**) [KNH$_2$/NH$_3$ (prepared *in situ*), substrate↓, KNO$_3$↓, 20°C, sealed, 8 days: 56%;[173] likewise, 50°C: high yield;[643] KNH$_2$/NH$_3$, substrate↓, 30 min; then KMnO$_4$↓ portionwise, 10 min: 40%].[425,cf.215] probably via the adduct (**14**).[353,425,998]

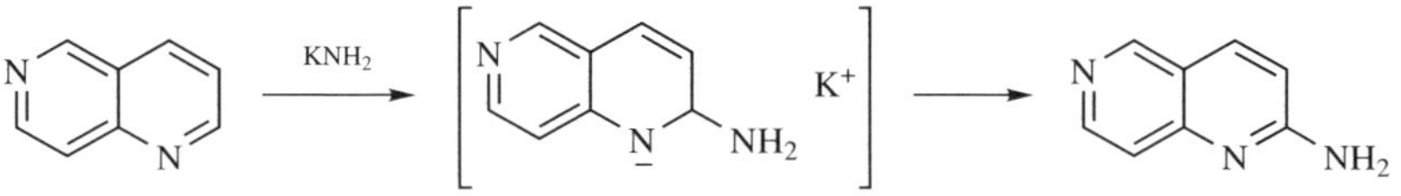

Cycloadditions

1,6-Naphthyridine with chloromethyl phenyl sulfone gave 1,7-bisphenylsulfonyl-6b,7,7a,7b-tetrahydro-1*H*-azirino [1,2-*a*]cyclopropa [*b*] [1,6]-naphthyridine (**16**) (KOH, Me$_2$SO, reactants↓ slowly, 20°C, tlc monitored: 59%).[393,803]

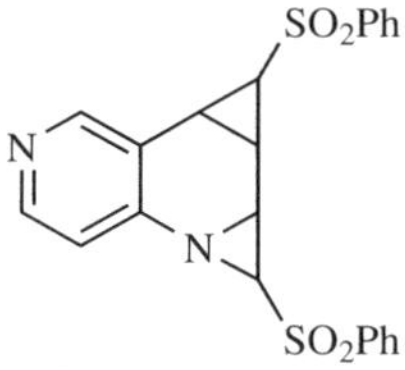

Also other additions.[322]

Deuteration

1,6-Naphthyridine gave mainly 5-*d*-1,6-naphthyridine (**17**) (substrate, D$_2$O, 170°C, sealed, 12 h: 95%)[302] or mainly perdeutero-1,6-naphthyridine (**18**) (Pt/asbestos, D$_2$O, 170°C, N$_2$, sealed, 24 h; then repeated: 92%).[629]

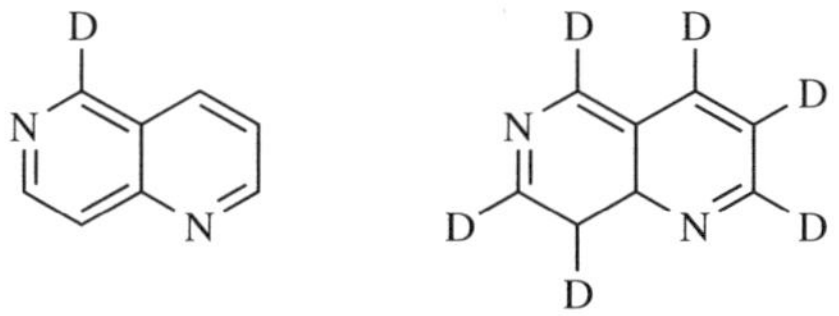

Halogenation

1,6-Naphthyridine gave a separable mixture of 3-bromo- (**19**, Q=Br, R=H), 8-bromo- (**19**, Q=H, R=Br), and 3,8-dibromo-1,6-naphthyridine (**19**, Q=R=Br) (substrate, Br_2, CCl_4, reflux, 1 h; then pyridine↓, reflux, 12 h: yields 68%, 22%, and 11%, respectively);[173] the same reactants in acetic acid appear to have given only 8-bromo-1,6-naphthyridine (**19**, Q=H, R=Br) (80°C, 12 h: 55%).[1042]

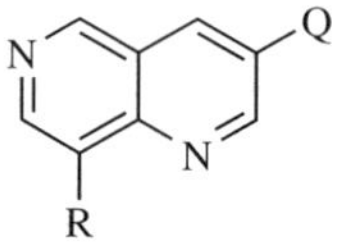

(**19**)

N-Oxidation

1,6-Naphthyridine (**20**) with limited *m*-chloroperoxybenzoic acid gave a separable mixture of its 1- (**21**) and 6-oxide (**22**) (reactants, $CHCl_3$, 20°C, 8 h: 12% and 38% after separation) but with an excess of oxidant gave the 1,6-dioxide (**23**) (likewise but reflux, 1 h: 53%);[1212] other oxidants proved less satisfactory[215,1011,1282] except for an excess of Na_2WO_4/H_2O_2, which gave the 1,6-dioxide (**23**) (reactants, 60°C, 7 h: 86%).[1282]

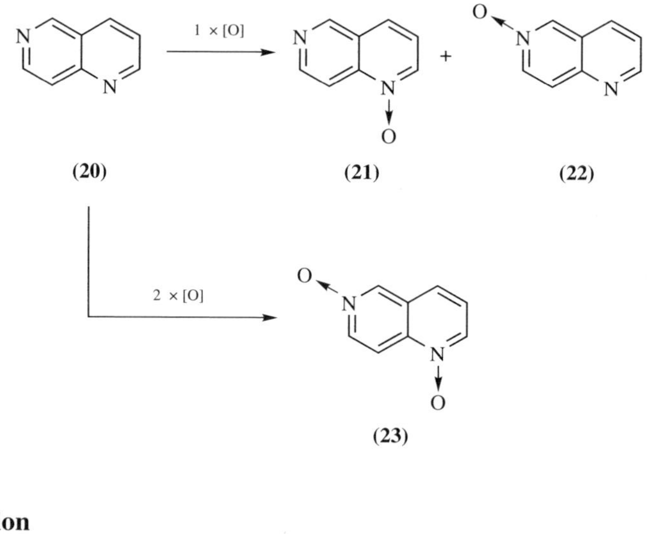

(**20**)   (**21**)   (**22**)

(**23**)

Reduction

Note: Examples have been given in Section 9.1.1.

Reissert Reactions

Note: 1,6-Naphthyridine appears to undergo Reissert-type additions only at its 5- and 6-positions; the products are useful intermediates for further elaboration.

1,6-Naphthyridine with diethyl malonate and acetic anhydride gave 6-acetyl-5-(diethoxycarbonylmethyl)-5,6-dihydro-1,6-naphthyridine (**24**) (neat reactants, 100°C, 50 h: 33%);[1207] with benzyltrimethylstannane and ethyl chloroformate, gave ethyl 5-benzyl-5,6-dihydro-1,6-naphthyridine-6-carboxylate (**25**) (reactants, CH_2Cl_2, ?°C, ? h: 43%);[197] with silver cyanide and benzoyl chloride, gave 6-benzoyl-5,6-dihydro-1,6-naphthyridine-5-carbonitrile (**26**, R=Ph) (reactants, $CHCl_3$, 15°C, 4 h: 40%);[1318] or with potassium cyanide and acetyl chloride, gave 6-acetyl-5,6-dihydro-1,6-naphthyridine-5-carbonitrile (**26**, R=Me) (substrate, KCN, CH_2Cl_2, 20°C; AcCl↓ during 20 min; 20°C, 5 h: 10%).[1375]

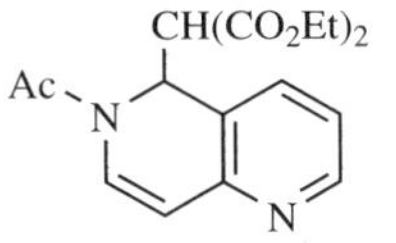

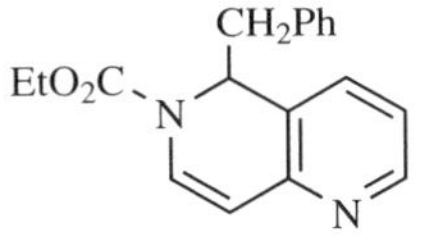

The same substrate with diketene and formic acid gave 5-acetonyl-5,6-dihydro-1,6-naphthyridine-6-carbaldehyde (**27**) (neat reactants, 25°C, 20 h: 60%);[1308,cf. 917] or with acetic anhydride alone gave 6-acetyl–5-carboxymethyl-5,6-dihydro-1,6-naphthyridine (**28**) (neat reactants, 45°C, 40 h: 20%).[962]

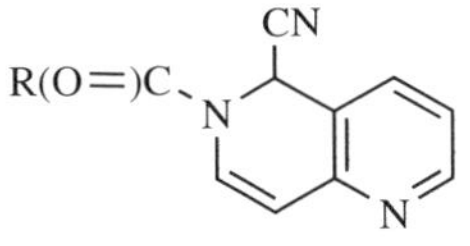

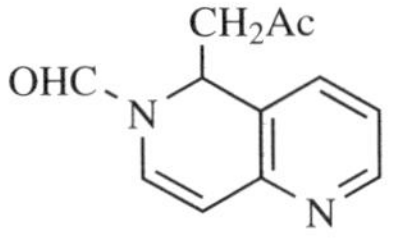

Also other examples.[336,392,791]

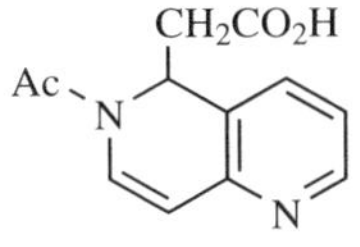

9.2. ALKYL- AND ARYL-1,6-NAPHTHYRIDINES

This section covers the insertion of alkyl or aryl groups into existing 1,6-naphthyridines and the reactions specific to the alkyl or aryl groups therein

The ^{1}HNMR spectra of *C*-methyl groups and the ^{13}CNMR spectra of *N*-methyl groups attached to 1,6-naphthyridine have been examined in some detail.[964,1319]

9.2.1. Preparation of Alkyl- and Aryl-1,6-naphthyridines

In almost all cases, the alkyl/aryl substituents on such naphthyridine derivatives have been put in place during *primary syntheses* (see Chapter 8). However, there are a few examples in which these substituents have been inserted subsequently, as illustrated here.

By C-Alkylation

Note: For additional C-alkylations, see Sections 9.1.3 and 11.1.2

1,6-Naphthyridine 6-oxide (**29**, R = H) gave 5-methyl-1,6-naphthyridine 6-oxide (**29**, R = Me) (NaH, Me$_2$SO, THF, −5°C; substrate/Me$_2$SO↓ dropwise, 5 min: 16%).[1202]

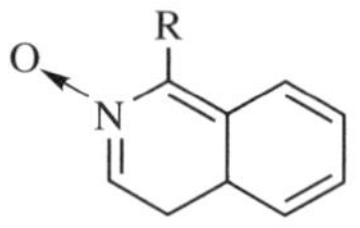

(**29**)

5-Benzyl-5,6,7,8-tetrahydro-1,6-naphthyridine (**30**, R = H) gave 5-benzyl-8-methyl-5,6,7,8-tetrahydro-1-6-naphthyridine (**30**, R = Me) [LiNPr$_2^i$/THF (made *in situ*), −70°C; substrate/THF↓ dropwise; MeI↓ dropwise; −55°C, 15 min: 87%].[1058]

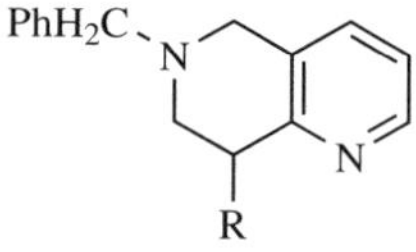

(**30**)

The Reissert product, 6-benzoyl-5,6-dihydro-1,6-naphthyridine-5-carbonitrile (**31**, R = H) gave gave 6-benzoyl-5-methyl-5,6-dihydro-1,6-naphthyridine-5-carbonitrile (**31**, R = Me) (MeI, 50% NaOH, phase transfer catalyst, 20°C, 2 h: 87%).[340]

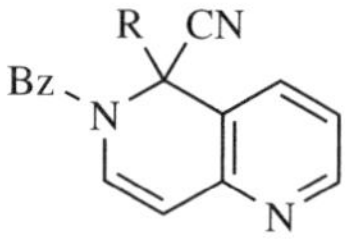

(**31**)

By N-Alkylation

5-Methyl- (**31a**, R = H) gave 1,5-dimethyl- (**31a**, R=Me) [sodio substrate (made *in situ*), NH$_3$/THF: MeI↓, 2 h: 50%] or 1-allyl-5-methyl-1,2,3,4-tetrahydro-1,6-naphthyridine (**31a**, R = CH$_2$CH : CH$_2$) likewise using ICH$_2$CH = CH$_2$: 50%].[1063]

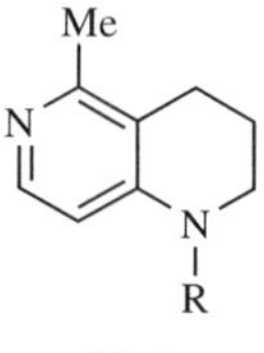

(**31a**)

From Halogeno-1,6-naphthyridines

8-Bromo-1,6-naphthyridine (**32**, R = Br) gave 8-(2-ethoxycarbonylvinyl)-1,6-naphthyridine (**32**, R = CH : CHCO$_2$Et) [substrate, H$_2$C=CHCO$_2$Et, Pd(OAc)$_2$, PPh$_3$, MeCN, 120°C, sealed, 24 h: 69%]; also analogous reactions from a chloro or iodo substrate.[1042]

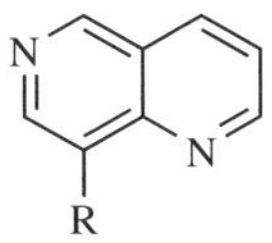

(**32**)

Also other examples.[244] See also Section 10.2.3.

By Grignard Addition

6-Benzyl-1,6-naphthyridin-6-ium bromide (**33**) with methylmagnesium bromide gave crude 6-benzyl-5-methyl-5,6-dihydro-1,6-naphthyridine (**34**) (MeMgBr, THF, 5°C, substrate↓ slowly, 5°C, 10 min:), characterized by borohydride reduction to 6-benzyl-5-methyl-5,6,7,8-tetrahydro-1,6-naphthyridine (substrate, MeOH, pH7 buffer, 5°C; NaBH$_4$↓ slowly, 5°C, 10 min: 92% overall).[1058]

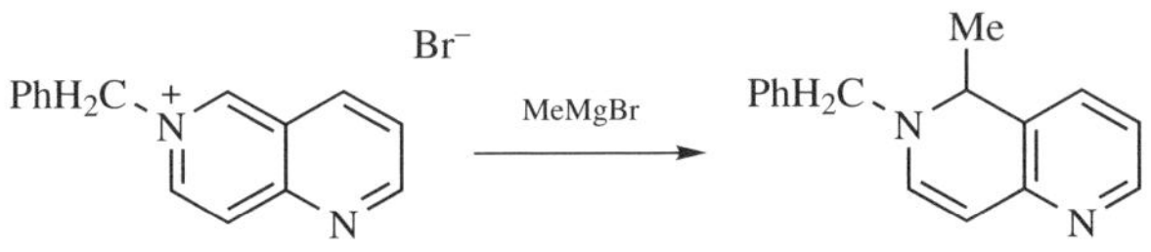

(**33**) → (**34**)

Also other examples.[340]

9.2.2. Reactions of Alkyl- and Aryl-1,6-naphthyridines

The reported reactions of the alkyl- and aryl-1,6-naphthyridines are illustrated by the following examples.

Alkylidenation

8-Acetyl-7-methyl- (**35**) with *tert*-butoxybis(dimethylamino)methane (Bredereck's reagent) gave 8-acetyl-7-(2-dimethylaminovinyl)-6-propoxy-1,6-

naphthyridin-5 (6*H*)-one (**36**) (reactants, dioxane, reflux, 5 h: 54%); likewise for the 8-benzoyl analog (46%).[183]

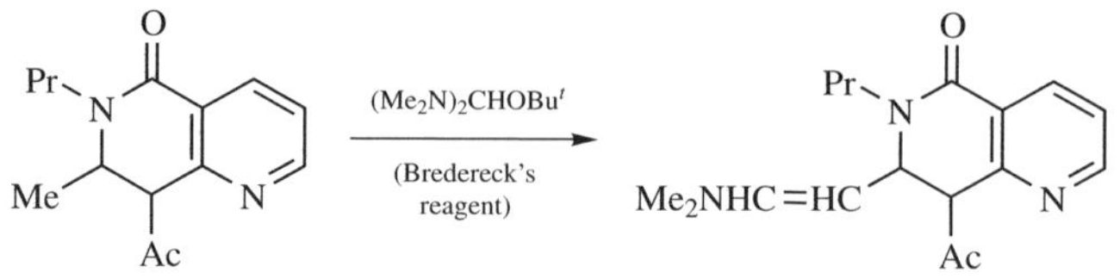

Cyclocondensation

The ylide, 1,6-naphthyridin-6-ium-6-dicyanomethanide (**37**) with dimethyl acetylenedicarboxylate, gave a separable mixture of dimethyl-8-cyanopyrrolo [2,1-f] [1,6]-naphthyridine-9,10-dicarboxylate (**39**) and its 8,8-dicyano-8,9-dihydro precursor (**38**) (44% and 23%, respectively; no other details).[901]

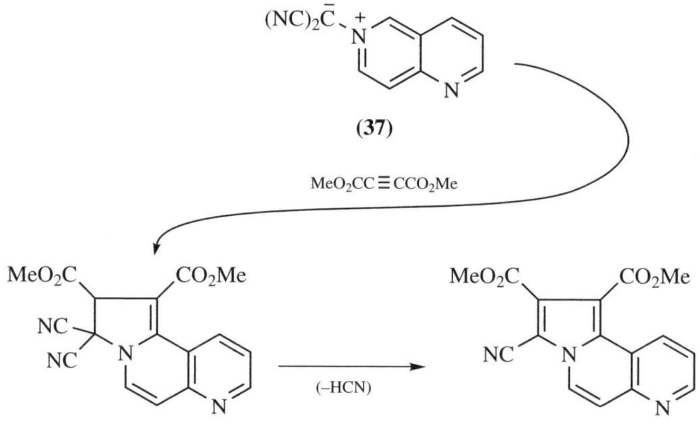

Dealkylation

Note: Both *N*- and *C*-demethylation have been reported, but the latter is rare.

6-Benzyl-3-methyl-4 (**40**, R = CH$_2$Ph) gave 3-methyl-5,6,7,8-tetrahydro-1,6-naphthyridine (**40**, R = H) (substrate·2HCl, Pd/C, H$_2$O, MeOH, H$_2$, 20°C: 61%);[277] 5-, 7- and 8-methyl-1,6-naphthyridine were made somewhat similarly from their 8-benzyl derivatives (Pd/C, AcOH, H$_2$: 95%, 92%, and 92% respectively; no further details).[1058]

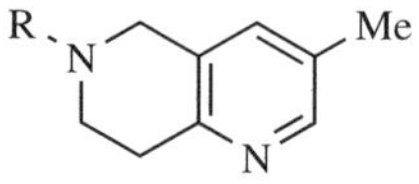

Ethyl 5-amino-7-benzylseleno-8-cyano-4-(fur-2-yl)-2-methyl-1,4-dihydro-1,6-naphthyridine-3-carboxylate (**41**) underwent deheterylation to give ethyl 5-amino-7-benzylseleno-8-cyano-2-methyl-1,6-naphthyridine-3-carboxylate (**42**) (AcOH, reflux, 1 h: then 20°C, 24 h: 72%).[959]

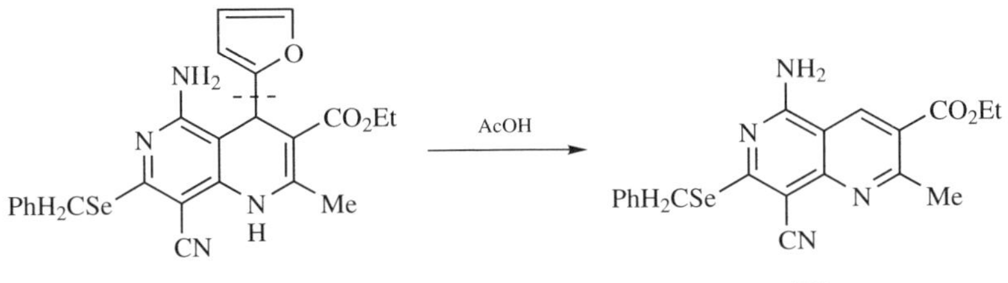

Note: An indirect dealkylation is illustrated in the next example.

Oxidative Reactions

5-Methyl-2-oxo-1,2-dihydro-1,6-naphthyridine-3-carbonitrile (**43**, R = Me) with selenium dioxide gave 2-oxo-1,2-dihydro-1,6-naphthyridine-3-carbonitrile (**43**, R = H), presumably by decarboxylation of an intermediate 5-carboxylic acid (**43**, R = CO_2H) (reactants, AcOH, reflux, 20 h: 55%).[1166]

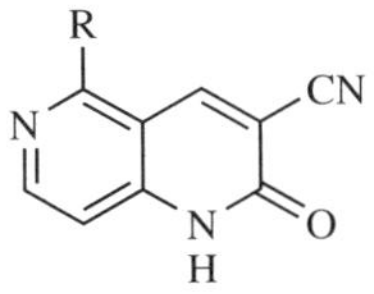

Methyl 1-acetyl-4-methoxycarbonylmethylene-1,2,3,4-tetrahydro-1,6-naphthyridine-5-carboxylate (**44**, X = : $CHCO_2Me$) underwent ozonolysis to give methyl 1-acetyl-4-oxo-1,2,3,4-tetrahydro-1,6-naphthyridine-5-carboxylate (**44**, X=O) (substrate, MeOH, −78°C, $O_3\downarrow$, 4 h: 74%).[1143]

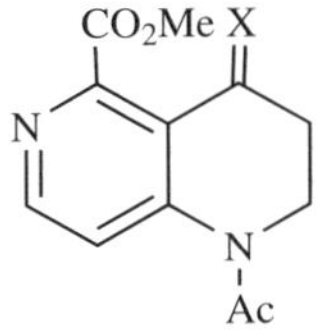

Reductive Reactions

Note: Only nuclear reduction of alkyl-1,6-naphthyridines has been reported.

5-Methyl-1,6-naphthyridine gave 5-methyl-1,2,3,4-tetrahydro-1,6-naphthyridine (**45**) (Pd/CaCO$_3$, MeOH, H$_2$, 20°C: 80%).[1063]

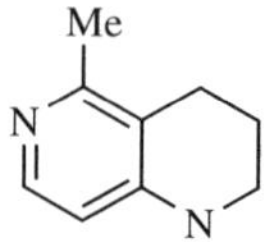

(**45**)

CHAPTER 10

Halogeno-1,6-naphthyridines

In contrast to the situation in halogeno-1,5-naphthyridines (see Chapter 3), a halogeno substituent on 1,6-naphthyridine will be activated at the 2-, 4-, 5-, and 7-positions and much less so at the 3- and 8-positions.

The structure of 7-amino-5-bromo-4-methyl-2-oxo–1,2,3,4-tetrahydro-1, 6-naphthyridine-3-carbonitrile (**1**), prepared by an ambiguous synthesis, has been confirmed by x-ray analysis.[779] Mass spectral fragmentation patterns for bromo-1, 6-naphthyridines to have been investigated.[546]

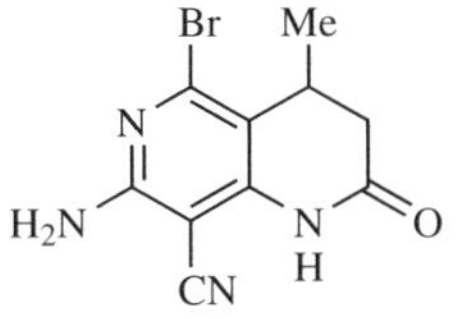

(**1**)

10.1. PREPARATION OF HALOGENO-1,6-NAPHTHYRIDINES

Some halogeno-1,6-naphthyridines have been made by *primary synthesis* (see Chapter 8), but several other procedures have been used, as explained in the following subsections.

10.1.1. By Direct Halogenation

This route may be used to prepare nuclear or extranuclear halogeno derivatives as indicated in Section 9.1.3 and in the following examples.

1,6-Naphthyridin-5(6*H*)-one (**2**, R = H) gave 8-iodo-1,6-naphthyridin-5(6*H*)-one (**2**, R = I) (I_2, 0.4M NaOH, 80°C, 3 h: 70%).[1042]

The Naphthyridines: The Chemistry of Heterocyclic Compounds, Volume 63, by D.J. Brown
Copyright © 2008 John Wiley & Sons, Inc.

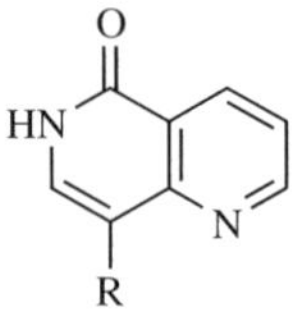

104 Halogeno-1,6-Naphthyridines

(2)

1,6-Naphthyridine-2-carboxylic acid (**3**, R = H) gave its 8-bromo derivative (**3**, R = Br) (Br$_2$, AcOH: 80%; no further details)[244,1411] or its 8-chloro derivative (**3**, R = Cl) (Cl$_2$, no details).[244]

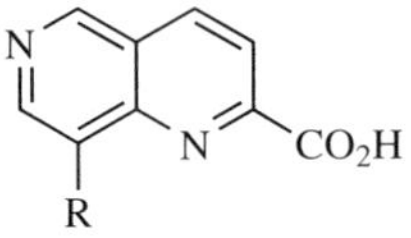

(3)

Ethyl 7-bromo-1-ethyl-1,2,3,4-tetrahydro-1,6-naphthyridine-3-carboxylate (**4**, R = H) underwent brominaion to its 3,7-dibromo analog (**4**, R = Br) (Br$_2$, CH$_2$Cl$_2$, crude), identified by dehydrobromination to ethyl 7-bromo-1-ethyl-1,4-dihydro-1,6-naphthyridine-3-carboxylate (**5**) (Et$_3$N, EtOH, reflux, 10 min: 97% Overall); analogs likewise.[1408]

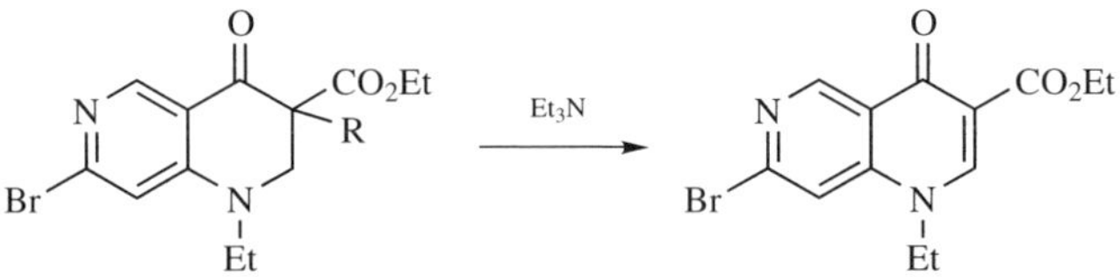

(4) → **(5)**

1,6-Naphthyridin-4(1H)-one gave its 3-chloro derivative (substrate, Ac$_2$O, AcOH, SO$_2$Cl$_2$, 50°C, 1 h: then 95°C, 10 h: 78%).[844]

Also other examples.[336,391,844]

10.1.2. By Halogenolysis of 1,6-Naphthyridinones or the Like

The conversion of a 1,6-naphthyridinone into the corresponding halogeno-1,6-naphthyridine is usually done by refluxing with phosphoryl chloride or bromide, with or without the addition of a tertiary base; alternatively, a phosphorus pentahalide can be used but additional *C*-halogenation may result. Nontautomeric 1,6-naphthyridinones also undergo (less satisfactory) halogenolysis. These possibilities are illustrated in the following typical examples.

1,6-Naphthyridin-5(6H)-one (**6**, R = H) gave 5-bromo-(**7**, R = H, X = Br) (POBr$_3$, 95°C, 1 h: 57%)[1067] or 5-chloro-1,6-naphthyridine (**7**, R = H, X = Cl) (POCl$_3$, 130°C, sealed, 20 h: ~90%);[1285] 7-methyl-1,6-naphthyridin-6(5H)-one (**6**, R = Me) gave 5-chloro-7-methyl-1,6-naphthyridine (**7**, R = Me, X = Cl) (likewise: 82%).[1072,1285]

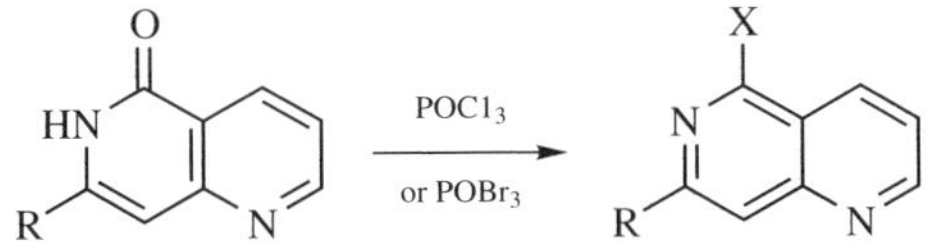

6-Benzyl-4,7-dimethyl-1,6-naphthyridine-2,5(1H, 6H)-dione (**8**) gave 6-benzyl-2-chloro-4,7-dimethyl-1,6-naphthyridin-5(6H)-one (**9**) (POCl$_3$, 95°C, 3 h: 84%; note the unreactivity of the fixed oxo substituent).[1074]

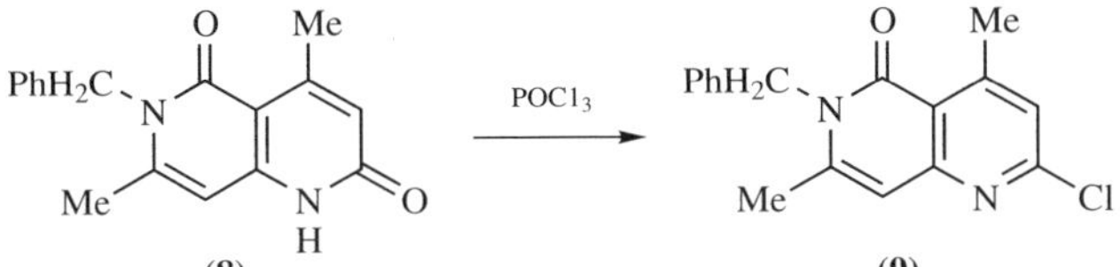

3-Nitro-1,6-naphthyridin-2(1H)-one gave 2-chloro-3-nitro-1,6-naphthyridine (**10**) (POCl$_3$, reflux, 3 h: 44%);[819] 4-chloro-8-nitro-1,6-naphthyridine (75 min: 68%)[48] and 5-chloro-8-iodo-1,6-naphthyridine (likewise: 64%)[1042] were made similarly from the corresponding 1,6-naphthyridinones.

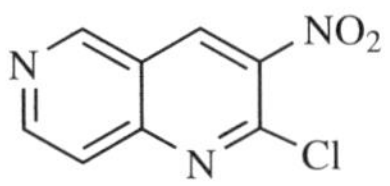

6-Methyl-5,6,7,8-tetrahydro-1,6-naphthyridin-2(1H)-one gave 2-chloro-6-methyl-5,6,7,8-tetrahydro-1,6-naphthyridine (**11**) (substrate, POCl$_3$, reflux; PCl$_5$↓ during 1 h; reflux, 2 h: ~90%).[180]

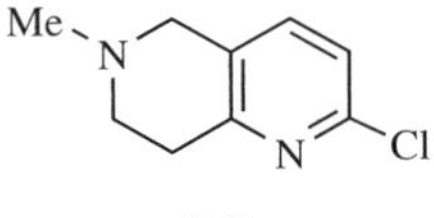

5-Methyl-1,6-naphthyridin-2(1H)-one gave 2-chloro-5-methyl-1,6-naphthyridine (12) POCl$_3$, PCl$_5$, reflux, 7 h: 77%;[490] 2-chloro-1,6-naphthyridine-3-carbonitrile was made by a similar procedure (1 h: 89%).[247]

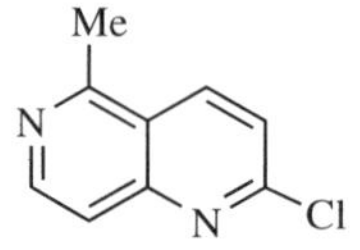

(12)

5,7-Dimethyl-1,6-naphthyridin-2(1H)-one gave 2-chloro-5,7-dimethyl-1,6-naphthyridine (13) (POCl$_3$, PhNMe$_2$, CHCl$_3$, reflux, 20 min: 50%);[1255] tripropylamine has been used as a tertiary base in a somewhat similar halogenolysis to afford 5-chloro-1,6-naphthyridine (40%).[1282]

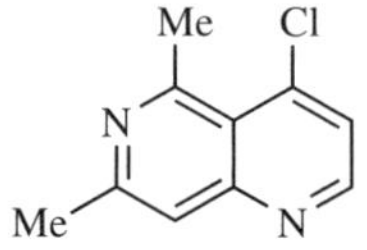

(13)

1,6-Naphthyridin-4(1H)-one (14) underwent halogenolysis, and C-halogenation with phosphorus pentabromide to give 3,4,8-tribromo-1,6-naphthyridine (15) (reactants, 95°C, 1 h: 21%);[1067] somewhat similar halogenations have been reported using phosphorus pentachloride.[950]

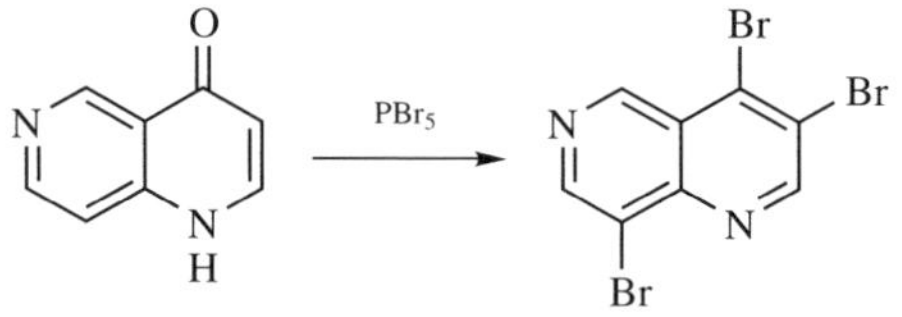

(14) → PBr$_5$ → (15)

The nontautomeric oxo substrate, 6-methyl-1,6-naphthyridin-5(6H)-one (16) with phosphoryl chloride gave only 5-chloro-1,6-naphthyridine (17) (reactants, 170°C, sealed, 20 h: 63%; reflux, 12 h: 0%) but with phosphorus pentachloride and phosphoryl chloride gave a separable mixture of the same product (17), 5,8-dichloro-1,6-naphthyridine (18), and 8-chloro-6-methyl-1,6-naphthyridin-5(6H-one (19) (reactants, reflux, 24 h: 2%, 24%, and 14%, respectively).[299]

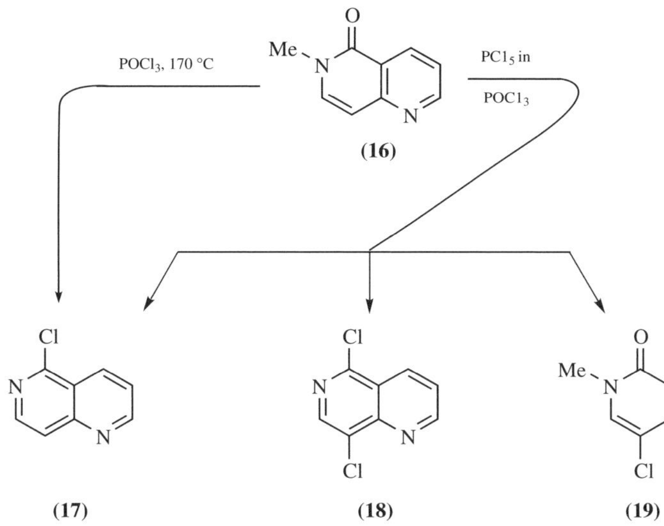

The extranuclear hydroxy substrate, ethyl 1-(2-hydroxyethyl)-7-(4-methylpiperazin-1-yl)-4-oxo-1,4-dihydro-1,6-naphthyridine-3-carboxylate, gave its ethyl 1-(2-chloroethyl) analog (SOCl$_2$, CHCl$_3$, reflux, ~2 h: 56%; analogs likewise).[1049]

Also other examples.[6,85,108,289,390,463,490,575,928]

10.1.3. By Other Methods

Halogeno-1,6-naphthyridines have also been made by the Meissenheimer reaction from 1,6-naphthyridine *N*-oxides and by a (copperless) Sandmeyer-type reaction from 1,6-naphthyridinamines. However, neither route has been developed into a practical preparative procedure, as indicated in the following examples.

1,6-Naphthyridine 1-oxide (**20**) and and phosphoryl chloride gave a separable mixture of 2-, 3-, and 4-chloro-1,6-naphthyridine (**21**) and unsubstituted 1,6-naphthyridine (reactants, 0°C→reflux, 10 min: 50%, 9%, 15%, and 2%, respectively).[232]

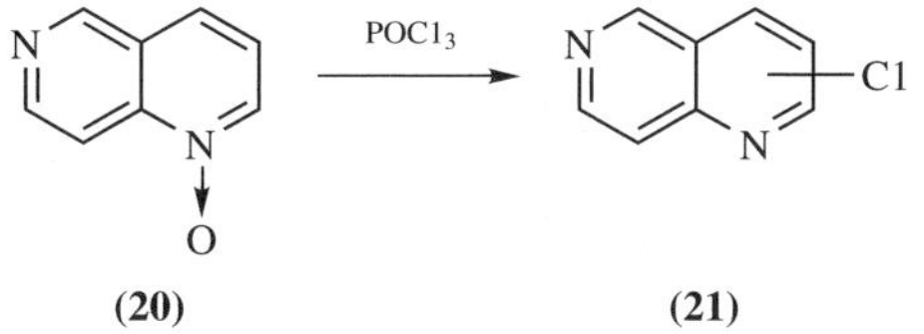

1,6-Naphthyridine 1,6-dioxide (**22**) likewise gave a mixture of chlorinated products from which the main constituents, 2,5- and 3,5-dichloro-1, 6-naphthyridine (**23**), appear to have been isolated.[1069]

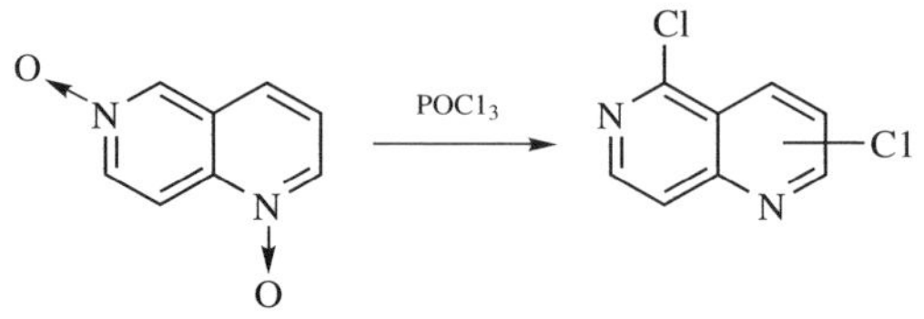

(**22**) → POCl₃ → (**23**)

Diazotization of 3-phenyl-1,6-naphthyridine-2,7-diamine (**24**) in an excess of hydrochloric acid gave mainly 7-chloro-3-phenyl-1,6-naphthyridin-2-amine (**25**) along with three separable byproducts (10M HCl, −10°C, HCl gas↓ substrate↓, −10°C, 5 h; 20°C, 8 h; 4°C, 2 days: 61%); analog likewise.[564]

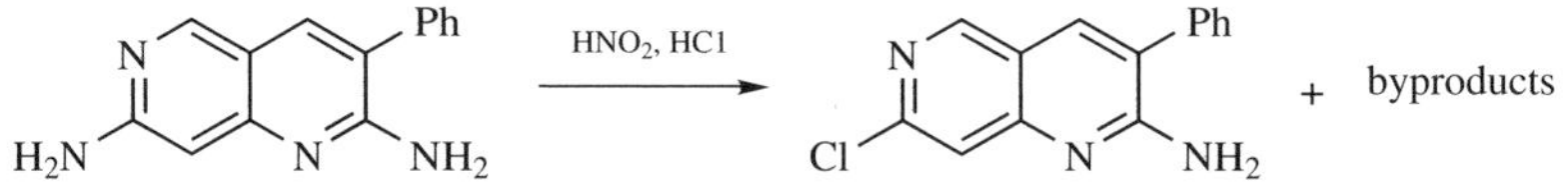

(**24**) → HNO₂, HCl → (**25**) + byproducts

10.2. REACTIONS OF HALOGENO-1,6-NAPHTHYRIDINES

Halogeno-1,6-naphthyridines undergo a variety of useful reactions, as illustrated in the following subsections.

10.2.1. Alcoholysis or Phenolysis of Halogeno-1,6-naphthyridines

Halogeno substituents undergo ready alcoholysis/phenolysis by alkoxide or phenoxide ion, especially when they occupy the more activated positions or are further activated by other electron-withdrawing substituents. Examples follow.

2-Chloro- (**26**, R = Cl) gave 2-methoxy-5-methyl-1,6-naphthyridine (**26**, R = OMe) (MeONa, MeOH, 20°C, 12 h: 79%),[490]

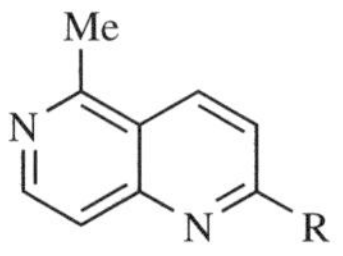

(**26**)

2-Chloro- (**27**, R = Cl) gave 2-ethoxy-3-nitro-1,6-naphthyridine (**27**, R = OEt) (EtONa, EtOH, 20°C, 3 h: 60%);[819] 2-chloro- (**28**, R = Cl) gave 2-methoxy-1,6-naphthyridine-3-carbonitrile (**28**, R = OMe) (MeONa MeOH, reflux, 1 h: 87%).[247] Note the activation by an electron-withdrawing group in these examples.

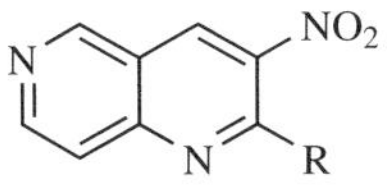

(**27**) 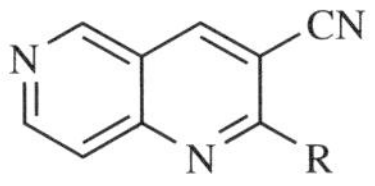   (**28**)

4-Chloro- (**29**, R = Cl) gave 4-phenoxy-1,6-naphthyridine (**29**, R = OPh) (PhOH, KOH, 150°C, 3 h: 55%).[1067]

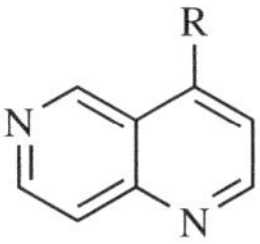

(**29**)

5-Chloro- (**30**, R = Cl) gave 5-methoxy-7-methyl-1,6-naphthyridine (**30**, R = OMe) (MeONa, MeOH, reflux, 15 min: 80%);[1072] 5-chloro- (**31**, R = Cl) underwent only monoalcoholysis to give 8-iodo-5-methoxy-1,6-naphthyridine (**31**, R = OMe) (MeONa, MeOH, reflux, 1 h: >95%).[1042]

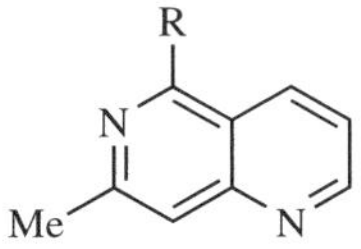

(**30**) 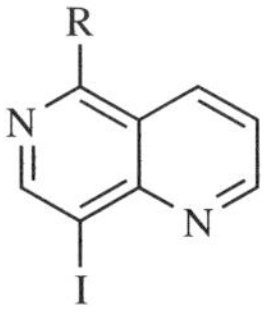   (**31**)

5-Amino-7-bromo- (**32**, R = Br) gave 5-amino-7-methoxy-2,4-dimethyl-1,6-naphthyridine-8-carbonitrile (**32**, R = OMe) (MeONa, MeOH, reflux, 20 h: 70%).[395,cf. 994]

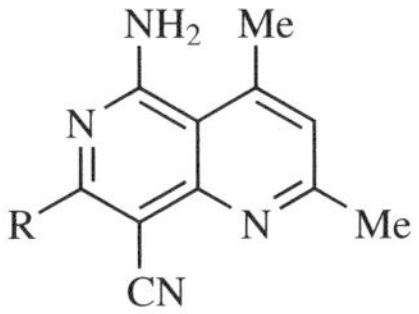

(**32**)

Also other examples.[299,367]

10.2.2. Aminolysis of Halogeno-1,6-naphthyridines

As with alcoholysis, halogeno substrates undergo aminolysis with vigor that depends on the positions they occupy in the naphtyridine system and on the type of amine used. The following examples illustrate the general picture.

2-Chloro- (**33**, R = Cl) gave 2-methylamino-1,6-naphthyridine-3-carbonitrile (**33**, R = NHMe) (substrate, PrOH, 20°C, MeNH$_2$ gas↓, 1 h: 92%);[247] 2-chloro- (**34**, R = Cl) gave 2-hydrazino-5-methyl-1,6-naphthyridine (**34**, R = NHNH$_2$) (H$_2$NNH$_2$·H$_2$O, MeOH, 25°C, 36 h: 65% as dihydrochloride).[490]

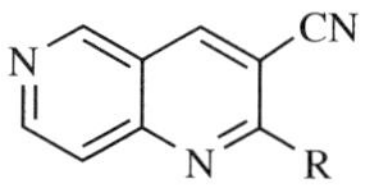

(33) 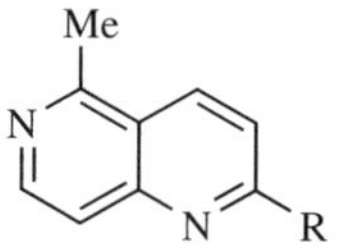   (34)

4-Bromo- (**35**, R = Br) with dimethylamine gave 4-dimethylamino-1,6-naphthyridine (**35**, R = NMe$_2$) (neat reactants, 135°C, sealed, 16 h: ?%)[1057] but 4-chloro- (**35**, R = Cl) gave 4-piperidino-1,6-naphthyridine [**35**, R = N(CH$_2$)$_5$] apparently under milder conditions (neat reactants, 95°C, 5 min: ?%)[1067] 4-chloro-3-nitro-1,6-naphthyridine (**36**, R = Cl) gave 3-nitro-1,6-naphthyridinamine (**36**, R = NH$_2$) (NH$_3$/MeOH, 110°C, sealed, 4 h: 75%).[819]

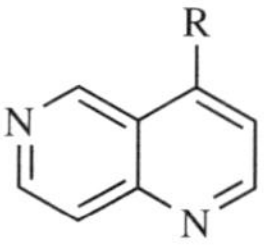

(35) 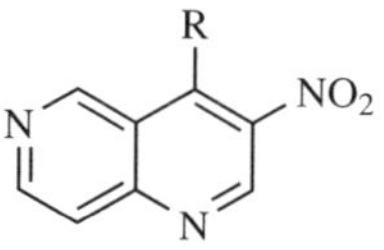   (36)

5-Chloro- (**37**, R = Cl) gave 5-benzylamino-7-methyl-1,6-naphthyridine (**37**, R = NHCH$_2$Ph) (neat PhCH$_2$NH$_2$, 145°C, N$_2$, 4 h: ∼40%).[1072]

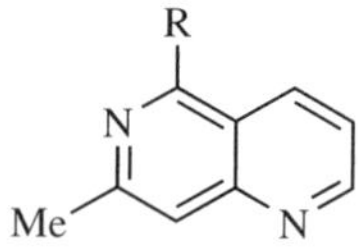

(37)

3-(2,6-Dichlorophenyl)-7-fluoro- (**38**, R = F) gave 7-amino-3-(2,6-dichlorophenyl)-1-methyl-1,6-naphthyridin-2(1*H*)-one (**38**, R = NH$_2$) (NH$_3$, Pr^{i}OH, NH$_3$↓, 110°C, sealed, 3 days: 88%); substituted-amino analogs similarly but in much shorter timeframes.[239]

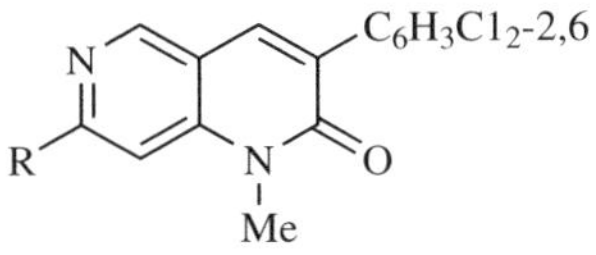

(38)

5-Amino-7-chloro- (**39**, R = Cl) gave 5-amino-7-hydrazino-2,4-diphenyl-1,6-naphthyridine-8-carbonitrile (**39**, R = NHNH$_2$) (H$_2$NNH$_2$.H$_2$O, dioxane, reflux, 3 h: 81%); analogs similarly.[895]

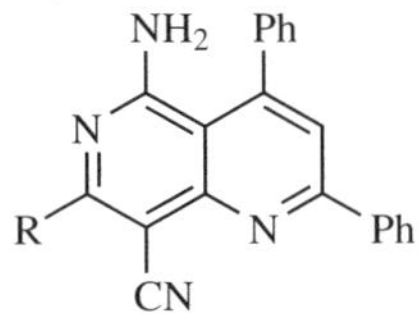

(39)

3-Bromo-, 3-chloro-, 4-bromo-, and 4-chloro-1,6-naphthyridines (**40**) all reacted separately with potassium amide in liquid ammonia to give a separable mixture of 1,6-naphthyridin-3 (and 4)-amines (**41**), albeit in different proportions [KNH$_2$/NH$_3$ (prepared *in situ*), substrate/Et$_2$O ↓ during 5 min, then 5 h: 50–80% before separation; the mechanism probably involved the intermediate shown][849,cf. 289]

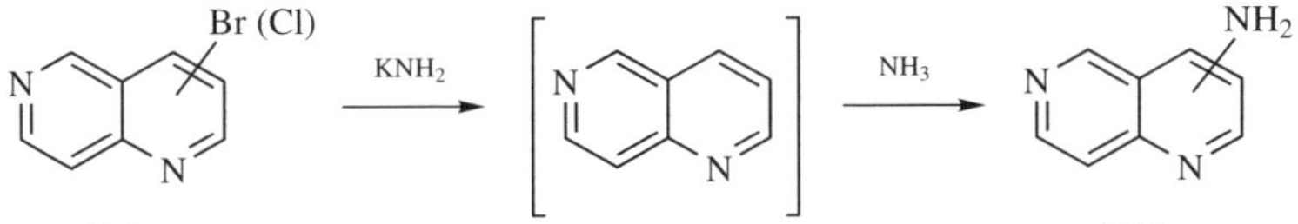

(40) → [intermediate] → (41)

Also other examples.[331,374,564,648,928,1049,1175]

10.2.3. Dehalogenation of Halogeno-1,6-naphthyridines

Dehalogenation of halogeno-1,6-naphthyridines has been done directly by hydrogenolysis, indirectly via a hydrazino analog, or by dehydrohalogenation of extranuclear halogeno substrates. Direct hydrogenolysis may be accompanied by nuclear reduction, especially when the reaction mixture is allowed to become acidic, so the indirect procedure is often used. Some simple examples have been given in Section 9.1.1; other examples follow here.

6-Benzyl-2-chloro- (**42**, R = Cl) gave 6-benzyl-4,7-dimethyl-1,6-naphthyridin-5 (6H)-one (**42**, R = H) (Pd/C, MeOH, H_2, 20°C, 75%).[1014]

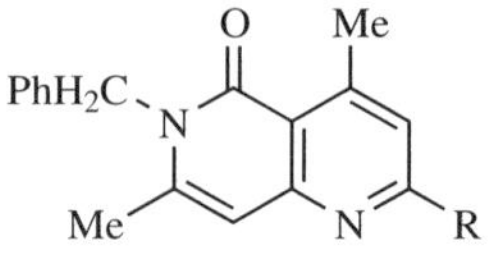

(**42**)

5-Chloro-7-methyl-1,6-naphthyridine (**43**) underwent direct hydrogenolysis to give 7-methyl-1,2,3,4-tetrahydro-1,6-naphthyridine (**44**) (Pd/CaCO$_3$, MeOH, H_2, 20°C ?%), but the same substrate (**43**) gave 4-hydrazino-7-methyl-1, 6-naphthyridine (**45**) ($H_2NNH_2·H_2O$, MeOH, reflux, 10 min: >95%) and thence 7-methyl-1,6-naphthyridine (**46**) (AcOH, H_2O, CuSO$_4$, reflux, 10 min: 65%).[1285]

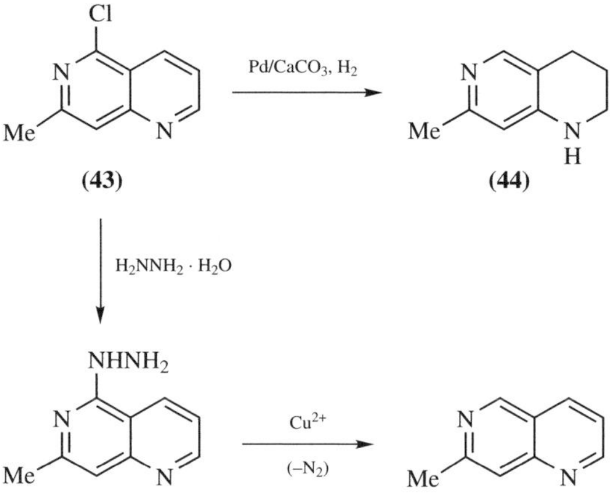

2-Chloro- (**47**, R = Cl) with tosylhydrazide gave 2-nitro-2-tosylhydrazinol-6-naphthyridine (**47**, R = NHNHTs) (TsNHNH$_2$, CHCl$_3$, 20°C, 3 days: 92% as crude hydrochloride) and thence 3-nitro-1,6-naphthyridine (**47**, R = H) (Na$_2$CO$_3$, H$_2$O, glycol, 100°C, 2 h: 11%);[48,cf. 819] 8-nitro-1,6-naphthyridine was made somewhat similarly.[48]

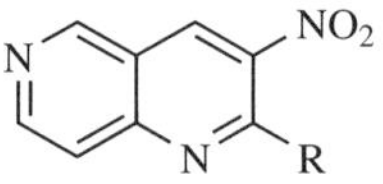

(**47**)

Ethyl 1-(2-chloroethyl)-7-(4-methylpiperazin-1-yl)-4-oxo-1,4-dihydro-1,6-naphthyridine-3-carboxylate (**48**, Q = CH$_2$CH$_2$Cl, R = Et) gave 7-(4-methylpiperazin-1-yl)-4-oxo-1-vinyl-1,6-naphthyridine-3-carboxylic acid (**48**, Q = CH:CH$_2$, R = H) (MKOH, reflux, 30 min: 69%); analogs likewise.[1049]

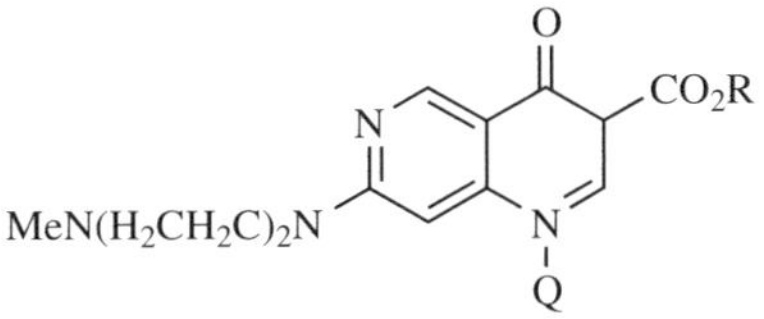

(**48**)

Also other examples.[85,180,575,844,1255]

10.2.4. Other Reactions of Halogeno-1,6-naphthyridines

A few rarely used reactions of halogeno-1,6-naphthyridines are exemplified here.

Hydrolysis

Note: It is quite remarkable that very few deliberate hydrolyses of these halogenonaphthyridines have been reported.

3-(3,5-Dimethyoxyphenyl)-2,7-difluoro-1,6-naphthyridine gave 3-(3,5-dimethoxyphenyl)-7-fluoro-1,6-naphthyridin-2(1*H*)-one (**49**) (NaOH, H$_2$O, THF, 53°C, 3 days: >95%).[564]

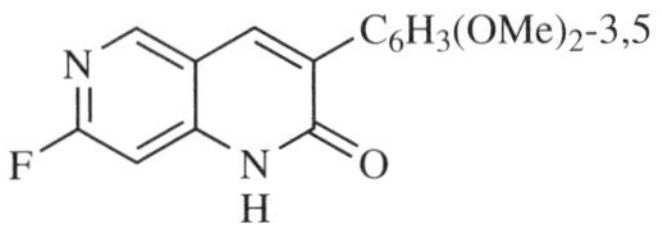

(**49**)

2-Chloro-5-dichloromethyl-1,6-naphthyridine gave 5-dichloromethyl-1,6-naphthyridin-2(1*H*)-one (**50**) (6M HCl, 95°C, 90 min: 67%).[490]

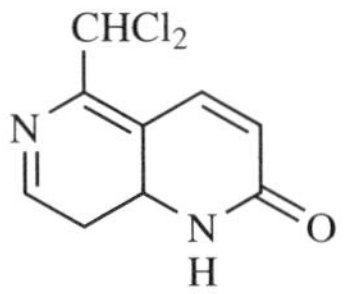

(**50**)

Thiolysis

2-Chloro-5-methyl-1,6-naphthyridine (**51**) with thioacetic acid gave 5-methyl-1,6-naphthyridine-2(1*H*)-thione (**52**), presumably via the 2-acetylthio intermediate (reactants, K_2CO_3, 20°C, 12 h: 74%).[490]

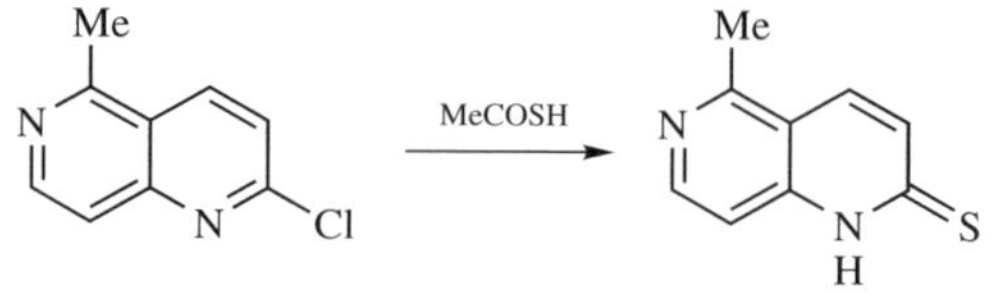

(51) → MeCOSH → (52)

Cyclocondensation

Note: A typical cyclocondensation of a halogeno-1,6-naphthyridine is illustrated in this example.

3-Bromo-5-methyl-1,6-naphthyridin-2-amine (**53**) with potassium *O*-ethylxanthate gave 8-methylthiazolo [4,5-b] [1,6]naphthyridin-2(3*H*)-thione (**54**) (reactants, 1-methylpyrrolidin-2-one, 165°C, 7 h: >95%); the 3-phenyl analog was made similarly.[928]

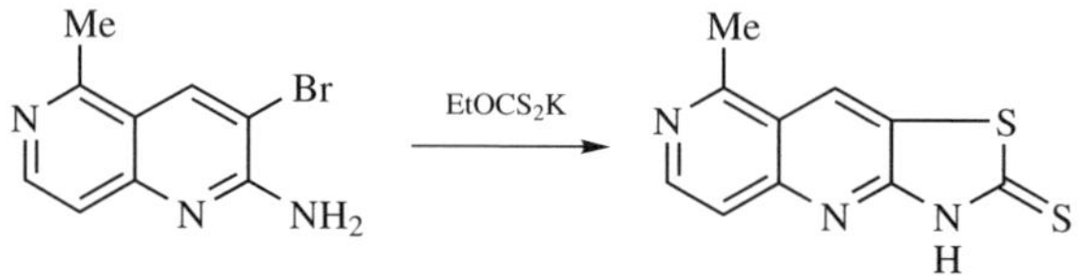

(53) → EtOCS$_2$K → (54)

CHAPTER 11

Oxy-1,6-naphthyridines

The term *oxy-1,6-naphthyridine* includes the tautomeric and nontautomeric 1,6-naphthyridinones, extranuclear hydroxy-1,6-naphthyridines, alkoxy/aryloxy-1,6-naphthyridines, and 1,6-naphthyridine *N*-oxides.

As well as their chemical importance, appropriate oxy-1,6-naphthyridines have shown appreciable (but not outstanding) biological activities as metal-binding antibacterials,[110] or as antineoplastic agents.[572,cf. 843] Medorinone, 5-methyl-1,6-naphthyridin-2(1H)-one, has been used as a cardiotonic/vasodilatory agent.[490, 716,1224,cf. 320] Oxy derivatives of fused/unfused 1,6-naphthyridines have been isolated from marine sponges.[214,408]

11.1. TAUTOMERIC/NONTAUTOMERIC 1,6-NAPHTHYRIDINONES AND EXTRANUCLEAR HYDROXY-1,6-NAPHTHYRIDINES

With the exception of 1,6-naphthyridin-8-ols, nuclear hdyroxy-1,6-naphthyridines would be expected to exist as the corresponding naphthyridinones. This has been confirmed by infrared,[1035] ultraviolet,[1026] and nuclear magnetic resonance spectral means.[639] Other physicochemical studies of these oxy-1,6-naphthyridines include IR stretching frequencies[1019] and ionization constants[1040] of 1,6-naphthyridin-8-ols; the mass spectra of 1,6-naphthyridinones; the effect of solvents on the UV spectrum of an 8-hydroxy-1,6-naphthyridine derivative;[773] and X-ray analysis of the dilactam, 4-hexyl-3,4,4a,5-tetrahydro-1,6-naphthyridine-2,7(1*H*,6*H*)-dione (**1**), in connection with its crystallization as self-assembled hydrogen-bonded polymers.[549]

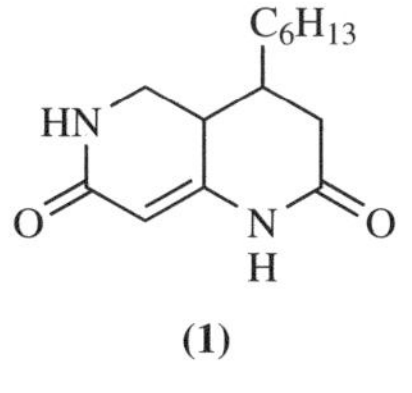

(1)

The Naphthyridines: The Chemistry of Heterocyclic Compounds, Volume 63, by D.J. Brown
Copyright © 2008 John Wiley & Sons, Inc.

11.1.1. Preparation of 1,6-Naphthyridinones and the Like

Many tautomeric and nontautomeric naphthyridinones have been made by *primary synthesis* (see Chapter 8) and a few by *ozonolysis of alkylidene-1,6-naphthyridines* (see Section 9.2.2) or by *hydrolysis of halogeno-1,6-naphthyridines* (see Section 10.4.2). Other preparative routes are illustrated by the following examples.

From 1,6-Naphthyridinamines (or Imines)

1,6-Naphthyridin-2-amine (**2**) gave 1,6-naphthyridin-2(1*H*)-one (**3**) (80% H_2SO_4, reflux, 50 h: 80%).[819]

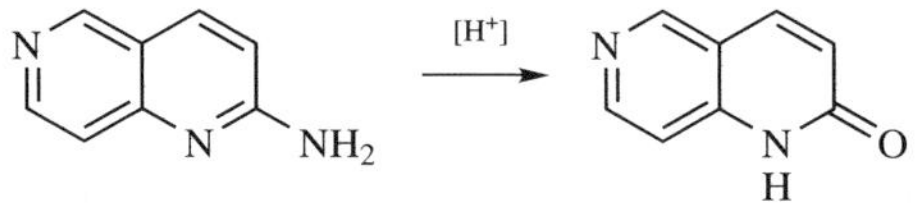

3-Amino- (**4**, R = NH_2) gave 3-hydroxy-5-methyl-1,6-naphthyridin-2(1*H*)-one (**4**, R = OH) (0.5M NaOH, 95°C, 10 h: 62%).[1166]

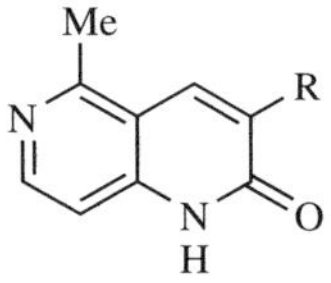

Ethyl 5-amino-7-benzylseleno-8-cyano-2-methyl-1,6-naphthyridine-3-carboxylate gave ethyl 7-benzylseleno-8-cyano-2-methyl-5-oxo-5,6-dihydro-1,6-naphthyridine-3-carboxylate (**5**) (substrate, AcOH, reflux, NaOH↓, slowly, then reflux, 30 min: 47%).[959]

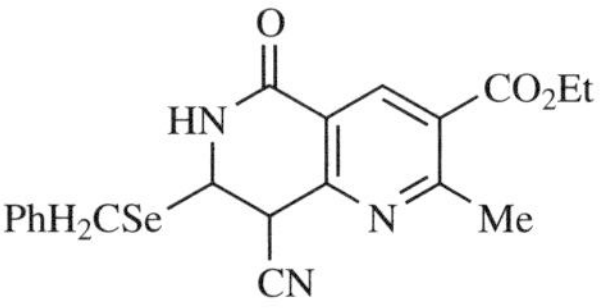

6-Methyl-1,6-naphthyridin-5(6*H*)-imine (**6**, X = NH) gave 6-methyl-1,6-naphthyridin-5(6*H*)-one (**6**, X=O) (2.5M NaOH, reflux, 3 h: ~65%).[372]

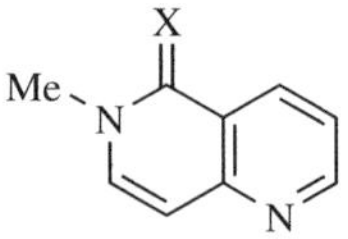

Also other examples.[215,590]

From Alkoxy-1,6-naphthyridines

2-Methoxy-1,6-naphthyridine (**7**) gave 1,6-naphthyridin-2(1H)-one (**8**) (10M HCl, 140°C, sealed, 8 h: 67%).[1318]

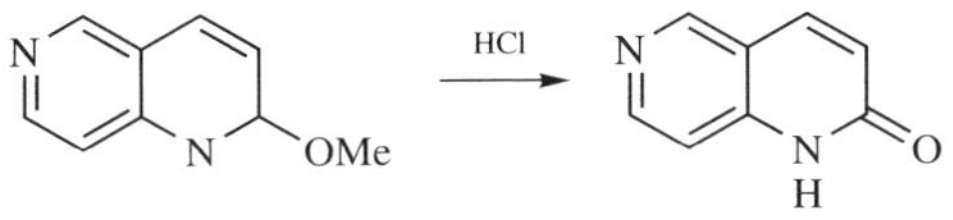

Also other examples.[1116]

From 1,6-Naphthyridine N-Oxides

1,6-Naphthyridine 1-oxide (**9**) gave 1,6-naphthyridin-2(1H)-one (**9a**) (Ac$_2$O, 160°C, sealed?, 3 h: 10%).[1282]

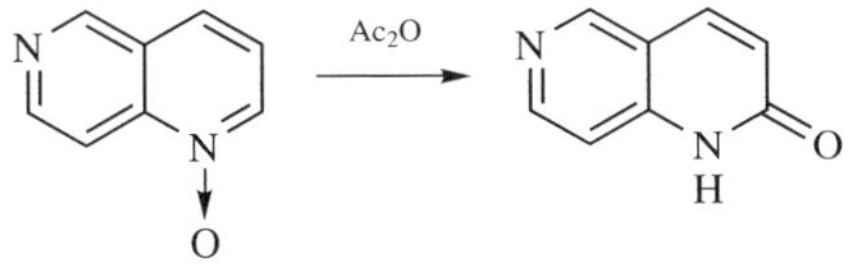

In contrast, 5-methyl-1,6-naphthyridin-2(1H)-one 6-oxide (**10**) gave 5-hydroxymethyl-1,6-naphthyridin-2(1H)-one (**10a**) (Ac$_2$O, 20°C, 16 h; 95°C, 1 h; then K$_2$CO$_3$, H$_2$O, 95°C, 1 h: 57%).[490]

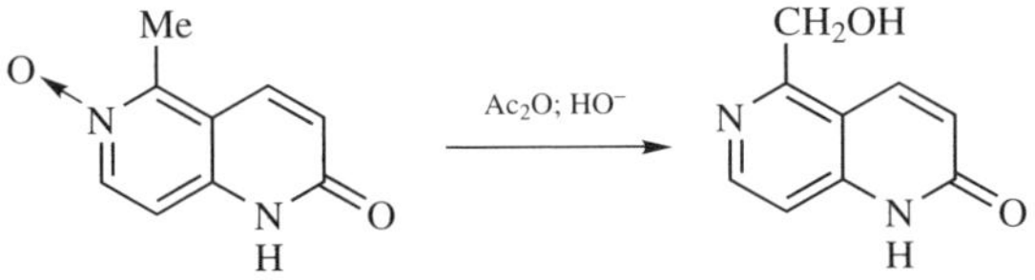

Irradiation of 1,6-naphthyridine 1,6-dioxide in water gave 1,6-naphthyridin-2(1H)-one 6-oxide (no details).[216]

From Quaternary N-Alkyl-1,6-naphthyridinium Salts

Note: This oxidative process provides nontautomeric naphthyridinones. 1,6-Naphthyridine 6-methiodide (**11**, R = H) underwent oxidation to 6-methyl-1,6-naphthyridin-5(6H)-one (**12**, R = H) [substrate, H$_2$O, 0°C, NaOH/H$_2$O↓ during 5 min; K$_3$Fe(CN)$_6$/H$_2$O↓ during 30 min; 0°C, 90 min; 20°C, 5 h:

57%];[1214,cf. 155,1275] the 2-phenyl substrate (**11**, R = Ph) likewise gave 6-methyl-2-phenyl-1,6-naphthyridin-5(6*H*)-one (**12**, R = Ph) (80%).[1033]

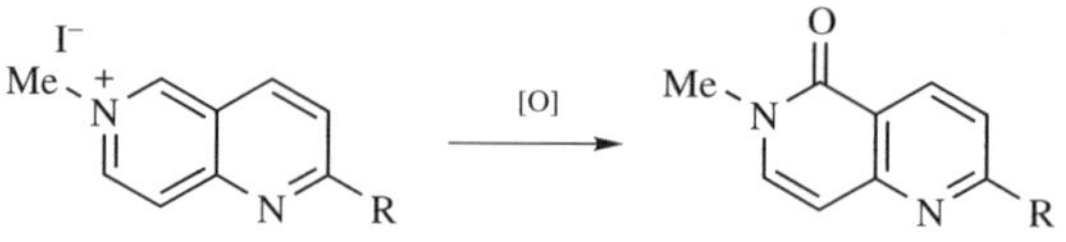

(**11**) → [O] → (**12**)

Tautomeric to Nontautomeric 1,6-Naphthyridinones

Note: This process is exemplified in Section 11.1.2.

11.1.2. Reactions of 1,6-Naphthyridinones and the Like

The *halogenolysis* of 1,6-naphthyridinones and extranuclear hydroxy-1,6-naphthyridines has been covered in Section 10.1.2. Other reactions are illustrated in the following classified examples.

Deoxygenation

Note: This is usually done via the corresponding halogeno derivative, but direct reduction can give a nuclear reduced product.

1,6-Dibenzyl-4-hydroxy-7-methyl-1,6-naphthyridine-2,5(1*H*,6*H*)-dione (**13**) gave 1,6-dibenzyl-4-hydroxy-7-methyl-5,6-dihydro-1,6-naphthyridin-2(1*H*)-one (**14**) (B_2H_6, THF: 79%).[950]

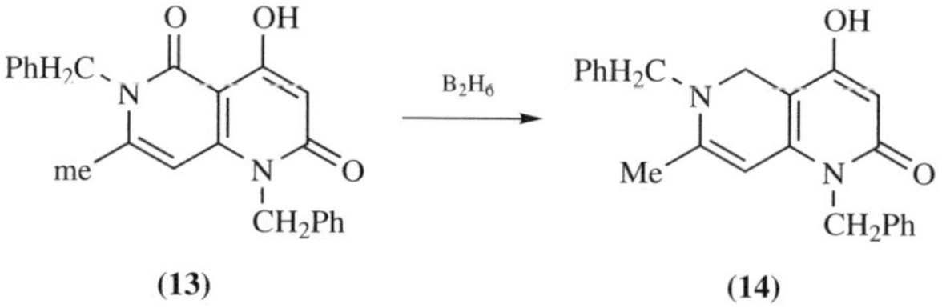

(**13**) → B_2H_6 → (**14**)

Alkylation or Arylation

Note: Alkylation or arylation of any tautomeric 1,6-naphthyridinone probably affords a mixture of *O*- and *N*-derivatives, but the latter usually predominates.

1,6-Naphthyridin-4(1*H*)-one (**15**, R = H) gave 1-methyl-1,6-naphthyridin-4(1*H*)-one (**15**, R = Me) (Me_2SO_4, 0.2M NaOH, 10°C, 2 h: 38%).[1067]

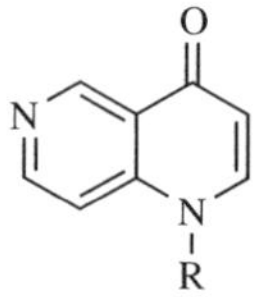

(**15**)

3-(2,6-Dichlorophenyl)-7-fluoro-1,6-naphthyridin-2(1H)-one (**16**, R = H) gave a separable mixture of 3-(2,6-dichlorophenyl-1-methyl-7-fluoro-1,6-naphthyridin-2(1H)-one (**16**, R = Me) and 3-(2,6-dichlorophenyl)-7-fluoro-2-methoxy-1,6-naphthyridine (**17**) (substrate, Me$_2$NCHO, 0°C, NaH↓, MeI↓ 2 h: 90% and 3%, respectively).[239]

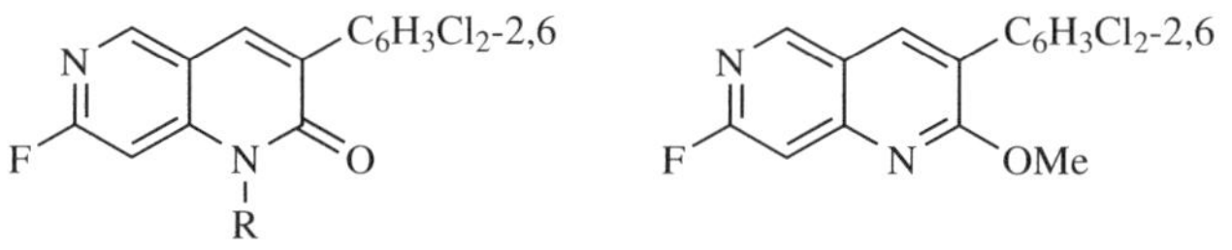

(**16**)   (**17**)

7-Phenyl-1,6-naphthyridin-5(6H)-one (**18**, R = H) gave (3,4-dimethoxybenzenesulfonyl)-7-phenyl-1,6-naphthyridin-5(6H)-one [**18**, R = C$_6$H$_3$(OMe)$_2$-3,4] [NaH, Me$_2$NCHO, 3,4-(OMe)$_2$C$_6$H$_3$SO$_2$Cl] (no details!).[576]

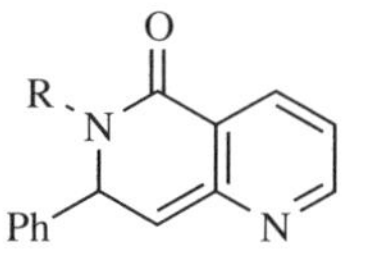

(**18**)

Also other examples.[490,924,951]

Aminolysis

Note: Only a one-pot indirect aminolysis has been reported. 3-Bromo-5-ethyl-1,6-naphthyridin-2(1H)-one (**19**) gave 3-bromo-5-ethyl-1,6-naphthyridin-2-amine (**20**) (POCl$_3$, reflux, 16 h: crude chloro compound; NH$_3$/EtOH, 100°C, sealed, 16 h: 92%; analogs likewise).[590]

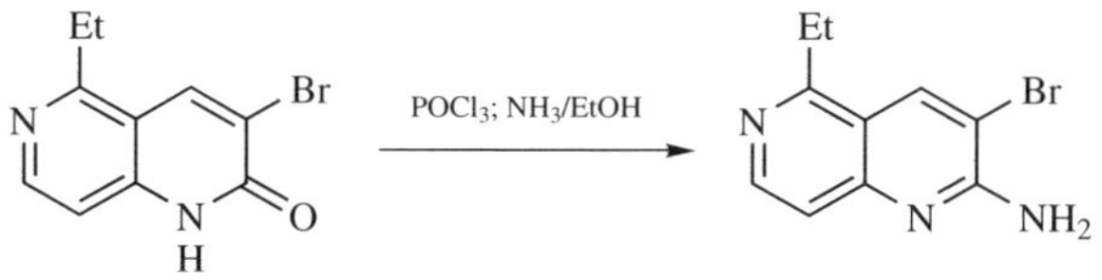

(**19**)   POCl$_3$; NH$_3$/EtOH   (**20**)

C-**Alkylation**

Note: This reaction does not involve the oxo/hydroxy substituent but probably depends on their presence.

4-Hydroxy-1,6-naphthyridin-2(1H)-one with dimethyl sulfoxide gave bis(4-hydroxy-2-oxo-1,2-dihydro-1,6-naphthyridin-3-yl)methane (**21**) (for details, see original).[412]

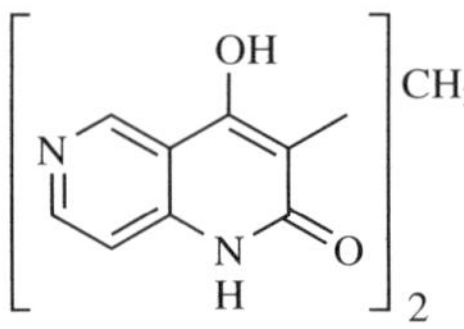

(**21**)

Cyclization Reactions

Ethyl 6-o-aminophenyl-2-methyl-5-oxo-5,6-dihydro-1,6-naphthyridine-3-carboxylate (**22**) gave ethyl 3-methylbenzimidazo[2,1-f] [1,6]naphthyridine-2-carboxylate (**23**) (POCl$_3$, reflux, 2 h: 98%).[1157]

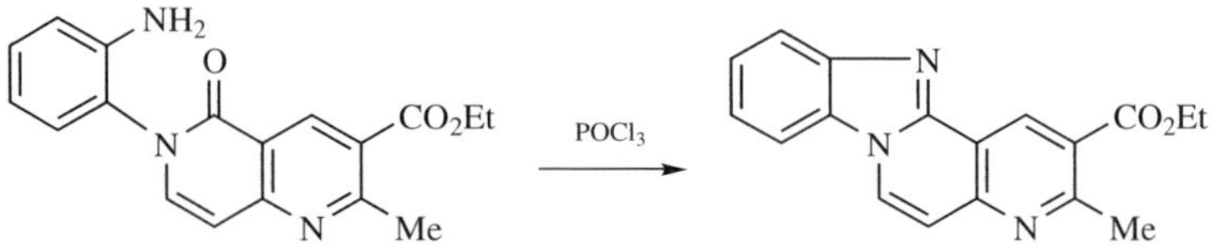

(**22**)   (**23**)

Also other examples.[750,1070]

Oxidation

The extranuclear hydroxy derivative, 5-(1-hydroxyethyl)- (**24**, R = CHOHMe), underwent a Swern oxidation to afford 5-acetyl-1,6-naphthyridin-2(1H)-one (**24**, R = Ac) (36%; for details, see original).[490]

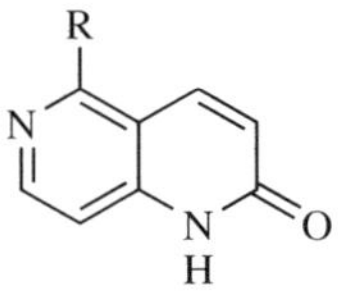

(**24**)

11.2. ALKOXY- AND ACYLOXY-1,6-NAPHTHYRIDINES

Not many such derivatives have been reported. Their methods of preparation and any subsequent reactions are summarized here.

Preparation

Note: All reported alkoxy-1,6-naphthyridines appear to have been made by *primary synthesis* (see Chapter 8), *alcoholysis of halogeno-1,6-naphthyridines* (see Section 10.2.1), or *O-alkylation of 1,6-naphthyridinones* (see Section 11.1.2).

Reactions

Note: For the *hydrolysis of alkoxy-1,6-naphthyridines*, see Section 11.1.1.

The acyloxy derivative, 4-acetoxy-6,7-dimethyl-1,6-naphthyridine-2,5(1H, 6H)-dione (**25**, Q = Ac, R = H), underwent *rearrangement* to 3-acetyl-4-hydroxy-6,7-dimethyl-1,6-naphthyridine-2,5(1H, 6H)-dione (**25**, Q = OH, R = Ac) (AlCl$_3$, 47%: no details).[950]

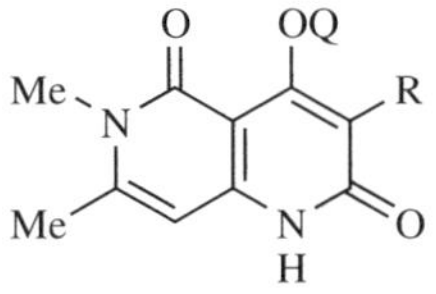

(25)

11.3. 1,6-NAPHTHYRIDINE N-OXIDES

The available information on such oxides is illustrated by the following examples.

Preparation of 1,6-Naphthyridine N-Oxides

Note: These oxides have been made either by *primary synthesis* (see Chapter 8) or, more frequently, by *oxidation* of the corresponding 1,6-naphthyridines (see Section 9.1 for some simple oxides as well as the examples given here).

4-Chloro-1,6-naphthyridine gave its 6-oxide (**26**) (*m*-ClC$_6$H$_4$CO$_3$H, CHCl$_3$, reflux, 25 min: 45%).[1069]

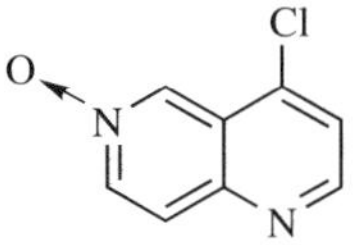

(26)

5,5,8,8-Tetramethyl-2-phenyl-5,6,7,8-tetrahydro-1,6-naphthyridine suffered similar *N*-oxidation to give a pseudo-*N*-oxide, formulated as the stable 6-nitroxyl radical (**27**) (H$_2$O$_2$, Na$_2$WO$_4$, MeOH, 20°C, dark, 3 days: 76%; homologs likewise).[933]

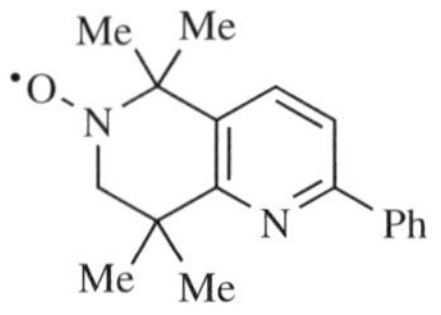

(**27**)

Also other examples.[215,1318]

Reactions of 1,6-Naphthyridine *N*-oxides

Note: The conversion of such oxides to *C*-chloro-1,6-naphthyridines by the *Meissenheimer* reaction has been covered in Section 10.1.3; the conversion of such *oxides into 1,6-naphthyridinones or extranuclear hydroxy-1,6-naphthyridines* has been exemplified in Section 11.1.1. Other reactions are illustrated here.

1,6-Naphthyridine 6-oxide underwent selective *quaternization* by methyl iodide to give its 1-methiodide (**28**) (reactants, MeOH, 23°C, 18 h: >95%).[1053]

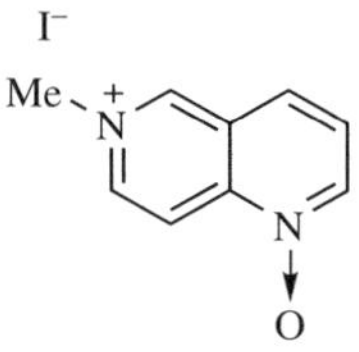

(**28**)

5-Methyl-1,6-naphthyridine 1-oxide (**29**) underwent *deoxygenation* by phosphorus trichloride to give 6-methyl-1,6-naphthyridine (**30**) (substrate, CHCl$_3$, PCl$_3$ in, slowly with cooling; then 1 h: ~65%).[1285]

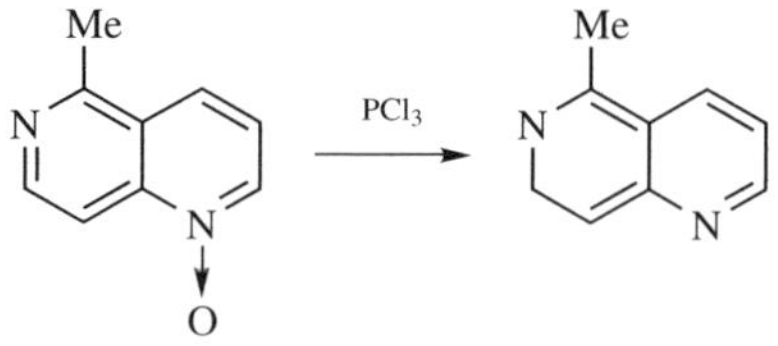

(**29**)   (**30**)

Hydrogenation of 1,6-naphthyridine 1,6-dioxide (**31**) gave a separable mixture of of 1,6-naphthyridine 6-oxide (**32**) and 1,6-naphthyridine (Raney Ni, H$_2$, 20°C: 30% and 12%, respectively).[1282]

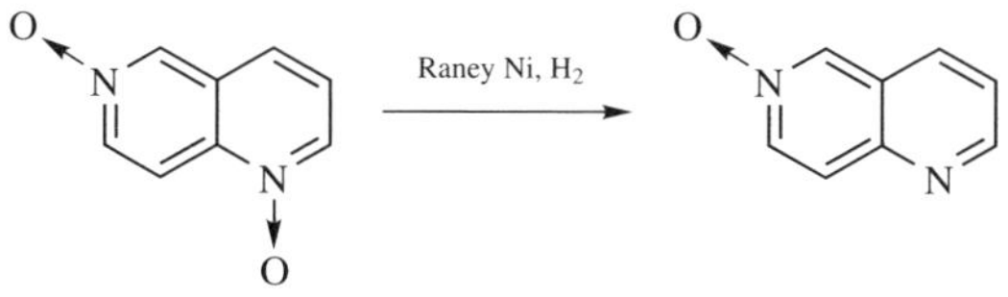

1,6-Naphthyridine 6-oxide (**33**) underwent *deoxidative cyanation* to give 1,6-naphthyridine-5-carbonitrile (**34**) (substrate KCN, H$_2$O, 20°C, ; BzCl in slowly; then 4 h: 63%);[1375] 1,6-naphthyridine 1-oxide and 1,6-dioxide also underwent *tele*-cyanation.[1318]

CHAPTER 12

Thio-1,6-naphthyridines

This chapter should summarize data on 1,6-naphthyridines that bear substituents joined directly or indirectly to the nucleus through a sulfur atom; for a list of such derivatives, see Chapter 5. In fact, very few thio-1,6-naphthyridines have been reported. Accordingly, available information is simply listed here.

Preparation of Thio-1,5-naphthyridines

Three *primary syntheses of 1,6-naphthyridinethiones* have been covered already: 8-benzylidene-6-isopropyl-4-phenyl-2-thioxo-1,2,5,6,7,8-hexahydrol-6-naphthyridine-3-carbonitrile[946] (Section 8.3.3); 1-benzyl-3-ethyl-4-hydroxy-7,7-dimethyl-5-thioxo-5,6,7,8-tetrahydro-1,6-naphthyridin-2(1H)-one[679] (Section 8.3.3); and 6-methyl-4-(thien-2-yl)-2-thioxo-1,2,5,6,7,8-hexahydro-1,6-naphthyridine-3-carbonitrile[1429] (Section 8.4). The *thiolysis of a halogeno-1,6-naphthyridine* to afford 5-methyl-1,6-naphthyridine-2(1H)thione has been described[490] (Section 10.2.4).

Reactions of Thio-1,5-naphthyridines

The *S-alkylation of a 1,6-naphthyridinethione* is exemplified by the conversion of 8-benzylidene-6-isopropyl-4-phemyl-2-thioxo-1,2,5,6,7,8-hexahydro-1,6-naphthyridine-3-carbonitrile (**1**) into 3-benzylidene-6-isopropyl-2-methylthio-5,6,7,8-tetrahydro-1,6-naphthyridine-3-carbonitrile (**2**) (MeI, AcONa, EtOH, reflux, 1 h: 96%) or analogous *S*-(substituted-alkyl) derivatives.[946]

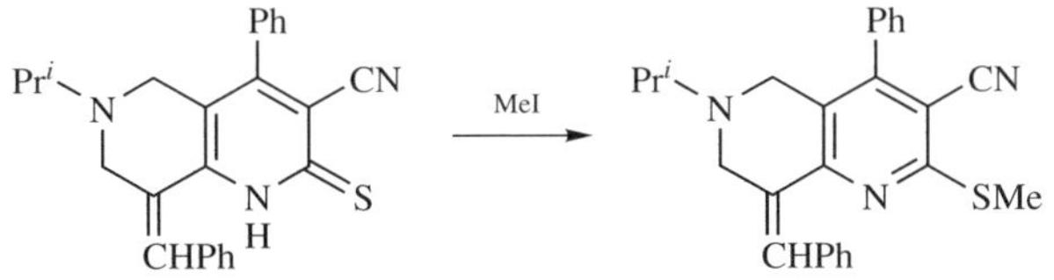

The Naphthyridines: The Chemistry of Heterocyclic Compounds, Volume 63, by D.J. Brown
Copyright © 2008 John Wiley & Sons, Inc.

Cyclocondensations involving thioxo or alkylthio substituents are represented by the conversion of 6-methyl-4-(thien-2-yl)-2-thioxo-1,2,5,6,7,8-hexahydro-1,6-naphthyridine-3-catbonitrile into 4-hydroxy-9-methyl-4-(thien-2-yl)-7,8,9,10-tetrahydropyrido[2′,3′:4,5]thieno[2.3-b][1,6]naphthyridin-2(1H)-one (**3**) (ClCH$_2$COCH$_2$CO$_2$Et, EtOH, EtONa, reflux, 10 min: 88%);[1429] also of 2-acetonylithio-8-benzylidene-6-isopropyl-4-phenyl-5,6,7,8-tetrahydro-1,6-naphthyridine-3-carbonitrile into 2-acetyl-8-benzylidene-6-isopropyl-4-phenyl-5,6,7,8-tetrahydrothieno [2,3-b][1,6]naphthyridin-3-amine (**4**) (EtONa, EtOH, 20°C,, 15 min: 75%).[946]

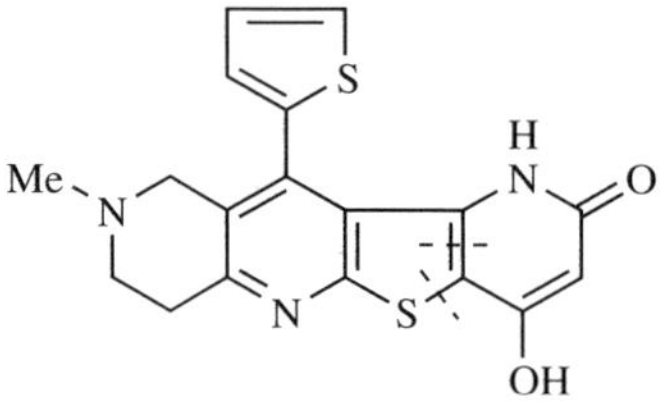

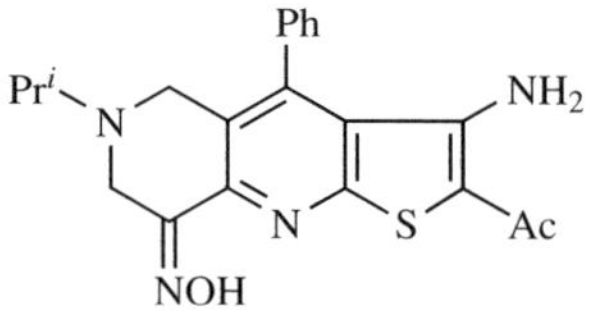

Alkylseleno-1,6-naphthyridines

A somewhat complicated alkylseleno-1,6-naphthyridine has been made by primary synthesis.[1448]

CHAPTER 13

Nitro-, Amino, and Related 1,6-Naphthyridines

This chapter covers 1,6-naphthyridines that bear nitrogenous substituents joined to the nucleus through their nitrogen atoms. Only nitro and amino derivatives appear to have been reported.

13.1. NITRO-1,6-NAPHTHYRIDINES

Most nitro-1,6-naphthyridines have been used for reduction to 1,6-naphthyridinamines. A comprehensive review of all nitronaphthyridines appeared in 2000.[1273] It includes information on those within the 1,6-series. The mass spectra of nitro derivatives in the 1,5-, 1,6-, and 1,8-naphthyridines have been reviewed.[492]

13.1.1. Preparation of Nitro-1,6-naphthyridines

Some of these nitro derivatives have been made by *primary synthesis* (see Chapter 8) and others by methods illustrated in the following examples.

By Nitration

1,6-Naphthyridin-2(1H)-one (**1**, R = H) gave 3-nitro-1,6-maphthyridin-2(1H)-one (**1**, R = NO$_2$) (fuming HNO$_3$, oleum, 95°C, 4 h: 96%),[819] somewhat similarly, 1,6-naphthyridin-4(1H)-one (**2**, R = H) gave 3-nitro-1,6-naphthyridin-4(1H)-one (**2**, R = NO$_2$) (fuming HNO$_3$, reflux, 75 min: 71%).[726]

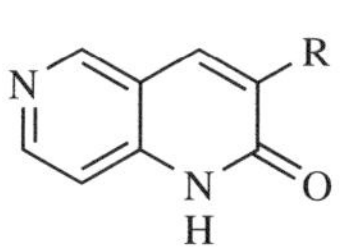

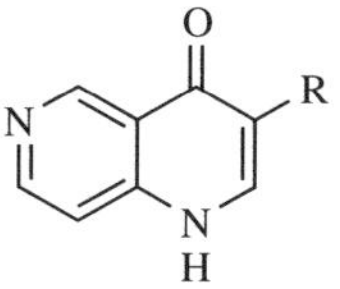

The Naphthyridines: The Chemistry of Heterocyclic Compounds, Volume 63, by D.J. Brown
Copyright © 2008 John Wiley & Sons, Inc.

A peculiar extranuclear nitration involving diazotization has also been described.[590]

By Oxidation of Dimethylsulfimido-1,6-naphthyridines

Note: This procedure offers a route from naphthyridinamines to nitro-1,6-naphthyridines.

1,6-Naphthyridin-2-amine (**3**, R = NH$_2$) gave 2-dimethylsulfimido-1,6-naphthyridine (**3**, R = N:SMe$_2$) (48%) and thence 2-nitro-1,5-naphthyridine (**3**, R = NO$_2$) 26%) (details same as for the 1,5-naphthyridine analogs; see Section 5.1.1).[692]

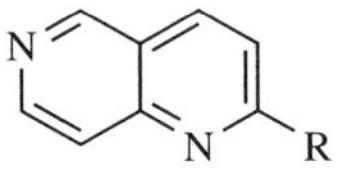

(3)

13.1.2. Reactions of Nitro-1,6-naphthyridines

Only two reactions of nitro-1,6-naphthyridines have been reported, as illustrated in the following examples.

Reduction

Note: Judging from the available data, stannous chloride reduction appears to have provided better yields than has hydrogenation over Pd, Pt, or Raney nickel.

6-Methyl-8-nitro- (**4**, R = NO$_2$) gave 8-amino-6-methyl-1,6-naphthyridin-5(6*H*)-one (**4**, R = NH$_2$) (SnCl$_2$, HCl, heat: 91%);[1116] similarly, 3-bromo-5-(*p* nitrophenyl)- (**5**, R = NO$_2$) gave 5-*p*-aminophenyl-3-bromo-1,6-naphthyridin-2 (1*H*)-one (**5**, R = NH$_2$) (SnCl$_2$, 6M HCl, 95°C, 12 h: 78%).[590]

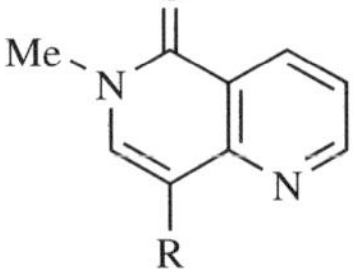

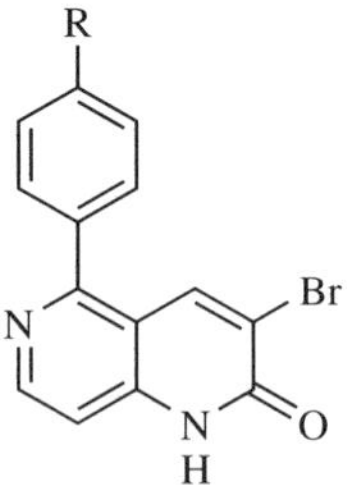

(4) (5)

3-Nitro- (**6**, R = NO$_2$) gave 3-amino-1,6-naphthyridin-4 (1*H*)-one (**6**, R = NH$_2$) (Raney Ni, NH$_4$OH, H$_2$O, H$_2$, 40°C 80 atm, 4 h: ~60%);[726] 5-*p*-nitrophenyl-

(7, R = NO$_2$) gave 5-*p*-aminophenyl-1,6-naphthyridin-2(1*H*)-one (7, R = NH$_2$) (PtO$_2$, AcOH, H$_2$, 3 atm: 54%);[490]

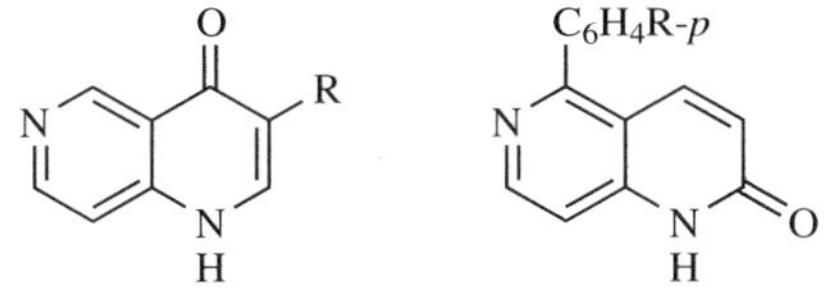

(6)

(7)

Also other examples.[204]

Rearrangement

6-Methyl-8-nitro-1,6-naphthyridin-5(6*H*)-one (8) with hydrazine hydrate gave 8-methylpyrido [2.3-*d*]pyrazin-5(6*H*)-one (9) (neat reactants, 135°C, 4 h: 94%; a mechanism was proposed).[1171]

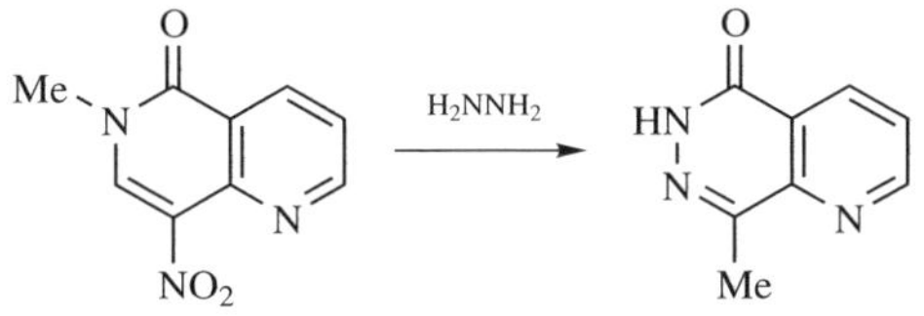

(8)

(9)

13.2. AMINO- AND (SUBSTITUTED-AMINO)-1, 6-NAPHTHYRIDINES

The ionization constants,[331] NMR spectra,[174] and mass spectra[1227] of representative amino-1,6-naphthyridines have been compared with those of related amino derivatives. Complexes of 1,6-naphthyridin-2-amines with tetracyanoquinodimethane have been studied.[416] 4-(3-Dimethylamino-1-methylpropylamino)-1,6-naphthyridine showed significant antimalarial activity.[236] X-ray analyses of a number of 5-amino-1,6-naphthyridines have been reported in connection with their antibacterial activities; for example, ethyl 5-amino-7-benzylseleno-8-cyano-4-(fur-2-yl)-1,2-dimethyl-1,4-dihydro-1,6-naphthyridine-3-carboxylate (10) proved to be a mixture of two symmetric conformers in each crystal;[772] the other related compounds appear to be composed of single conformers.[558,560,562,759,781,1289]

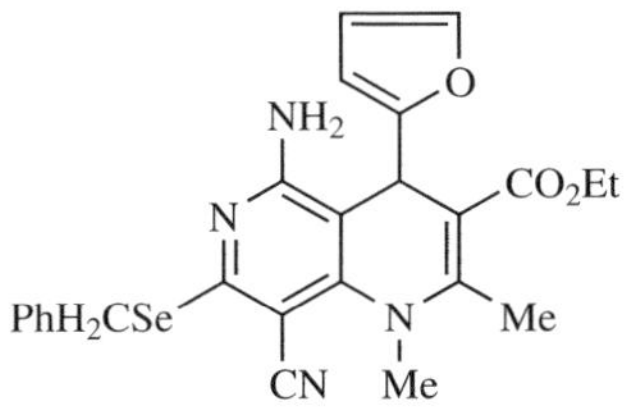

(10)

13.2.1. Preparation of Amino-1,6-naphthyridines

These aminonaphthyridines have been made by *primary synthesis* (see Chapter 8) or *aminolysis of halogeno-1,6-naphthyridines* (see Section 10.2.2). Other preparative routes are illustrated by the following examples.

By Amination

Note: The amination of unsubstituted 1,6-naphthyridine has been discussed in Section 9.1.3. Reviews that include the amination of 1,6-naphthyridines have appeared previously.[514,1160,1268]

3-Nitro-1,6-naphthyridine (**11**, R = H) gave 3-nitro-1,6-naphthyridin-4-amine (**11**, R = NH$_2$) (KMnO$_4$, liq NH$_3$, 55%; no other details).[514]

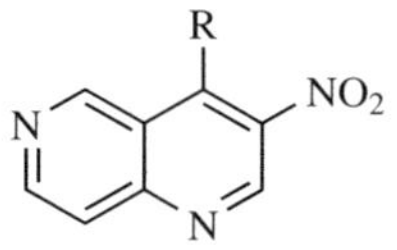

(**11**)

1,6-Naphthyridine 6-methiodide (**12**) with ammonia and permanganate gave 6-methyl-1,6-naphthyridin-5(6*H*)-imine hydriodide (**13**) (substrate, liq NH$_3$, KMnO$_4$ in slowly; 20 min: less than 5% after separation from other products).[372]

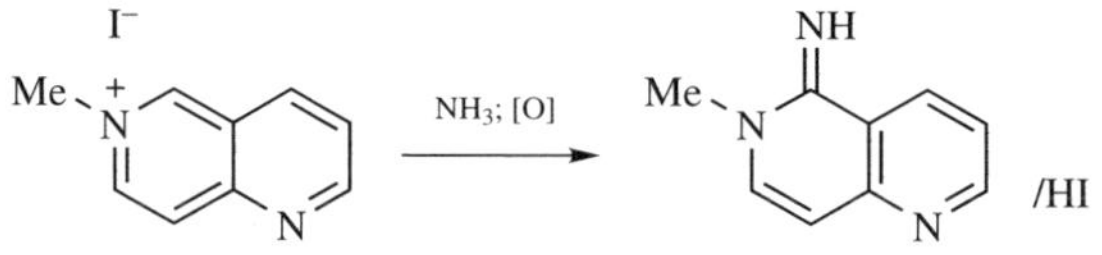

(**12**) → (**13**)

By the Curtius or Hofmann Reaction

5-Methyl-2 oxo-1,2-dihydro-1,6-naphthyridine-3-carbohydrazide (**14**) gave 3-amino-5-methyl-1,6-naphthyridin-2(1*H*)-one (**15**) (substrate, HCl, 5°C, NaNO$_2$/H$_2$O in during 30 min; 5°C, 3 h; then 20–95°C, 5 h: 50%).[1166]

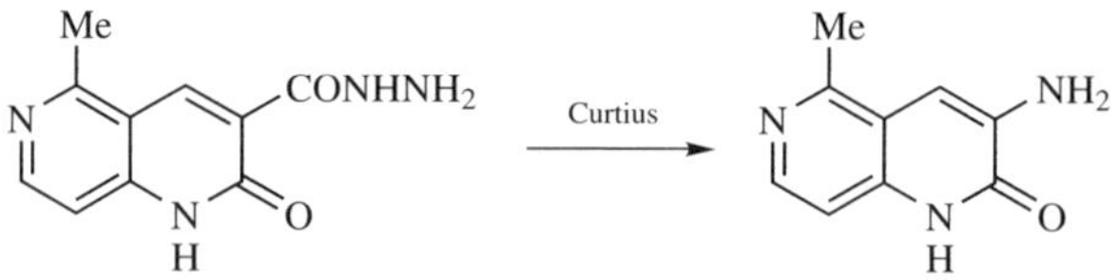

(**14**) —Curtius→ (**15**)

Also other examples.[105,310]

By Modification of the Amino Group

2-Hydrazino-3-methyl-1,6-naphthyridine (**16**, R = NH$_2$) gave 5-methyl-1,6-naphthyridin-2-amine (**16**, R = H) (Raney Ni, Me$_2$NCHO, limited H$_2$, 3 atm, 65°C, 71%).[490]

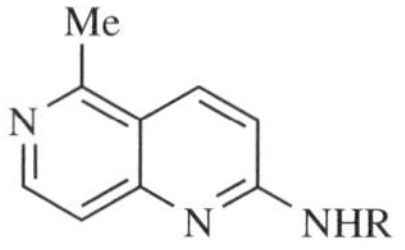

(**16**)

Note: For alkylation/alkyludenation of existing amines, see Section 13.2.2.

13.2.2. Reactions of Amino-1,6-naphthyridines

Already covered are the *halogenolysis* (Section 10.1.3) and *hydrolysis* of these naphthyridinamines (Section 11.1). Other reported reactions are illustrated in the following examples.

Alkylation

8-Amino-6-methyl-1,6-naphthyridin-5(6*H*)-one (**17**) with *N*-(4-iodopentyl) phthalimide gave 6-methyl-8-(1-methyl-4-phthalimidobutyl)amino-1,6-naphthyridin-5(6*H*)-one (**18**, R = phthalimido) (substrate, Me$_2$NCHO, 80°C, A, synthon + NEt$_3$ in during 4 h; 85°C, 2 h: ~45%) and thence, by deprotection with hydrazine, 8-(4-amino-1-methylbutyl)amino-6-methyl-1,6-naphthyridin-5(6*H*)-one (**18**, R = NH$_2$).[1116]

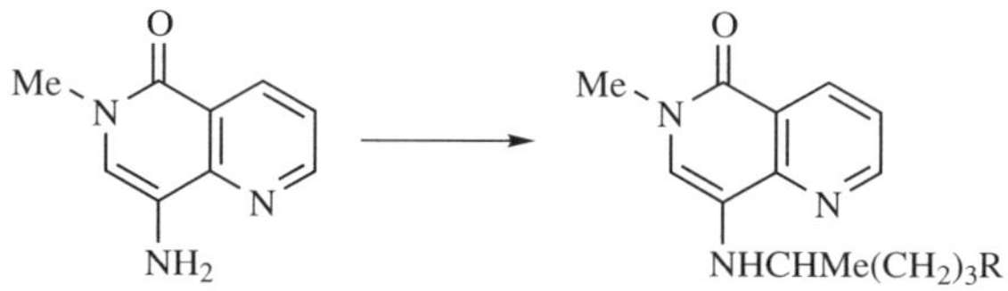

(**17**)   (**18**)

3-(3,5-Dimethoxyphenyl)-1,6-naphthyridine-2,7-diamine (**19**, R = H) and trityl chloride gave selectively 7-tritylamino-3-(3,5-dimethoxyphenyl)-1,6-naphthyridin-2-amine (**19**, R = CPh$_3$) (reactants, NEt$_3$, THF, 50°C, sealed, 20 h: more synthon + NEt$_3$ added, 50°C, 48 h: 80%).[220]

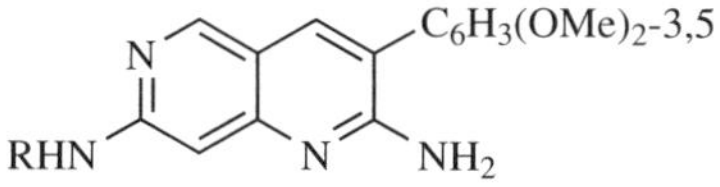

(**19**)

5-Phemyl-5,6,7,8-tetrahydro-1,6-naphthyridine (**20**, R = H) gave its 6-methyl derivative (**20**, R = Me) (CH$_2$O, HCO$_2$H, H$_2$O, reflux, N$_2$, 4 h: 58%).[1215]

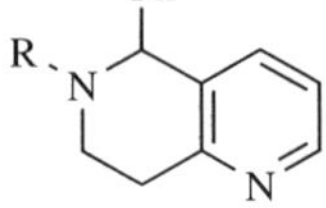

(**20**)

1,6-Naphthyridin-2-amine with diethyl ethoxymethylenemalonate gave 2-(2,2-diethoxycarbonylvinyl)amino-1,6-naphthyridine (**21**) (reactants, EtOH, reflux, 12 h: 51%);[1068] in contrast, 7-amino-1,2,3,4-tetrahydro-1,6-naphthyridine-8-carbonitrile with dimethylformamide diethylacetal gave 7 dimethylamino methyleneamino-1,6-naphthyridine-8-carbonitrile (**22**) (57%, no other details).[792]

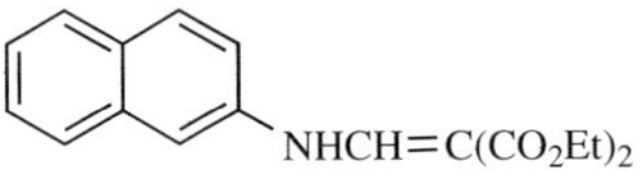

(**21**) 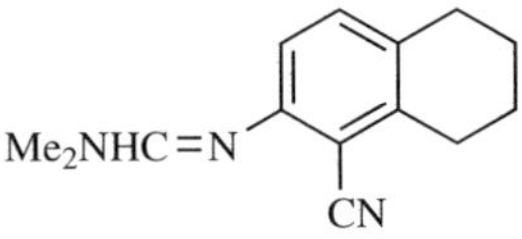   (**22**)

Also other examples.[1062,1323]

Acylation

6,8,8-Trimethyl-5,6,7,8-tetrahydro-1,6-naphthyridin-3-amine (**23**, R=H) gave 3-benzamido-6,8,8-trimethyl-5,6,7,8-tetrahydro-1,6-naphthyridine (**23**, R=Bz) (BzCl, NEt$_3$, CH$_2$Cl$_2$, 25°C, 6 h: 70%; analogs likewise).[204]

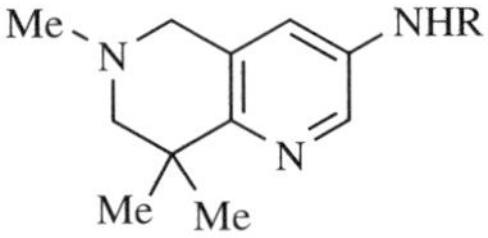

(**23**)

Ethyl 5-amino-7-benzylseleno- (**24**, R = H) underwent diacetylation to give 5-acetamido-1-acetyl-7-benzylseleno-5-cyano-4-(fur-2-yl)-2-methyl-1,4-dihydro-1,6-naphthyridine-3-carboxylate (**24**, R = Ac) (Ac$_2$O, pyridine, reflux, 90 min: 92%).[959]

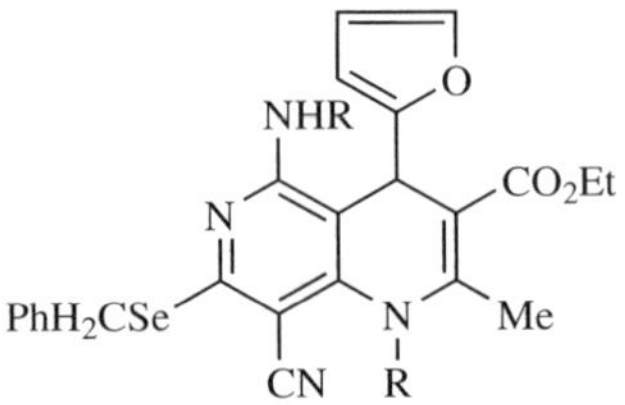

(**24**)

Carbamoylation to Ureido Derivatives

2-Phenyl-1,6-naphthyridine-1,7-diamine gave selectively 2-(N'-*tert*-butylureido)-3-phenyl-1,6-naphthyridin-7-amine (**25**) (substrate, NaH, Me$_2$NCHO, N$_2$, 20°C, 10 min; Bu^{t}NCO/Me$_2$NCHO added dropwise; 20°C, 4 h: 66%); analogs likewise.[220]

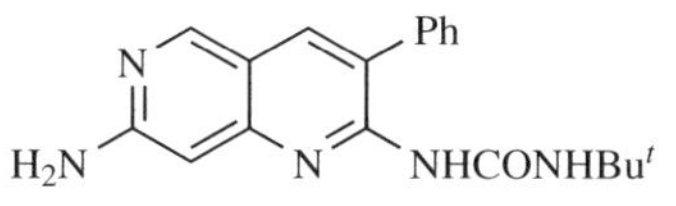

(**25**)

Diazotization

3-Amino-1,5-naphthyridin-4 (1H)-one gave crude 3-diazonio-1,6-naphthyridin-4(1H)-one chloride (HCl, NaNO$_2$, 5°C, 12 h) and thence 3-diazonio-1,6-naphthyridin-4-olate (**26**) (NaHCO$_3$, Et$_2$O: 70% overall);[726] subsequent irradiation caused ring contraction to pyrrolo [3,2-*d*] pyridine-3-carboxylic acid (**27**) (40%).[726]

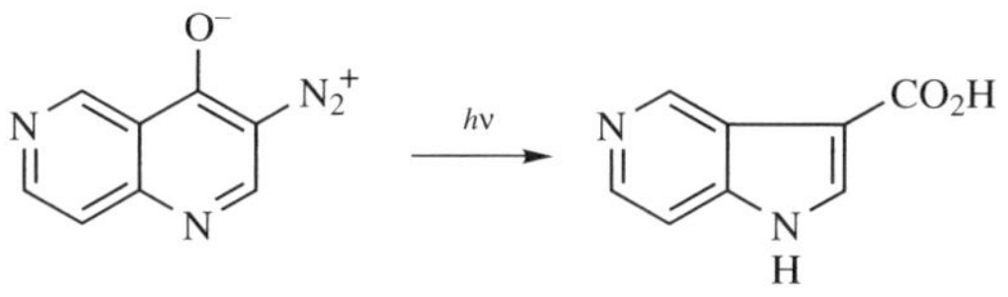

(**26**)   (**27**)

Cyclization Reactions

Note: Only a typical example is given here. Crude 5-amino-7-hydrazino-2,4-dimethyl-1,6-naphthyridine-8-carbonitrile (**28**) gave 2,4-dimethyl-7H-pyrazolo-[3,4-*h*] [1,6]naphthyridine-2,9-diamine (**29**) (EtOH, reflux, 6 h: 72%); analogs likewise.[895]

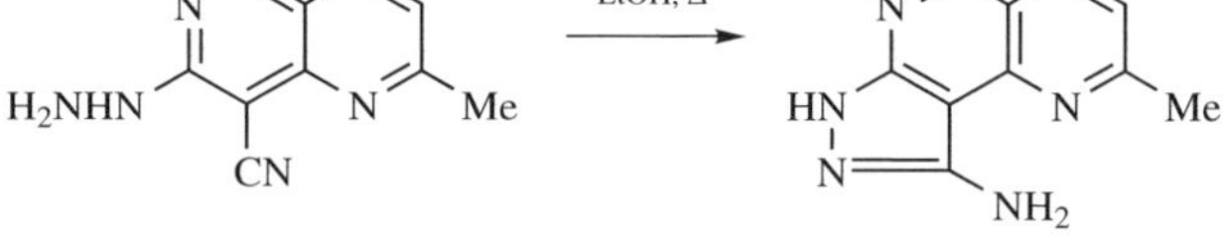

(**28**)   (**29**)

CHAPTER 14

1,6-Naphthyridinecarboxylic Acids and Related Derivatives

This chapter covers reported information on 1,6-naphthyridines that bear functional groups joined to the nucleus through their carbon atoms: carboxylic acids, esters, amides, nitriles, aldehydes, and ketones.

14.1. 1,6-NAPHTHYRIDINECARBOXYLIC ACIDS

A brief review of antibacterial naphthyridinecarboxylic acids has appeared.[542]

14.1.1. Preparation of 1,6-Naphthyridinecarboxylic Acids

These carboxylic acids have been made by *primary synthesis* (see Chapter 8), *oxidation of alkyl-1,6-naphthyridines* (see Section 9.2.2), or the routes illustrated in the following classified examples.

By Hydrolysis of Esters

Note: The choice between acidic or alkaline hydrolysis may depend on the other substituents present in the molecule.

Ethyl 7-chloro-1-cyclopropyl-8-fluoro-4-oxo-1,4-dihydro-1,6-naphthyridine-3-carboxylate (**1**, R = Et) gave 7-chloro-1-cyclopropyl-8-fluoro-4-oxo-1,4-dihydro-1,6-naphthyridine-3-carboxylic acid (**1**, R = H) (HCl, H$_2$O, EtOH, reflux, 9 h: 81%);[1175] analogs likewise.[648]

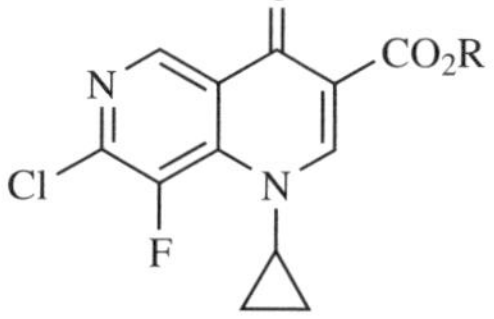

(1)

The Naphthyridines: The Chemistry of Heterocyclic Compounds, Volume 63, by D.J. Brown
Copyright © 2008 John Wiley & Sons, Inc.

2-Amino-6-*p-tert*-butoxycarbonylphenyl- (**2**, R = Bu^t) gave 2-amino-6-p-carboxyphenyl-5,6,7,8-tetrahydro-1,6-naphthyridine-3-carbonitrile (**2**, R = H) (substrate, MeNO$_2$, 0°C, HCl gas in, 5 min; 20°C, 77%).[378]

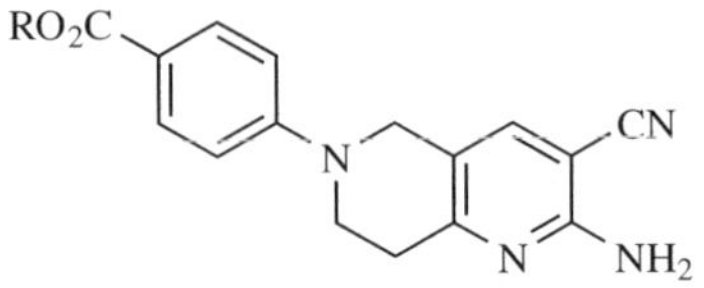

(**2**)

Ethyl 7-chloro-1-ethyl-4-oxo-1,4-dihydro-1,6-naphthyridine-3-carboxylate (**3**, R = Et) gave 7-chloro-1-ethyl-4-oxo-1,4-dihydro-1,6-naphthyridine-3-carboxylic acid (**3**, R = H) (2M KOH, 90°C, 14 min: 93%).[1284]

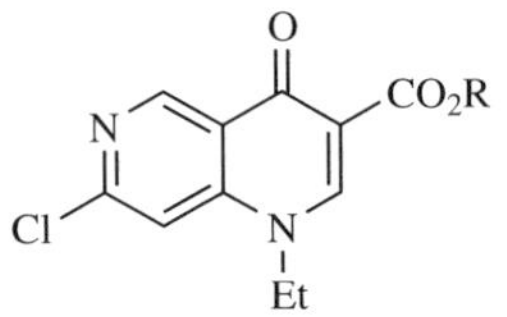

(**3**)

Ethyl 5,7-dimethyl-4-oxo-1,4-dihydro-1,6-naphthyridine-3-carboxylate (**4**, R = Et) gave 5,7-dimethyl-4-oxo-1,4-dihydro-1,6-naphthyridine-3-carboxylic acid (**4**, R = H) (M NaOH, reflux, 2 h: 65%).[1255]

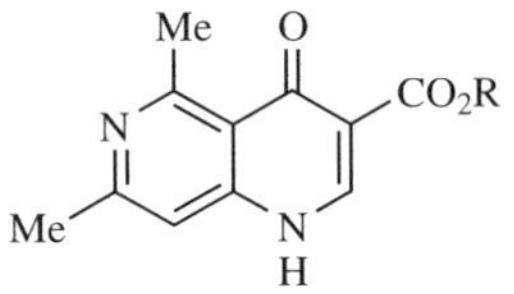

(**4**)

Also other examples.[48,108,391,575] including the hydrolysis of an extranuclear diethoxyphosphinyl to a dihydroxyphosphinyl derivative of 1,6-naphthyridine.[924]

By Hydrolysis of Nitriles

5-Methyl-2-oxo-1,2-dihydro-1,6-naphthyridine-3-carbonitrile (**5**, R = CN) gave 5-methyl-2-oxo-1,2-dihydro-1,6-naphthyridine-3-carboxylic acid (**5**, R = CO$_2$H) (50% H$_2$SO$_4$, reflux, 18 h: 90%);[1166] somewhat similarly, 6-methyl-2-oxo-1,2,5,6,7,8-hexahydro-1,5-naphthyridine-3-carbonitrile

(**6**, R = CN) gave the corresponding 3-carboxylic acid (**6**, R = CO$_2$H) (10M HCl, reflux, 15 h: 75% as hydrochloride).[180]

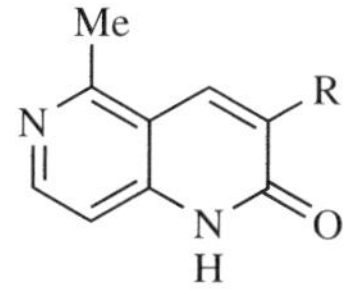

2-Oxo-5-(pyridin-4-yl)-1,2-dihydro 1,6-naphthyridine-3-carbonitrile (**7**, R = CN) gave the corresponding 3-carboxylic acid (**7**, R = CO$_2$H) (M NaOH, 95°C, 5 h: 88%).[490]

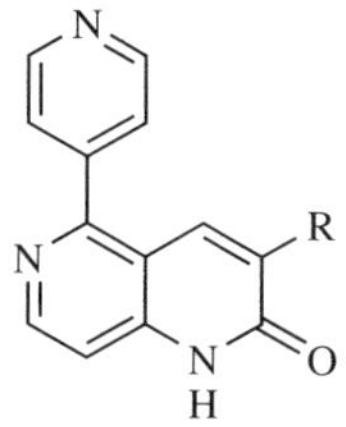

Also other examples.[247,340]

14.1.2. Reactions of 1,6-Naphthyridinecarboxylic Acids

Only decarboxylation, amide formation, and esterification are represented in the 1,6-naphthyridine literature. Examples follow.

Decarboxylation

6-Methyl-5-oxo-5,6-dihydro-1,6-naphthyridine-8-carboxylic acid (**8**, R = CO$_2$H) gave 6-methyl-1,6-naphthyridin-5(6H)-one (**8**, R = H) (neat substrate, 250°C, ? min: 77%);[876] 2-oxo-5-(pyridin-4-yl)-1,2-dihydro-1,6-naphthyridine-3-carboxylic acid (**9**, R = CO$_2$H) gave 5-(pyridin-4-yl)-1,6-naphthyridin-2(1H)-one (**9**, R = H) (neat substrate, 370°C, 3 min: 72%).[490]

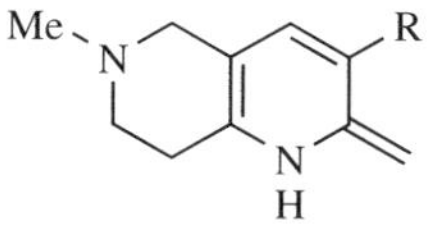

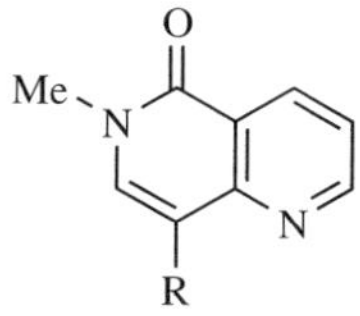

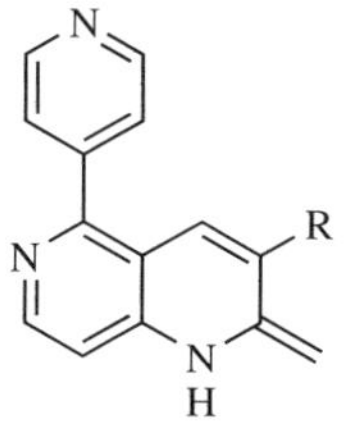

5,7-Dimethyl-4-oxo-1,4-dihydro-1,6-naphthyridine-3-carboxylic acid (**10**, R = CO_2H) gave 5,7-dimethyl-1,5-naphthyridin-4(1H)-one (**10**, R = H) (substrate, Cu powder, 260°C, 10 mmHg: 50% as sublimate).[1255]

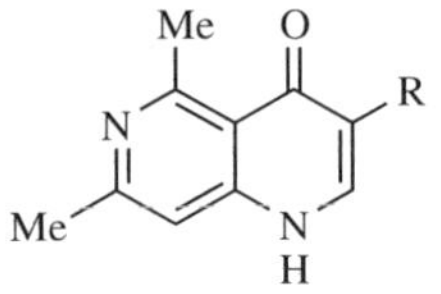

(**10**)

8-Nitro-4-oxo-1,4-dihydro-1,6-naphthyridine-3-carboxylic acid (**11**, R = CO_2H) gave 8-nitro-1,6-naphthyridin-4 (1H)-one (**11**, R = H) (substrate, quinoline, 190°C, N_2, 25 min: 77%).[48]

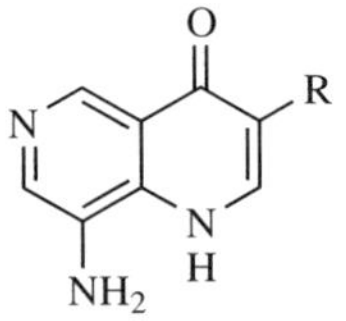

(**11**)

Also other examples.[180,391,1166]

Conversion to Carboxamides

1,6-Naphthyridine-2-carboxylic acid (**12**, R = H) gave 1,6-naphthyridine-2-carbanilide (**12**, R = NHPh) [$PhNH_2$, 1-hydroxybenzotriazole hydrate, 1-(3-dimethylaminopropyl)-3-ethyl-carbodimide, Me_2NCHO, no details];[244] many analogs were made similarly, some with cytomegalovirus activities.[249,318,1411]

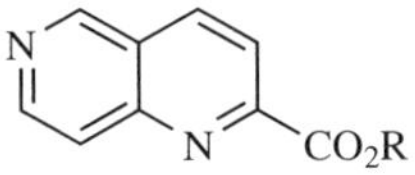

(**12**)

Esterification

2-Amino-1,6-naphthyridine-3-carboxylic acid (**13**, R = H gave methyl 2-amino-1,6-naphthyridine-3-carboxylate (**13**, R = Me) (substrate, H_2SO_4; H_2SO_4/MeOH in dropwise; reflux, 3 h with additional reagent every 45 min: 70%).[247]

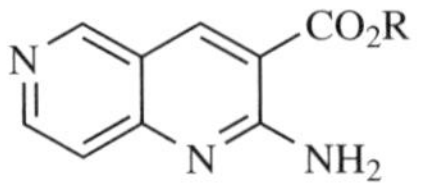

(**13**)

14.2. 1,6-NAPHTHYRIDINECARBOXYLIC ESTERS

The available information on these esters is summarized in the following paragraphs.

1,6-Naphthyridinecarboxylic Esters—Preparation

Note: Most of these esters have been made by *primary synthesis* (see Chapter 8); others, by the *Reissert reaction* (see Section 9.1.3), *esterification* (see Section 14.1.2), or by *passenger introduction*.

1,6-Naphthyridinecarboxylic Esters—Reactions

Note: The *hydrolysis of esters* has been covered in Section 14.1.1; other reactions are exemplified here.

Methyl 2-amino-1,6-naphthyridine-3-carboxylate (**13**, R = Me) underwent *aminolysis* to give 2-amino-1,6-naphthyridine-3-carbohydrazide (**14**, R = NH$_2$) (neat H$_2$NNH$_2$ · H$_2$O, reflux, 30 min: 83%) or 2-amino-*N*-amidino 1,6-naphthyridine-3-carboxamide [**14**, R = C(=NH)NH$_2$] [HN=C(NH$_2$)$_2$, MeOH, reflux, 1 h: 80%];[247] another example of such amide formation.[108,946]

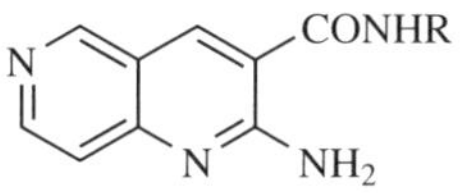

(**14**)

8-(2-Ethoxycarbonylethyl)-5-methoxy-1,2,3,4-tetrahydro-1,6-naphthyridine (**15**) underwent *cyclization* to give 1-methoxy-4,5,9,10-tetrahydro-6*H*, 8*H*-pyrido[3,2,1-*ij*] [1,6]-naphthyridin-6-one (**16**) (MeONa, MeOH, reflux, 30 min: 95%).[1042]

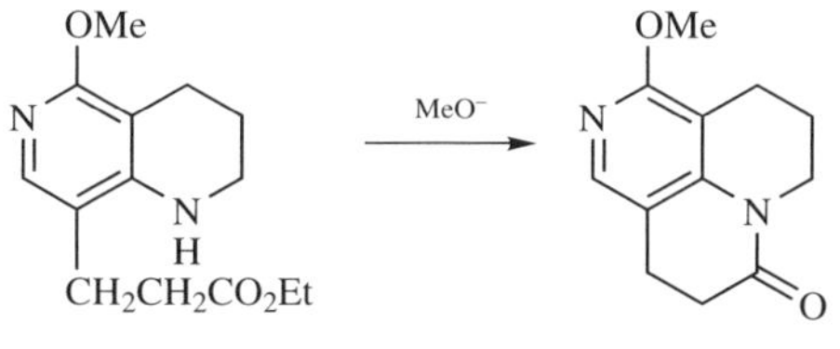

(**15**)   (**16**)

14.3. 1,6-NAPHTHYRIDINECARBOXAMIDES

The available information on these carboxamides, carbohydrazides, and the like are summarized in the following paragraphs. Some do show bioactivity. e.g.,[1437]

1,6-Naphthyridinecarboxamides—Preparation

Note: Most such carboxamides have been made directly by *primary synthesis* (see Chapter 8); others from *1,6-naphthyridinecarboxylic acids or esters* (see Sections 14.1.2 and 14.2, respectively or by minor routes exemplified here.

2-Methylamino-1,6-naphthyridine-3-carbonitrile (**18**, R = Me) underwent limited *hydrolysis* of its cyano group to afford 2-methylamino-1,6-naphthyridine-3-carboxamide (**17**) (KOH, H$_2$O, EtOH, reflux, 5 min: 79%); the related substrate, 2-amino-1,6-naphthyridine-3-carbonitrile (**18**, R = H), underwent *thiolysis* of the cyano group to afford 2-amino-1,6-naphthyridine-3-carbothioamide (**19**) (substrate, NEt$_3$, pyridine, 20°C, H$_2$S in, 2 h: 97%) or *hydrazinolysis* of the cyano group to afford 2,N diamino -1,6-naphthyridine-3-carboxamidine (**20**) (H$_2$NNH$_2$.H$_2$O, reflux, 5 min: 65%).[247]

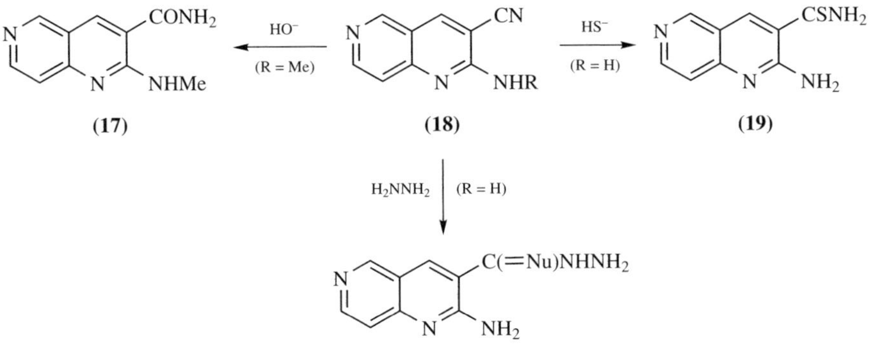

5-Methyl-2-oxo-1,2-dihydro-1,6-naphthyridine 3-carboxamide (**21**, R = H) underwent *transamination* of the amide group to furnish the corresponding 3-carbohydrazide (**21**, R = NH$_2$) (H$_2$NNH$_2$.H$_2$O, 95°C, 18 h: 80%).[1166]

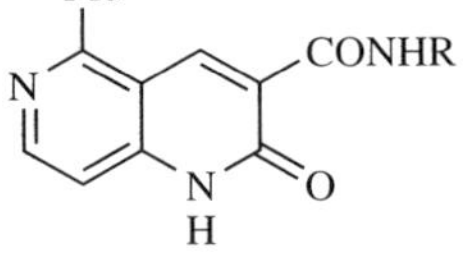

1,6-Naphthyridinecarboxamides—Reactions

Note: Apart from N-alkylations,[1434] an amide has been reported to undergo *intramolecular cyclization*.

8-(2-Carbamoylethyl)-6-methyl-3,4,5,6,7, 8-hexahydro-1,6-naphthyridin-2(1H)-one (**22**) underwent reductive cyclization to 2-methylperhydropyrido

[3,4,5-*ij*]quinolizine (**23**) (copper chromite, dioxane, H$_2$, 125 atm, 250°C, 1 h: 86%).[127]

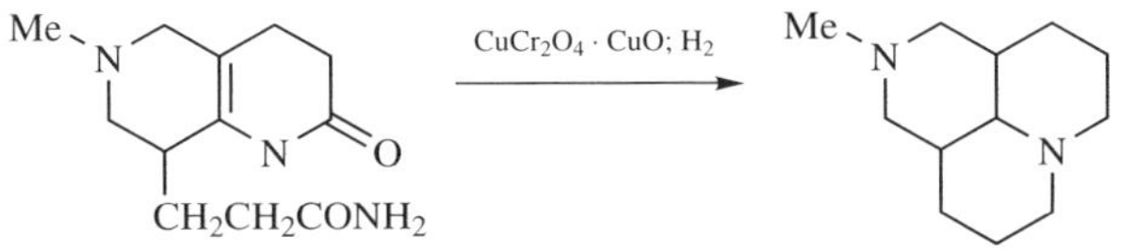

14.4. 1,6-NAPHTHYRIDINECARBONITRILES, CARBALDEHYDES, AND KETONES

Available information on these entities is summarized here.

1,6-Naphthyridinecarbonitriles

Preparation. Most such nitriles have been made by *primary synthesis* (see Chapter 8). However, 1,6-naphthyridine *N*-oxides undergo *oxidative cyanation*, as illustrated here. 1,6-Naphthyridine 1-oxide (**24**) gave 1,6-naphthyridine-2-carbonitrile 1-oxide (**25**) [substrate, K$_3$Fe(CN)$_6$, KCN, H$_2$O, 0°C, 3 h: 58%]; 1,6-naphthyridine-5-carbonitrile 6-oxide (29%) and a separable mixture of 1,6-naphthyridine-2-carbonitrile 1,6-dioxide (30%), 1,6-naphthyridine 5-carbonitrile 1,6-dioxide (14%), and 1,6-naphthyridine-2,5-dicarbonitrile 1,6-dioxide (17%) were made similarly.[224,1317]

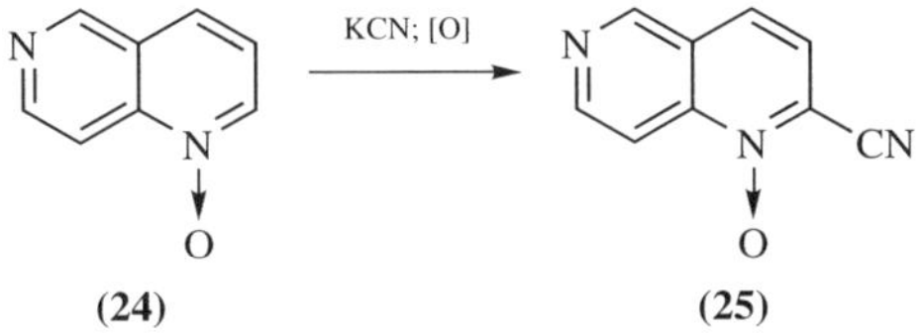

Reactions. These nitriles have been hydrolyzed to carboxylic acids or carboxamides (see Sections 14.1.1 and 14.3, respectively).

1,6-Naphthyridinecarbaldehydes and Ketones

Preparation. Both types of derivative have been prepared by *primary synthesis* (see Chapter 8) or by *N-acylation* (see Section 13.2.2).

Reactions. A case of *C-deacylation* appears to be the only reported reaction of these derivatives. A Ru complex has been made.[1262] 3-Acetyl-1,5-dibenzyl-

(**26**, R = Ac) gave 1,5-dibenzyl-4,7-dimethyl-1,6-naphthyridine-2,5(1*H*, 6*H*)-dione (**26**, R = H) (KOH, H$_2$O, EtOH, reflux, 7 h: 93%).[1014]

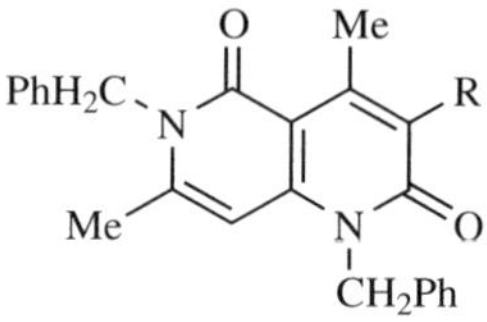

(**26**)

CHAPTER 15

Primary Syntheses of 1,7-Naphthyridines

The general approaches to primary synthesis of 1,7-naphthyridines have been much more limited than those for 1,5- or 1,6-naphthyridines. Thus 1,7-naphthyridines have been made by cyclization of a single aliphatic substrate, by cyclization of an appropriately substituted pyridine substrate, by condensation of a pridine substrate with a synthon that provides one or more ring atoms in the product, or from a small variety of other heterocyclic precursors by rearrangement or other elaborative processes. A Meillard reaction between acetyllysine and glucose gave 1,7-naphthyridines.[942]

Some previous reviews contain limited information and/or examples of such methods of synthesis.[49–53,57,58,61,210,265,407,897,1260,1357,1430,1432]

15.1. FROM A SINGLE ALIPHATIC SUBSTRATE

A mixture if *cis*- and *trans*-isomers of diethyl 2-amino-3-(2-aminoethyl)hexanedioate (**1**) underwent thermal cyclization to give a separable mixture of *cis*- and *trans*-3,4,4a,5,6,8a-hexahydro-1,7-naphthyridine-2,8(1*H*,7*H*)-dione (**2**) (by sublimation at ∼200°C/0.1 mmHg: 69%).[1087]

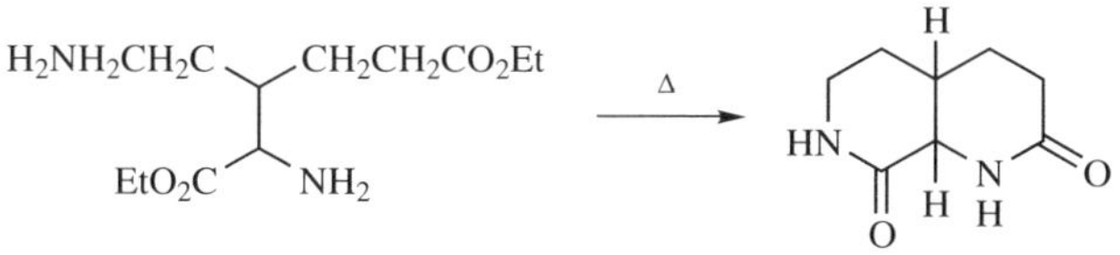

15.2. FROM A SINGLE PYRIDINE SUBSTRATE

An appropriately substituted pyridine substrate may be cyclized to a 1,7-naphthyridine by completion of any one of the 10 peripheral bonds in the product.

The Naphthyridines: The Chemistry of Heterocyclic Compounds, Volume 63, by D.J. Brown
Copyright © 2008 John Wiley & Sons, Inc.

Seven such possibilities have been used in in the literature, and the examples that follow are classified according to which bond is formed during the synthesis.

By Completion of the 1,2-Bond

4-Ethoxalylmethyl-2-methoxy-5-nitropyridine (3, R = H) gave 3-hydroxy-6-methoxy-3,4-dihydro-1,7-naphthyridin-2(1*H*)-one (**4**, R = H) (PtO_2, EtOH, H_2, 3 atm, 2 h: 75%),[152,234] that underwent subsequent dehydration to afford 6-methoxy-1,7-naphthyridin-2(1*H*)-one (TsCl, pyridine, 150°C, sealed?, 4 h: 79%);[234] similarly, 4-α-ethoxybenzyl-2-methoxy-5-nitropyridine (**3**, R=Ph) gave 3-hydroxy-6-methoxy-4-phenyl-3,4-dihydro-1,7-naphthyridin-4(1*H*)-one (**4**, R = Ph) (34%).[245]

(**3**) (**4**)

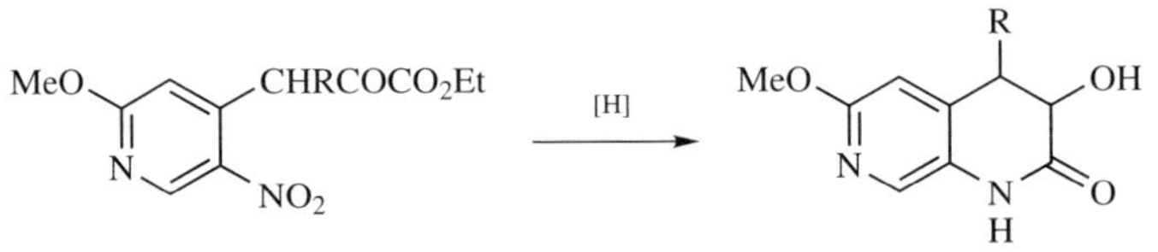

4-(2-Ethoxycarbonylvinyl)-2,6-dimethyl-3-pyridinamine (**5**) gave 6,8-dimethyl-1,7-naphthyridin-2(1*H*)-one (**6**) (EtONa, EtOH, reflux, 1 h: 78%).[1044]

(**5**) (**6**)

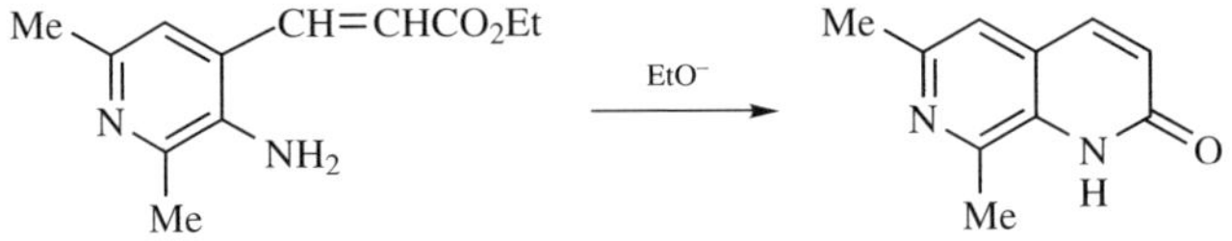

3-(2-Cyanobut-3-enyl)-2-pyridinecarbonitrile (**7**) underwent addition of piperidine and cyclization to give 5-allyl-8-piperidino-1,7-naphthyridin-6-amine (**8**) [$LiN(CH_2)_5$, (made *in situ*), THF, substrate in, −70°C, 30 min, then −20°C, 2 h: 98%]; analogs likewise.[1335]

(**7**) (**8**)

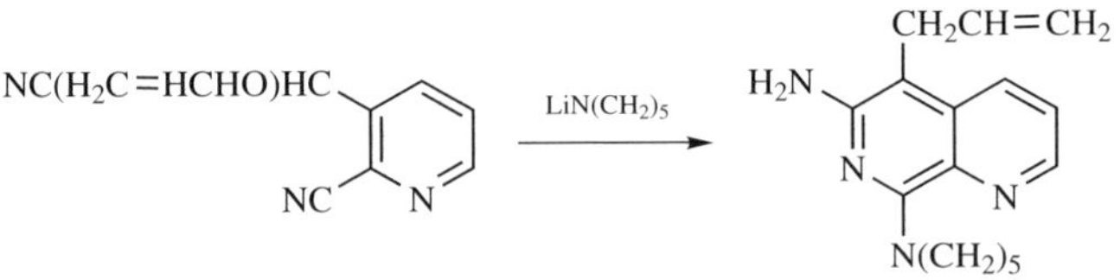

Also other examples.[100,165,477,1271]

By Completion of the 1,8a-Bond

Note: Only a fused 1,7-naphthyridine derivative has been made in this way.[604]

By Completion of the 3,4-Bond

3-(α-Benzylaminophenethyliden)amino-4-pyridinecarbonitrile (**9**) gave 2-diethylamino-3-phenyl-1,7-naphthyridin-4-amine (**10**) (BuONa, BuOH, reflux, 2.5. h: 60%); analogs likewise.[1141]

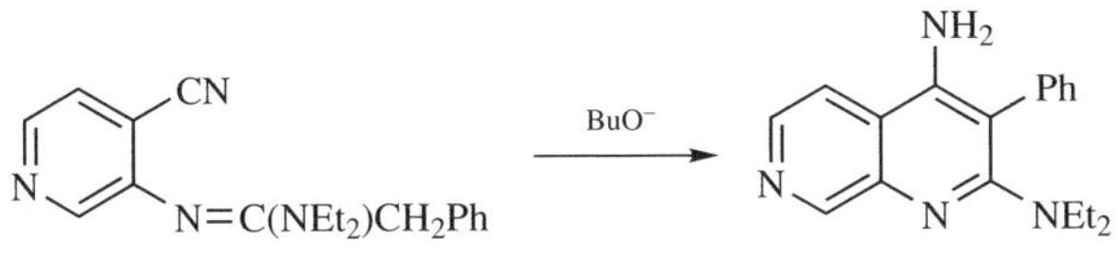

By Completion of the 4,4a-Bond

3-(2,2-Diethoxycarbonylvinyl)amino-1-methyl-2(1H)-pyridinone (**11**) gave ethyl 7-methyl-4,8-dioxo-1,4,7,8-tetrahydro-1,7-naphthyridine-3-carboxylate (**12**) (Dowtherm A, reflux, N_2, 25 min: 83%).[1116]

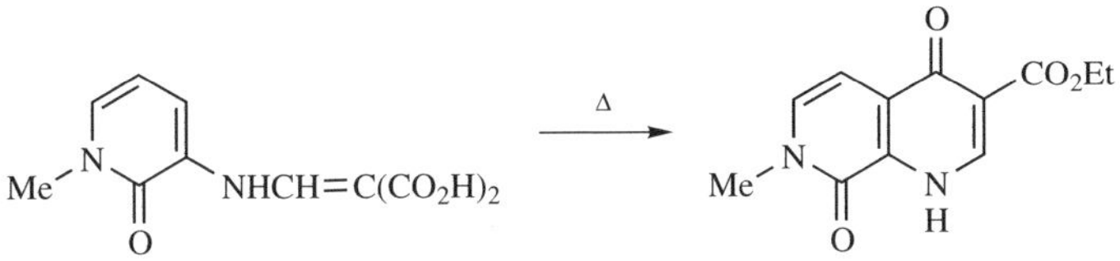

Also other analogous examples,[93,258,865,1130] including the formation of fused 1,7-naphthyridines.[828]

By Completion of the 5,6-Bond

Ethyl 2-(N-ethoxycarbonylmethyl-N-tosylaminomethyl)-3-pyridinecarboxylate (**13**) gave ethyl 5-oxo-1,5-dihydro-1,7-naphthyridine-6-carboxylate (**14**) (EtONa, EtOH, 95°C, 5 h: 84%).[744]

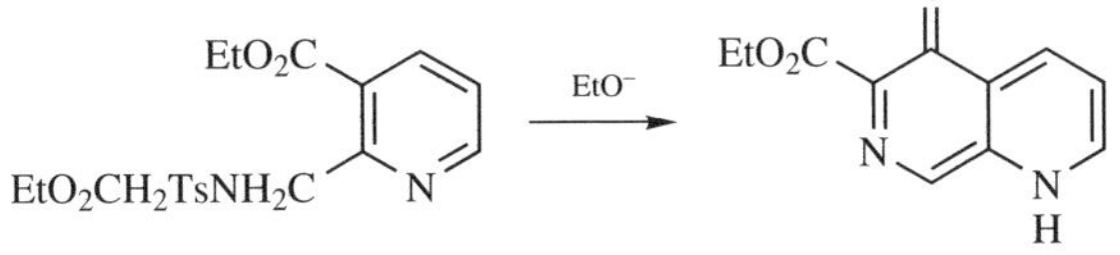

The analogous substrate, ethyl 2-(N-benzyl-N-ethoxycarbonylmethylaminomethyl)-3-pyridinecarboxylate (**15**), gave only a nucleus-reduced product, ethyl 7-benzyl-5-hydroxy-7,8-dihydro-1,7-naphthyridine-6-carboxylate (**16**)

(EtONa, PhH, 95°C, 4 h: 66%).[1307]

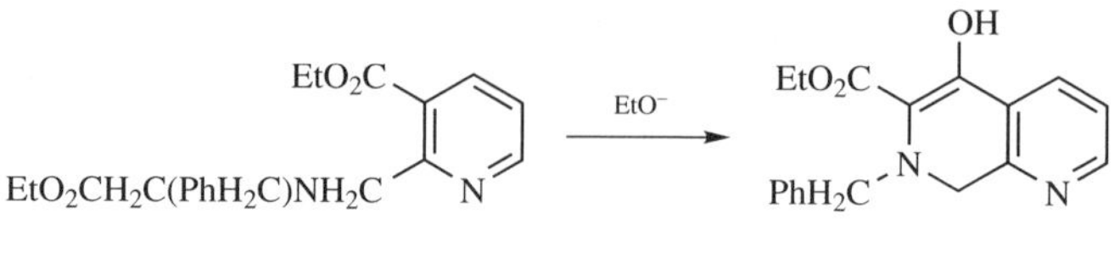

(15) **(16)**

Also other examples.[593]

By Completion of the 6,7-Bond

4,6-Dimethyl-3-phenylethynyl-2-pyridinecarboxamide (**17**) gave 2,4-dimethyl-6-phenyl-1,7-naphthyridin-8(7*H*)-one (**18**) (EtONa, EtOH, reflux, 3 h: 72%).[1055]

(17) **(18)**

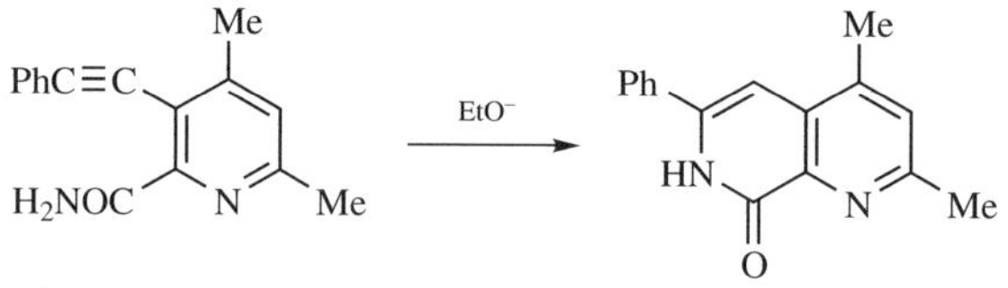

3-Cyanomethyl-2-pyridinecarbonitrile (**19**) gave 8-bromo-1,7-naphthyridin-6-amine (**20**) (substrate, HBr/AcOH, 30°C; then 20°C, 1 h: 74%).[198,cf. 187]

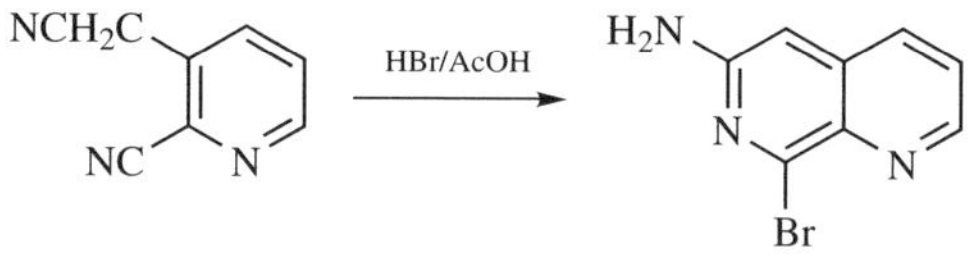

(19) **(20)**

Also other examples.[339,710,747,1451]

By Completion of the 7,8-Bond

Aminolysis of 3-[α-chloro-α-(cyclohexylimino)acethyl]-2-pyridinecarbonyl chloride (**21**) is reported to afford 7-cyclohexyl-6-phenylimino-6,7-dihydro-1,7-naphthyridine-5,8-quinone (**22**) (substrate, Et_2O, 5°C, $PhNH_2$ in slowly at 20°C; 12 h: 45%; note that the amine does not contribute to the ring atoms of the product).[731]

(21) **(22)**

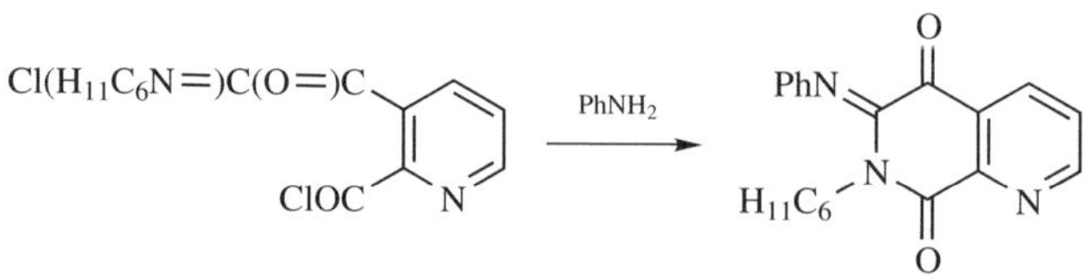

By Completion of the 8,8a-Bond

Note: Efforts to cyclize 2-[(2,2-diethoxyethyl)iminomethyl]pyridine (**23**) to 1,7-naphthyridine (**24**) failed when using a variety of reagents and conditions.[46]

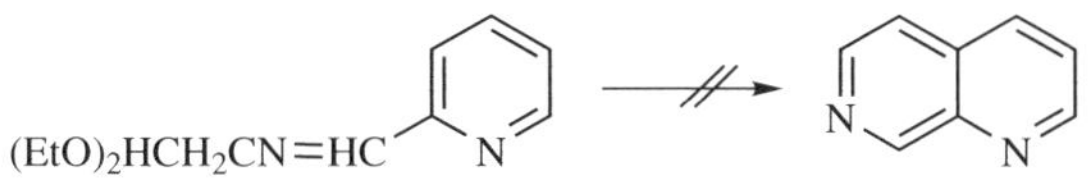

15.3. FROM A PYRIDINE SUBSTRATE WITH A SYNTHON

A pyridine substrate may be converted into a 1,7-naphthyridine by condensation with an aliphatic synthon that provides one or more adjacent ring atoms in the product. Of the many possibilities, only seven types have been used in the literature, as illustrated in the following classified examples.

Provision of N1 by the Synthon

1-Benzyl-4-[2-(1,3-dioxolan-2-yl)ethyl]piperidin-3-one (**25**) with hydroxylamine hydrochloride gave 7-benzyl-5,6,7,8-tetrahydro-1,7-naphthyridine (**26**) (reactants, EtOH, 70°C, 16 h: 34%).[634]

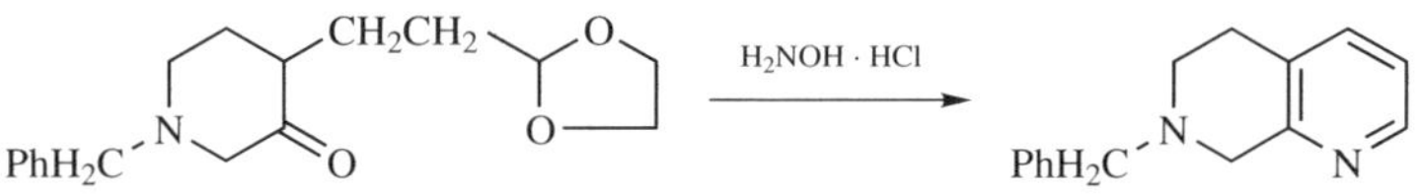

Also other examples.[178]

Provision of C2 by the Synthon

3-*tert*-Butoxycarbonylamino-4-cyanoacetyl-6-fluoropyriidne (**27**) with dimethylformamide dimethylacetal gave 6-fluoro-4-oxo-1,4-dihydro-1,7-naphthyridine-3-carbonitrile (**28**) (neat reactants, 20°C, 5 h: 85%).[355]

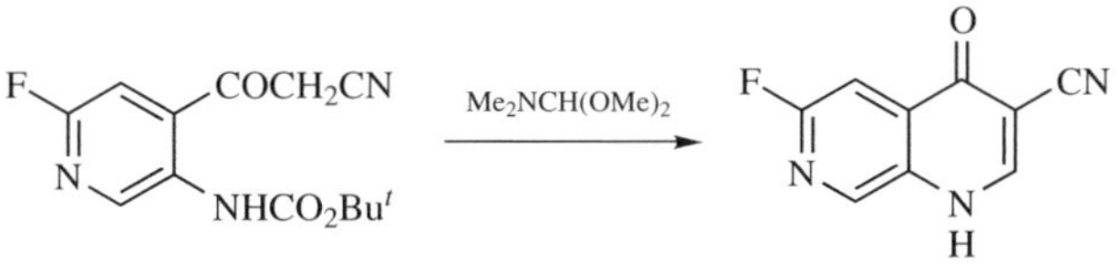

Provisions of C2 + C3 by the Synthon

3-Amino-4-pyridinecarbaldehyde (29) reacted with acetone to give 2-methyl-1,7-naphthyridines (30, R = Me) (reactants, NaOH, EtOH, reflux, 1 h: 80%), with acetophenone to give 2-phenyl-1,7-naphthyridine (30, R = Ph) (likewise, 24 h: 80%), or with other such synthons to give appropriate analogs;[1103] the Schiff base, 4-p-tolyliminomethyl-3-pyridinamine, has also been used as a substrate to give 2-phenyl-1,7-naphthyridine (30, R = Ph) (BzMe, AcOH, H_2O, EtOH, reflux, 8 h: 84%).[100]

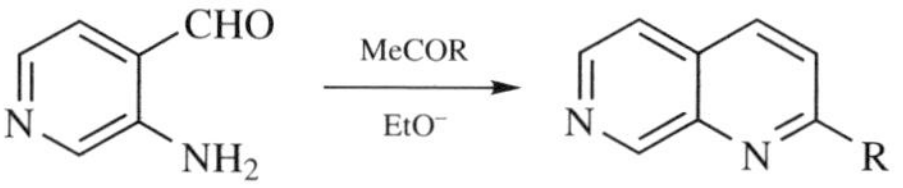

3-Amino-4-pyridinecarbonitrile (31) with diethyl malonate gave ethyl 4-amino-2-oxo-1,2-dihydro-1,7-naphthyridine-3-carboxylate (32) (reactants, EtONa, EtOH, reflux, 5.5 h: 71%).[725]

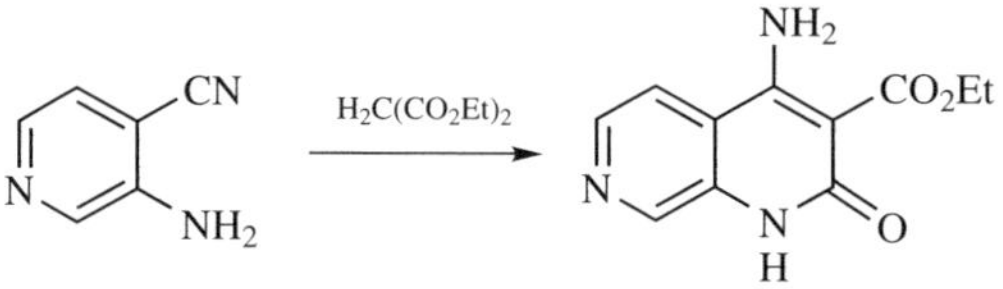

Also other examples.[197,266,575,1302]

Provision of N1 + C2 + C3 by the Synthon

3-Bromo-4-pyridinecarbaldehyde (33) with phenylacetamide gave 3-phenyl-1,7-naphthyridin-2(1H)-one (34) [reactants, tris(dibenzylideneacetone)dipalladium, 4,5-bis(diphenylphosphino)-9,9-dimethylxanthene, Cs_2CO_3, PhMe, N_2, 100°C, 18 h: 75%].[117]

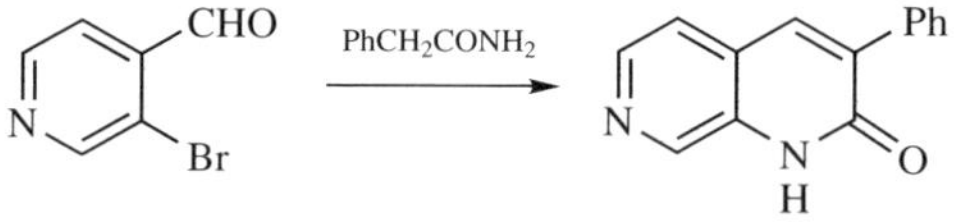

Provision of C2 + C3 + C4 by the Synthon

2-Phenylthio-3-pyridinamine (35) with diethyl α-ethoxymethylenemalonate gave ethyl 4-oxo-8-phenylthio-1,4-dihydro-1,7-naphthyridine-3-carboxylate

(**36**) [neat reactants, 120°C, 2 h: then poly(phosphoric acid) in 150°C, 1 h: 35%].[157]

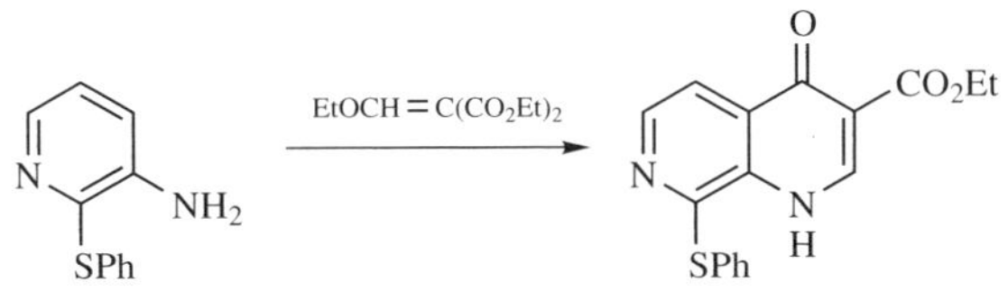

3-Amino-2(1H)-pyridinone (**37**) underwent a modified Skraup reaction to give 1,7-naphthyridin-8(7H)-one (**38**) (see original for considerable detail: (20%).[1048]

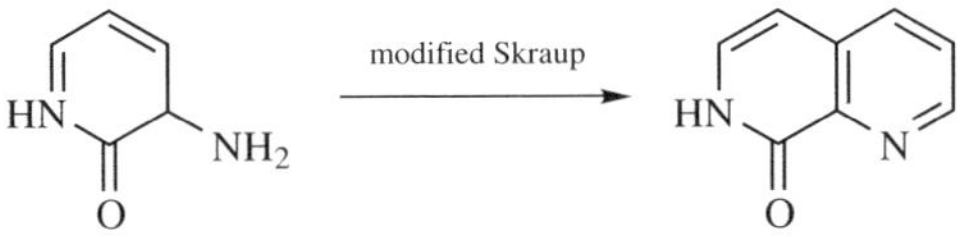

Also other examples.[165]

Provision of C6 + N7 + C8 by the Synthon

3-Acetyl-2-chlorophyridine (**39**, Q = Ac) with *N*-benzylidenebenzylamine (as its anion) gave 5-methyl-6,8-diphenyl-1,7-naphthyridine (**40**, R = Me) (reactants, LiNPr$_2^i$: 52%; see original for the complicated procedure); likewise, 2-chloro-3-pyridinecarbonitrile (**39**, Q = CN) gave 6,8-diphenyl-1,7-naphthyridin-5-amine (**40**, R = NH$_2$) (72%) and ethyl 2-chloro-3-pyridinecarboxylate (**39**, Q = CO$_2$Et) gave 6,8-diphenyl-1,7-naphthyridin-5-ol (**40**, R = OH) [65%; probably better formulated as 5,8-diphenyl-1,7-naphthyridin-5(1H)-one]; several homologs were made similarly.[1146]

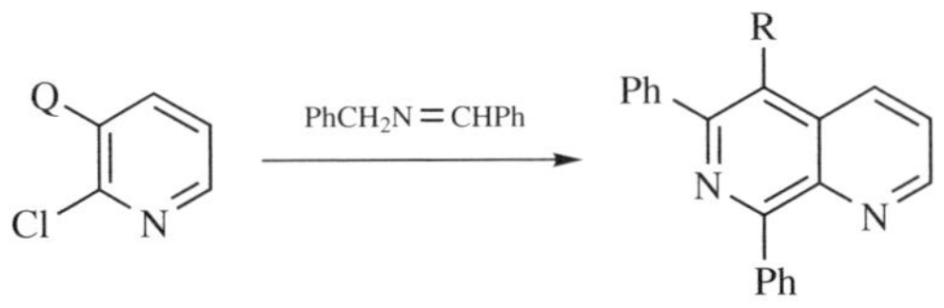

Provision of N7 by the Synthon

3-Phenylethynyl-2-pyridinecarbaldehyde (**41**, R = Ph) with ammonia gave 6-phenyl-1,7-naphthyridine (**42**, R = Ph) (NH$_3$/EtOH, 80°C, sealed, 2 h: 31%); 3-(oct-1-ynyl)-2-pyridinecarbaldehyde (**41**, R = C$_6$H$_{13}$) likewise gave

6-hexyl-1,7-naphthyridine (**42**, R = C$_6$H$_{13}$) (59%).[621] A somewhat similar procedure gave 1,7-naphthyridine.[197]

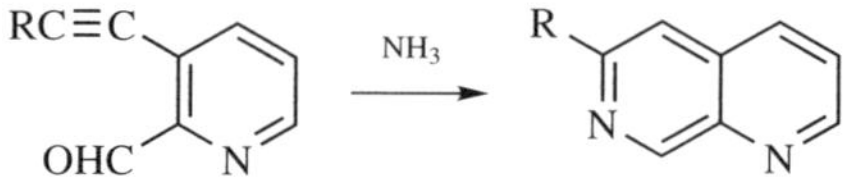

Provision of C8 by the Synthon

3-[2-Ethoxycarbonyl-2-(triphenylphosphoranylidenamino)vinyl]pyridine (**43**) with phenyl isocyanate gave a separable mixture of ethyl 8-anilino-1,7-naphthyridine-6-carboxylate (**44**) and an isomeric 2,6-naphthyridine derivative (formed by an alternative cyclocondensation) (reactants, PhMe, 20°C, 2 h: then 180°C, sealed, 72 h: 23% and 22%, respectively, after separation); analogs likewise.[442]

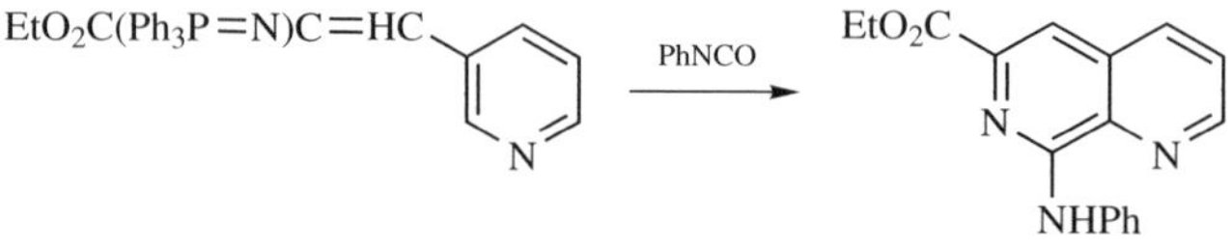

15.4. FROM OTHER HETEROCYCLIC SYSTEMS

Only a few heterocyclic systems other than pyridine have been used for the synthesis of 1,7-naphthyridines or their fused[161] derivatives; even fewer such procedures have appreciable synthetic potential. Examples follow.

3,10-Diazatricyclo[5.3.1.0^{3,8}]undecane as Substrates

1-Phenyl-3,10-diazatricyclo[5.3.1.0^{3,8}]undecane-2,9-dione with methanesulfonic acid gave methyl 8-oxo-6-phenyldecahydro-1,7-naphthyridine-6-carboxylate (**46**) (reactants, MeOH, reflux, 12 h: 97%).[124,195]

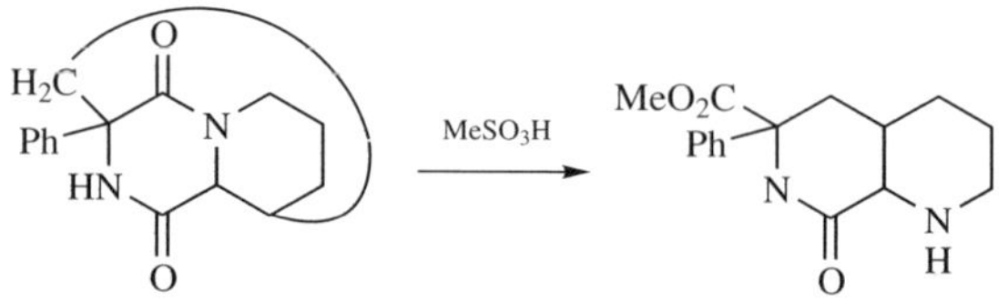

Cyclopenta[*b*]pyridines (Pyrindines) as Substrates

Note: A useful route to 1,7-naphthyridine (48%) and its 8-alkyl derivatives is offered by this synthesis; see Section 8.5.

Pyrano[3,4-*b*]pyridines as Substrates

N-(3,5-Bis(trifluoromethyl)benzyl]-8-oxo-5-*p*-totyl-8*H*-pyrano[3,4-*b*]pyridine-6-carboxamide (**47**, X=O) and ethanolamine gave *N*-[3,5-bis(trifluoromethyl)benzyl]-7-(2-hydroxyethyl)-8-oxo-5-*p*-totyl-7,8-dihydro-1,7-naphthyridine-6-carboxamide (**47**, X = NCH$_2$CH$_2$OH) (reactants, MeOH, THF, 20°C, 16 h; then to dryness; solid, diazabicycloundecane, MeCN, PhMe, reflux, 1 h: 70%); analogs likewise.[243,944, cf. 556,1252]

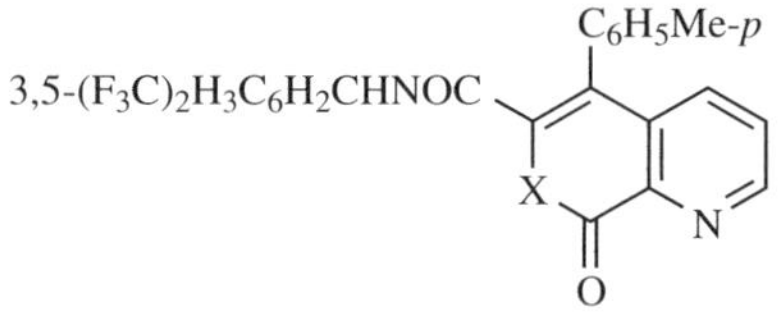

(**47**)

Pyrazines as Substrates

1-Benzyl-6-[*N*-(but-3-ynyl)-*N*-propionylaminomethyl]-5-chloro-3-methoxy-2(1*H*)-pyrazinone (**48**) underwent thermolysis to afford 1-benzyl-3-methoxy-7-propionyl-5,6,7,8-tetrahydro-1,7-naphthyridin-2(1*H*)-one (**49**) (PhBr, reflux, 2 h: 53%, after separation from a second product); the proposed intermediate was not isolated.[602]

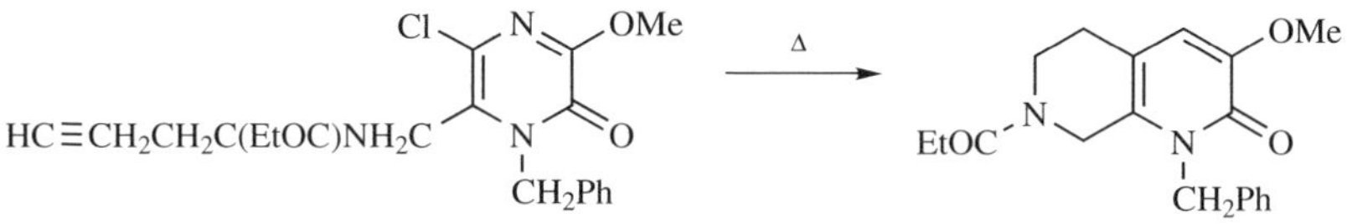

(**48**)   (**49**)

Pyrrolo[3,4-*b*]pyridines as Substrates

Note: This ambiguous but useful synthesis[1048,1221] has been exemplified in Section 8.5.

CHAPTER 16

1,7-Naphthyridine, Alkyl-1, 7-Naphthyridines, and Aryl-1, 7-naphthyridines

The scope of this chapter is similar to that on the corresponding 1,5-naphthyridines; see the introduction to Chapter 3.

The general chemistry of 1,7-naphthyridines has been reviewed from time to time but usually somewhat briefly.[49–53,55–58,61,265,407,417,897,1260,1357,1430,1432]

16.1. 1,7-NAPHTHYRIDINE

This section also includes information on the unsubstituted hydro-1,7-naphthyridines.

16.1.1. Preparation of 1,7-Naphthyridine

1,7-Naphthyridine has been made[197] by a *primary synthesis* (see Section 15.3), but the best preparative route is probably the first of the following indirect examples.

4-Hydrazino-1,7-naphthyridine (**1**) gave 1,7-naphthyridine (**2**) ($CuSO_4$, H_2O, AcOH, 95°C, 10 min: 67%[1297] or $CuSO_4$, H_2O, kieselguhr, reflux, 16 h: 60%);[45] similar treatment of 8-hydrazino- or 6,8-dihydrazino-1,7-naphthyridine afforded the same product (**2**) but in poor yield.[155,187]

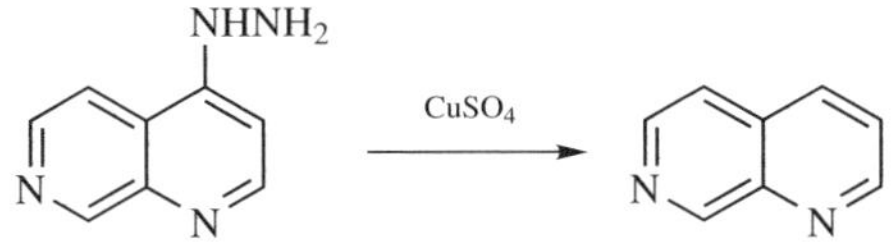

The Naphthyridines: The Chemistry of Heterocyclic Compounds, Volume 63, by D.J. Brown
Copyright © 2008 John Wiley & Sons, Inc.

8-Tosylhydrazino-1,7-naphthyridine also afforded the parent compound (**2**) (Na_2CO_3, H_2O, 100°C, 3 h: 66%).[42]

7-Benzyl-5,6,7,8-tetrahydro-6,7-naphthyridine (**3**, R = CH_2Ph) underwent reductive debenzylation to give 5,6,7,8-tetrahydro-1,7-naphthyridine (**3**, R = H) (Pd/C, trace HCl, MeOH, H_2, 3 atm, 20°C, 78% as dihydrochloride).[634]

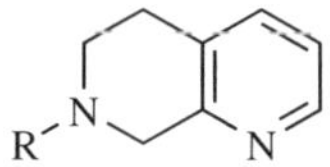

(**3**)

1,7-Naphthyridine (**4**) underwent hydrogenation to give a separable mixture of 1,2,3,4-tetrahydro- (**5**) and 5,6,7,8-tetrahydro-1,7-naphthyridine (**6**) (Pd/C, EtOH, H_2, 1 atm, 20°C, 8 h: 7% and 4%, respectively, after separation as picrates and reformation of the free bases);[47,cf. 1243] successive reduction of the same substrate (**4**) under Bouvault–Blanc conditions and by hydrogenation gave *trans*-decahydro-1,7-naphthyridine (**7**) (EtOH, Na, reflux, 5 h; then crude product, EtOH, PtO_2, H_2, 1 atm, 20°C, 1 h: 67%).[47]

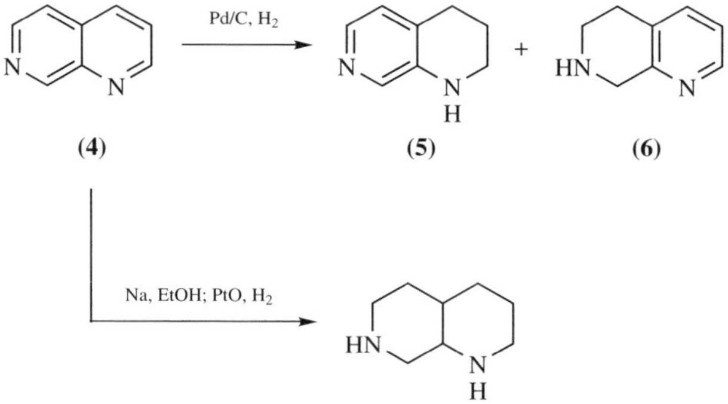

16.1.2. Properties of 1,7-Naphthyridine

Reported physical data on 1,7-naphthyridine can be found from its entry in Table A3. Any more extensive studies on such material are listed here.

Electron Density/Resonance Energy. Such phenomena have been calculated for 1,7-naphthyridine, mainly for comparison with those of related systems, to reveal the effect of ring nitrogen atoms on aromaticity, and for correlation with observed electronic spectra.[676,840,1126,1173]

Electron Spin Resonance. The ESR spectra for anions from 1,7- and related naphthyridines have been studied in some detail.[1083]

Infrared Spectra. The IR spectra of 1,7-naphthyridine derivatives have been compared with those of related molecules.[44,363,1251]

Ionization. Theoretical studies on the ionization of 1,7- and other naphthyridines have been reported.[806,813]

Mass Spectra. The MS of 1,7-naphthyridine have been reported and compared with those of 1,5-, 1,6-, and 1,8-isomers.[975] Those of perhydro derivatives have been studied.[1368]

Nuclear Magnetic Resonance Spectra. The ^{1}HNMR spectra of 1,7-naphthyridine and related molecules have been reported,[42] especially in relation to long-range spin–spin coupling,[1140] the enhancing effect of lanthanide reagents on chemical shifts,[43] the site(s) of covalent hydration,[1170] the nature of adducts formed with sodium amide/ammonia,[302] and the effects of various substituents.[296] The ^{13}CNMR spectra for 1,7- and other naphthyridines have been compared with calculated values[923] and used to measure pK_a values for mono- and diprotonation.[492]

Polarography. The electrochemical reduction of 1,7-naphthyridine and related systems has been examined.[829,1329]

Ultraviolet Spectra. Calculated electronic spectra for 1,7-naphthyridine and related systems have been obtained by several methods and subsequently compared with experimental values.[820,878,1181,1312,1486]

16.1.3. Reactions of 1,7-Naphthyridine

The nuclear *reduction of* 1,7-naphthyridine has been covered in Section 16.1.1. Other reported reactions are illustrated in the examples that follow.

C-Alkylation/Arylation

1,7-Naphthyridine (**8**, R=H) gave 2-phenyl-1,7-naphthyridine (**8**, R=Ph) [PhLi (made *in situ*), Et$_2$O, substrate in, 0°C, 2.5 h, N$_2$: 22%].[1033,cf. 998]

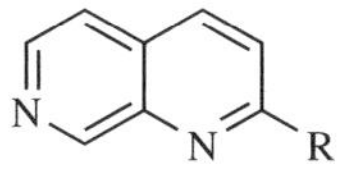

(**8**)

Amination

1,7-Naphthyridine (**9**, X = Y = Z = H) in potassium amide/liquid ammonia with subsequent addition of oxidant gave a separable mixture of 1,7-naphthyridin-2-amine (**9**, X = NH$_2$, Y = Z = H), its 4-isomer (X = Z = H, Y = NH$_2$), and its 8-isomer (**9**, X = Y = H, Z = NH$_2$) (substrate, KNH$_2$/NH$_3$, 10 min; KMnO$_4$ in slowly, 10 min: 25%, 10%, and 19%, respectively, after chromatographic separation),[925,cf. 353,1046] probably via appropriate intermediate adducts such as the dihydroamine (**10**).[54]

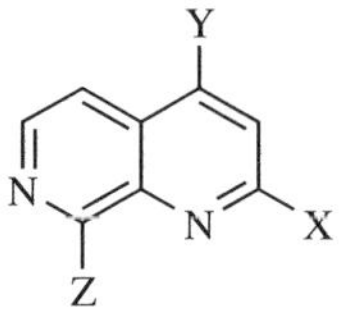

(**9**) 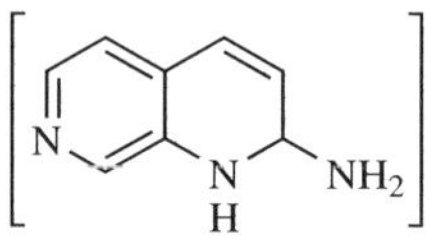   (**10**)

Cycloadditions

1,7-Naphthyridine with chloromethyl phenyl sulfone gave 1,7-bisphenylsulfonyl-6b,7,7a,7b-tetrahydro-1*H*-azirino[1,2-*a*]cyclopropa[*c*][1,7]naphthyridine (**11**) (KOH, Me$_2$SO, reactants in slowly, 20°C, tlc monitored: 77%).[803,cf. 393]

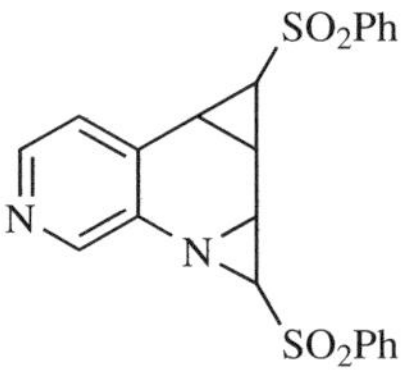

(**11**)

Halogenation

1,7-Naphthyridine (**12**, X=Y=H) underwent bromination to give mainly 3,5-dibromo-1,7-naphthyridine (**12**, X=Y=Br) accompanied by 3- (**12**, X=Br, Y=H) and 5-bromo-1,7-naphthyridine (**12**, X=H, Y=Br) [substrate, as hydrobromide, PhNO$_2$, 175°C, Br$_2$ (2.5 equiv) in slowly, 6 h: 74%, 1%, and 3%, respectively, after separation; use of Br$_2$ (1.1 equiv) still gave predominantly the dibromo product).[985]

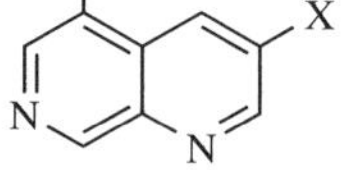

(**12**)

N-Oxidation

1,7-Naphthyridine underwent mono-*N*-oxidation to give a separable mixture of the 1- (**13**) and 7-oxide (**14**) [substrate, *m*-ClC$_5$H$_4$CO$_3$H (1 equiv) CHCl$_3$,

20°C, 8 h: 3% and 45%, respectively] or di-*N*-oxidation to give the dioxide (**15**) [substrate, *m*-ClC₆H₄CO₃H (2 equiv), CHCl₃, reflux, 1 h: 47%].[1212]

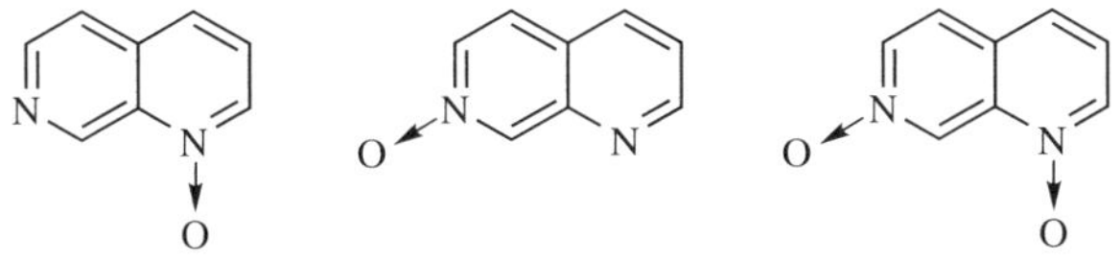

(13)   (14)   (15)

Quaternization

1,7-Naphthyridine gave its 7-methiodide (**16**) (substrate, excess MeI, PhMe, reflux, 12 h: 90%),[1214,cf. 993] or its bismethofluorosulfonate (**17**) (substrate, excess FSO₃Me, initially exothermic, then 95°C, 1 h; more synthon in, 95°C, 30 min: 90%).[1061]

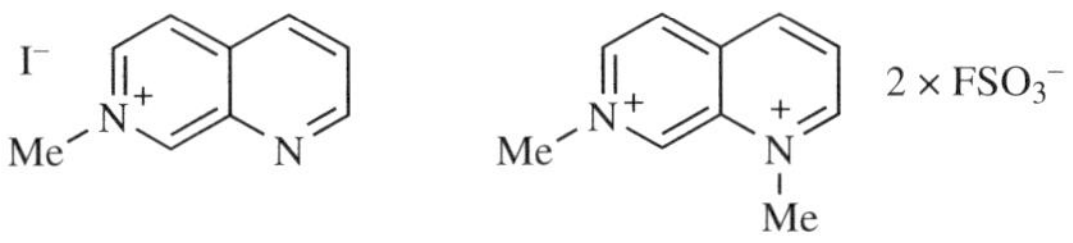

(16)   (17)

Reissert Reactions

1,7-Naphthyridine underwent Reissert addition to give 7-benoyl-7,8-dihydro-1,7-naphthyridin-8-carbonitrile (**18**) (substrate, KCN, H₂O; BzCl in slowly, 20°C, 4 h: 18%).[1033]

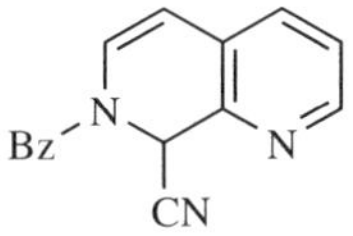

(18)

16.2. ALKYL- AND ARYL-1,7-NAPHTHYRIDINES

This section covers the insertion of alkyl/aryl groups into existing 1,7-naphthyridines and the reactions specific to such groups attached to 1,7-naphthyridines.

16.2.1. Preparation of Alkyl and Aryl-1,7-naphthyridines

In almost all cases, the alkyl/aryl group in such naphthyridines has been put there by *primary synthesis* (see Chapter 15). However, such groups have occasionally been inserted subsequently, as illustrated by the following examples.

By C-Alkylation/Arylation

Note: An example of such an arylation has been given in Section 16.1.3.

By N-Alkylation (Quaternization)

Note: See Section 16.1.3 for an example.

From Halogeno-1,7-naphthyridines

8-Bromo-1,7-naphthyridin-6-amine (**19**, R = Br) gave 8-*m*-nitrophenyl-1,7-naphthyridin-6-amine (**19**, R = $C_6H_4NO_2$-*m*) [substrate, PhMe, Me_2NCHO; $(HO)_2BC_6H_4NO_2$-*m*, bis(dibenzylideneacetone)palladium, Ph_3P, H_2O; reflux, 4 h: 90%); analogs likewise].[198]

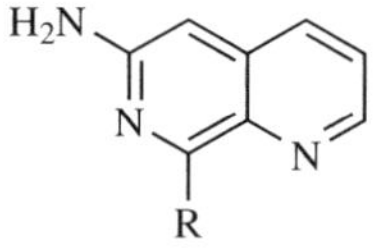

From Trifluoromethanesulfonyloxy-1,7-naphthyridines

8-*m*-Nitrophenyl-6-(trifluoromethanesulfonyloxy)-1,7-naphthyridine (**20**) with 3-(tributylstannyl)pyridine (**21**) gave 8-*m*-nitrophenyl-6-(pyridin-3-yl)-1,7-naphthyridine (**22**) [reactants, bis(dibenzylideneacetone)palladium, Ph_3P, Me_2NCHO, LiCl, 110°C, 16 h: 75%]; analogs likewise.[198]

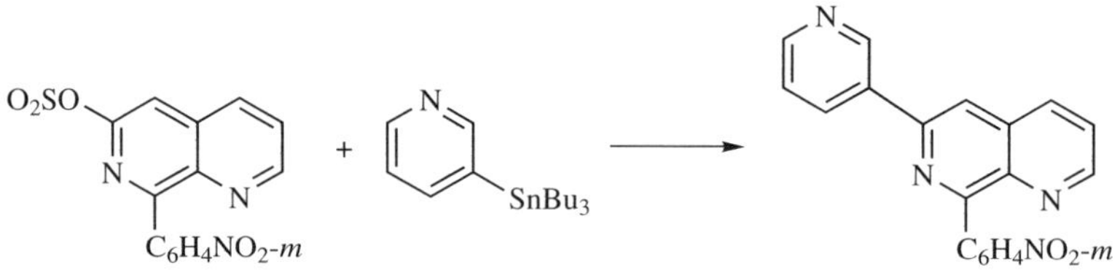

16.2.2. Reactions of Alkyl- and Aryl-1,7-naphthyridines

There appear to be no reported reactions that are specific to C-alkyl/aryl groups attached to 1,7-naphthyridines, but the effect of methiodide formation on the ^{13}CNMR spectrum of 1,7-naphthyridine has been studied by comparison with similar treatment of analogous substrtates.[1319] In addition, several N-debenzylations from hydro-1,7-naphthyridines have been recorded; see the following examples.

7-Benzyl-5,6,7,8-tetrahydro-1,7-naphthyridine (**23**, R = CH$_2$Ph) gave 5,6,7,8-tetrahydro-1,7-naphthyridine (**23**, R = H) (Pd/C, HCl, MeOH, dioxane, H$_2$, 3 atm, 20°C, 6 h: 78% as dihydrochloride).[534] In much the same way, debenzylation of appropriate substrates afforded ethyl 5-hydroxy-5,6,7,8-tetrahydro- (**24**) (77–95%) and the isomeric ethyl 5-hydroxy-1,2,3,4-tetrahydro-1,7-naphthyridine-2-carboxylate (45%).[732,1307]

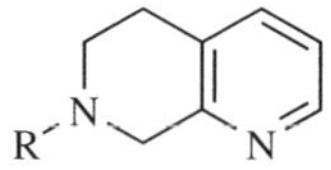

(**23**)

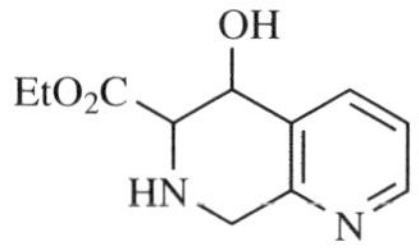

(**24**)

CHAPTER 17

Halogeno-1,7-naphthyridines

Because of the location of ring nitrogen atoms, a halogeno substituent on 1,7-naphthyridine could be expected to be most activated at the 8-position; slightly less so at the 2-, 4-, and 6-positions; and least so at the 3- and 5-positions. This postulate appears to be borne out by the limited aminolytic, alcoholysis, and other such data available to date (see Section 7.2$^{\text{cf. 60}}$).

17.1. PREPARATION OF HALOGENO-1,7-NAPHTHYRIDINES

Some halogeno-1,7-naphthyridines have been made by *primary syntheses* (see Chapter 15), but most have been prepared by indirect routes, illustrated in the following examples.

By Direct Halogenation

Note: The bromination of unsubstituted 1,7-naphthyridine has been summarized in Section 16.1.3.

1,7-Naphthyridin-8-amine (**1**, R = H) underwent halogenation to give 5-bromo- (**1**, R = Br) (substrate, AcOH, 50°C; Br$_2$ in slowly, 95°C, reflux, 2.5 h: 72%) or 4-chloro-1,7-naphthyridin-8-amine (**1**, R = Cl) (substrate, HCl; KClO$_4$/H$_2$O in dropwise, 40°C; then 95°C, 2 h: 57%).[1046]

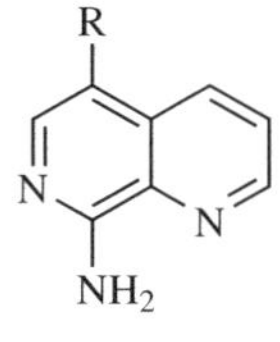

(**1**)

The Naphthyridines: The Chemistry of Heterocyclic Compounds, Volume 63, by D.J. Brown
Copyright © 2008 John Wiley & Sons, Inc.

From 1,7-Naphthyridinones

Note: This route may be used to place a chloro or bromo substituent at the 2-, 4-, 6-, or 8-position of 1,7-naphthyridine.

6-Methoxy-1,7-naphthyridin-2(1H)-one (**2**) gave 2-chloro-6-methoxy-1,7-naphthyridine (**3**) (POCl$_3$, reflux, 12 h: 63%).[234]

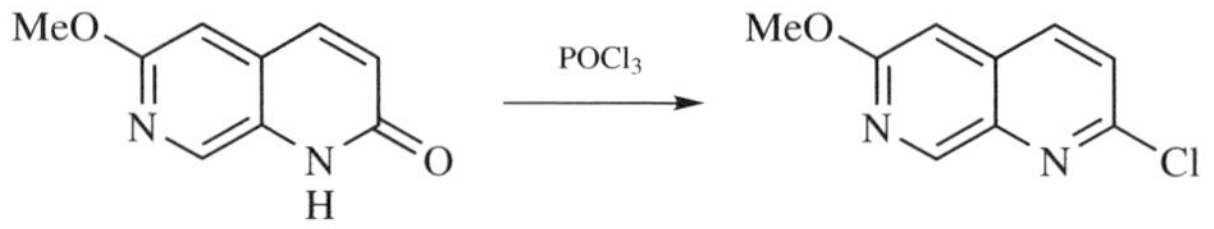

1,7-Naphthyridin-4(1H)-one gave 4-chloro-1,7-naphthyridine (**4**, R = H) (POCl$_3$, 95°C, 90 min: ~60%);[93] 4-chloro-6,8-dimethyl-1,7-naphthyridine (**4**, R = Me) was made somewhat similarly (reflux, 60 min: ~80%).[1098]

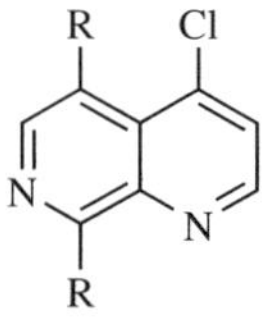

3-Bromo-1,7-naphthyridin-4(1H)-one (**5**) gave 3,4-dibromo-1,7-naphthyridine (**6**) (neat POBr$_3$, 125°C, 3 h: 78%).[844]

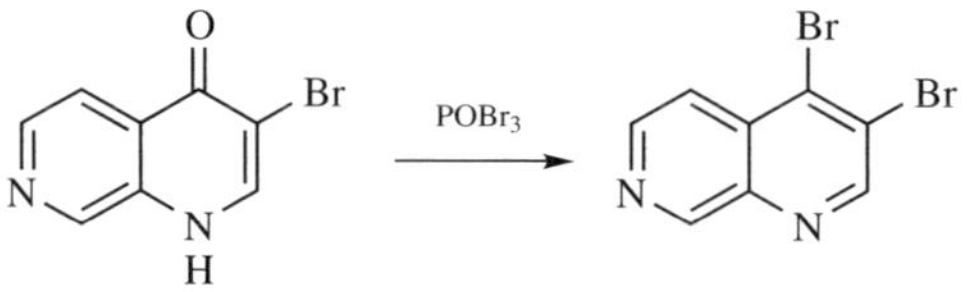

1,7-Naphthyridin-8(7H)-one (**7**, R = H) gave 8-chloro-1,7-naphthyridine (**8**, R = H) (POCl$_3$, reflux, 12 h: 58%;[155] POCl$_3$, reflux, 75 min: 72%);[42] 5-chloro-1,7-naphthyridin-8(7H)-one (**7**, R = Cl) likewise gave 5,8-dichloro-1,7-naphthyridine (**8**, R = Cl) (POCl$_3$, reflux, 3 h: 67%).[1046]

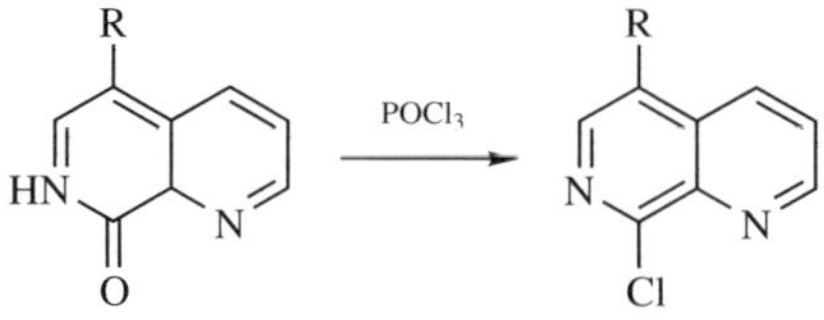

Also other examples.[45,85,266,355,1116]

By the Meissenheimer Reaction from *N*-Oxides

1,7-Naphthyridine 1-oxide (**9**) in refluxing phosphoryl chloride for 1 h gave a mixture of 2-, 4-, and 3-chloro-1,7-naphthyridine (in a ratio of 19 : 12 : 1 by gas chromatography) from which the first two (**10**, X = Cl, Y = Z = H) and (**10**, X = Y = H, Z = Cl) were isolated in low yield.[225]

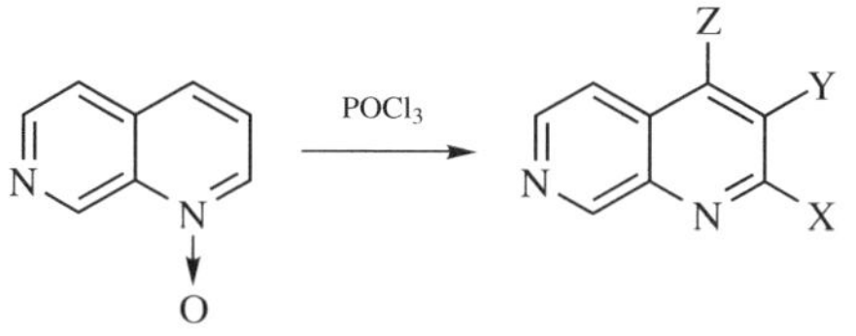

17.2. REACTIONS OF HALOGENO-1,7-NAPHTHYRIDINES

The aminolysis of halogeno-1,7-naphthyridines is fairly well represented in the literature, but other potential reactions have seldom been used. Reported reactions are illustrated in the following examples.

Alcoholysis

4-Chloro-1,7-naphthyridine (**11**) gave 4-methoxy-1,7-naphthyridine (**12**) (MeONa, MeOH, reflux, 8 h: 25%).[225]
Also other examples.[355,445,1130]

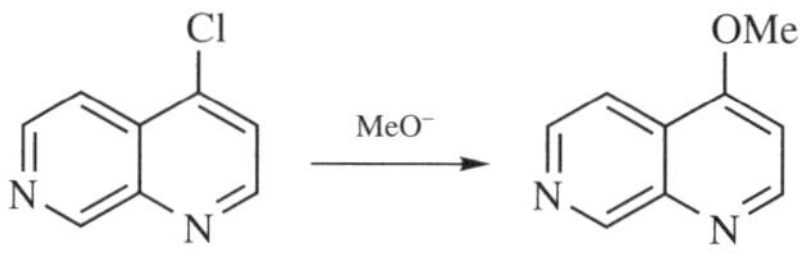

Aminolysis

2-Chloro- (**13**, R = Cl) gave 2-hydrazino-1,7-naphthyridine (**13**, R = NHNH$_2$) (neat H$_2$NNH$_2$. H$_2$O, 100°C, 1 h: 85%).[234]

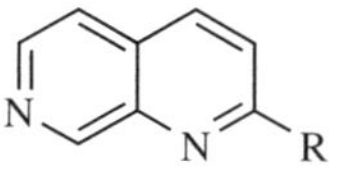

4-Chloro-1,7-naphthyridine with 4-diethylamino-1-methylbutylamine gave 4-(4-diethylamino-1-methylbutylamino)-1,7-naphthyridine (**14**) (neat reactants,

160°C, 6 h: 73%);[175] likewise, 4-chloro-7-methyl-1,7-naphthyridin-8(7H)-one with the same amine gave 4-(4-diethylamino-1-methylbutylamino)-7-methyl-1,7-naphthyridin-8(7H)-one (15) (reacants, Me$_2$NCHO, reflux ?, 8 h: 38%).[1116]

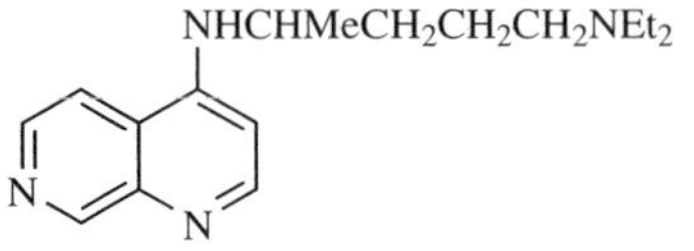

(14) 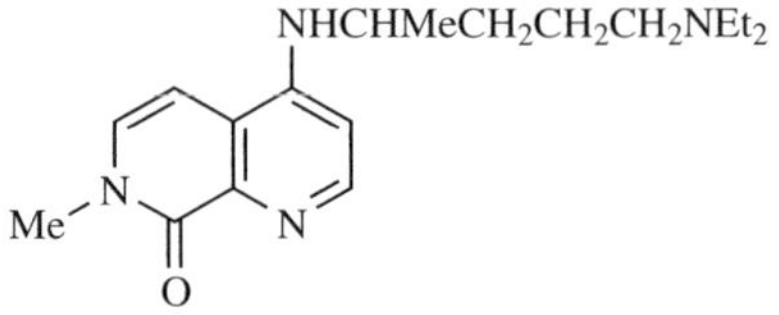 (15)

4-Chloro-1,7-naphthyridine gave 4-hydrazino-1,7-naphthyridine (H$_2$NNH$_2$·H$_2$O, EtOH, reflux, 3 h: 90%).[45,cf. 1297]

Ethyl 5,8-difluoro-4-oxo-1,4-dihydro-1,7-naphthyridine-3-carboxylate (16, R = F) with an excess of morpholine gave ethyl 6-fluoro-8-morpholino-4-oxo-1,4-dihydro-1,7-naphthyridine-3-carboxylate [16, R = N(CH$_2$CH$_2$)$_2$O] (reactants, Me$_2$NCHO, 60°C, 1 h: 83%).[1130]

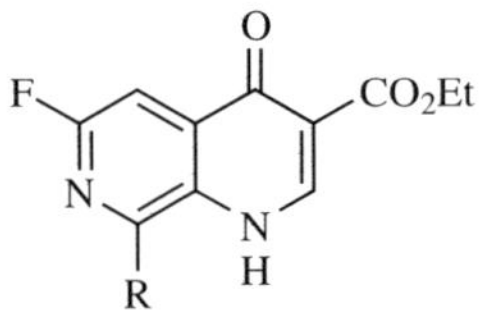

(16)

8-Chloro-1,7-naphthyridine gave 8-hydrazino-1,7-naphthyridine (H$_2$NNH$_2$·H$_2$O, EtOH, reflux, 10 min: 99%).[155]

8-Bromo-1,7-naphthyridin-6-amine with hydrazine hydrate (17) gave 6,8-dihydrazino-1,7-naphthyridine (18) (reactants, dioxane, reflux, 1 h: 65%; note the additional transamination).[187]

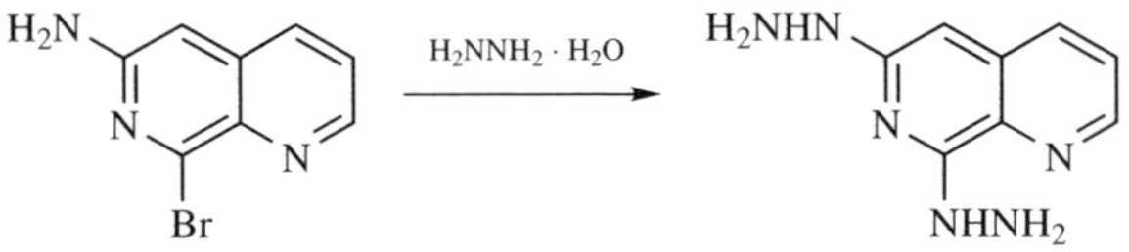

(17) (18)

Also other examples.[265,445,503,1146]

Note: Potassium amide in liquid ammonia can be used for aminolysis of halogeno-1,7-naphthyridines, but it also induces *tele*-aminolysis.[54] For example, 2-chloro-1,7-naphthyridine gave a separable mixture of 1,7-naphthyridin-2-amine and both the 4- and 8-isomers (KNH$_2$, NH$_3$, Et$_2$O, −33°C, 4 h: 70% yield of the mixture) from which each product was isolated in low yield;[834] also analogous examples.[332,808,834]

Dehalogenation

Note: Reductive dehalogenation appears to have been avoided for fear of concomitant nuclear reduction. Accordingly, the indirect route via hydrazino derivatives has been used to make the parent 1,7-naphthyridine (see Section 16.1.1) and other products exemplified here.

3,4-Dibromo-1,7-naphthyridine (**19**) underwent hydrazinolysis to give a mixture of 3-bromo-4-hydrazino- (**20**, R = Br) and 3,4-dihydrazino-1,7-naphthyridine (**20**, R = NHNH$_2$) (H$_2$NNH$_2 \cdot$ H$_2$O, EtOH, 20°C, 24 h: 70%). Oxidation of the crude mixture then gave a separable mixture of 1,7-naphthyridine (**21**, R = H) and its 3-bromo derivative (**21**, R = Br) CuSO$_4$, AcOH, H$_2$O, reflux, 15 min: 30% and 6%, respectively after separation).[985]

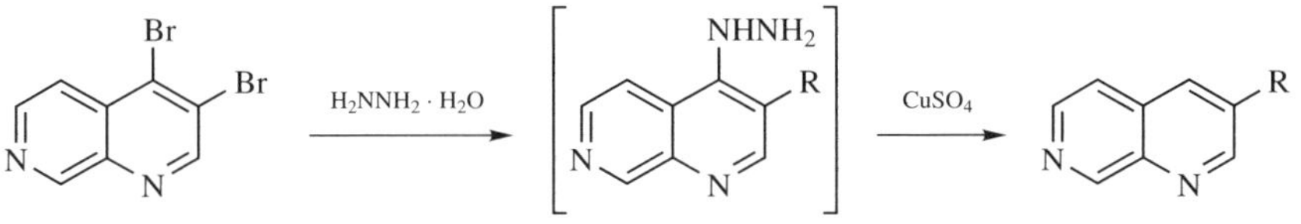

Ring Fission

4-Chloro-6,8-dimethyl-1,7-naphthyridine (**22**) underwent ring fission by hydrazine to give 2,6-dimethyl-4-(pyrazol-3-yl)-3-pyridinamine (**23**) (H$_2$NNH$_2 \cdot$ H$_2$O, 150°C, sealed, 5 h: 65%), confirmed in structure by cyclization with triethyl orthoformate to 7,9-dimethyllpyrazolo[1,5-c]pyrido[4,3-e]pyrimidine (**24**) (reactants, EtOH, reflux, ~3 h: 60%).[1098]

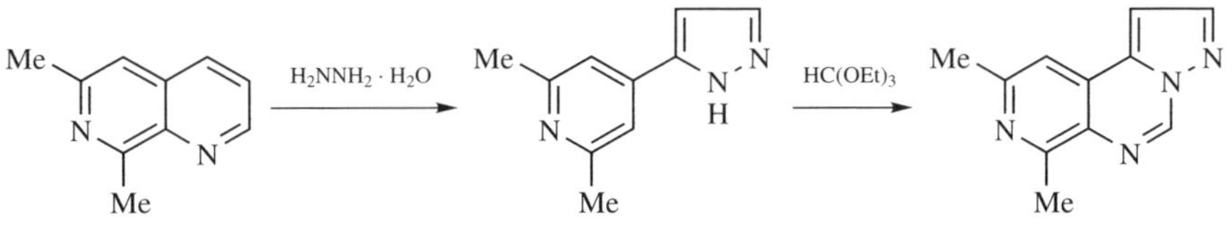

CHAPTER 18

Oxy-1,7-Naphthyridines

This chapter summarizes what has been reported on the 1,7-naphthyridinones, the alkoxy- or acyloxy-1,7-naphthyridines, and the 1,7-naphthyridine N-oxides.

18.1. 1,7-NAPHTHYRIDINONES

The IR spectra of 1,7-naphthyridin-8(7H)-one clearly point to its existence as such rather than as 1,7-naphthyridin-8-ol, at least in the solid state and in nonhydroxylic solvents,[1035] and (by analogy with related systems) it appears safe to assume that the other isomeric 1,7-naphthyridinones will exist as their keto tautomers. General studies that include the ^{1}HNMR,[639] UV,[173] and mass spectra[1253] of 1,7-naphthyridinones have appeared. The antibacterial activity of 8-quinolinol, possibly related to its chelation of heavy metals, is evident to a lesser degree in its aza analogs, including 1,7-naphthyridin-8(7H)-one.[99,110,1040,cf.76] Other activities are evident.[385]

18.1.1. Preparation of 1,7-Naphthyridinones

Most known tautomeric and nontautomeric 1,7-naphthyridinones have been made by *primary synthesis* (see Chapter 15). Of the many potential indirect preparative routes, only two appear to have been used, as indicated in the following examples.

By *C*-Oxidation

Note: This example affords only a nontautomeric naphthyridinone.

1,7-Naphthyridine 7-methiodide (**1**) underwent oxidation to give 7-methyl-1,7-naphthyridin-8(7H)-one (**2**) [substrate, H_2O, 0°C; $NaOH/H_2O$ + $K_3Fe(CN)_6$

The Naphthyridines: The Chemistry of Heterocyclic Compounds, Volume 63, by D.J. Brown
Copyright © 2008 John Wiley & Sons, Inc.

in slowly, then 15 min: isolated as picrate][1214,cf. 153]

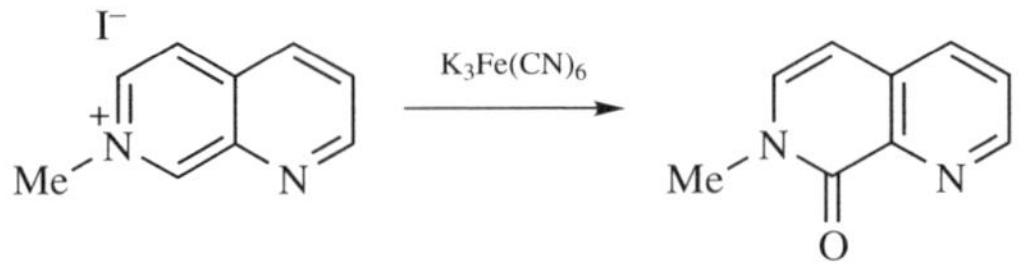

From 1,7-Naphthyridinamines

1,7-Naphthyridin-8-amine (**3**) underwent slow hydrolysis to give 1,7-naphthyridin-8(7*H*)-one- (**4**) (70% H_2SO_4, reflux, 4 days: 81%).[1046]

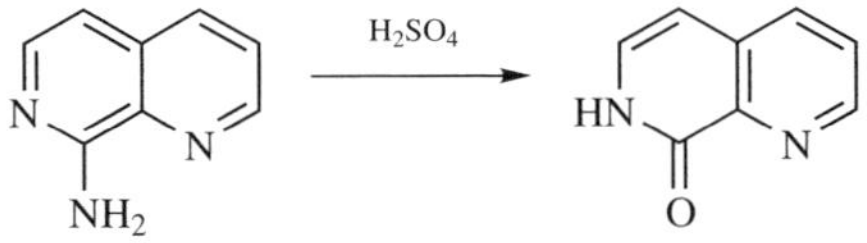

18.1.2. Reactions of 1,7-Naphthyridinones

The most important reaction of tautomeric 1,7-naphthyridinones, *halogenolysis*, has been covered in Section 17.1. Only a few of the other possible reactions have been reported, as illustrated by the following examples.

N-Alkylation

Ethyl 4-amino-2-oxo-1,2-dihydro- (**5**, R = H) with ethyl bromoacetate gave ethyl 4-amino-1-ethoxycarbonylmethyl-2-oxo-1,2-dihydro-1,7-naphthyridine-3-carboxylate (**5**, R = CH_2CO_2Et) (reactants, K_2CO_3, AcEt, reflux, 18 h: 66%).[725]

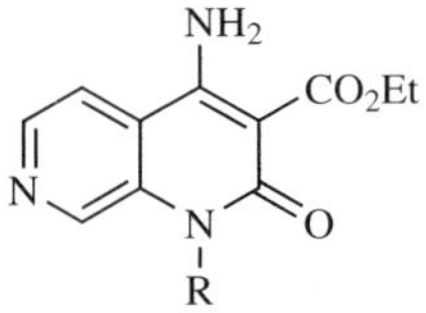

Ethyl 6-chloro-8-morpholino-4-oxo-1,4-dihydro- (**6**, R = H) gave ethyl 5-chloro-1-ethyl-8-morpholino-4-oxo-1,4-dihydro-1,7-naphthyridine-3-carboxylate (**6**, R = Et) (substrate, K_2CO_3, Me_2NCHO, 100°C, 30 min; then EtI in,

100°C, 6 h; then more K_2CO_3 + EtI in, 100°C, 6 h: 44%); analogs likewise.[1130]

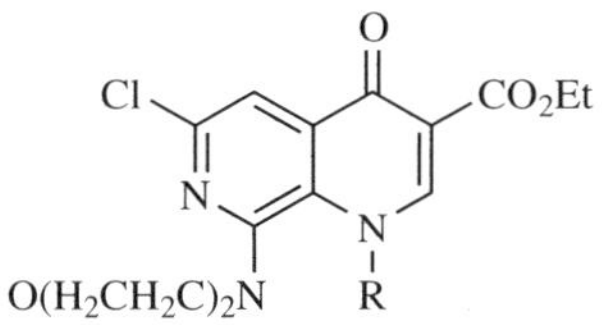

(6)

N-Dealkylation

Note: This reaction amounts to the conversion of a nontautomeric to a tautomeric 1,7-naphthyridinone.

1-Benzyl-5-hydroxy-3,4-dihydro- (**7**, R = CH_2Ph) underwent reductive debenzylation to give 5-hydroxy-3,4-dihydro-1,7-naphthyridin-2(1H)-one (**7**, R = H) (substrate, liquid NH_3, −50°C, Na in during 30 min, −50°C, 30 min: 44%).[1272]

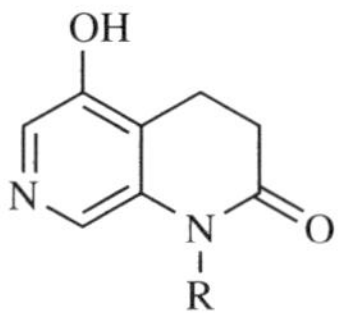

(7)

Cyclization Reactions

7-Cyclohexyl-6-phenylimino-6,7-dihydro-1,7-naphthyridine-5,8-quinone (**8**) with benzonitrile oxide or *N*-phenylbenzonitrileimide gave the spiro derivatives **9**(X = O) and **9**(X = NPh), respectively (reactants, Et_3N, solvent: 70% and 91%, respectively; for fine detail, see original).[750]

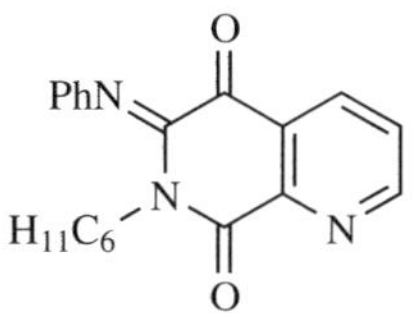

(8) 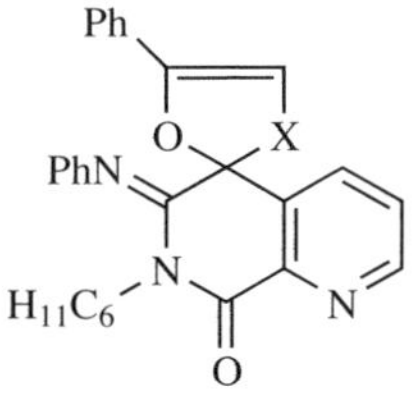   (9)

18.2. ALKOXY- AND ACYLOXY-1,7-NAPHTHYRIDINES

Only a few such derivatives of 1,7-naphthyridine have been made. Available information is listed briefly here.

Preparation of Alkoxy/Aryloxy Derivatives

Note: These derivatives have been made either by *primary synthesis* (see, Chapter 15) or by *alcoholysis of halogeno-1,7-naphthyridines* (see Section 17.2).

Preparation of Acyloxy- and Arylsulfonyloxy-1,7-naphthyridines

6-Acetyl-1-benzyl-5-hydroxy-3,4,7,8-tetrahydro-1,7-naphthyridin-2(1*H*)-one (**10**) underwent catalytic dehydrogenation and rearrangement to give 5-acetoxy-3,4-dihydro-1,7-naphthyridin-2(1*H*)-one (**11**) (Pd/C, xylene, 135°C, 30 h: 45%).[1272]

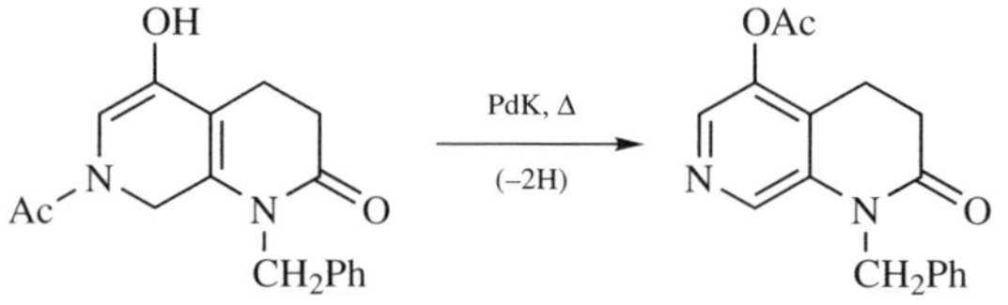

8-*m*-Nitrophenyl-1,7-naphthyridin-6-amine (**12**, R = NH$_2$) underwent diazotization in the presence of trifluoromethanesulfonic acid to give 8-*m*-nitrophenyl-6-trifluoromethanesulfonyloxy-1,7-naphthyridine (**12**, R = OSO$_2$CF$_3$) (reactants, Me$_2$NCHO, NaNO$_2$ in slowly, 25°C, 3 h: 60%; analogs likewise).[198]

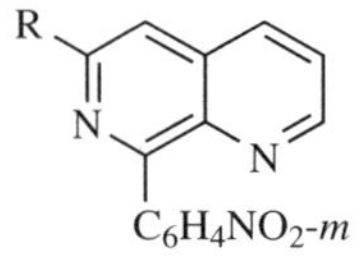

Reactions of Alkoxy/Aryloxy Derivatives

Note: No reactions specific to the ether group appear to have been reported. However, the otherwise reasonably stable proton pump inibitor, 4-(α-methylbenzylamino)-3-butyryl-8-ethoxy-1,7-naphthyridine, is decomposed by light.[1327]

Reactions of Arenesulfonyloxy-1,7-naphthyridines

Note: The *arenolysis* of such a derivative was covered in Section 16.2.1.

8-*m*-Nitrophenyl-6-trifluoromethanesulfonyloxy-1,7-naphthyridine (**13**) underwent *deacyloxylation* to give 8-*m*-nitrophenyl-1,7-naphthyridine (**14**) {substrate, Et₃SiH, [1,1'-bis(diphenylphosphino)ferrocene]dichloropallasium(II), Me₂NCHO, 50°C, 3 h: more Pd complex in, 60°C, 1 h: 92%} or *amidolysis* to give 6-(4-methylpiperazin-1-ylcarbonyl)-8-*m*-nitrophenyl-1,7-naphthyridine (**15**) [substrate, MeN(CH₂CH₂)₂NH, bis(dibenzylideneacetone)palladium, PPh₃, Et₃N, Me₂NCHO, CO in, 80°C, 16 h: 61%].[198]

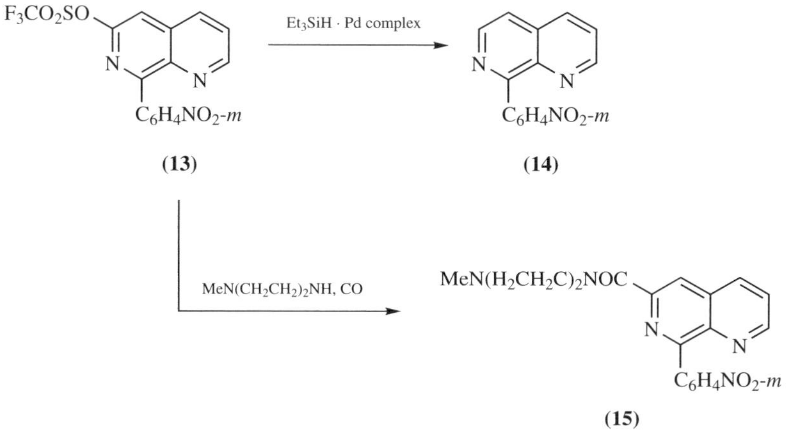

18.3. 1,7-NAPHTHYRIDINE *N*-OXIDES

The very limited data on such compounds are summarized here.

Preparation of 1,7-Naphthyridine *N*-Oxides

Note: A few such oxides have been made by *primary synthesis* (see Chapter 15), and the *oxidation* of unsubstituted 1,7-naphthyridine to its mono- and di-*N*-oxides has been covered in Section 16.1.3.

Reactions of 1,7-Naphthyridine *N*-Oxides

Note: The conversion of such oxides to halogeno-1,7-naphthyridines by the *Meissenheimer reaction* has been exemplified in Section 17.1.

Ethyl 4-oxo-1,4-dihydro-1,7-naphthyridine-3-carboxylate 6-oxide underwent *N-deoxygenation* to give the 4-oxo-1,4-dihydro-1,7-naphthyridine-3-

carboxylic acid (**17**) (Fe powder, AcOH, pyridine, 20°C, 90 min: then NaOH, H$_2$O, reflux, 1 h: ~80%).[93]

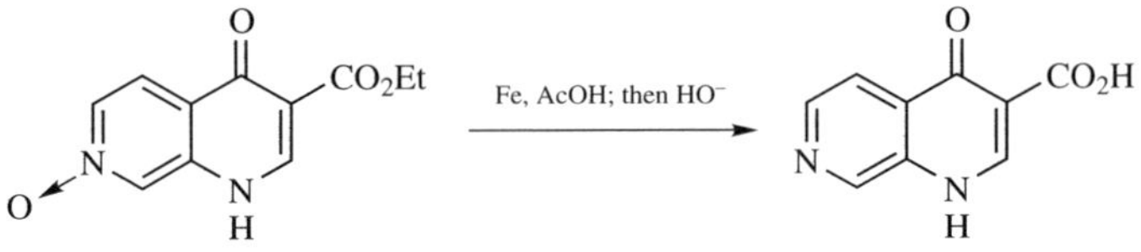

CHAPTER 19

Thio-1,7-naphthyridines

Thio-1,7-naphthyridines appear to be represented by only two thioethers, 4-oxo-8-phenylthio-1,4-dihydro-1,7-naphthyridine-3-carboxylic acid (**1**, R = H) and its ethyl ester (**1**, R = Et) (see Section 15.3),[157] and a few peripheral sulfur-containing 1,7-naphthyridines (see Section 18.2).

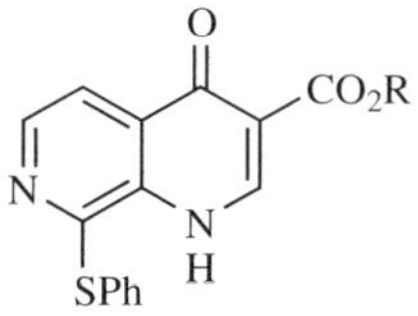

(1)

Accordingly, the whole area is almost a virgin field for research and the present chapter has been inserted simply to emphasize this fact.

The Naphthyridines: The Chemistry of Heterocyclic Compounds, Volume 63, by D.J. Brown

CHAPTER 20

Nitro-, Amino-, and Related 1,7-Naphthyridines

Information on such nitrogenous derivatives is not plentiful in the literature. Available data are summarized (mainly by cross-references and examples) in the sections that follow,

20.1. NITRO-1,7-NAPHTHYRIDINES

A review of nitronaphthyridines (in all six series) is particularly valuable for comparative purposes.[1273]

Preparation of Nitro-1,7-naphthyridines

Note: These nitro compounds have been made by *primary synthesis* (see Chapter 15) and by *nitration* as exemplified here.

1,7-Naphthyridin-4(1*H*)-one (**1**, R = H) gave its 3-nitro derivative (**1**, R = NO_2) (70% HNO_3, reflux, 24 h: 21%;[723] 98% HNO_3, reflux, 1 h: 74%).[1017] Somewhat similarly, 1,7-naphthyridin-8(7*H*)-one (**2**, R = H) gave its 5-nitro derivative (**2**, R = NO_2) (70% HNO_3, 95% H_2SO_4, 95°C, 1 h: 42%).[1046]

(1)

(2)

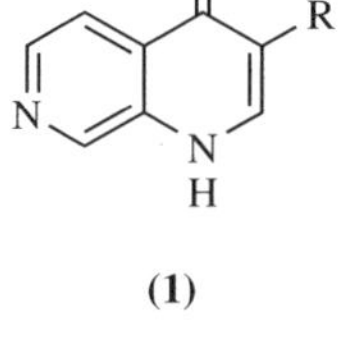

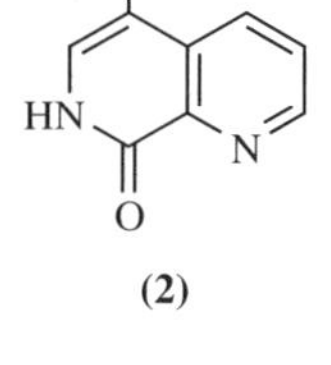

Reactions of Nitro-1,7-naphthyridines

Note: The only reported reaction of nitro-1,7-naphthyridines appears to be their *reduction* to the corresponding naphthyridinamines.

The Naphthyridines: The Chemistry of Heterocyclic Compounds, Volume 63, by D.J. Brown

3-Nitro-1,7-naphthyridin-4(1H)-one (**1**, R = NO$_2$) gave 3-amino-1,7-naphthyridin-4(1H)-one (**1**, R = NH$_2$) (substrate, NaOH, H$_2$O, Raney Ni, H$_2$: 70% as dihydrochloride;[723] or substrate, MeOH, Raney Ni, H$_2$, 20°C: 75% as dihydrochloride).[1017]

20.2. AMINO-1,7-NAPHTHYRIDINES

This section includes the chemistry of all types of amino group attached to 1,7-naphthyridine; aspects already covered are indicated by cross-references.

Preparation of Amino-1,7-naphthyridines

Note: These amino compounds have been made by *primary synthesis* (see Chapter 15), by *aminolysis of halogeno-1,7-naphthyridines* (see Section 17.2), by *reduction of nitro-1,7-naphthyridines*, (see Section 20.1), or by *direct amination* as illustrated in Section 16.1.3 and by further examples here.

5,8-Dichloro-1,7-naphthyridine (**3**, X = Cl) gave 5,8-dichloro-1,7-naphthyridin-2-amine (**4**, X = Cl) (substrate, KNH$_2$/NH$_3$, 5 min; then KMnO$_4$ in slowly; 15 min: 52%); likewise the dibromo derivative (**3**, X = Br) gave 5,8-dibromo-1,7-naphthyridin-2-amine (**4**, X = Br) (33%).[819]

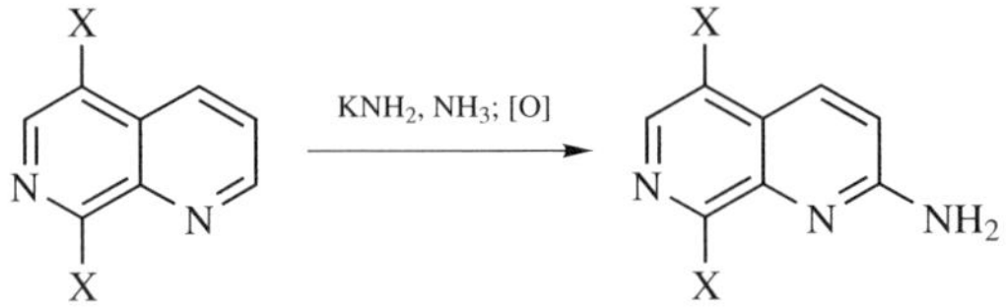

1,7-Naphthyridine 7-methiodide (**5**) underwent addition of ammonia and subsequent oxidation to give a mixture from which were isolated the hydriodide of 7-methyl-1,7-naphthyridin-8(7H)-imine (**6**, X = NH), 7-methyl-1,7-naphthyridin-8(7H)-one, and two ring contraction products (substrate, liquid NH$_3$, −33°C; KMnO$_4$ in slowly; 20 min: each in <5% yield).[372,cf. 460]

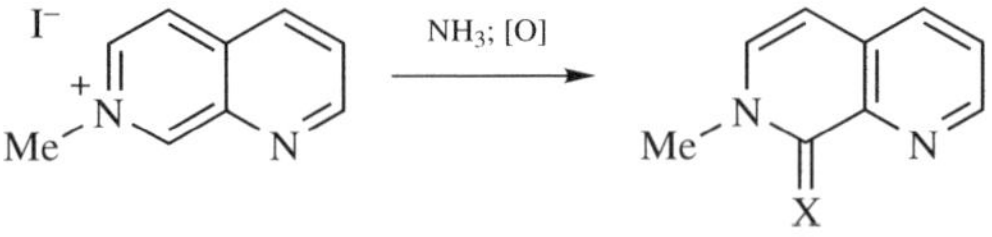

Reactions of Amino-1,7-naphthyridines

Note: The oxidative *removal of hydrazino groups* has been exemplified in Section 16.1.1 as well as here, *hydrolysis to naphthyridinones* in Section 18.1.1, and *transamination* in Section 17.1 under "aminolysis."

1,7-Naphthyridin-4-amine gave a charge transfer complex with tetracyanoquinodimethane.[416]

8-Butyl-5-formylmethyl-1,7-naphthyridin-6-amine (**7**) underwent intramolecular *cyclization* to give 5-butyl-3*H*-pyrrolo[3,2-*f*][1,7]naphthyridine (**8**) (acidic conditions: 72%; no details).[132]

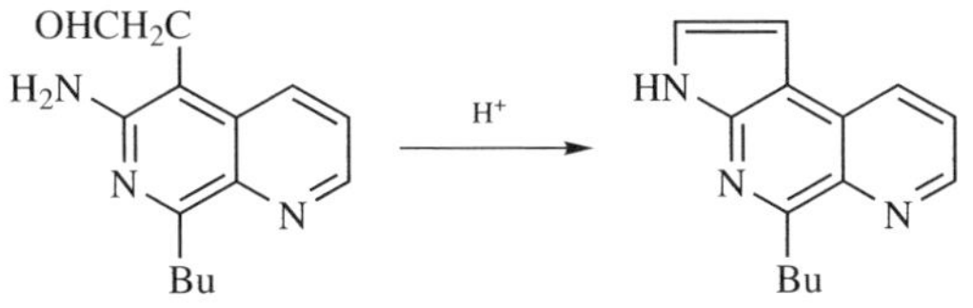

2-Hydrazino-6-methoxy-1,7-naphthyridine (**9**, R = NHNH$_2$) underwent *oxidative deamination* to give 6-methoxy-1,7-naphthyridine (**9**, R = H) (CuSO$_4$, AcOH, H$_2$O, 95°C, 1 h: 45%).[234]

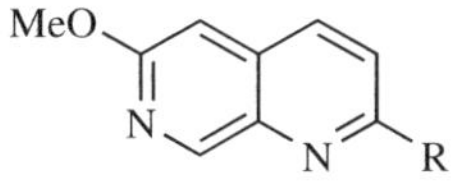

3-Amino-1,7-naphthyridin-4(1*H*)-one (**10**) underwent *diazotization* to give 4-oxo-3,4-dihydro-1,7-naphthyridin-3-ylidendiazonium chloride hydrochloride (**11**) (NaNO$_2$, HCl, 0°C: 80%) and subsequent ring contraction on irradiation to afford pyrrolo[2,3-*c*]pyridine-3-carboxylic acid (**12**) (AcOH, H$_2$O, *hv*, 20°C, 3 h: 90%).[1097,cf. 723]

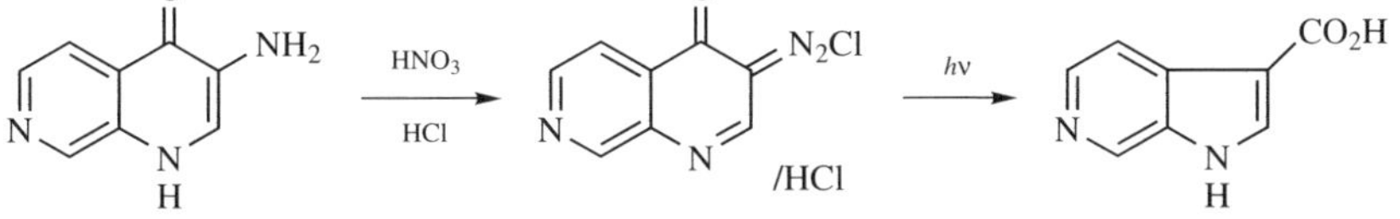

CHAPTER 21

1,7-Naphthyridinecarboxylic Acids and Related Derivatives

This chapter summarizes any published information on 1,7-naphthyridinecarboxylic acids, esters, amides, nitriles, aldehydes, ketones, and the like.

21.1. 1,7-NAPHTHYRIDINECARBOXYLIC ACIDS

21.1.1. Preparation of 1,7-Naphthyridinecarboxylic Acids

A few such acids have been made directly by *primary synthesis* (see Chapter 15) but most, by indirect routes illustrated in the following examples.

By Hydrolysis of Esters

Ethyl 5-oxo-1,5-dihydro-1,7-naphthyridine-6-carboxylate (**1**, R = Et) gave 5-oxo-1,5-dihydro-1,7-naphthyridine-6-carboxylic acid (**1**, R = H) (M NaOH, 95°C, 1 h: 93%);[744] similarly, ethyl 4-oxo-8-phenylthio-1,4-dihydro-1,7-naphthyridine-3-carboxylate (**2**, R = Et) gave 4-oxo-8-phenylthio-1,4-dihydro-1,7-naphthyridine-3-carboxylic acid (**2**, R = H) (5M NaOH, 20°C, 3 h: 30%).[157]

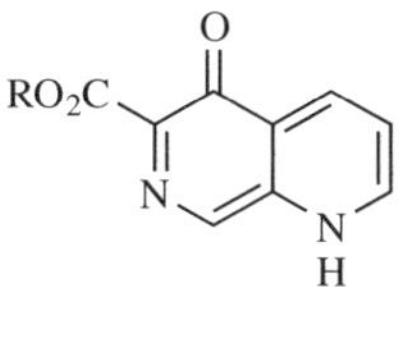

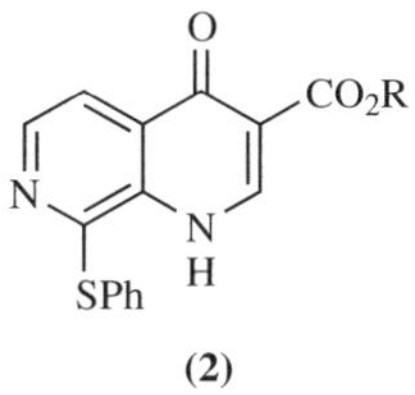

Ethyl 7-methyl-4,8-dioxo-1,4,7,8-tetrahydro-1,7-naphthyridine-3-carboxylate (**3**, R = Et) gave 7-methyl-4,8-dioxo-1,4,7,8-tetrahydro-1,7-naphthyridine-

---

The Naphthyridines: The Chemistry of Heterocyclic Compounds, Volume 63, by D.J. Brown
Copyright © 2008 John Wiley & Sons, Inc.

3-carboxylic acid (**3**, R = H) (2M HCl, reflux, 12 h: 79%).[1116]

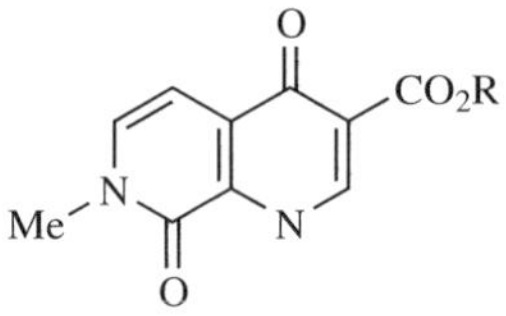

(**3**)

Also other examples.[93,593,723,732]

By Hydrogenolysis of Benzyl Esters

Benzyl 7-methyl-8-oxo-5-phenyl-7,8-dihydro-1,7-naphthyridine-6-carboxylate (**4**, R = CO$_2$CH$_2$Ph) gave 7-methyl-8-oxo-5-phenyl-7,8-dihydro-1,7-naphthyridine-6-carboxylic acid (**4**, R = CO$_2$H) (Pd/C, H$_2$O, MeOH, H$_2$, 20°C, 1 h: 89%).[593]

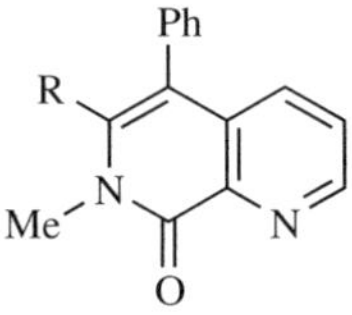

(**4**)

From Carboxamides

7-Methyl-8-oxo-5-phenyl-7,8-dihydro-1,7-naphthyridine-6-carboxamide (**4**, R = CONH$_2$) also afforded the foregoing acid (**4**, R = CO$_2$H) (substrate, HCl, NaNO$_2$ in slowly, 20°C, 15 h: 82%).[593]

By Oxidation of Carbaldehydes

7-Methyl-8-oxo-5-phenyl-7,8-dihydro-1,7-naphthyridine-6-carbaldehyde (**4**, R = CHO) also gave the acid (**4**, R = CO$_2$H) (substrate, NaOH, Bu^{t}OH, H$_2$O; KMnO$_4$ in slowly; 20°C, 40 min: 91%).[593]

21.1.2. Reactions of 1,7-Naphthyridinecarboxylic Acids

Of the many potential reactions of these acids, only two have been reported. They are illustrated by the examples that follow.

Decarboxylation

7-Methyl-4,8-dioxo-1,4,7,8-tetrahydro-1,7-naphthyridine-3-carboxylic acid (**5**, R = CO$_2$H) gave 7-methyl-1,7-naphthyridine-4,8(1H,7H)-dione (**5**, R = H) (mineral oil, 290°C, 20 min: 80%).[1116]

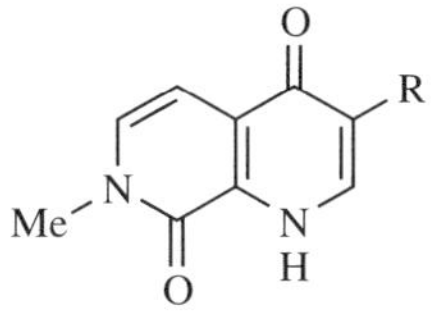

(5)

4-Oxo-1,4-dihydro-1,7-naphthyridine-3-carboxylic acid (**6**, R = CO$_2$H) gave 1,7-naphthyridin-4(1H)-one (**6**, R = H) (quinoline, 200°C, 1 h; then 230°C, 15 min: ~78%).[723,cf. 93]

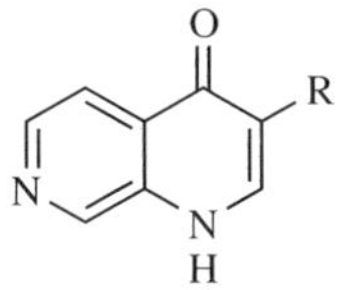

(6)

Also other examples.[732,744]

Conversion to Amides

Note: This has been done directly and also indirectly via the uncharacterized 1,7-naphthyrieinecarbonyl chlorides.

6,8-Dimethyl-4-oxo-1,4-dihydro-1,7-naphthyridine-3-carboxylic acid with 1-(4-aminobutyl)-4-diphenylmethylpiperazine gave N-[4-(4-diphenylmethylpiperazin-1-yl)butyl]-6,8-dimethyl-1,7-naphthyridin-4(1H)-one (**7**) (substrate, N,N-carbonyldiimidazole, Me$_2$NCHO, 60°C, 5 h; then substrate in, 60°C, 1 h: 50%).[1204]

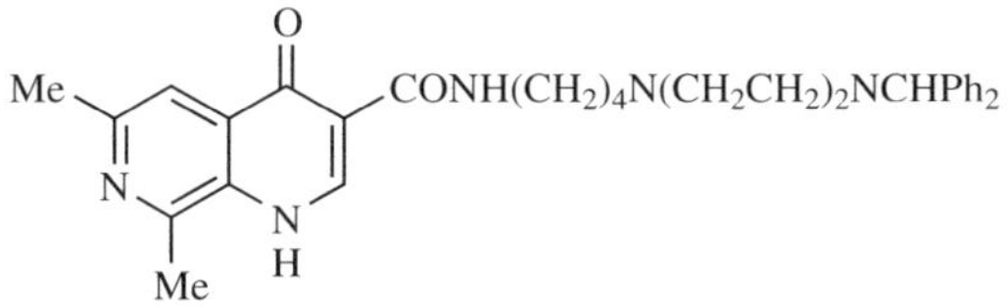

(7)

7-Methyl-8-oxo-5-*p*-tolyl-7,8-dihydro-1,7-naphthyridine-6-carboxylic acid was converted into the corresponding 6-carbonyl chloride (SOCl$_2$, Me$_2$NCHO, CH$_2$Cl$_2$, reflux, 3 h), which then gave 7-methyl-N-(α-methylbenzyl)-8-oxo-6-*p*-tolyl-7,8-dihydro-1,7-naphthyridine-6-carboxamide (**8**) (PhMeCHNH$_2$, Et$_3$N, ClCH$_2$CH$_2$Cl, 20°C, 1 h: 67% overall); analogs likewise.[578]

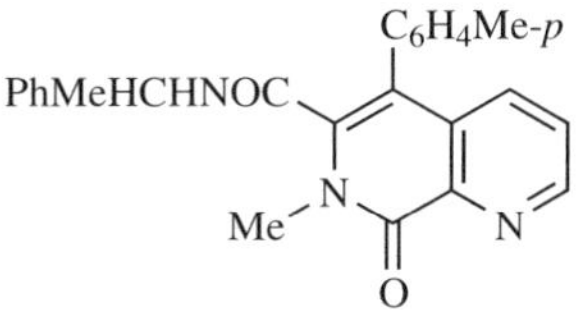

(8)

21.2. 1,7-NAPHTHYRIDINECARBOXYLIC ESTERS

Virtually all known examples of such esters have been prepared by *primary synthesis* (see Chapter 15), and their only reported reaction appears to be *hydrolysis (or hydrogenolysis)* to the corresponding acids (see Section 21.1.1).

21.3. 1,7-NAPHTHYRIDINECARBOXAMIDES

Available data on these amides are summarized briefly in the following list.

Preparation of Amides

Note: Most amides have been made by *primary synthesis* (see Chapter 15) or from *1,7-naphthyridinecarboxylic acids* (see Section 21.1.2). In addition, some have been made by *hydrolysis of nitriles* as illustrated here.

7-Methyl-8-oxo-5-phenyl-7,8-dihydro-1,7-naphthyridine-6-carbonitrile (**9**, R=Me) gave the carboxamide (**10**, R=Me) (NaOH, H$_2$O, EtOH, reflux, 40 min: 97%);[593] less effectively, the 7-isopropyl substrate (**9**, R=Pr^i) gave the amide (**10**, R=Pr^i) (KOH, Pr—OH, reflux, 90 min: 15%; perhaps [H$_2$O] was too low?).[1397]

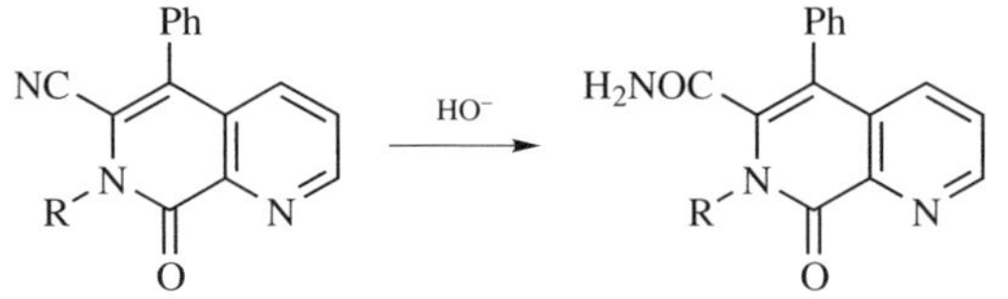

21.4. 1,7-NAPHTHYRIDINECARBONITRILES

All such nitriles appear to have been made by *primary synthesis* (see Chapter 15) or by the *Reissert reaction* (see Section 16.1.3); their only reported reaction involves *hydrolysis to amides* (see Section 21.3).

21.5. 1,7-NAPHTHYRIDINECARBALDEHYDES AND KETONES

A few such derivatives have been made by *primary synthesis* (see Chapter 15 and reference a 1995 paper by Natsugari and colleagues[593]); aldehydes have been *oxidized* to carboxylic acids (see Section 21.1).

CHAPTER 22

Primary Syntheses of 1,8-Naphthyridines

Like the primary syntheses of other naphthyridines, the primary synthesis of 1,8-naphthyridines may be done by cyclization of appropriate aliphatic substrates, with or without auxiliary synthons, by cyclization of appropriately substituted pyridines with or without synthons, or from other heterocyclic substrates by several processes.

Some reviews of 1,8-naphthyridine chemistry contain information on these primary syntheses.[49–52,55,57,58,61,265,407,419,670,1357,1430,1432] Other papers appear to be mainly of historical interest.[5,9,11,23,25,28,35–38,62,63]

22.1. FROM AN ALIPHATIC SUBSTRATE

Such syntheses are more suited to the preparation of hydro rather than aromatic 1,8-naphthyridines and may require auxiliary synthons. A few typical examples follow.

Using a Substrate Alone

2,2,4-Trichloro-4,6-dicyanohexanoyl chloride (**1**) gave 3,7-dichloro-1,8-naphthyridin-2(1*H*)-one (**2**) (HCl, Bu$_2$O, 140°C: 78%).[715]

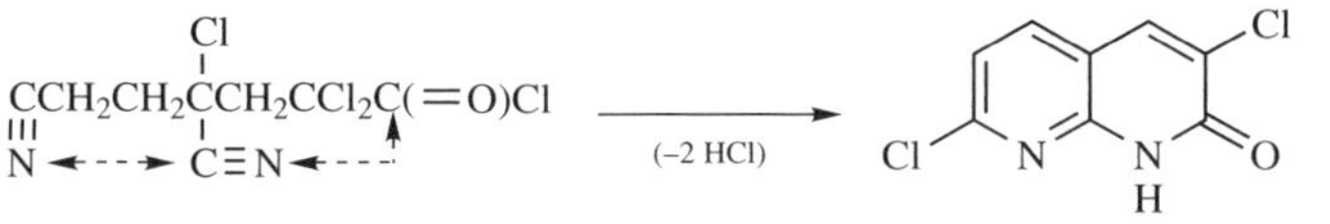

The carboxamidine (**3**) gave 2,2,4a,7,7a-pentamethyl-1,2,3,4,4a,5,6,7-octahydro-1,8-naphthyridine (**4**) (HCl/CH$_2$Cl$_2$, 200°C, 80%).[711]

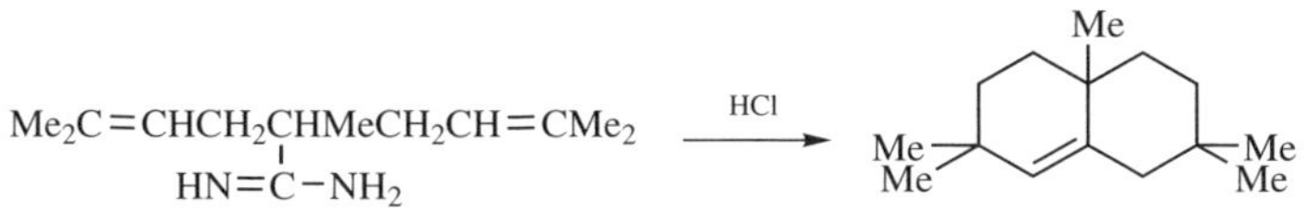

Also other examples.[212,519,734,824,939,1080,1150,1325]

The Naphthyridines: The Chemistry of Heterocyclic Compounds, Volume 63, by D.J. Brown
Copyright © 2008 John Wiley & Sons, Inc.

Using a Substrate and Synthon(s)

1-Amino-1,2,2-tricyanoethylene (malononitrile dimer) with 1-(1-acetyl-2-dimethylaminovinyl)benzotriazole gave 4-amino-6-(benzotriazol-1-yl)-5-methyl-2-oxo-1,2-dihydro-1,8-naphthyridine-3-carbonitrile (**5**) with loss of dimethylamine (EtO$^-$: 75%).[644]

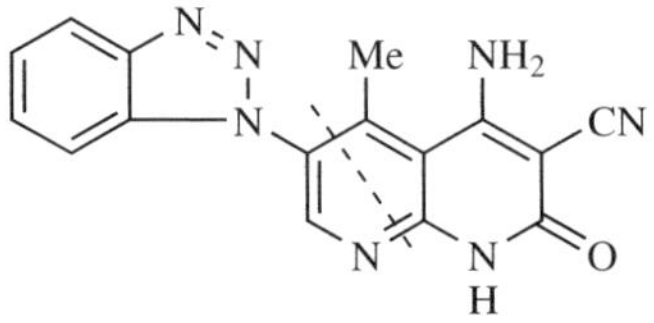

(**5**)

N-(2-Benzoylethyl)-N,N-dimethylammonium chloride (2 mol) and cyanoacetamidinium chloride (made *in situ*) gave 2,7-diphenyl-1,4,4a,5-tetrahydro-1,8-naphthyridine-4a-carbonitrile (**6**) (HCl: 52%).[1402]

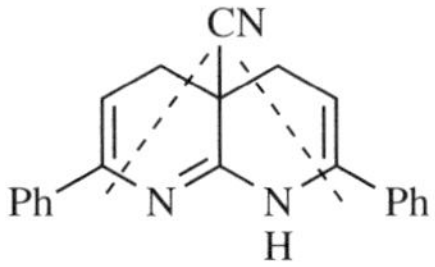

(**6**)

22.2. FROM A SINGLE PYRIDINE SUBSTRATE

The intramolecular cyclization of a pyridine derivative to a 1,8-naphthyridine can involve completion of the 1,2-, 1,8a-, 2,3-, 3,4-, or 4,4a-bond. The following examples are so grouped.

By Formation of the 1,2-Bond

5-Methyl-3-(*m*-tolylethynyl)-2-pyridinamine (**7**) gave 4-ethoxy-6-methyl-2-*m*-tolyl-1,8-naphthyridine (**8**) (EtONa/EtOH: 90%; note the addition of ethanol).[700]

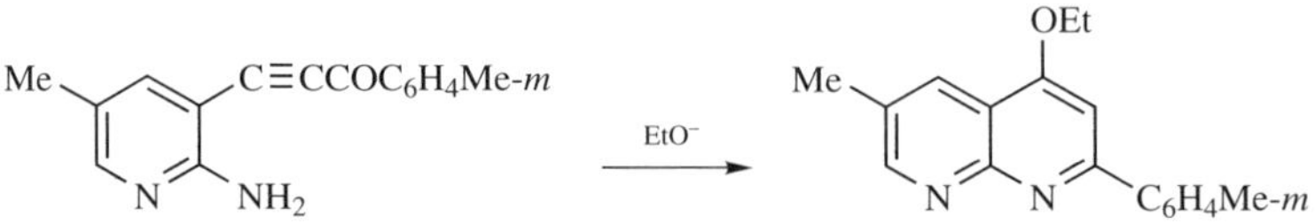

(**7**) → (**8**)

3-(2-Ethoxycarbonylvinyl)-2-pyridinamine (**9**) gave 1,8-naphthyridin-2(1*H*)-one (**10**) (EtONa/EtOH: 72%),[1044]

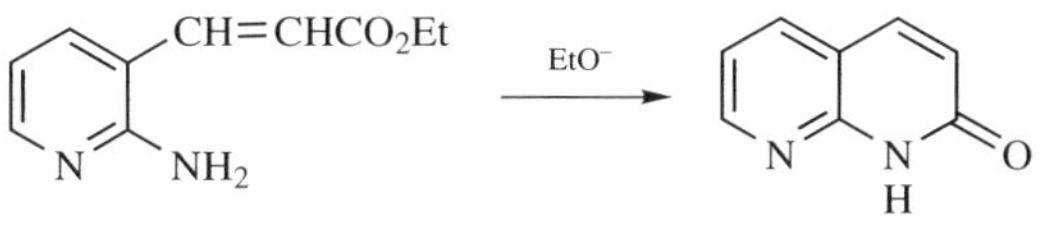

Also other examples.[95,477,588,1039,1056]

By Formation of the 1,8a-Bond

2,5-Bis(3-aminopropyl)pyridine (**11**) gave 2-(3-aminopropyl)-1,2,3,4-tetrahydro-1,8-naphthyridine (**12**) (NaNH$_2$/PhMe: 90%).[223]

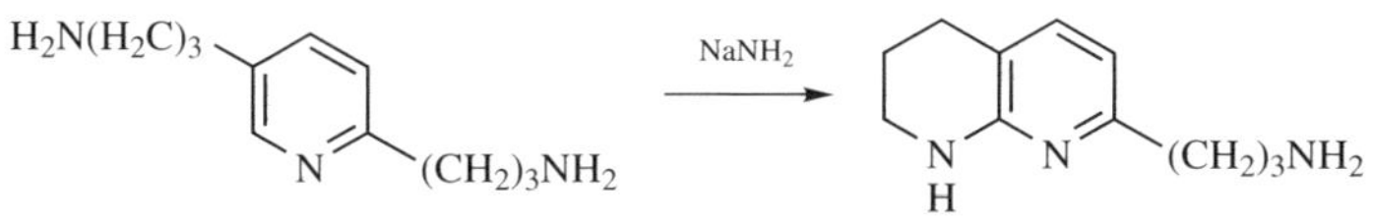

2,6-Dichloro-3-[2-ethoxycarbonyl-3-(thiazol-2-ylamino)acryloyl]-5-fluoropyridine (**13**) gave ethyl 7-chloro-6-fluoro-4-oxo-1-(thiazol-2-yl)-1,4-dihydro-1,8-naphthyridine-3-carboxylate (**14**) (Bu^{t}OK/dioxane: 63%).[143]

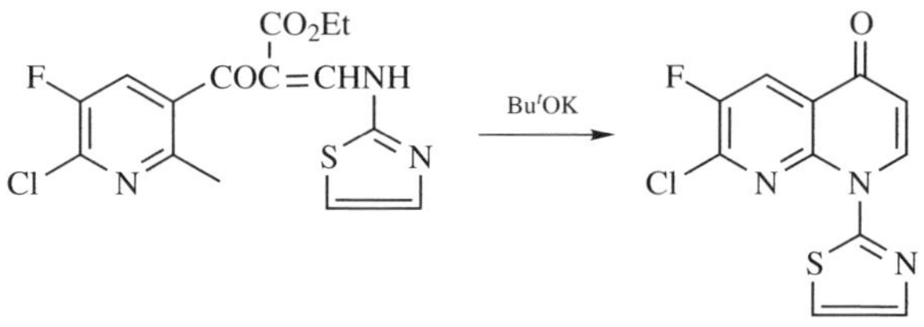

Also other examples.[1,139,166,167,334,355,454,458,493,613,696,1122,1172,1216,1223,1234,1300,1346]

By Formation of the 2,3-Bond

Note: Only fused 1,8-naphthyridines appear to have been made by this type of synthesis.[138]

By Formation of the 3,4-Bond

Methyl 2-[*N*-phenyl-*N*-(pyrrolidin-1-ylacetyl)amino]-2-pyridinecarboxylate (**15**) gave 4-hydroxy-1-phenyl-3-pyrrolidin-l-yl)-1,8-naphthyridin-2(1*H*)-one (**16**)

($Bu^tOK/MeOBu$: ~85%).[717]

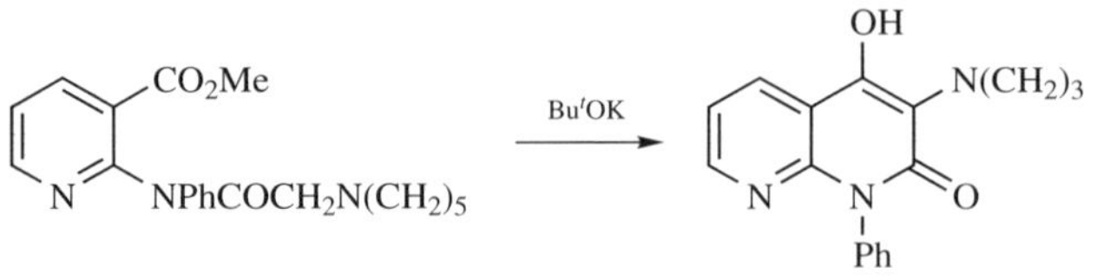

[Structure: (15) pyridine with CO$_2$Me and NPhCOCH$_2$N(CH$_2$)$_5$ substituents → Bu^{t}OK → (16) 1,8-naphthyridine with OH, N(CH$_2$)$_3$, =O, N-Ph]

2-[N-Ethyl-N-(2-ethoxycarbonylethyl)amino]-3-fluoro-6-p-tolylthiopyridine-3-carbonitrile (17) gave ethyl 4-amino-1-ethyl-6-fluoro-7-p-tolylthio-1,2-dihydro-1,8-naphthyridine-3-carboxylate (18) (NaH/PhMe: 91%).[1162]

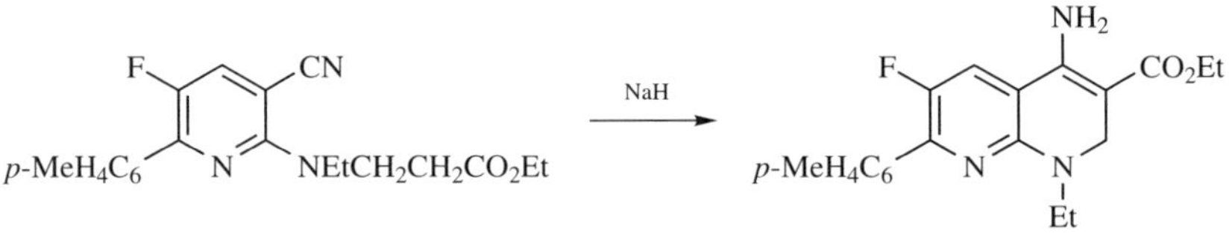

[Structure: (17) → NaH → (18)]

Also other examples.[10,213,260,311,1236]

By Formation of the 4,4a-Bond

6-(2-Acetyl-1-methylethylidene)amino-2-pyridinamine (19) gave 5,7-dimethyl-1,8-naphthyridin-2-amine (20) (H_3PO_4, 100°C: 84%).[104]

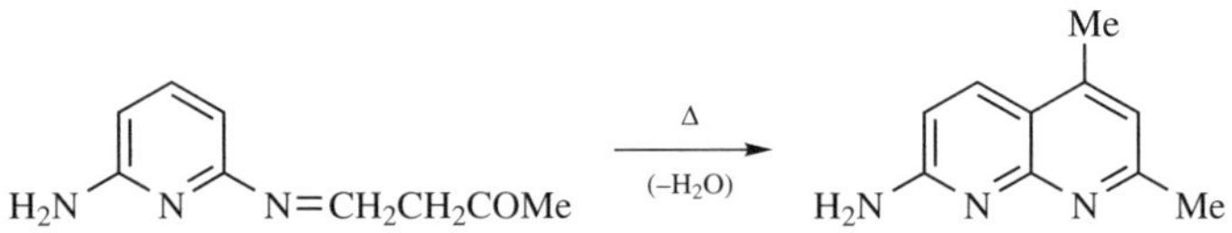

[Structure: (19) H$_2$N-pyridine-N=CH$_2$CH$_2$COMe → Δ (−H$_2$O) → (20) 5,7-dimethyl-1,8-naphthyridin-2-amine]

2-(2-Carboxyethylamino)-6-methylpyridine (21) gave 7-methyl-3,4-dihydro-1,8-naphthyridin-4(1H)-one (22) (P_2O_5/H_3PO_4 or H_2SO_4, heat: ~50%).[1383]

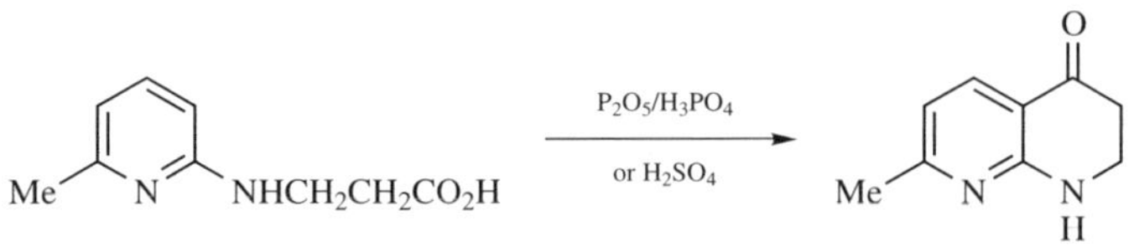

[Structure: (21) Me-pyridine-NHCH$_2$CH$_2$CO$_2$H → P_2O_5/H_3PO_4 or H_2SO_4 → (22)]

2-(2,2-Diethoxycarbonylvinylamino)-6-fluoro-7-methylpyridine (23) gave ethyl 6-fluoro-7-methyl-4-oxo-1,4-dihydro-1,8-naphthyridine-3-carboxylate (24) (Dowtherm, heat: 80%);[1091] such a cyclization has also been done under different conditions to give ethyl 1-cyclopropyl-4-oxo-7-phenylthio-6-trifluoromethyl-1,4-dihydro-1,8-naphthyridine-3-carboxylate H_2SO_4/Ac_2O: 45%).[1163]

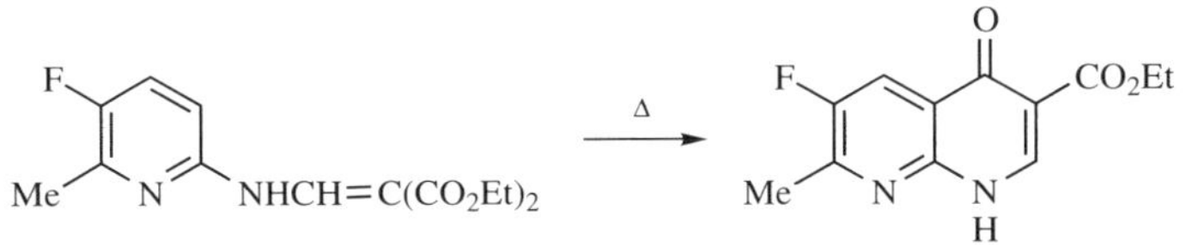

2-[2,2-Dicyano-1-(o-methoxyphenyl)vinylamino]-1-methylpiperidine (**25**) gave the 4-amino-2-(p-methoxyphenyl)-8-methyl-5,6,7,8-tetrahydro-1,8-naphthyridine-3-carbonitrile (**26**) (NaH/Me_2NCHO: 85%).[1177]

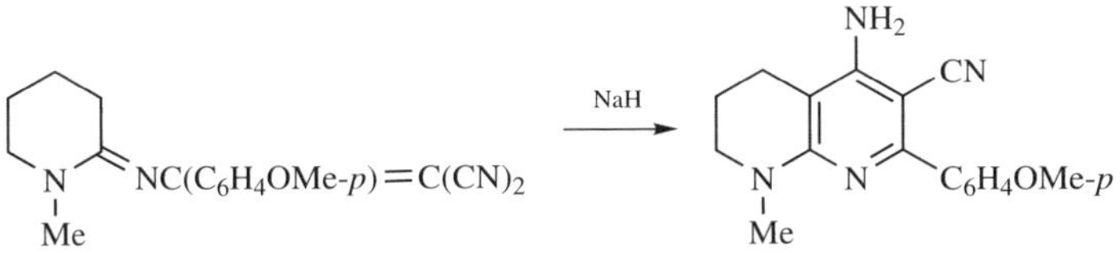

Also other examples.[7,31,102,103,154,179,687,701,702,720,835,841,847,863,1050,1060,1076,1117,1120, 1235,1249,1284,1385,1388,1405]

22.3. FROM A PYRIDINE SUBSTRATE AND SYNTHON(S)

Of the many possibilities within this type of synthesis, only nine are exemplified in the literature: where one synthon supplies N1, C2, C2 + C3, C4a + C8a, N1 + C2 + C3, C2 + C3 + C4, or N1 + C2 + C3 + C4; or when two synthons supply N1 and C2 or C2 and C3 + C4. The following examples illustrate each of these primary synthetic routes.

Where the Synthon Supplies N1

2-Chloro-3-(2-ethoxycarbonylvinyl)pyridine (**27**) gave 1,8-naphthyridin-2(1H)-one (**28**) (NH_3/EtOH: 74%).[1044]

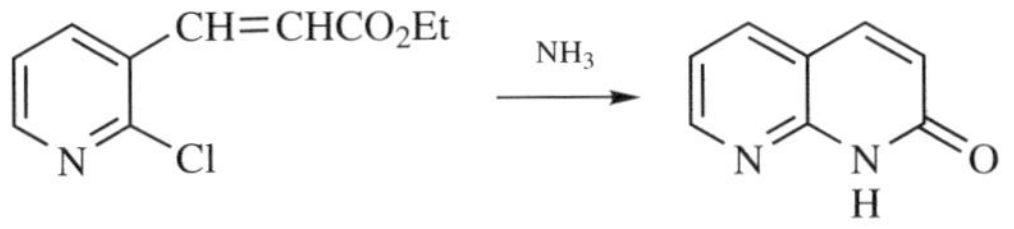

Also other examples.[252]

Where the Synthon Supplies C2

3-(2-Nitrovinyl)-2-pyridinamine (**29**) gave 3-nitro-2-phenyl-1,8-naphthyridine (**30**) (PhCHO, xylene: 75%).[1278]

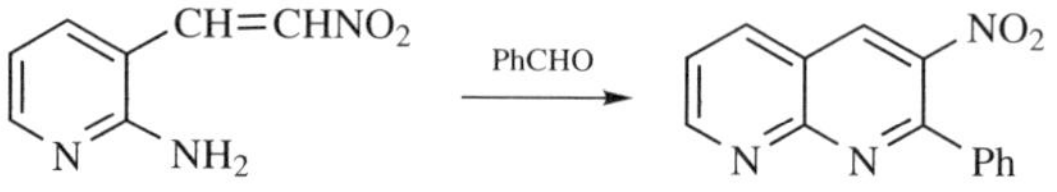

2-Chloro-6-methyl-3-(sulfamoylacetyl)pyridine gave 1-ethyl-7-methyl-4-oxo-1,4-dihydro-1,8-naphthyridine-3-sulfonamide (**31**) [HC(OEt)$_3$, heat: 44%].[1045]

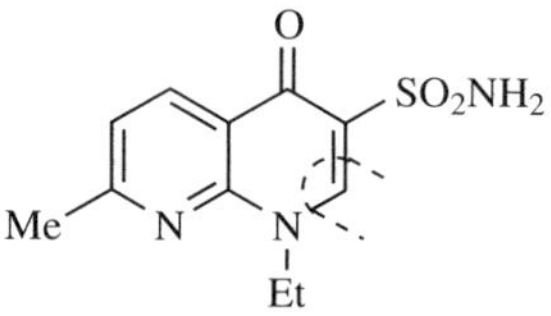

Also other examples.[1332]

Where the Synthon Supplies C2 + C3

Note: A great many such syntheses have been reported, most of them using 2-amino-3-pyridinecarbaldehyde or related substrates.

2-Amino-5,6-diphenyl-3-pyridinecarbaldehyde (**32**) with α-benzoyltoluene gave 2,3,6,7-tetraphenyl-1,8-naphthyridine (**33**) (EtOH, trace KOH: 90%).[284]

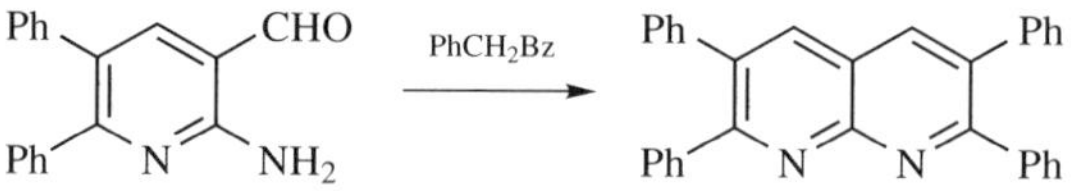

2-Amino-3-pyridinecarbonitrile (**34**) with *m*-chlorobenzyl cyanide gave 3-*m*-chlorophenyl-1,8 naphthyridin-2-amine (**35**) (neat reactants, trace KOH/H$_2$O, microwave: 98%);[77] the same substrate (**34**) with methyl cyanoacetate gave 2-oxo-1,2-dihydro-1,8-maphthyridine-3-carbonitrile (**36**) (piperidine, EtOH: 81%).[367]

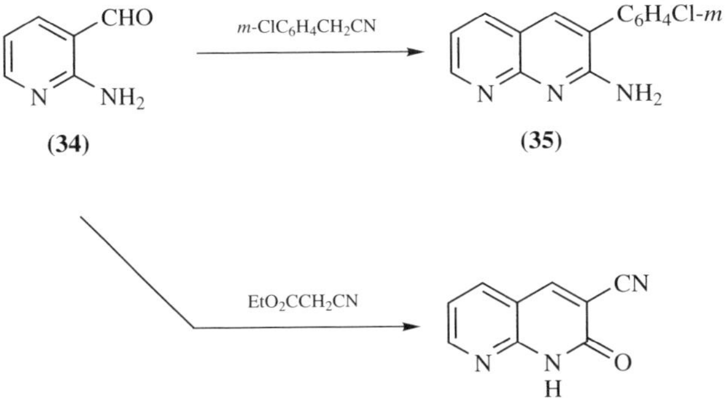

2-Benzylamino-3-pyridinecarboxylic acid (**37**) and acetic anhydride gave 4-acetoxy-1-benzyl-1,8-naphthyridin-2(1H)-one (**38**) (Ac$_2$O/AcOH: 63%).[160]

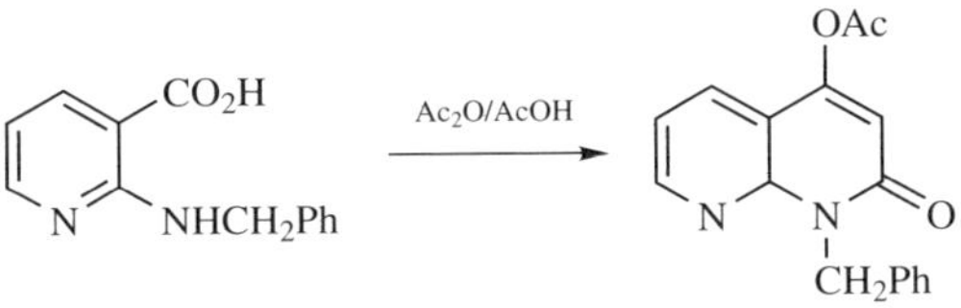

Also other examples.[8,18,33,65,78,121,123,130,136,142,145,162,183,184,189,194,199,208,239,264,276, 278,283,306,312,384,468,482,531,579,582,607,618,622,650,653,658,659,666,674,707,709,712,727,753, 796,801,851,856,858,859,872,877,896,900,983,1005,1009,1041,1099,1100,1112,1139,1195,1196,1217, 1258,1274,1291,1328,1347,1400,1419,1425,1439,1440]

Where the Synthon Supplies N1 + C2 + C3

2 Bromo-3-pyridinecarbaldehyde (**39**) with 2-phenylacetamide gave 2-phenyl-1,8-naphthyridin-2(1H)-one (**40**) (reactants, Cs$_2$CO$_3$, P- and Pd-catalysts: 91%).[117]

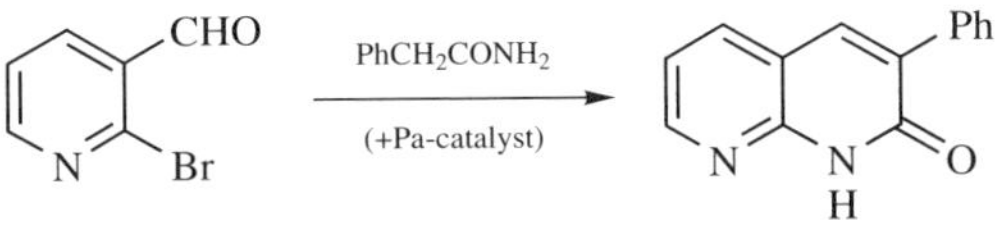

Where the Synthon Supplies C2 + C3 + C4

Note: This synthesis involves cyclocondensation of the 2-pyridinamine with a 1,5-dicarbonyl synthon or equivalent. Although the classical Skraup procedures may be used,[172,348,755] yields are usually poor.

2,6-Pyridinediamine (**41**) with α-phenylmalondialdehyde gave 7-phenyl-1,8-naphthyridin-2-amine (**42**) (85% H_3PO_4, 95°C: 59%)[338] or with bis(trifluoroacetyl)methane gave 5,7-bis-(trifluoromethyl)-1,8-naphthyridin-2-amine (**44**) (likewise: 67%).[1121] Analog (**43**) was made similarly.[33]

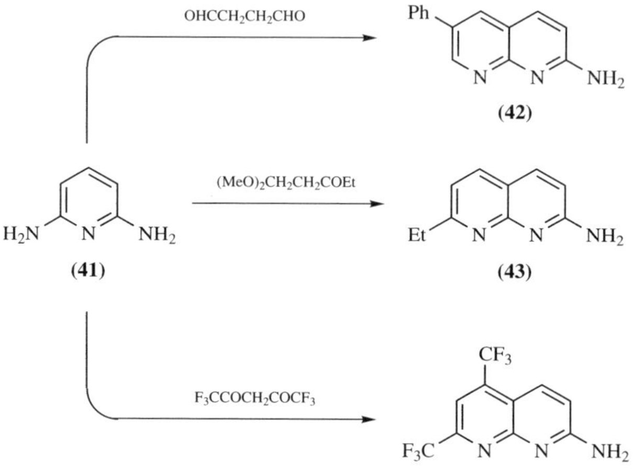

6-Methyl-2-pyridinamine (**45**, R=Me) with diethyl (1-ethoxyethylidene)malonate gave ethyl 2,7-dimethyl-4-oxo-1,4-dihydro-1,8-naphthyridine-3-carboxylate (**46**) (Dowtherm, 240°C: 22%);[249] 2-pyridinamine (**45**, R=H) with diethyl malonate gave 4-hydroxy-1,8-naphthyridin-2(1*H*)-one (**47**) (neat reactants, heat: 85%);[693] and 2,6-pyridinediamine (**45**, R=NH_2) with malic acid gave 7-amino-1,8-naphthyridin-2(1*H*)-one (**48**) (H_2SO_4, heat: 85–97%).[389,681,1018]

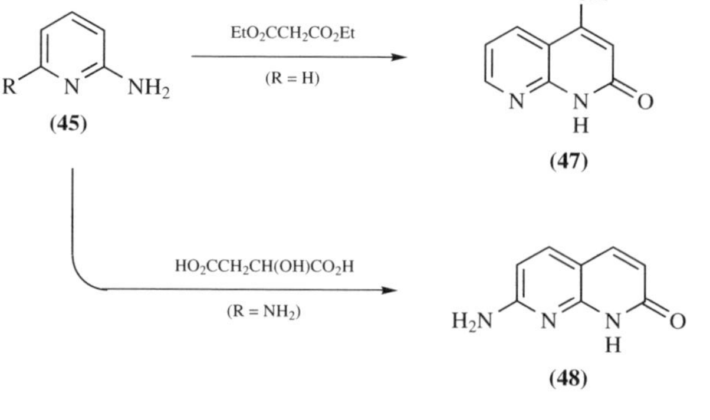

Also other examples.[34,67,70,71,73,88,114,115,201,345,350,428,505,538,686,690,718,721,755,763,771, 810,864,880,964,1059,1071,1073,1075,1156,1159,1218,1326,1333,1399,1441]

Where the Synthon Supplies N1 + C2 + C3 + C4

Note: As might be expected, this is a rare type of synthesis. During the preparation of diethyl 2,6-dimethyl-3,5-pyridinedicarboxylate (**49**) from ethyl acetoacetate and hexamethylene tetramine, triethyl 2,7,8a-trimethyl-1,4,4a,5,8,8a-hexahydro-1,8-naphthyridine-3,6-dicarboxylate (**51**) was isolated in 2% yield; it appears to have arisen from addition of the intermediate (**50**) to the 2,3-bond of the main product (**49**).[1147]

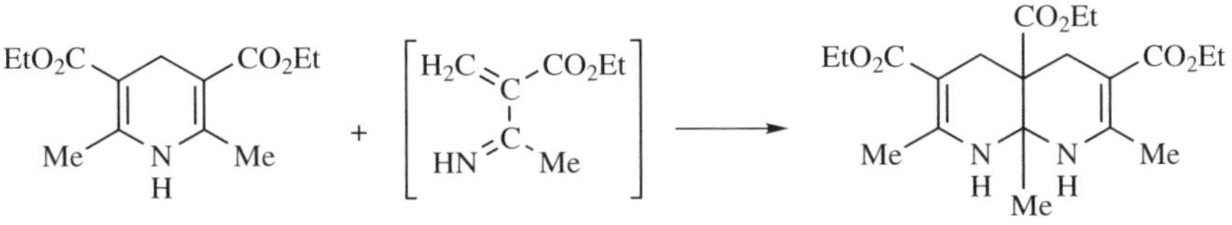

Also other examples.[3,510]

Where Two Synthons Supply N1 and C2

4,6-Dichloro-3-ethoxycarbonylacetyl-5-fluoropyridine (**52**) with triethyl orthoformate and subsequently cyclopropylamine gave 7-chloro-1-cyclopropyl-6-fluoro-4-oxo-1,4-dihydro-1,8-naphthyridine-3-carboxylic acid (**53**) (∼60%; note hydrolysis of the ester group).[487]

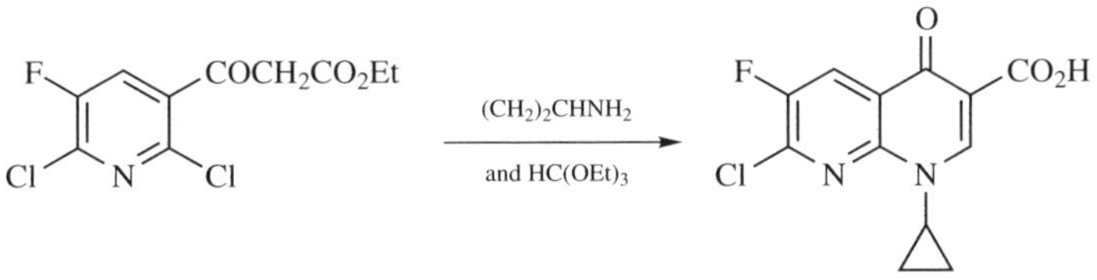

Where Two Synthons Supply C2 and C3 + C4

2-Pyridinamine (**54**) with benzaldehyde and subsequently pyruvic acid gave 2-phenyl-1,8-naphthyridine-4-carboxylic acid (**55**) (EtOH, heat: ?%).[68]

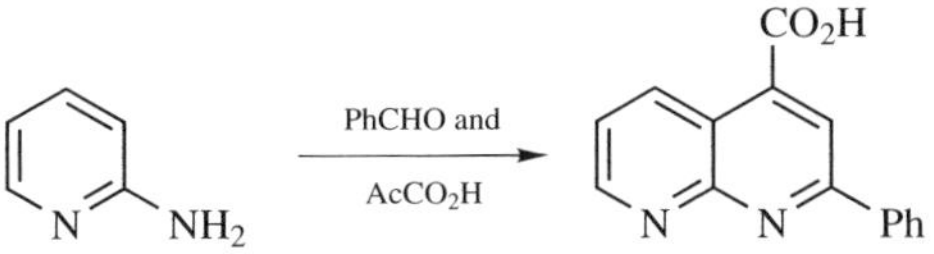

22.4. FROM OTHER HETEROCYCLIC SUBSTRATES

1,8-Naphthyridines may be made from a variety of heteromono-, heterodi-, or heterotricyclic substrates as illustrated by the following examples. Fused 1,8-naphthyridine derivatives may also be so made. e.g.,[567]

Isoxazolo[4,3-c][1,8]naphthyridines as Substrates

8-Chloro-5-methyl-3,3a,4,5-tetrahydroisoxazolo[4,3-c][1,8]naphthyridine (**56**) gave 6-chloro–3-hydroxymethyl-1-methyl-2,3-dihydro-1,8-naphthyridin-4(1H)-one (**57**) (Raney Ni: 88%; note the additional hydrolysis of =NH to =O).

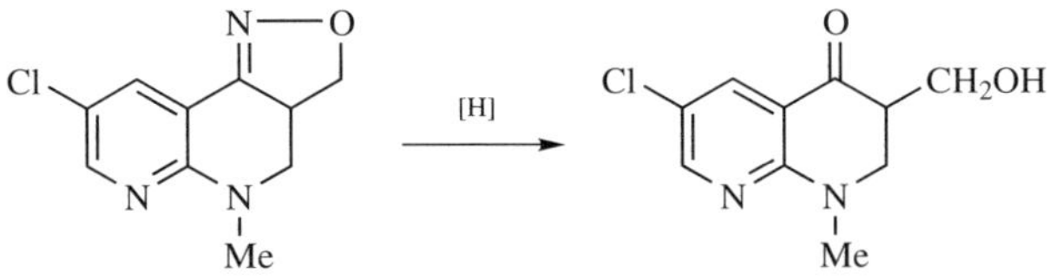

Oxazoles as Substrates

4-Methyl-2-(N-pent-4-enylacetamido)oxazole (**58**) gave 1-acetyl-7-methyl 1,2,3,4-tetrahydro-1,8-naphthyridine (**59**) (diazabicycloundecene, PhH, 180°C: 39%; a rational mechanism is suggested).[574]

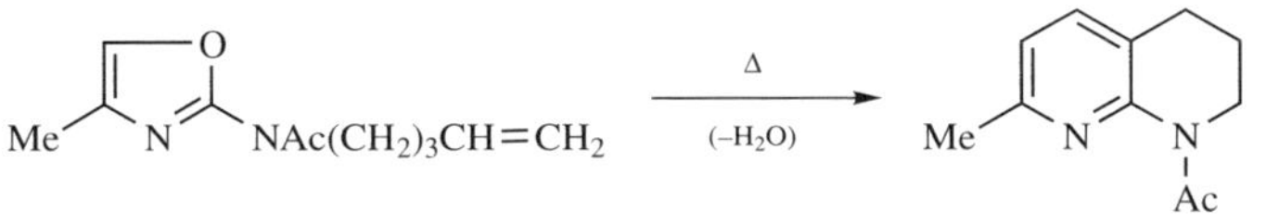

Pyrans as Substrates

Ethyl 6-amino-4-p-chlorophenyl-5-cyano-2-methylpyran-3-carboxylate (**60**) with 2-cyanoacetamide gave ethyl 5-amino-6-cyano-2-methyl-7-oxo-7,8-dihydro-1,8-naphthyridine-3-carboxylate (**61**) (piperidine, 130°C: 80%).[685]

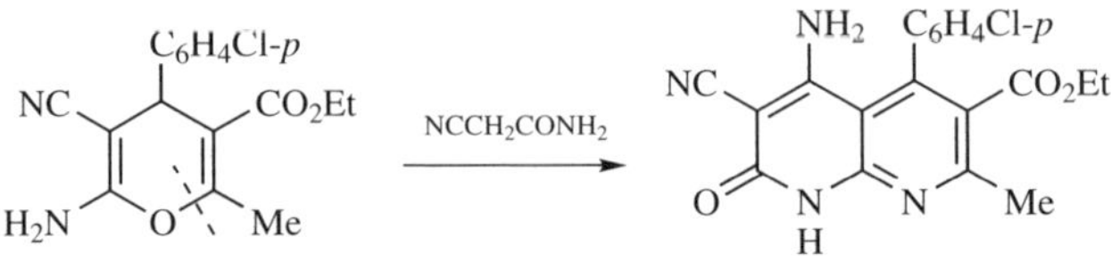

Pyrido[2,3-*b*][1,8]naphthyridines as Substrates

4,8-Dimethylpyrido[2,3-*b*][1,8]naphthyridine-2,6(1*H*,9*H*)-dione (**62**) underwent oxidation to give 2-amino-5-methyl-7-oxo-7,8-dihydro-1,8-naphthyridine-3-carboxylic acid *N*-oxide, probably the 1-oxide (**63**) (MeCO$_3$H: ~30%).[748]

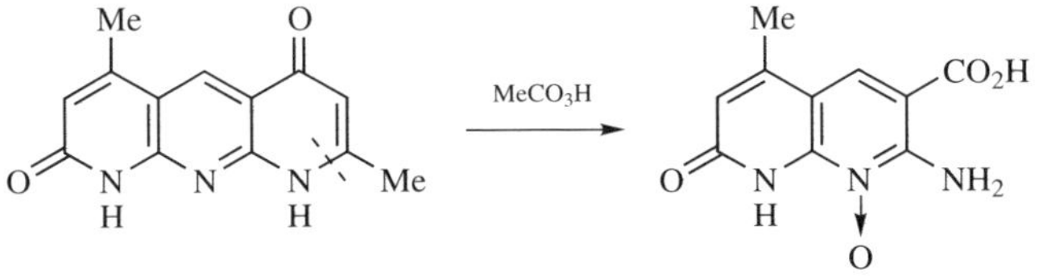

Pyrido[2,3-*d*][1,3]oxazines as Substrates

1-Phenyl-2*H*-pyrido[2,3-*d*][1,3]oxazine-2,4(1*H*)-dione (**64**) with diethyl malonate gave ethyl 4-hydroxy-2-oxo-1-phenyl-1,2-dihydro-1,8-naphthyridine-3-carboxylate (**65**) (reactants, NaH, 150°C: ~85%);[453,407] the same substrate (**64**) with ethyl nitroacetate gave 4-hydroxy-3-nitro-1-phenyl-1,8-naphthyridin-2(1*H*)-one (**66**) (NaH/MeCONMe$_2$, 100°C: 77%).[1191]

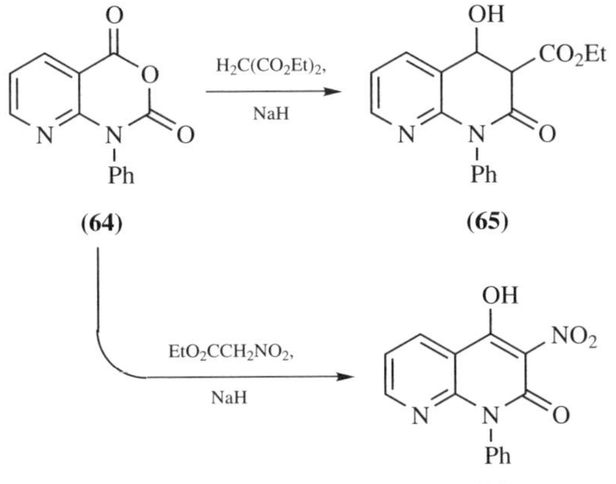

Also other examples.[140,471,816,912,1127,1225]

Pyrido[1,2-*a*]pyrimidines as Substrates

2-Benzyl-6-methyl-4*H*-pyrido[1,2-*a*]pyrimidin-4-one (**67**, R=CH$_2$Ph) underwent thermal rearrangement to give 2-benzyl-7-methyl-1,8-naphthyridin-4(1*H*)-one (**68**, R=CH$_2$Ph) (Dowtherm, 220°C: 42%;[263] the homologous substrate (**67**, R=Ph) likewise gave 7-methyl-2-phenyl-1,8-naphthyridin-4(1*H*)-one (**68**, R=Ph) (reflux: 76%).[1390]

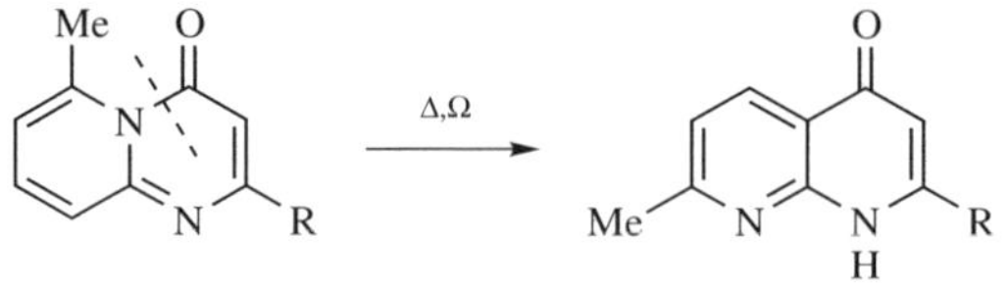

6-Amino-2-trifluoromethyl-4*H*-pyrido[1,2-*a*]pyrimidin-4-one (**69**) gave 7-amino-2-trifluoromethyl-1,8-naphthyridin-4(1*H*)-one (**70**) (Ph$_2$O, reflux: 90%).[533]

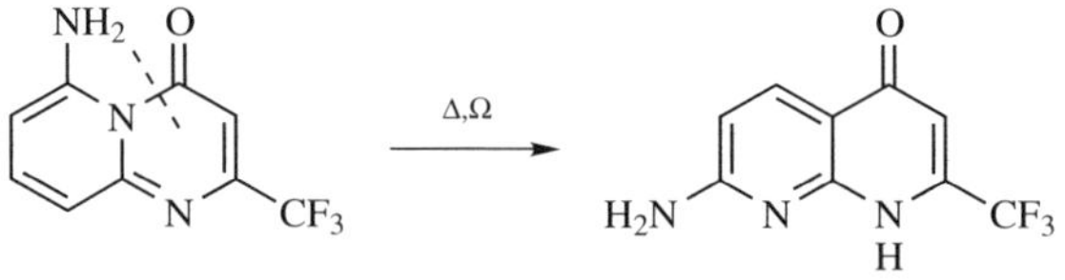

Also other examples.[524,565,569,909,971,1006,1082,1085,1185,1186,1269]

Pyrido[2,3-*d*]pyrimidines as Substrates

Pyrido[2,3-*d*] pyrimidine (**71**) with malononitrile gave 2-amino-1,8-naphthyridine-3-carbonitrile (**72**, R=CN) (reactants, MeOH, 20°C: 75%) or with ethyl cyanoacetate gave ethyl 2-amino-1,8-naphthyridine-3-carboxylate (**72**, R=CO$_2$Et) (likewise: 8%); other activated methylene synthons also gave products but in very poor yield, even at appreciably increased temperatures.[1030]

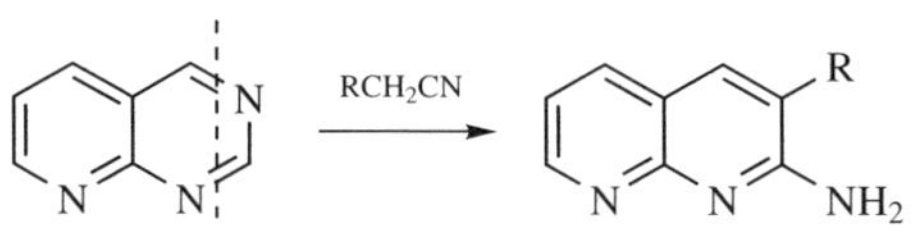

Also other examples.[1013,1054]

Pyrimido[1,2-*a*]naphthyridines as Substrates

Diethyl 4-hydroxy-6-nitro-10-oxo-7,10-dihydropyrimido-[1,2-*a*][1,8]naphthyridine-3,9-dicarboxylate (**73**) underwent alkaline degradation and saponification to give 7-amino-6-nitro-4 oxo-1,4-dihydro-1,8-naphthyridine-3-carboxylic acid (**74**) (NaOH/EtOH: above 80%).[758]

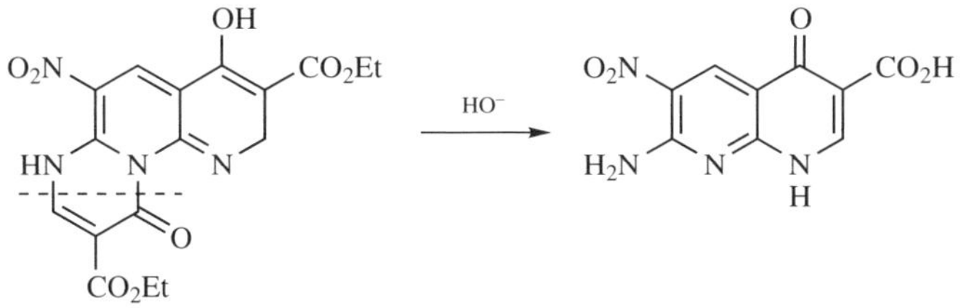

Also other examples.[748]

Pyrrolo[2,3-*b*]pyridine as Substrates

Pyrrolo[2,3-*b*]pyridine (**75**) underwent pyrolysis in chloroform to give a separable mixture of 1,8-naphthyridine (**76**, R=H) and 3-chloro-1,8-naphthyridine (**76**, R=Cl) (vapor phase, 550°C: low yields).[1145]

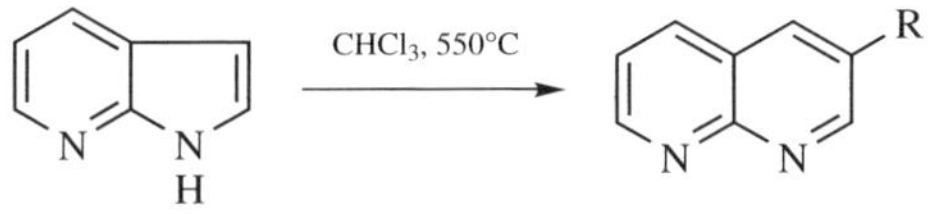

Tetrazolo[2,4,5-*ij*][1,8]naphthyridines as Substrates

4,4,9,9-Tetramethyl-5,6,7,8-tetrahydro-4*H*,9*H*-tetrazolo[2,4,5-*ij*][1,8]naphthyridine (**77**) lost nitrogen on irradiation to give 2,2,7,7-tetramethyl-1,2,3,5,6,7-hexahydro-1,8-naphthyridine (**78**) (*h*ν, $C_6D_5CD_3$: 100% by NMR).[534]

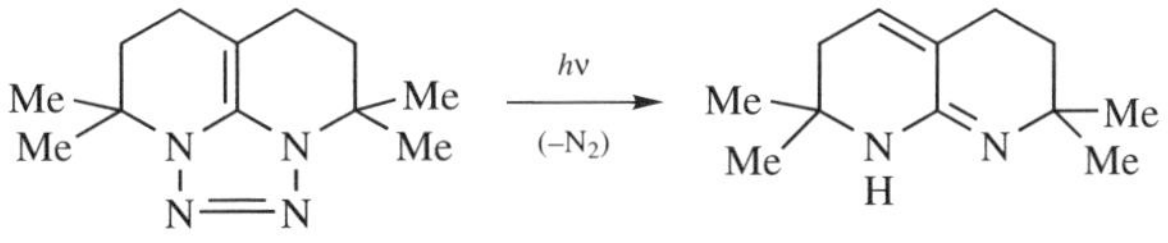

1,2,4-Triazines as Substrates

3-(Pent-4-ynylamino)-6-trifluoromethyl-1,2,4-triazine (**79**) underwent thermal cyclization to give 3-trifluoromethyl-5,6,7,8-tetrahydro-1,8-naphthyridine

(**80**) with loss of nitrogen and formation of the 4a,8a- and 5,6-bonds (Ph$_2$O, 240°C: 67%.[1404]

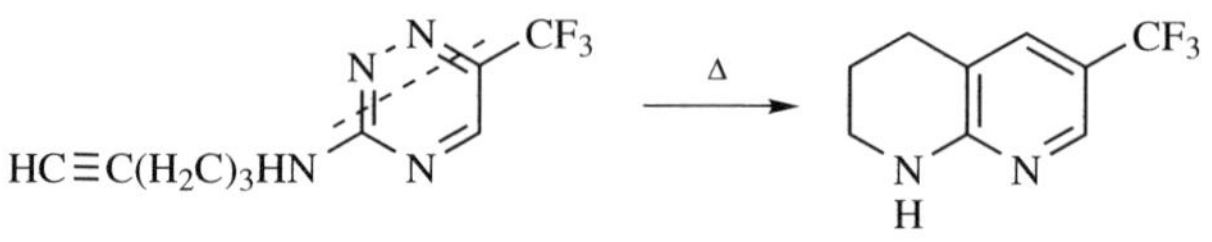

(**79**) → (**80**)

Also other examples.[1396]

CHAPTER 23

1,8-Naphthyridine, Alkyl-1, 8-naphthyridines, and Aryl-1, 8-naphthyridines

The scope of this chapter with respect to 1,8-naphthyridines is analogous to that of Chapter 2 in terms of 1,5-naphthyridines. Much more work has been done on the 1,8-naphthyridines than of any of the other naphthyridine systems; indeed, the word "naphthyridine" was used originally to cover only the 1,8-system.[1] Its symmetry and the contiguity of its nitrogen atoms in space have made 1,8-naphthyridine especially attractive to theoretical chemists and for the construction of metallic and other complexes; moreover, a variety of its derivatives have shown useful biological activities. e.g.,[356,357,397,1263,1422]

The general chemistry of 1,8-naphthyridines, usually in association with those of related systems, has been reviewed from time to time but not always in any depth.[49,51,55,58,61,265,407,419,670,1260,1357,1430,1432]

23.1. 1,8-NAPHTHYRIDINE AND HYDRO DERIVATIVES

23.1.1. Preparation of 1,8-Naphthyridine

The parent 1,8-naphthyridine was first made by catalytic dechlorination of 2,4-dichloro-1,8-naphthyridine (no details);[30,33] it is probably best made by a Skraup procedure (33%)[250] or by a modified Friedländer procedure (80%).[1441] Other routes have been mentioned in Chapter 22.

The preparation of several unsubstituted hydro-1,8-naphthyridimes has been described as follows.

1,2-Dihydro-1,8-naphthyridine (**2**) by catalytic hydrogenlysis of 2,4-dichloro-1,8-naphthyridine (Pd/C, H_2: ?%)[31] or from 3-methyl-2-pyridinamine (**1**)

The Naphthyridines: The Chemistry of Heterocyclic Compounds, Volume 63, by D.J. Brown

with 2-bromoacetaldehyde diethyl acetal (250°C: ?%).[8]

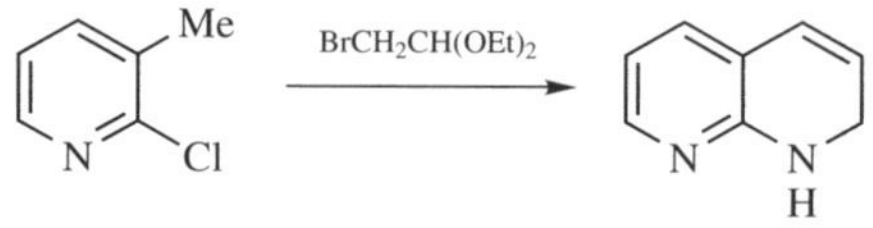

1,2,3,4-Tetrahydro-1,8-naphthyridine (3) by reduction of 1,8-naphthyridine (B_2H_6, 20°C: 44%;[411] or Pd/C, H_2: 35%);[47] or by cyclization of 3-(3-aminopropyl)pyridine (4) (Na/PhMe: 30%).[1122]

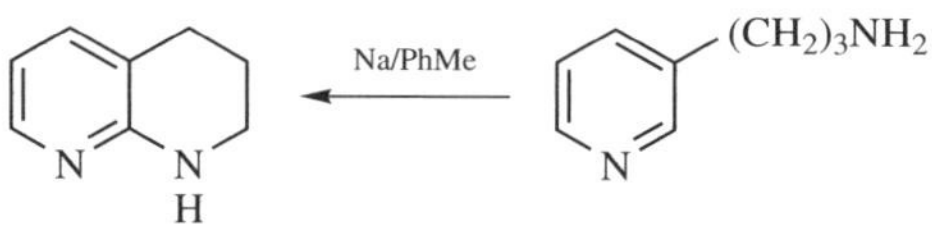

1,2,3,4,4a,5,6,7-Octahydro-1,8-naphthyridine (5) by oxidation of decahydro-1,8-naphthyridine [N-bromosuccinimide: 55%; or Hg(OAc)$_2$: 18%][627,cf. 1088] or by thermal bicyclization of 2,2-bis(3-aminopropyl)acetic acid (6) (dry distillation: ?%).[2]

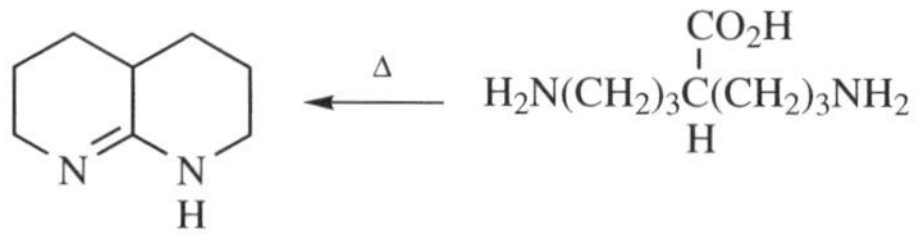

Decahydro-1,8-naphthyridine (7) by reduction of 1,8-naphthyridine (Na, EtOH: 70%;[627] or with an additional hydrogenation: 50%).[47,cf. 64] Also by a reductive dehalogenation of 2,4-dichloro-1,8-naphthyridine (?%).[64]

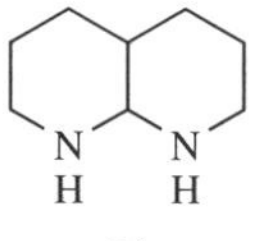

23.1.2. Properties of 1,8-naphthyridine

The bald physical properties of 1,8-naphthyridine may be found from its entry in Table A4. More extensive information or studies, usually involving comparisons with related systems and/or alkyl derivatives, are referenced here.

Aromaticity[676]
Dipole moment[488]

Electron spin resonance[1079,1083]
Infrared spectra[44,343,689,948,1124,1247]
Ionization[45,528,806,813,1176,1192]
Mass spectra[366,975]
Nuclear magnetic resonance
    ^{13}C spectra[913,1174,1176,1192]
    ^{1}H spectra[42,43,296,922]
    ^{14}N spectra[365]
    ^{15}N spectra[1128,1189]
Nuclear quadrapole resonance spectra[882]
Polarography[829,1303]
Resonance energy, etc.[209,812,840,998,1081,1101,1126,1173,1241]
Tautomerism[1242]
Ultraviolet spectra[45,759,807,820,1016,1181,1312,1426]
X-ray structural studies[809,832]

23.1.3. 1,8-Naphthyridine Complexes

As mentioned earlier, 1,8-naphthyridine and its derivatives are admirably suited to the preparation of metallic or other complexes. Although the details of such complexes are beyond the scope of this book, selected papers in this area are listed here according to the element or other entity involved.

Ag[325,459]
Au[254]
B[671]
Ba[255]
Ca[255]
Cd[256,261,1369]
Co[256,261,652,748,1369]
Cr[616,937,991,1238]
Cu[256,257,261,298,459,704,775,783,790,984,1369]
Fe[256,314,348,935,955,1362,1365,1369,1372]
Hg[1149,1206,1209]
La, etc.[512,662,1287,1355,1367]
Mn[256,261,652,934,1369]
Mo[661,904,1358]
Ni[128,256,261,315,652,739,1369]
Pd[256,957]

Pr[297,958]
Pt[280,cf. 319]
Re[675]
Rh[398,430,1155,1156,1436]
Ru[279,287,399,638,898,940,1125,1222,1370]
Sn[678]
Sr[255]
U[341]
Yb[321]
Zn[256,261,652,1369]
Organic entities[424,506,606,665,927,1160]

23.1.4. Reactions of 1,8-Naphthyridine

The nuclear *reduction* of 1,8-naphthyridine has been covered in Section 23.1.1. Other reactions are illustrated by the following examples, but for *deuteration*, see a 1989 paper by Woźniak.[511]

C-Amination

1,8-Naphthyridine (**8**) gave 1,8-naphthyridin-2-amine (**10**) (KNH$_2$/NH$_3$; then KMnO$_4$: 10%);[425] the intermediate dihydro adduct (**9**) was clearly involved on NMR evidence.[302,353]

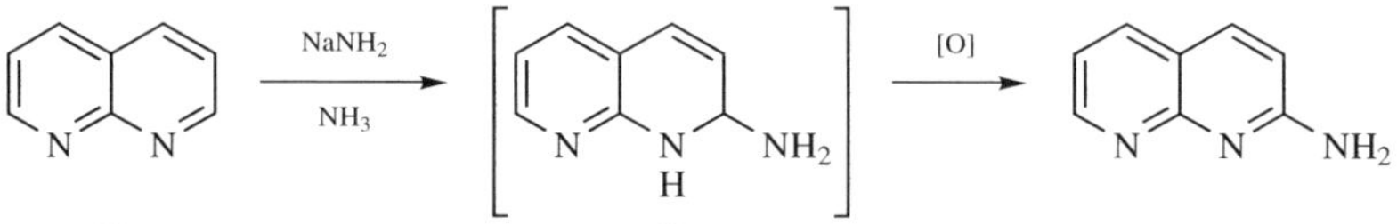

N-Amination

1,8-Naphthyridine with *O*-mesitylenesulfonylhydroxylamine gave the quaternary product, 1-amino-1,8-naphthyridin-1-ium mesitylenesulfonate (**11**) (reactants, CH$_2$Cl$_2$, 20°C: 72%), converted by benzoyl chloride into the zwitterionic product (**12**) (neat reactants, 90°C: 60%).[1003,cf. 1004,1028]

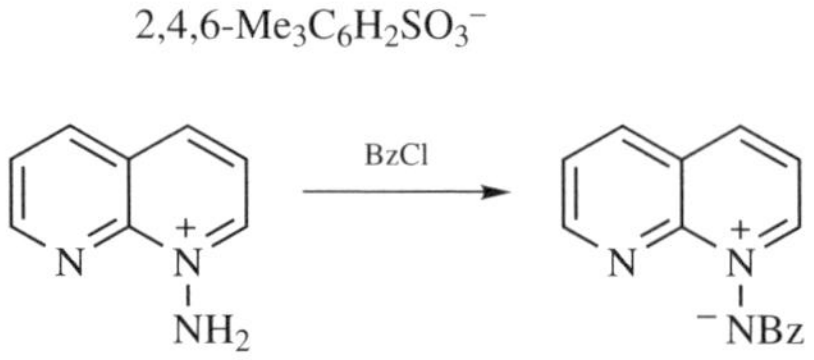

Halogenation

1,8-Naphthyridine hydrobromide gave a separable mixture of 3-bromo- (**13**, R = H) and 3,6-dibrimo-1,8-naphthyridine (**13**, R = Br) (Br$_2$, PhNO$_2$, 175°C: ~30% each; excess of Br$_2$ gave 72% of the dibromo product).[905,cf. 173]

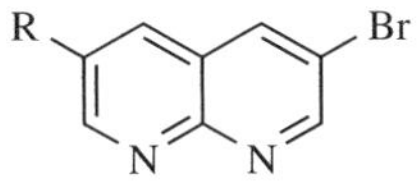

(**13**)

Cycloaddition

1,8-Naphthyridine with chloromethyl phenyl sulfone gave 5,6-bis(phenylsulfonyl)-5,5a,5b,6-tetrahydro-4bH-azirino[1,2-a]cyclopropa[c][1,8]naphthyridine (**14**) (reactants, KOH, Me$_2$SO, 20°C: 66%).[803,cf. 393]

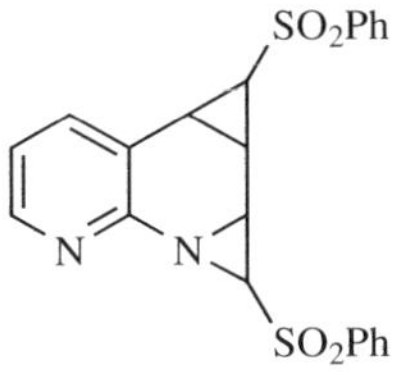

(**14**)

N-Oxidation

1,8-Naphthyridine gave its 1-oxide (**15**) (PhCO$_3$H, CHCl$_3$: 47%;[1212] or H$_2$O$_2$, Na$_2$WO$_4$: 5%).[1011]

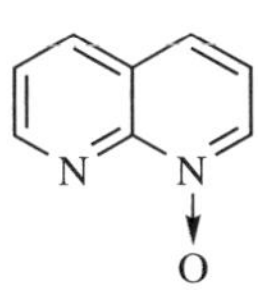

(**15**)

Quaternization

1,8-Naphthyridine gave its 1-methiodide (**16**) (MeI, AcMe, reflux: ~90%),[1214] its 1,8-bismethofluorosulfonate (**17**) (neat FSO$_3$Me, 100°C: 88%),[1061] or

1,2-dihydroimidazo[1,2,3-*ij*][1,8]-naphthyridinediium dibromide (**18**) (BrCH$_2$CH$_2$Br, reflux: 90%).[1144]

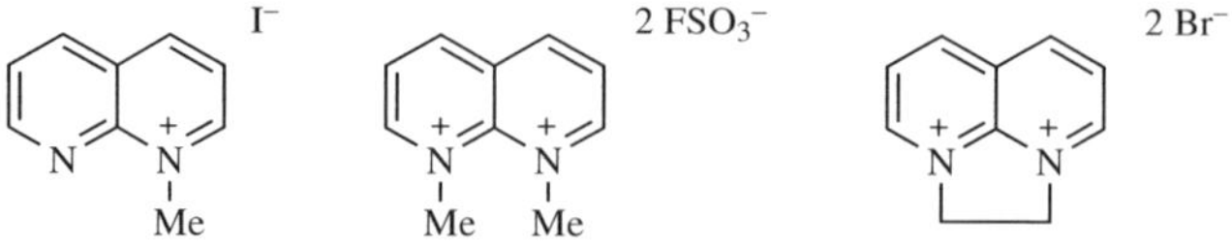

In contrast to the last example, 1,8-naphthyridine with excess *o*-bis(bromomethyl)benzene gave *o*-bis(1,8-naphthyridiniomethyl)benzene dibromide (**19**) (*o*-Cl$_2$C$_6$H$_4$, 140°C: 41%).[817]

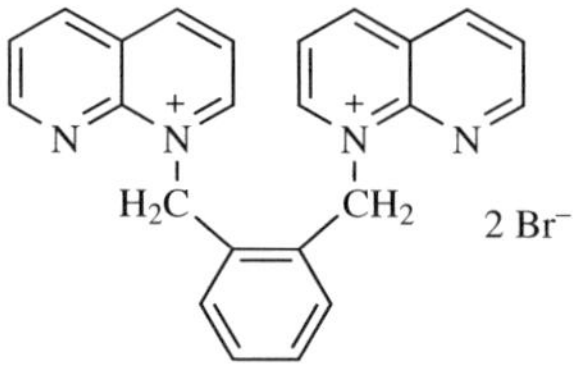

Also other examples, including miscellaneous information such as kinetic data.[59,377,906,961,993,1002,1137,1306]

Reissert Reactions

1,8-Naphthyridine with potassium cyanide and benzoyl chloride gave 1-benzoyl-1,2-dihydro-1,8-naphthyridine-2-carbonitrile (**20**) (reactants, 20°C: 21%).[1095]

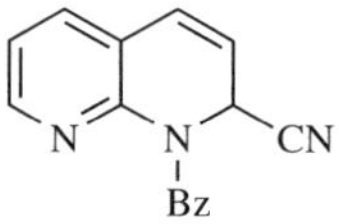

1,8-Naphthyridine with benzoyl chloride and indole gave 1-benzoyl-2-(indol-3-yl)-1,2-dihydro-1,8-naphthyridine (**21**) (20°C: 32%) or 1,8-dibenzoyl-2,7-bis(indol-3-yl)-1,2,7,8-tetrahydro-1,8-naphthyridine (**22**) (90°C: 30%).[804,cf. 791,842]

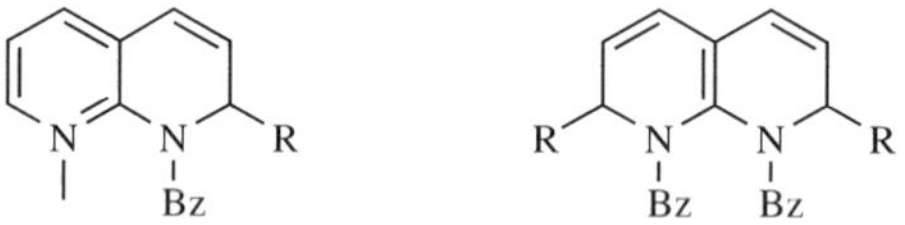

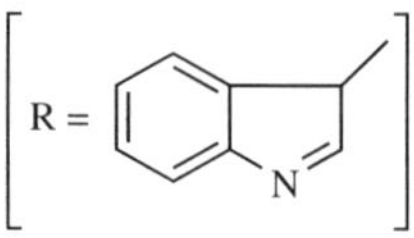

23.2. ALKYL- AND ARYL-1,8-NAPHTHYRIDINES

23.2.1. Preparation of Alkyl- and Aryl-1,8-naphthyridines

Many alkyl or aryl-1,8-naphthyridines have been made by *primary synthesis* (see Chapter 22). Other procedures involve the introduction of substituted or unsubstituted alkyl or aryl groups into existing 1,8-naphthyridines or the interconversion of such groups already in place. The following examples illustrate the methods used.

By Direct Alkylation/Arylation

1,8-Naphthyridine underwent self-arylation to give 2,2′-bi-1,8-naphthyridine (LiBu, dithiane; then O_2: 69%).[737]

2-Methyl- gave 2,7-dimethyl-1,8-naphthyridine (MeLi, Et_2O, −50°C: 84%).[473]

Ethyl 7-chloro-1-cyclopropyl-6-fluoro-4-oxo-1,2,3,4-tetrahydro-1,8-naphthyridine-3-carboxylate (23, R = H) gave its 5-methyl derivative (23, R = Me) (Bu^{t}Li: then MeI: 86%).[1158]

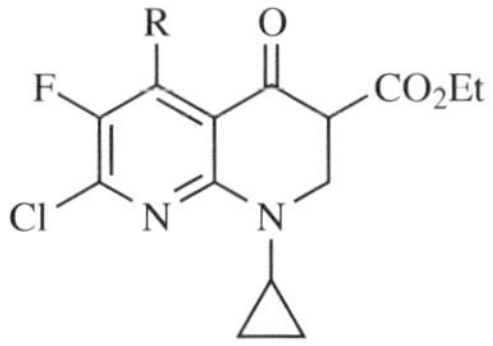

(23)

3-Nitro-1,8-naphthyridine (24, R = H) gave 3-nitro-4-phenylsulfonylmethyl-1,8-naphthyridine (24, R = CH_2SO_2Ph) ($ClCH_2SO_2Ph$, NaOH, Me_2SO: 82%).[1164]

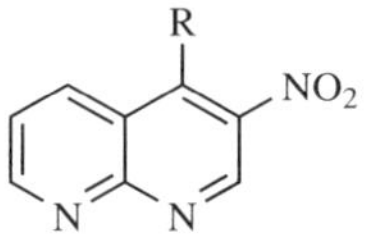

(24)

Also other examples.[86,194,409,1075,1102]

From Halogeno-1,8-naphthyridines

1-Butyl-4-chloro-3-nitro-1,8-naphthyridin-2(1*H*)-one (25, R = Cl) gave 1-butyl-3-nitro-4-phenyl-1,8-naphthyridin-2(1*H*)-one (25, R = Ph) [PhB(OH)$_2$, Pd(PPh$_3$)$_4$, Cs_2CO_3: 86%, analogs likewise].[160]

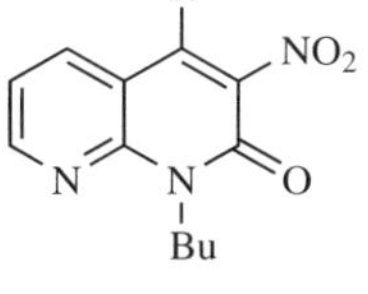

(25)

2,7-Dichloro-1,8-naphthyridine with *tert*-butyl 2-ethynylpyrrole-1-carboxylate gave 2,7-bis(1-*tert*-butoxycarbonylpyrrol-2-yl-ethynl)-1,8-naphthyridine (**26**) [reactants, Et$_3$N, CuI, Pd(PPh$_3$)$_2$Cl$_2$, THF: 784].[122,cf. 118]

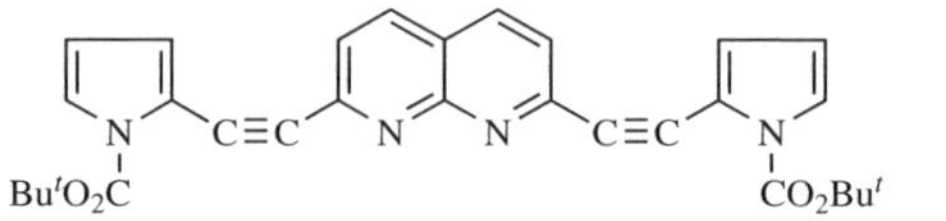

(**26**)

Also other examples.[176,288,444,465,466,905,1158,1184]

From 1,8-Naphthyridine *N*-Oxides

1,8-Naphthyridine 1-oxide (**27**) with *o*-methoxyphenyllithium gave 2-*o*-methoxyphenyl-1,8-naphthyridine (**28**) (reactants, Et$_2$O: 13%).[404]

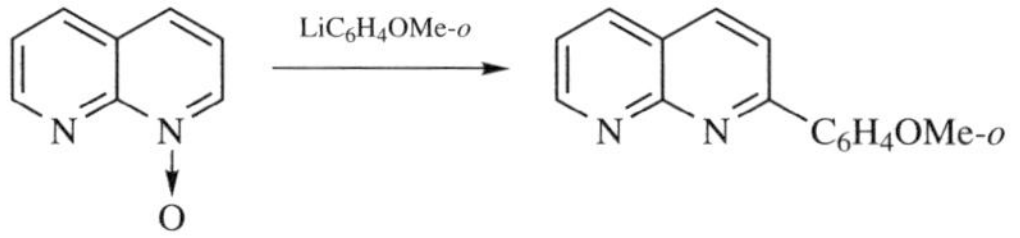

(**27**)   (**28**)

By Modification of Existing Alkyl Groups

Ethyl 1-(2-chloroethyl)-4-oxo-7-(pyrrolidin-1-yl)-1,4-dihydro-1,8-naphthyridine-3-carboxylate (**29**) gave 4-oxe-7-(pyrrolidin-1-yl)-1-vinyl-1,4-dihydro-1,8-naphthyridine 3-carboxylic acid (**30**) (NaOH: 69%); note the additional hydrolysis of the ester group].[1049]

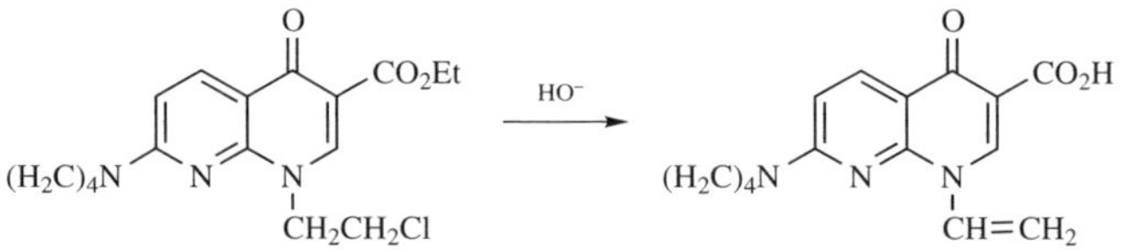

(**29**)   (**30**)

Ethyl 2,7-dimethyl- gave ethyl 2-methyl-7-[2-(5-nitrofuran-2-yl)vinyl]-4-oxo-1,4-dihydro-1,8-naphthyridine-3-carboxylate (**31**) (5-nitrofuran-2-carbaldehyde, Ac$_2$O/AcOH: 27%)[249]

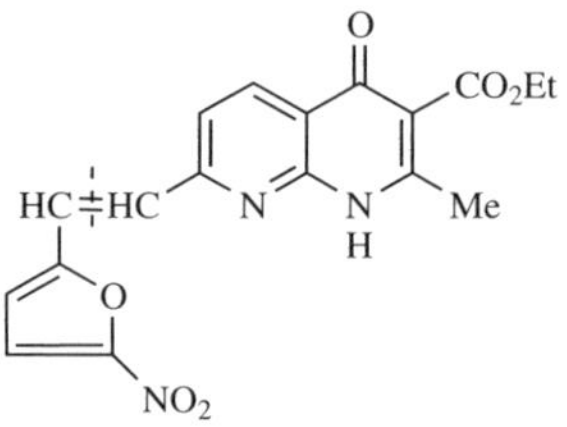

(**31**)

Also other examples.[667,926,938,1134,1163,1182,1407]

Note: Other types of modification are exemplified in Section 23.2.2.

23.2.2. Reactions of Alkyl- and Aryl-1,8-naphthyridines

Only reactions specific to the alkyl substituent(s) are covered here; nuclear reactions and those affecting aryl groups (e.g., nitration) will be found in appropriate chapters. Examples follow.

Halogenation of Alkyl Substituents

2,7 Dimethyl-1,8 naphthyridine (**32**, R = Me) with 2,4,6-trichloro-1,3,5-triazine gave 2,7-bis(chloromethyl)- (**32**, R = CH_2Cl) ($CHCl_3$: 33%)[794] or with *N*-chlorosuccinimide and benzoyl peroxide gave either 2,7-bis(dichloromethyl)- (**32**, R = $CHCl_2$) (4 mol NCS, CCl_4: 97%) or 2,7-bis(trichloromethyl)-1,8-naphthyridine (**32**, R = CCl_3) (NCS 8 mol, CCl_4: 89%).[473]

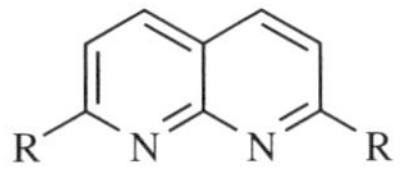

(**32**)

1-Ethyl-6-fluoro-7-methyl-4-oxo- (**33**, R = Me) gave 1-ethyl-6-fluoro-4-oxo-7-trichloromethyl-1,4-dihydro-1,8-naphthyridine-3-carboxylic acid (**33**, R = CCl_3) (neat $SOCl_2$, 60°C: 90%).[1091]

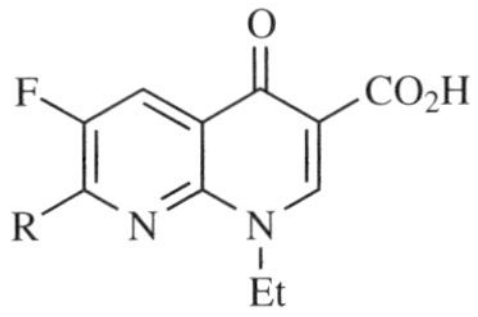

(**33**)

Ethyl 6-ethoxy-7-(2-ethoxycarbonylvinyl)-(**34**) gave ethyl 7-(1,2-dibromo-2-ethoxycarbonylethyl)-6-ethoxy-1-ethyl-4-oxo-1,4-dihydro-1,8-naphthyridine-3-carboxylate by addition of bromine to the exocyclic double bond (Br_2/AcOH: 91%).[1060]

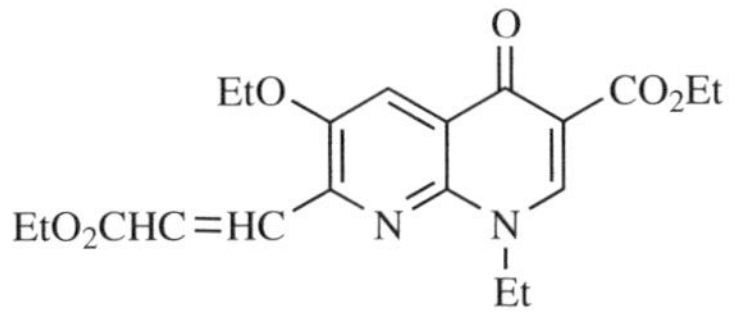

(**34**)

Also other examples.[112,924,1064,1135]

Oxidation of Alkyl Substituents

2,7-Dimethyl-1,8-naphthyridine (**35**, R = H) gave 1,8-naphthyridine-2,7-dicarbaldehyde (**36**, R = H) (SeO_2, dioxane: 59%);[1133,cf. 724] likewise, 2,7-dimethyl-4-octyloxy-1,8-naphthyridine (**35**, R = OC_8H_{17}) gave 4-octyloxy-1,8-naphthyridine-2,7-dicarbaldehyde (**36**, R = OC_8H_{17}) (82%).[1376]

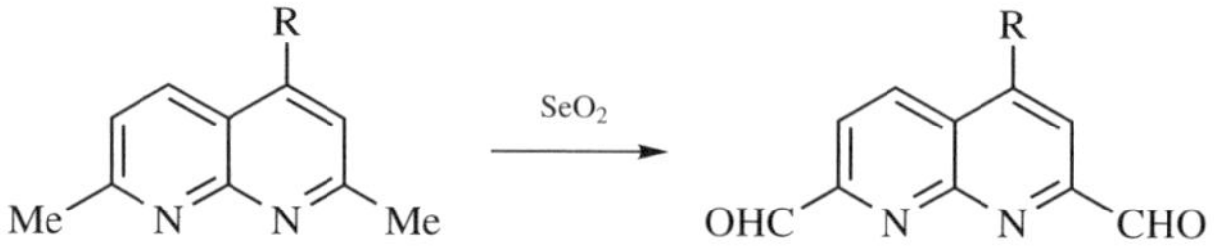

2-Styryl-1,8-naphthyridine gave 1,8-naphthyridine-2-carboxylic acid (CrO_3, AcOH, H_2O: 50%).[874]

7-Methyl-1,8-naphthyridin-4(1H)-one (**37**) on nitration gave 6-nitro-5-oxo-5,8-dihydro-1,8-naphthyridine-2-carboxylic acid (**38**) (HNO_3, heat: ~55%).[726]

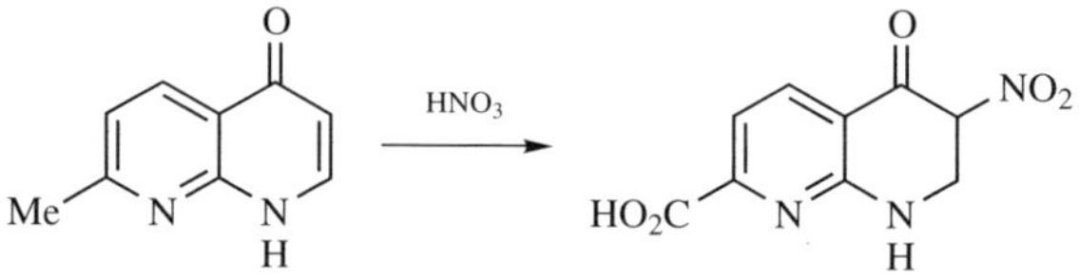

Also other examples.[350,395,483,1060,1084,1163,1441,1445,1450]

Acylation of Alkyl Substituents

5,7-Dimethyl-1,8-naphthyridin-2(1H)-one with butyllithium followed by benzonitrile and then hydrochloric acid gave 5-methyl-7-phenacyl-1,8-naphthyridin-2(1H)-one (**39**); in contrast, the same substrate treated similarly, but with sodium amide/ammonia in place of butyllithium, gave the isomeric 7-methyl-5-phenacyl-1,8-naphthyridin-2(1H)-one (**40**) (49%).[401]

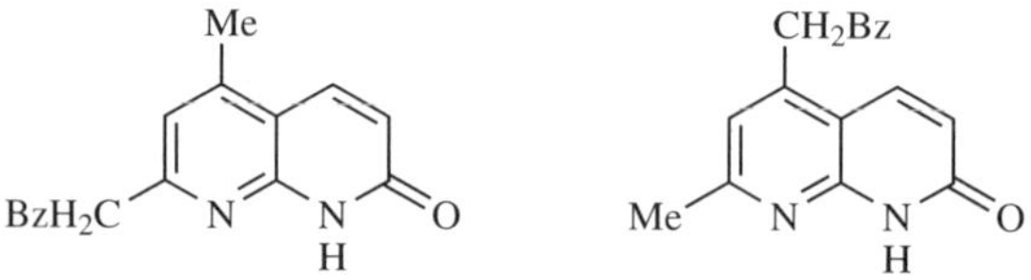

Also other examples.[938]

Cyclizations Involving Alkyl Substituents

2-Styryl-1,8-naphthyridine (**41**) with diazomethane gave 2-(4-phenylpyrazolin-3-yl)-1,8-naphthyridine (**42**) (55%); analogs likewise.[1131]

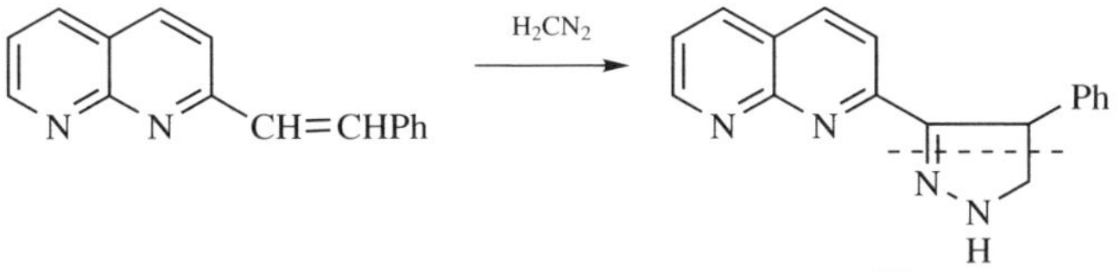

Also other examples.[1106]

CHAPTER 24

Halogeno-1,8-naphthyridines

Halogeno substituents at the 2-, 4-, 5-, and 7-positions of 1,8-naphthyridine will be activated toward aminolysis and similar reactions; those at the 3- and 6-positions will be only mildly so activated; and extranuclear halogeno substituents will resemble at best that in benzyl chloride.

Accordingly, halogeno-1,8-naphthyridines have proved to be valuable intermediates.[427] Moreover, such naphthyridines (with structures broadly based on that of nalidixic acid and usually with fluoro substituents) have shown remarkable antimicrobial and other bioactivities: for example, enoxacin (**1**),[406,431,457,489,563,780,1091,1236,1316] tosufloxacin (**2**),[451,452,461,484,497] trovafloxacin,[530,541,579,765,1314] and related compounds.[423,501,527,547,550,617,883]

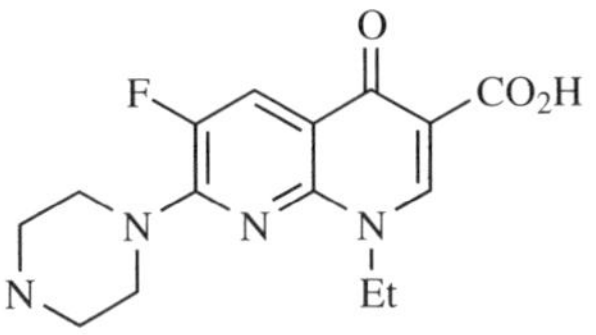

(1) 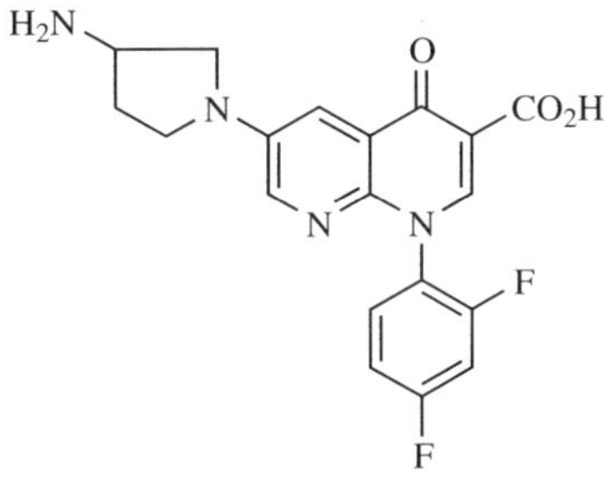   (2)

The X-ray structures, UV spectra, and mass spectra of various halogeno-1,8-naphthyridines have been studied.[302,546,800,1313]

24.1. PREPARATION OF HALOGENO-1,8-NAPHTHYRIDINES

This can be done in a variety of ways, including *primary synthesis* (see Chapter 22) and the *halogenation of alkyl groups* (see Section 23.2.2), which have been covered already. Other routes are illustrated by the following examples.

The Naphthyridines: The Chemistry of Heterocyclic Compounds, Volume 63, by D.J. Brown
Copyright © 2008 John Wiley & Sons, Inc.

By Direct Nuclear Halogenation

Note: Such halogenation of the parent 1,8-naphthyridine has been exemplified in Section 23.1.2.

1,8-Naphthyridin-2(1*H*)-one (**3**, R = H) gave 3-chloro-1,8-naphthyridin-2(1*H*)-one (**3**, R = Cl) (KClO$_3$, HCl, 50°C: 78%).[272]

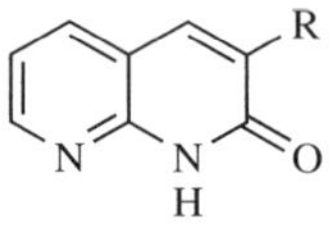

(**3**)

7-Methyl-1,8-naphthyridin-4(1*H*)-one (**4**, R = H) gave its 3-iodo derivative (**4**, R = I) (I$_2$, K$_2$CO$_3$, Me$_2$NCHO, H$_2$O: 82%).[905]

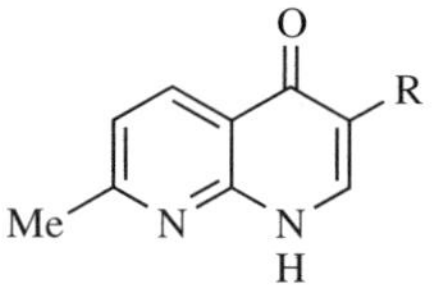

(**4**)

In contrast to the last example, 7-methyl-2,3-dihydro-1,8-naphthyridin-4(1*H*)-one (**5**, R = H) apparently gave its 7-bromo derivative (**5**, R = Br) (*N*-bromosuccinimide, CHCl$_3$: ~80%).[1383]

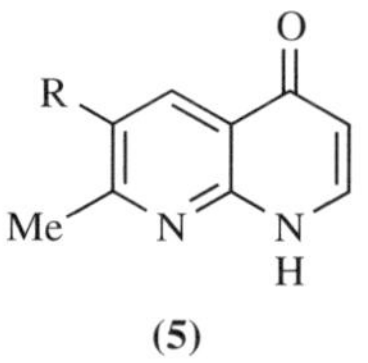

(**5**)

Also other examples.[379,693,982]

By Halogenolysis of 1,8-Naphthyridinones

Note: The chlorolysis of naphthyridinones or naphthyridinoles is usually done with neat phosphoryl chloride to which may be added phosphorus pentachloride or dimethylformamide in difficult cases; corresponding bromolysis is achieved by the use of phosphoryl bromide; and the chlorolysis of (extranuclear) hydroxyalkyl-1,8-naphthyridines may be done with thionyl chloride with or without added zinc chloride. Typical examples of these procedures are given here.

1,8-Naphthyridin-2(1*H*)-one (**6**, R = H) gave 2-chloro-1,8-naphthyridine (**7**, R = H) (neat POCl$_3$, reflux: 84%);[1100] 2-*m*-chlorophenyl-1,8-naphthyridin-

2(1*H*)-one (**6**, R = C$_6$H$_4$Cl-*m*) gave 2-chloro-3-*m*-chlorophenyl-1,8-naphthyridine (**7**, R = C$_6$H$_4$Cl-*m*) (POCl$_3$, microwave *hν*, 2.5 min: 94%).[77]

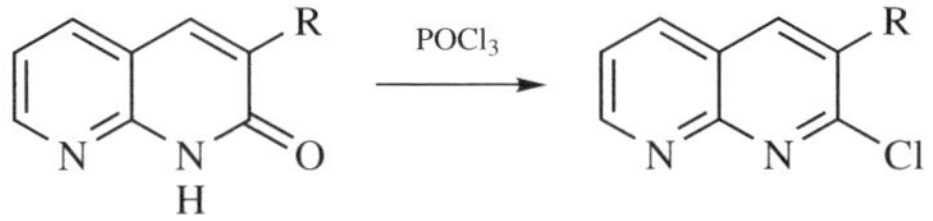

(**6**) (**7**)

3,6-Dinitro-1,8-naphthyridin-2(1*H*)-one gave 2-chloro-3,6-dinitro-1,8-naphthyridine (**8**) (POCl$_3$, reflux, 8 h: 60%).[1128]

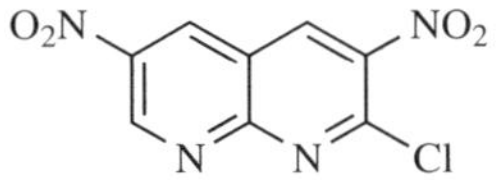

(**8**)

2,3-Dimethyl-1,8-naphthyridin-4(1*H*)-one gave 4-chloro-2,7-dimethyl-1,8-naphthyridine (**9**) (POCl$_3$, 90–95°C, 30 min: 85%).[1133]

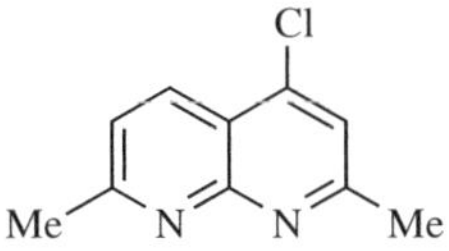

(**9**)

Also other examples of nuclear chlorolysis using neat phosphoryl chloride.[31, 79,85,88,89,101,153,239,418,471,533,543,551,600,686,690,701,718,720,728,743,763,764,771,850,856, 926,972,1091,1098,1186,1191,1221,1267,1293,1328,1384,1390,1401]

1,8-Naphthyridine-2,7(1*H*, 8*H*)-dione gave 2,7-dichloro-1,8-naphthyridine (**10**) (PCl$_5$, POCl$_3$, reflux: 81%)[389] 2,4-dichloro-1,8-naphthyridine was made similarly in comparable yield.[33]

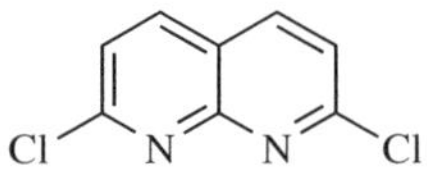

(**10**)

2-Oxo-1,2-dihydro-1,8-naphthyridine-3-carbonitrile gave 2-chloro-1,8-naphthyridine-3-carbonitrile (**11**) (PCl$_5$, POCl$_3$, reflux: 81%);[367] 2-chloro-7-ethoxy-5-phenyl-1,8-naphthyridine-3,6-dicarbonitrile (**12**) was made similarly (85%).[78]

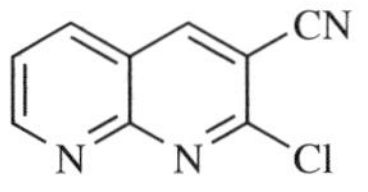

(**11**) 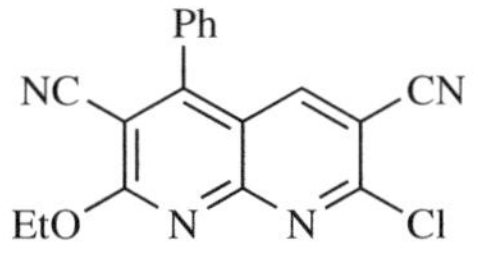 (**12**)

212   Halogeno-1,8-naphthyridines

Ethyl 1-(2-chloroethyl)-6-nitro-4,7-dioxo-1,4,7,8-tertahydro- gave ethyl 7-chloro-1-(2-chloroethyl)-6-nitro-4-oxo-1,4-dihydro-1,8-naphthyridine-3-carboxylate (**13**) [POCl$_3$, Me$_2$NCHO (i.e., a Vilsmeier reagent), 65°C: 82%].[1029]

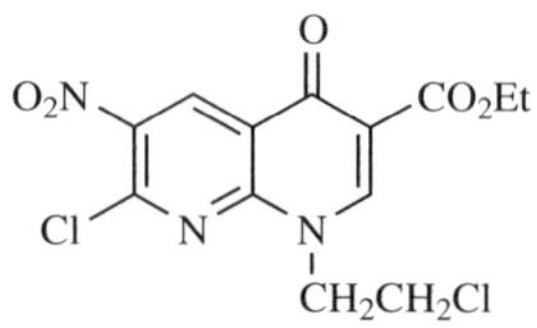

(**13**)

1,8-naphthyridin-4(1*H*)-one gave 4-bromo-1,8-naphthyridine (**14**) (POBr$_3$, 145°C: ~75%);[1186,] also analogous cases.[160,272,390]

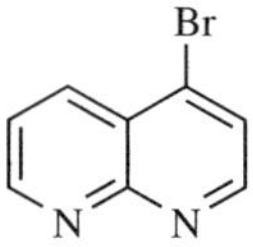

(**14**)

Ethyl 7-chloro-1-(2-hydroxyethyl)- (**15**, R = OH) gave ethyl 2-chloro-1-(2-chloroethyl)-4-oxo-1,4-dihydro-1,8-naphthyridine-3-carboxylate (**15**, R = Cl) (SOCl$_2$, CHCl$_3$, reflux: 70%);[1049] the analogous product, ethyl 7-chloromethyl-1-ethyl-4-oxo-1,4-dihydro-1,8-naphthyridine-3-carboxylate (**16**) was made somewhat similarly (SOCl$_2$, ZnCl$_2$, CHCl$_3$ 25°C: 83%).[1248]

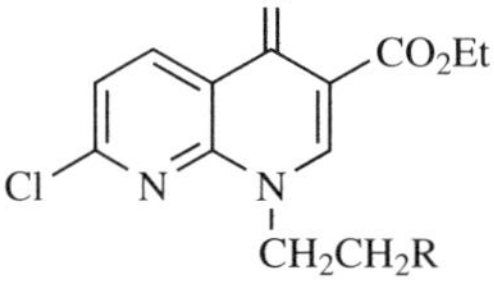

(**15**) 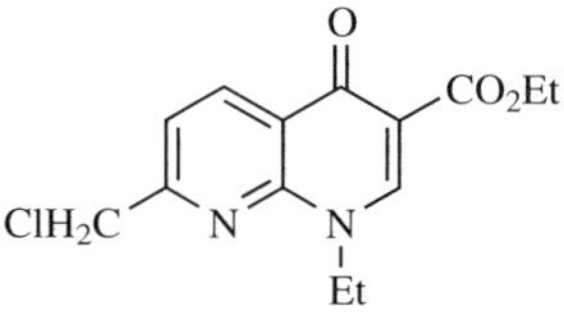   (**16**)

From 1,8-Naphthyridinamines

Ethyl 6-amino-1-ethyl- (**17**, R = Et, X = NH$_2$) gave ethyl 6-chloro-1-ethyl-7-methyl-4-oxo-1,4-dihydro-1,8-naphthyridine-3-carboxylate (**17**, R = Et, X = Cl) (HNO$_2$, 0°C; CuCl, 65°C: 64%);[387] an analogous substrate,(**17**, R = H, X = NH$_2$), likewise gave 1-ethyl-6-fluoro-7-methyl-4-oxo-1,4-dihydro-1,8-naphthyridine-3-carboxylic acid (**17**, R = H, X = F) (NaNO$_2$, HBF$_4$,

H_2O, $-5°C$; then HCl reflux: 36%).[395]

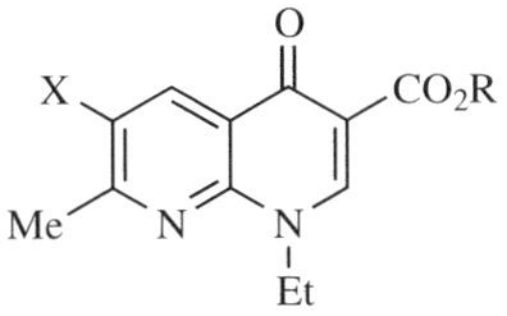

(17)

Also other examples.[893]

From 1,8-Naphthyridine N-Oxides

Note: At least in simple cases, the Meissenheimer reaction appears to be unsatisfactory for the production of halogeno-1,8-haphthyridines. Thus 1,8-naphthyridine 1-oxide with phosphoryl chloride gave a difficult-to-separate mixture of 2-, 3-, and 4-chloro-1,8-naphthyridine (in ratio 36 : 7 : 57);[225] phosphoryl bromide and bromine/acetic acid both gave even worse mixtures of brominated 1,8-naphthyridines.[317,448]

From 1,8-Naphthyridinecarboxylic Acids

1-Ethyl-7-methyl-4-oxo-1,4-dihydro-1,8-naphthyridine-3-carboxylic acid (**18**, R = CO_2H) gave 1-ethyl-3-iodo-7-methyl-1,8-naphthyridin-4(1*H*)-one (**18**, R = I) [substrate, I_2, $Pb(OAc)_4$, *hv*: 62%]; analogs likewise.[1213]

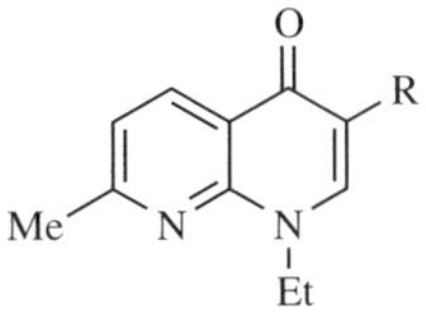

(18)

By Interchange of Halogeno Substituents

Note: Such interchange has been confined to the conversion of chloro to fluoro derivatives.

2,7-Dichloro- (**19**, X = Cl) gave 2,7-difluoro-1,8-naphthyridine (**19**, X = F) [KF, $O_2S(CH_2)_4$, 180°C: 50%];[918,cf. 821] hexachloro-gave hexafluoro-1,8-naphthyridine (likewise, 200°C: ?%).[918]

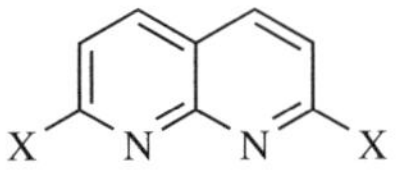

(19)

Ethyl 1-ethyl-4-oxo-1,4-dihydro-7-trichloromethyl- (**20**, X = Cl) gave ethyl 1-ethyl-4-oxo-1,4-trifluoromethyl-1,8-naphthyridine-3-carboxylate (**20**, X = F) (SbF$_5$, 30°C: ~40%).[1135]

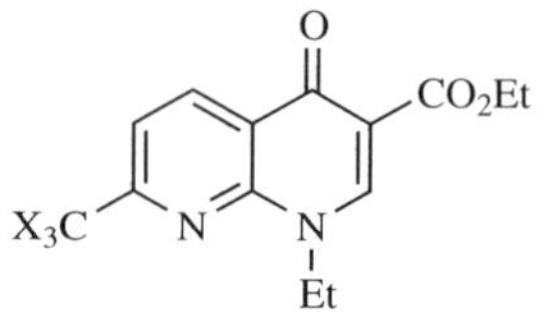

(**20**)

Also other examples.[1248]

24.2. REACTIONS OF HALOGENO-1,8-NAPHTHYRIDINES

The *interchange of halogeno substituents* has been covered at the end of Section 2.4.1. The many other important reactions of halogeno-1,8-naphthyridines are illustrated in the following classified examples.

Hydrolysis

Note: This may be done under either acidic or alkaline conditions.

2-Chloro 1,8-naphthyridine (**21**) gave 1,8-naphthyridin-2(1*H*)-one (**22**) (1.25M NaOH, reflux: 69%).[1008]

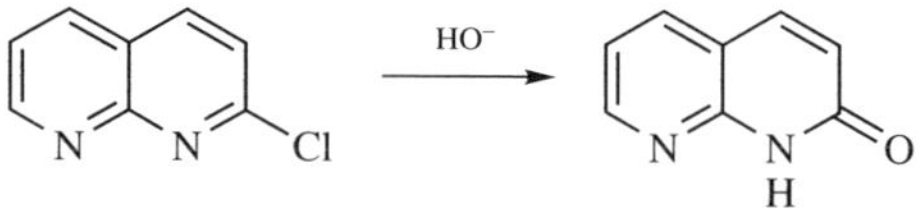

(**21**)   (**22**)

4-Bromo-1,8-naphthyridine gave 1,8-naphthyridin-4(1*H*)-one (HCl, reflux: 83%).[628]

3-Carboxymethyl-7-chloro-1,8-naphthyridin-2(1*H*)-one gave 3-carboxymethyl-1,8-naphthyridine-2,7(1*H*, 8*H*)-dione (**23**) 6M NaOH, reflux: 94%).[309]

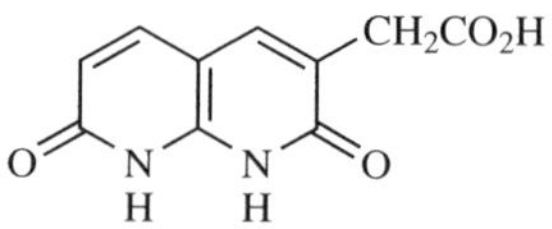

(**23**)

Also other examples.[263,568,660]

Alcoholysis/Phenolysis

2-Chloro- (**24**, R = H) gave 2-methoxy-1,8-naphthyridine (**25**, R = H) (MeONa/MeOH, reflux: 71%);[1100] likewise, 2-chloro-3-nitro- (**24**, R = NO$_2$) gave

2-methoxy-3-nitro- (**25**, R = NO$_2$) (66%)[682] or 2-ethoxy-3-nitro-1,8-naphthyridine (KOH/EtOH, 20°C: 47%).[1320]

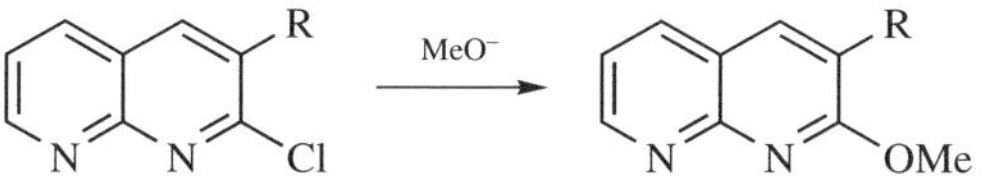

4-Chloro-2,7-dimethyl- gave 2,7-dimethyl-4-octyloxy-1,8-naphthyridine [C$_8$H$_{17}$ONa (made *in situ*), Me$_2$SO, 20°C: 70%].[1376]

4-Chloro-7-methyl-2-phenyl- gave 2-methyl-1,8-naphthyridine [PhONa (made *in situ*), Me$_2$NCHO, 80°C: 77%].[221]

Also other examples.[31,74,185,338,369,389,533,543,600,718,763,764,771,838,860,926,1029,1052,1076,1128,1364,1385,1406,1414,1445]

Aminolysis/Hydrazolysis and Acylaminolysis

Note: This reaction has been used heavily to afford primary, secondary, and tertiary amino-1,8-naphthyridines. A few simple typical examples are given here.

4-Bromo-1,8-naphthyridine (**26**, R = Br) gave 1,8-naphthyridin-4-amine (**26**, R = NH$_2$) (NH$_3$/PhOH, 170°C: 75%).[628] 2-chloro-3-nitro-1,8-naphthyridine (**27**, R = Cl) gave 3-nitro-1,8-naphthyridin-2-amine (**27**, R = NH$_2$) (NH$_3$/EtOH, 110°C: 70%).[1320]

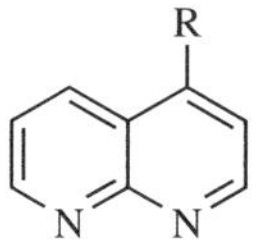

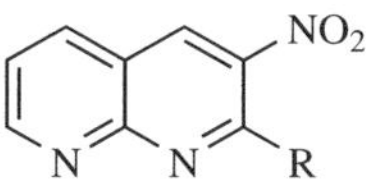

4-Chloro- (**28**, R = Cl) gave 4-methylamino-3-nitro-1-phenyl-1,8-naphthyridin-2(1*H*)-one (**28**, R = NHMe) MeNH$_2$/H$_2$O + THF, 20°C: 97%);[1191] 2,7-dichloro- (**29**, R = Cl) gave 2,7-bis(benzylamino)-1,8-naphthyridine (**29**, R = NHCH$_2$Ph) (neat PhCH$_2$NH$_2$, 150°C: 56%).[83]

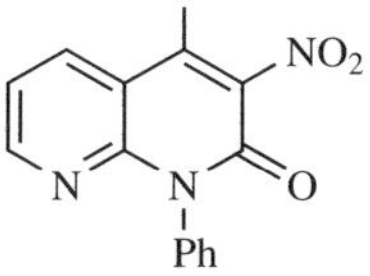

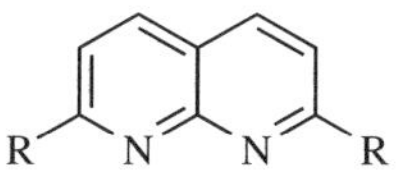

7-Chloro- gave 7-[bis(2-hydroxyethyl)amino]-1,8-naphthyridin-2(1H)-one (**30**) (diethanolamine, 120°C: 88%).[1081]

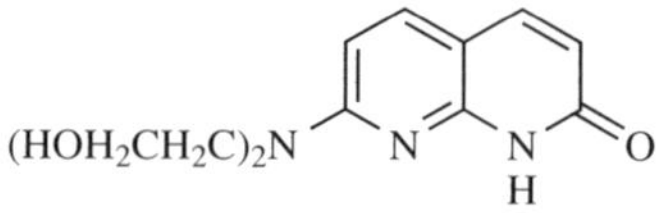

(HOH$_2$CH$_2$C)$_2$N

(**30**)

2-Chloro- (**31**, R = Cl) gave 2-hydrazino-3-p-methoxyphenyl-1,8-naphthyridine (neat H$_2$NNH$_2$·H$_2$O, reflux,: 88%).[1328]

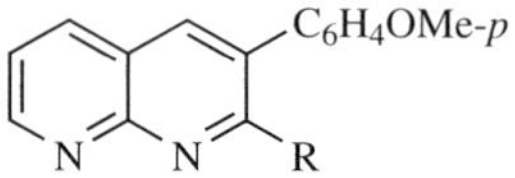

C$_6$H$_4$OMe-p

(**31**)

7-Chloro- (**32**, R = Cl) gave 7-(N-methylhydrazino)-4-oxo-1,4-dihydro-1,8-naphthyridine-3-carboxylic acid (**32**, R = NMeNH$_2$) (H$_2$NNHMe/EtOH, reflux: 83%).[1129]

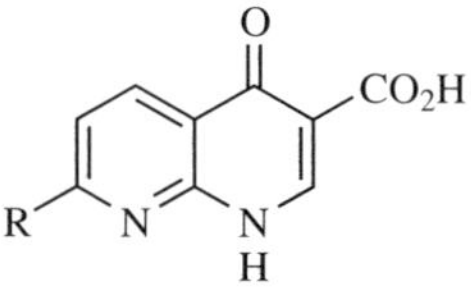

CO$_2$H

(**32**)

Also other examples.[74,77,78,80,101,143,156,158,166,263,335,346,353,364,386,397,444,447,449, 454,467,471,479,486,494,508,509,548,551,553,570,608,609,610,613,615,617,657,701,715,730,765,771, 850,856,860,925,1029,1049,1097,1107,1128,1177,1216,1223,1283,1293,1304,1345,1381,1401,1412, 1438]

Azidolysis

2-Chloro- (**33**, R = Cl) gave 2-azido-7-ethoxy-5-phenyl-1,8-naphthyridine (**33**, R = N$_3$), formulated as the tautomeric tetrazolo[1,5-a][1,8]naphthyridine (**33a**) (NaN$_3$, Me$_2$NCHO, reflux: ~80%).[718]

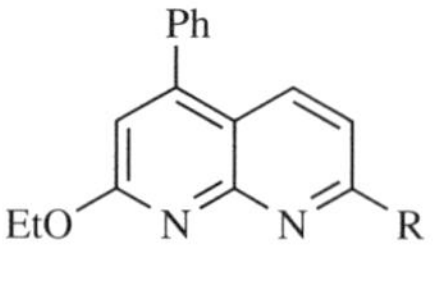

(**33**) 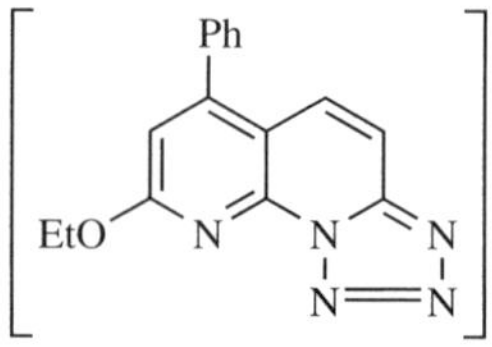   (**33a**)

Also other examples.[221,720,743,752,761,764,1384,1390,1391,1445]

Cyanolysis

1-Ethyl-3-iodo-7-methyl-1,8-naphthyridin-4(1*H*)-one (**34**, R = I) gave 1-ethyl-7-methyl-4-oxo-1,4-dihydro-1,8-naphthyridine-3-carbonitrile (**34**, R = CN) [KCN, CuI, MeN(CH$_2$)$_4$, reflux: 67%; note that the iodo substituent was situated at an unactivated position].[1213]

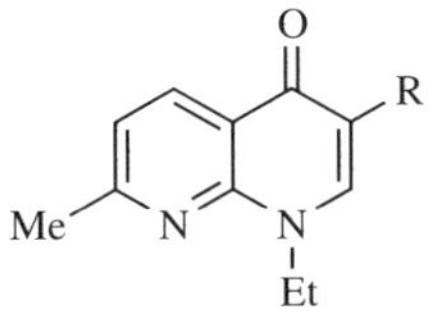

(**34**)

Cyclizations

2,4-Dichloro-*N,N*-diethyl-1,8-naphthyridine-3-carboxamide (**35**) with *N,N*-diethylisobutyrohydrazide gave 5-chloro-*N,N*-diethyl-9-isopropyl[1,2,4]triazolo[4,3-*a*][1,8]naphthyridine-1-carboxamide (**36**) (Dowtherm A, 155°C: 55%).[1423]

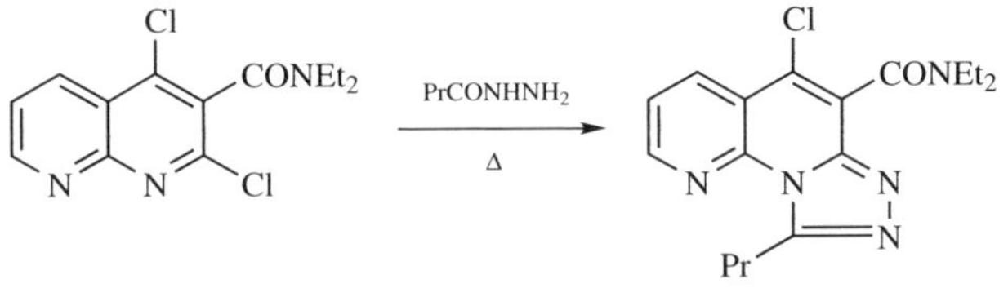

(**35**) → PrCONHNH$_2$, Δ → (**36**)

Also other examples.[1049]

Dehalogenation

Note: Dehalogenation may be done directly by hydrogenolysis under alkaline conditions or indirectly via a hydrazino-type derivative.

4-Chloro-2,7-dimethyl- (**37**, R = Cl) gave 2,7-dimethyl-1,8-naphthyridine (**37**, R = H) (Pd/CaCO$_3$, KOH, MeOH, H$_2$, 30°C: 92%).[1133]

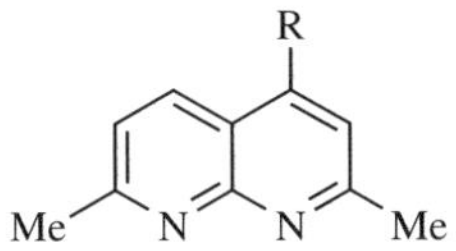

(**37**)

4-Chloro-7-methyl-1,8-naphthyridine (**38**, R = Cl) gave 2-methyl-1,8-naphthyridine (**38**, R = H), either by hydrogenolysis (Pd/CaCO$_3$, Pd/C, KOH, MeOH, H$_2$: 87%) or by conversion into 4-hydrazino-7-methyl-1,8-naphthyridine (**38**, R = NHNH$_2$) (H$_2$NNH$_2 \cdot$H$_2$O, EtOH: ~75%) and subsequent oxidation (CuSO$_4$/H$_2$O: ~50%).[153]

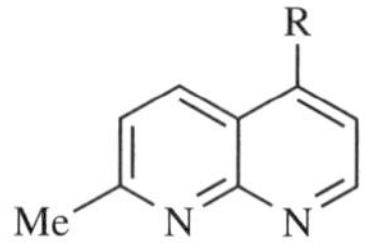

(**38**)

Also other examples.[65,129,185,343,628,690,980,1390]

Conversion into Esters and the Like

2,7-Dichloro-1,8-naphthyridine (**39**, R = Cl) gave dibutyl 1,8-naphthyridine-2,7-dicarboxylate (**39**, R = CO$_2$Bu) [substrate, Pd(PPh$_3$)$_2$Cl$_2$, BuOH, Bu$_3$N, CO, 120°C: 40%; analogs likewise];[92,cf. 536] by an unrelated procedure, 2,7-bis(trichloromethyl)-1,8-naphthyridine (**39**, R = CCl$_3$) gave dimethyl 1,8-naphthyridine-2,7-dicarboxylate (**39**, R = CO$_2$Me) (H$_3$PO$_4$, 170°C; then MeOH, reflux: 80%).[473]

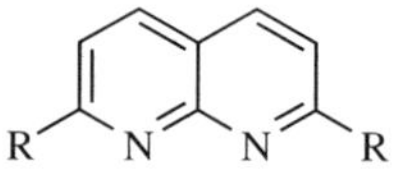

(**39**)

Also other examples.[1012]

Phosphinolysis

2-Chloro- (**40**, R = Cl) gave 2-(diphenylphosphino)-5,7-dimethyl-1,8-naphthyridine (**40**, R = PPh$_2$) (HPPh$_2$, BuLi, hexane; substrate added: 55%).[1200]

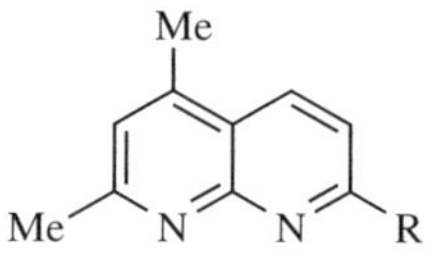

(**40**)

Also other examples.[270]

Thiolysis

Note: This process can be done directly with hydrosulfide ion or indirectly (via an unisolated thiouronium intermediate) using thicurea.

7-Bromo-2,3-dihydro-1,8-naphthyridin-4(1H)-one (**41**) gave 7-thioxo-2,3,7,8-tetrahydro-1,8-naphthyridin-4(1H)-one (**42**) (NaSH/EtOH, 20°C: 91%).[1385]

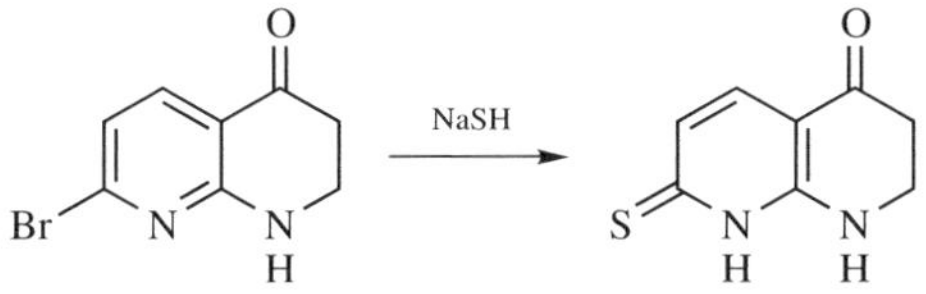

2-Chloro-7-methyl-1,8-naphthyridine gave 7-methyl-1,8-naphthyridine-2(1H)-thione (**43**) [(H$_2$N)$_2$CS, MeOH, reflux; then KOH: 91%].[520]

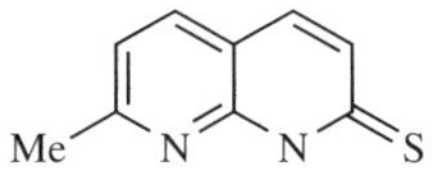

Also other examples.[1031]

Heteroarenolysis

2,7-Dichloro- gave 2,7-di(pyridin-2-yl)-1,8-naphthyridine [3-(Bu$_3$Sn)-pyridine, Pd(PPh$_3$)$_4$, PhMe, reflux: 84%; analogs likewise].[168]

CHAPTER 25

Oxy-1,8-naphthyridines

This chapter includes the 1,8-naphthyridinones, alkoxy- and aryloxy-1,8-naphthyridines, and 1,8-naphthyridine *N*-oxides. Much of the available material in these areas was stimulated by a search for analogs of nalidixic acid (**1**) with improved antimicrobial and other bioactivities. Some of the resulting compounds (usually bearing halogeno substituents) have been mentioned at the beginning of Chapter 24; others are included in numerous papers and reviews, mainly covering their bioactivities.[151,159,171,192,212,219,227,228,235,238,242,248,294,295,313,342,347,370,373,380, 405,438,581,592,596,799,868,915,1272,1296,1359,1382,1387,1413]

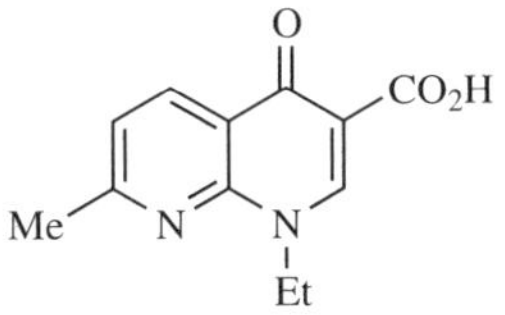

(**1**)

25.1. 1,8-NAPHTHYRIDINONES AND THE LIKE

1,8-Naphthyridinones are usually tautomeric with corresponding naphthyridinols, but they may be fixed (become nontautomeric) by *N*-alkylation. As in related series, 1,8-naphthyridinones prefer to exist as oxo rather than hydroxy derivatives.[91,591,826,881,1026]

The mass spectra of such naphthyridinones have been studied.[286]

25.1.1. Preparation of 1,8-Naphthyridinones and the Like

Two important routes to 1,8-naphthyridinones have been covered already: by *primary synthesis* (Chapter 22) and the *hydrolysis of halogeno-1,8-naphthyridines*

The Naphthyridines: The Chemistry of Heterocyclic Compounds, Volume 63, by D.J. Brown
Copyright © 2008 John Wiley & Sons, Inc.

(Section 24.2). Other preparative methods are illustrated in the following typical examples.

By Direct Oxidation

Note: The oxidation of nalidixic acid (**1**) to its 6-hydroxy derivative by microorganisms has been reported.[219] Oxidative routes to nontautomeric naphthyridinones have been described using chemical oxidants[64,1214] or aerial oxidation with or without irradiation.[344,949]

From Alkoxy- or Amino-1,8-naphthyridines

Ethyl 6-chloro-1-ethyl-7-methoxy-4-oxo-1,4-dihydro-1,8-naphthyridine-3-carboxylate (**2**) gave 6-chloro-1-ethyl-4,7-dioxo-1,4,7,8-tetrahydro-1,8-naphthyridine-3-carboxylic acid (**3**) (1M NaOH, 95°C: 95%; note the additional saponification of the ester group).[387]

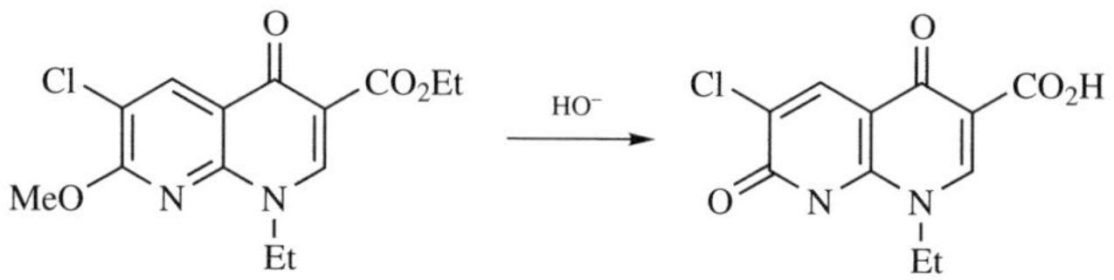

3-(2,6-Dichlorophenyl)-7-(2-ethoxyethoxy)-1,8-naphthyridin-2-amine gave 7-amino-6-(2,6-dichlorophenyl)-1,8-naphthyridin-2(1*H*)-one (**4**) (HCl/dioxane, reflux: 81%).[239]

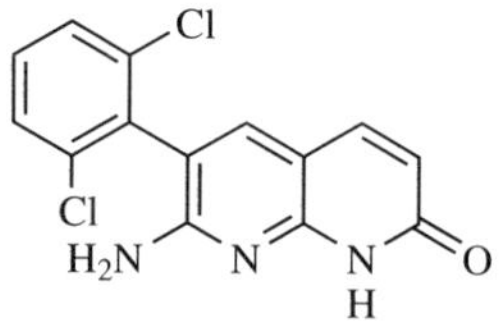

Ethyl 1-ethyl-7-phenyl-6-methoxy- (**5**, R = Me) gave ethyl 1-ethyl-7-phenyl-6-hydroxy-4-oxo-1,4-dihydro-1,8-naphthyridine-3-carboxylate (**5**, R = H) (AlCl$_3$, CH$_2$Cl$_2$: 81%; note that the unactivated methoxy group would be resistant to regular hydrolysis).[1061] Also other examples.[402,769]

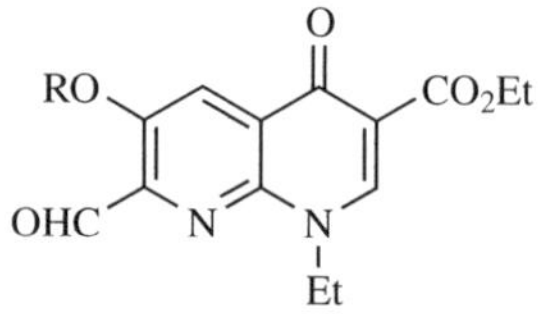

5,7-Dimethyl-1,8-naphthyridin-2-amine (**6**) gave 5,7-dimethyl-1,8-naphthyridin-2(1*H*)-one (**7**) (NaNO$_2$, H$_2$SO$_4$, H$_2$O, 0°C: 86%);[938] 7-amino-1,8-naphthyridin-2(1*H*)-one gave 1,8-naphthyridine-2,7(1*H*,8*H*)-dione (**8**) (H$_2$SO$_4$, solid NaNO$_2$, 20°C: 87%).[389]

Also other examples.[75,77,158,505,533,643,681,687,690,718,721,743,761,763,856,972,1100,1226,1392]

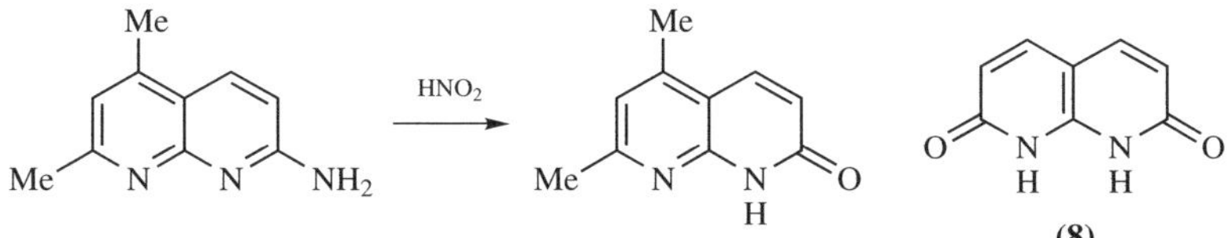

From 1,8-Naphthyridinecarboxylic Acids

Irradiation of 1-ethyl-7-methyl-4-oxo-1,4-dihydro-1,8-naphthyridine-3-carboxylic acid in methanol with oxygen caused decarboxylation and oxidative hydroxylation to afford 1-ethyl-3-hydroxy-7-methyl-1,8-naphthyridin-4(1*H*)-one (**9**) in isolable amount.[1010]

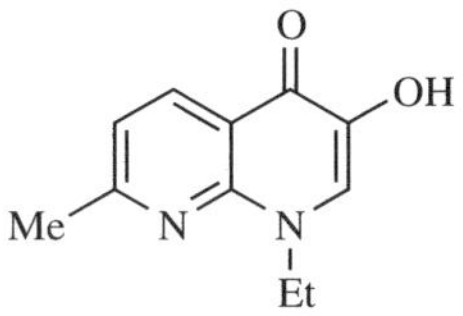

From Trichloromethyl-1,8-naphthyridines

1-Ethyl-6-fluoro-4-oxo-7-trichloromethyl-1,4-dihydro- (**10**) has been reported to afford 1-ethyl-6-fluoro-4,7-dioxo-1,4,7,8-tetrahydro-1,8-naphthyridine-3-carboxylic acid (**11**) (NaOH/H$_2$O, 100°C: 97%).[1091]

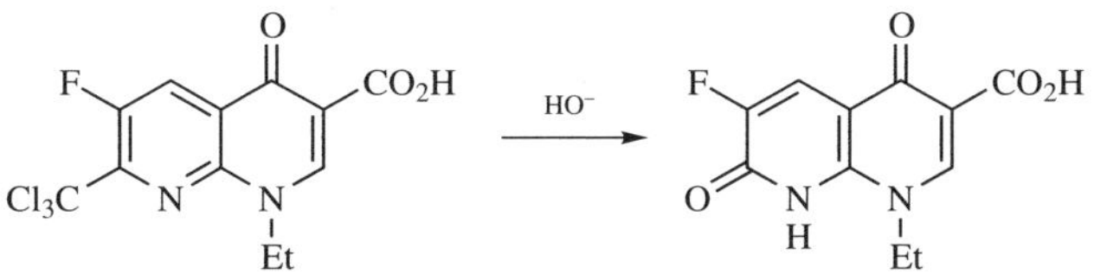

By Reduction of 1,8-Naphthyridinecarboxylic Esters or the Like

Note: This process affords (extranuclear) hydroxyalkyl-1,8-naphthyridines.

Dimethyl 1,8-naphthyridine-2,7-dicarboxylate gave 2,7-bis(hydroxymethyl)-1,8-naphthyridine (**12**) [NaB(OMe)$_3$H, THF, CH$_2$Cl$_2$, 25°C: 55%; or NaBH$_4$, EtOH, 25°C: 19%].[473]

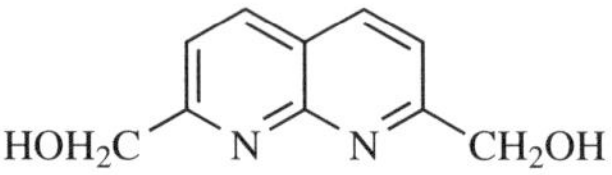

(**12**)

Carbaldehyde substrates may also be used.[1445]

25.1.2. Reactions of 1,8-Naphthyridinones and the Like

The *halogenolysis* of 1,8-naphthyridinones has been discussed in Section 24.1. Other reactions are illustrated by the following classified examples.

O-Alkylation

Note: Such alkylation of 1,8-naphthyridinones is seldom reported and then only in poor yield; in contrast, 1,8-naphthyridin-3/6-ols, which cannot undergo *N*-alkylation, readily form their *O*-alkylated derivatives in good yield.

Ethyl 1-ethyl-6-hydroxy- (**13**, R = H) gave ethyl 6-benzyloxy-1-ethyl-7-methyl-4-oxo-1,4-dihydro-1,8-naphthyridine-3-carboxylate (**13**, R = CH$_2$Ph) (PhCH$_2$Cl, K$_2$CO$_3$, AcNMe$_2$: 98%).[1060]

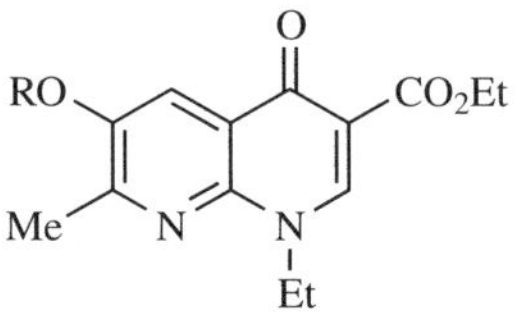

(**13**)

7-Bromo-2-phenyl-1,8-naphthyridin-4(1*H*)-one gave 2-bromo-5-ethoxy-7-phenyl-1,8-naphthyridine (**14**) (NaH, Me$_2$SO; then EtI, 80°C: 32%).[263]

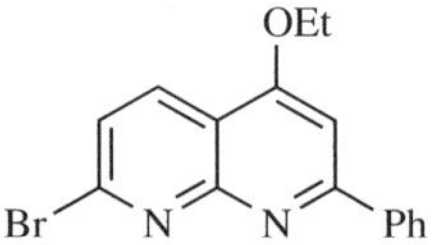

(**14**)

Also other examples.[402,418,815,1211]

N-Alkylation

Note: Alkylation of 1,8-naphthyridinones usually affords *N*-alkyl derivatives, at least predominantly.

3-Iodo (**15**, R = H) gave 1-ethyl-3-iodo-7-methyl-1,8-naphthyridin-4(1*H*)-one (**15**, R = Et) (K_2CO_3, EtI, Me_2NCHO: 84%).[905]

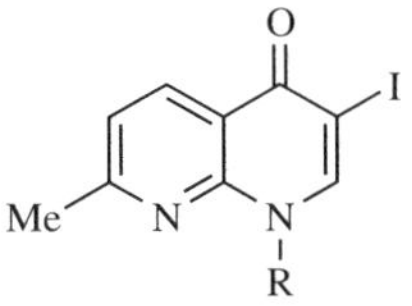

(**15**)

1,8-Naphthyridin-2(1*H*)-one (**16**, R = H) gave 1-pentyl-1,8-naphthyridin-2(1*H*)-one (**16**, R = C_5H_{11}) (KOH, EtOH, H_2O, $C_5H_{11}Br$, 60°C: 51%).[1298]

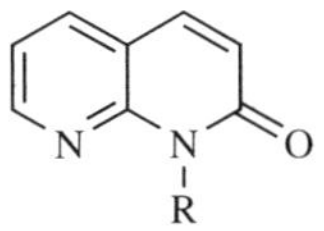

(**16**)

7-Methyl- (**17**, R = H) gave 1-ethyl-7-methyl-4-oxo-1,4-dihydro-1,8-naphthyridine-3-carbonitrile (**17**, R = Et) [K_2CO_3, excess $(EtO)_3PO$, reflux: 66%].[1405]

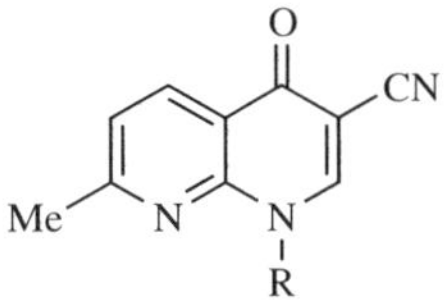

(**17**)

Also other examples.[76,158,249,305,378,412,421,460,740,1076,1102,1112,1216,1233,1253,1352,1393]

Aminolysis

Note: The conversion of 1,8-naphthyridinones into amino-1,8-naphthyridines is usually done via a halogeno intermediate, but direct aminolysis is possible sometimes.

4-Hydroxy- (**18**, R = OH) gave 4-amino-2-phenyl-1,8-naphthyridin-2(1*H*)-one (**18**, R = NH_2) (neat $AcONH_4$, 100°C: 90%).[892]

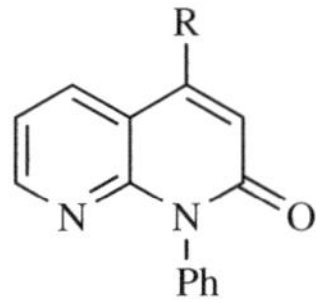

(**18**)

7-Chloro-2,3-dihydro-1,8-naphthyridin-4(1H)-one (**19**) gave 7-chloro-4-hydroxyamino-2,3-dihydro-1,8-naphthyridine (**20**) (or its hydroxyimino tautomer) ($H_2NOH \cdot HCl$, pyridine, EtOH, reflux: 91%; analogs likewise).[1403,cf. 1406]

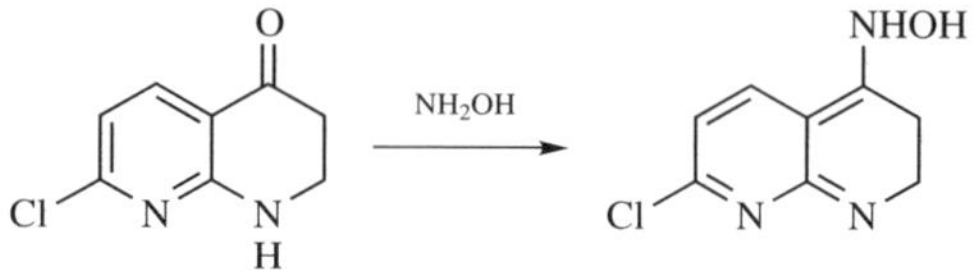

(**19**)   (**20**)

Thiation

Ethyl 7-methyl-4-oxo- (**21**, X = O) gave ethyl 7-methyl-4-thioxo-1,4-dihydro-1,8-naphthyridine-3-carboxylate (**21**, X = S) (P_2S_5, pyridine, reflux: 98%); analogs likewise.[1380] 1,8-Naphthyridine-2(1H)-thione similarly.[862]

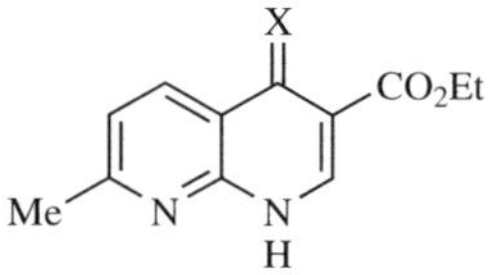

(**21**)

Cyclization Reactions

4-Hydroxy-1-phenyl-1,8-naphthyridin-2(1H)-one underwent cycloaddition by isobutene to give 8b-hydroxy-1,1-dimethyl-4-phenyl-1,2,2a,8b-tetrahydrocyclobuta[c][1,8]naphthyridin-3-(4H)-one (**22**) MeOH, hν: 95%); analogs likewise.[603]

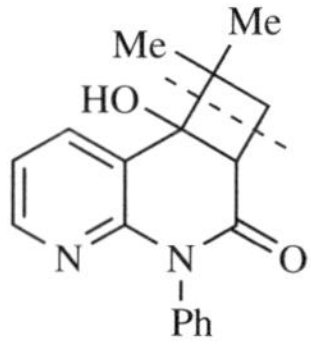

(**22**)

7-Methyl-2,3-dihydro-1,8-naphthyridin-4(1H)-one with o-aminoacetophenone underwent cyclocondensation to afford 3,7-dimethyl-5,6-dihydroquino[3,2-c][1,8]naphthyridine (**23**) (HCl/EtOH, 20°C: 61%).[1117]

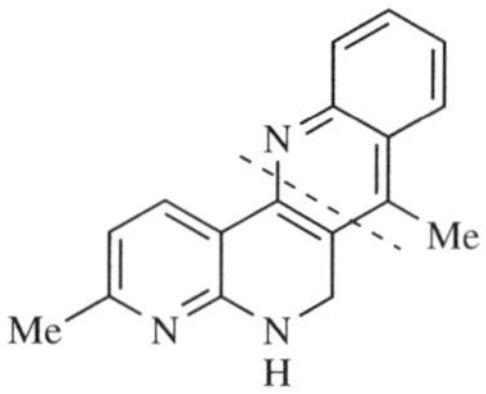

(**23**)

Also other examples.[304,979,1427]

Complex Formation

Note: References to the formation of some typical complexes between relatively simple 1,8-naphthyridinones and various metallic elements are listed here.

Ca[440,583]
Cd[583]
Co[645,1281]
Cr[583]
Cu[440,583,645,1264,1281]
Fe[440,583,1264,1281]
Hg[583]
Mg[440,583]
Mn[440,583,645]
Ni[583,645,1264,1281]
Pd[91,326,583]
Pt[91,326]
Rh[394,396,523,584]
U[1264]
Zn[440,583,645,1281]

25.2. ALKOXY- AND ARYLOXY-1,8-NAPHTHYRIDINES

Most available information on these ethers has been covered already.

Preparation

Note: Alkoxy/aryloxy-1,8-naphthyridines have been made by *primary synthesis* (see Chapter 22), by *alcoholysis/phenolysis of halogeno-1,8-naphthyridines* (see Section 24.2), by *alkylation of 1,8-naphthyridinones* (see Section 25.1.2), and by two minor procedures illustrated here.

2,7-Diazido-3-phenyl-1,8-naphthyridine (**24**, R = N$_3$) gave selectively 2-azido-7-methoxy-3-phenyl-1,8-naphthyridine (**24**, R = OMe) (MeONa/MeOH, reflux: 73%).[285]

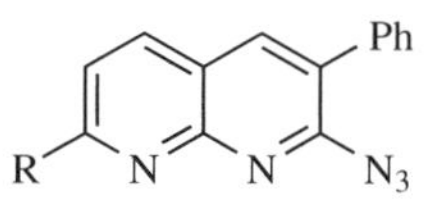

(**24**)

A substrate formulated as the benzylidene derivative (**25**) has been reported to afford the epoxy derivative (**26**) (H_2O_2, HO^-: 75%).[1102]

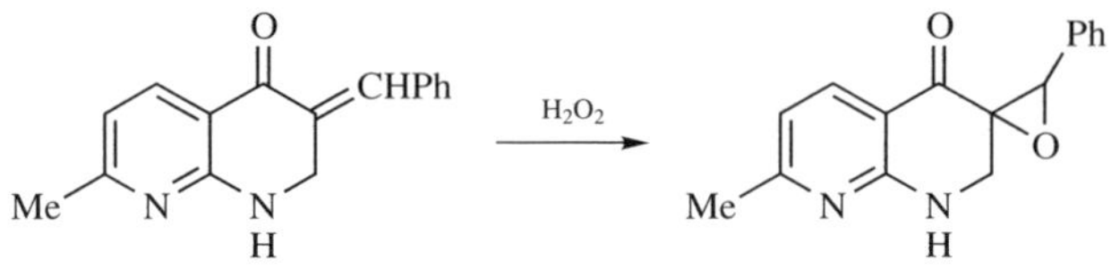

(**25**) →(H_2O_2)→ (**26**)

Reactions

Note: The *hydrolysis of alkoxy-1,8-naphthyridines* to 1,8-naphthyridinones has been exemplified in Section 25.1.1. No other reactions appear to have been reported.

25.3. 1,8-NAPHTHYRIDINE *N*-OXIDES

Despite a voluminous literature on 1,8-naphthyridines, their *N*-oxides have seldom been investigated. The limited information is summarized here.

Preparation

Note: Virtually all these oxides have been made by oxidation of existing 1,8-naphthyridines with peroxy acids, as illustrated in Section 23.1.4 and by the following example.

2-Ethoxy-5,7-dimethyl-1,8-naphthyridine gave its 8-oxide (**27**) ($MeCO_3H$: ~10% after purification).[1008,cf. 769]

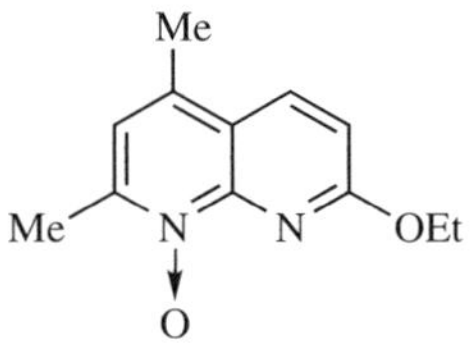

(**27**)

Reactions

Note: The *Meissenheimer conversion* of 1,8-naphthyridine *N*-oxides into halogeno-1,8-naphthyridines has been covered in Section 24.1; complexes of 1,8-naphthyridine 1-oxide with Cr,[555] Cu,[496] and Sr,[462] and the lanthanide compounds[429] have been reported; and deoxygenation of an *N*-oxide is exemplified here.

2-Amino-8-hydroxy-5-methyl-7-oxo-7,8-dihydro-1,8-naphthyridine-3-carboxylic acid (**28**, R = OH) (a tautomer of the 8-oxide) underwent reduction to 2-amino-5-methyl-7-oxo-7,8-dihydro-1,8-naphthyridine-3-carboxylic acid (**28**, R=H) (TiCl$_3$, HCl, H$_2$O, 20°C: ~75%).[748]

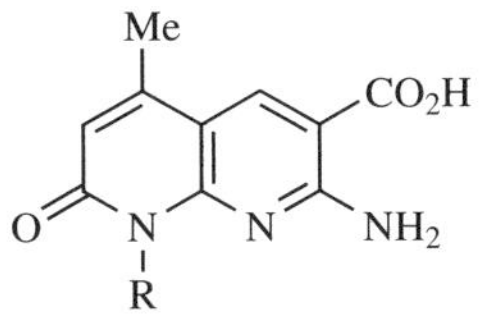

(**28**)

CHAPTER 26

Thio-1,8-naphthyridines

Little use has been made of 1,8-naphthyridinethiones, alkylthio-1,8-naphthyridines, or their respective oxidation products. Available information is summarized in the following brief sections.

26.1. 1,8-NAPHTHYRIDINETHIONES

A UV spectral study of 1,8-naphthyridine-2(1H)-thione and its 4(1H)-isomer suggests that they exist as thiones rather than the thiol tautomers.[591]

Preparation

Note: Reported preparative routes have been covered already: *primary synthesis* (Chapter 22), by *thiolysis of halogeno-1,8-naphthyridines* (at the end of Section 24.2), or by *thiation of 1,8-naphthyridinones* (Section 25.1.2).

Reactions

Note: Of the many possible reactions of 1,8-naphthyridinethiones, only three are available for exemplification here.

1,8-Naphthyridine-2(1H)-thione (**1**) underwent *S-alkylation* to give 2-carboxymethylthio-1,8-naphthyridine (**2**) (ClCH$_2$CO$_2$Na, KOH, H$_2$O: 84%).[862]

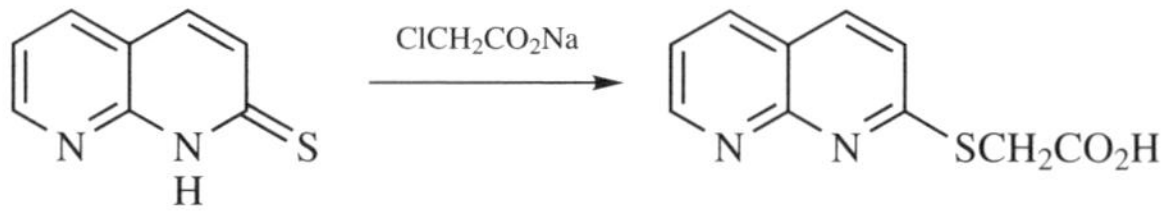

Also other examples.[263]

6-Amino-1-ethyl-4-oxo-7-thioxo-1,4,7,8-tetrahydro-1,8-naphthyridine-3-carboxylic acid (**3**) underwent *cyclocondensation* to give 5-ethyl-8-oxo-5,8-

---

The Naphthyridines: The Chemistry of Heterocyclic Compounds, Volume 63, by D.J. Brown
Copyright © 2008 John Wiley & Sons, Inc.

dihydrothiazolo[5,4-*b*][1,8]naphthyridine-7-carboxylic acid (**4**) (HCO$_2$H, reflux: 94%), analogs likewise.[1031]

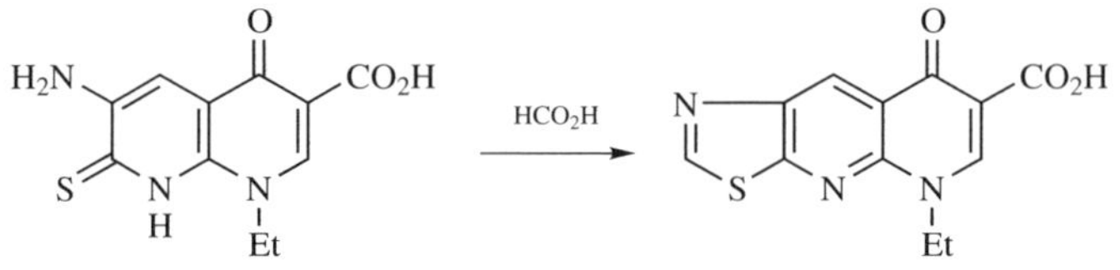

7-Methyl-1,8-naphthyridine-2(1*H*)-thione formed complexes with Mo(II) and Ru(II).[520]

26.2. ALKYLTHIO- AND ARYLTHIO-1,8-NAPHTHYRIDINES

Preparation

Note: Such thioethers have been made by *primary synthesis* (see Chapter 22) and by *alkylation of 1,8-naphthyridinethiones* (see Section 26.1).

Reactions

Note: The only reported reaction of these thioethers appears to be their *oxidation to sulfoxides or sulfones*, as illustrated here.

Ethyl 1-cyclopropyl-6-fluoro-4-oxo-7-*p*-tolylthio- (**5**) gave ethyl 1-cyclopropyl-6-fluoro-4-oxo-7-*p*-tolylsulfinyl- (**6**, $n = 1$) or ethyl 1-cyclopropyl-4-oxo-7-*p*-tolylsulfonyl-1,4-dihydro-1,8-naphthyridine-3-carboxylate (**6**, $n = 2$) [*m*-ClC$_6$H$_4$CO$_3$H (1 or 2 mol, respectively), CHCl$_3$: 90% or 89% respectively].[1162]

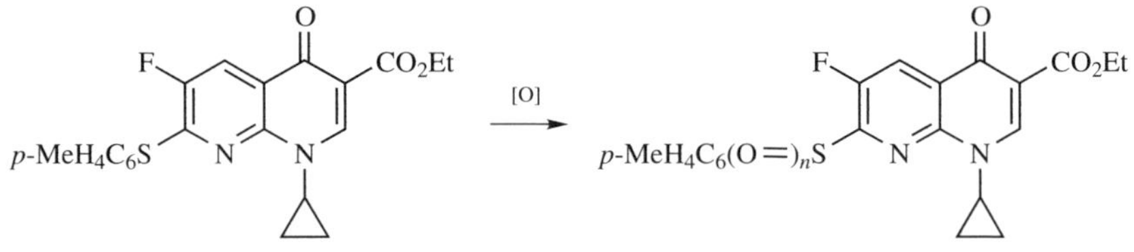

Also other examples.[418,1167]

26.3. 1,8-NAPHTHYRIDINE SULFOXIDES AND SULFONES

Preparation

Note: Both sulfoxides and sulfones have been made only by *oxidation of corresponding thioethers* (see Section 26.2).

Reactions

Note: Only *aminolysis* has been reported, as illustrated here.

Ethyl 1-cyclopropyl-4-oxo-7-phenylsulfonyl-6-trifluoromethyl-1,4-dihydro-1,8-naphthyridine-3-carboxylate (**7**) gave the crude ester (**8**, R = Et) (piperazine, MeCN, 20°C), characterized after saponification to 1-cyclopropyl-4-oxo-7-(piperazin-1-yl)-6-trifluoromethyl-1,4-dihydro-1,8-naphthyridine-3-carboxylic acid (**8**, R = H) (NaOH, EtOH: 37% overall).[1167]

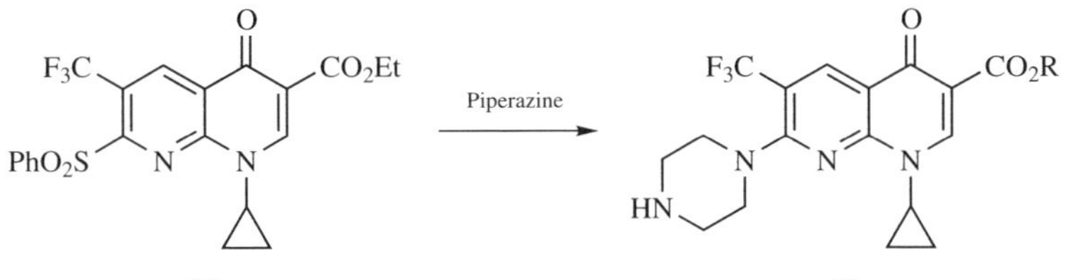

Also analogous aminolyses.[1162,1256]

26.4. 1,8-NAPHTHYRIDINESULFONIC ACIDS AND THE LIKE

The only reported sulfonic acid appears to be an *N*-sulfo derivative of a hydro-1,8-naphthyridine.

2,2,4a,7,7-Pentamethyl-1,2,3,4,4a,5,6,7-octahydro-1,8-naphthyridine (**9**, R = H) underwent *sulfonation* to give the corresponding 1,8-naphthyridine-1-sulfonic acid (**9**, R = SO$_3$H) (ClSO$_3$H, dioxane, 5–20°C: 61%).[1237,cf. 990]

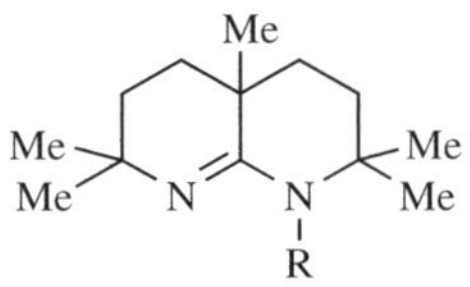

CHAPTER 27

Nitro-, Amino-, and Related 1,8-Naphthyridines

This chapter covers 1,8-naphthyridines bearing nitrogenous substituents that are joined directly or indirectly to the nucleus through their nitrogen atoms.

27.1. NITRO-1,8-NAPHTHYRIDINES

A review of such derivatives appeared in 2000,[1273] it is especially useful for the electronic effects of nitro groups on the reactivities of other substituents in the molecule. Mass spectra have also been reviewed.[492]

27.1.1. Preparation of Nitro-1,8-naphthyridines

Many of these nitro derivatives, both nuclear and extranuclear, have been prepared by *primary synthesis* (see Chapter 22). Other preparative routes are illustrated by the following classified examples.

By Direct Nitration

Note: Unsubstituted 1,8-naphthyridine resists nitration even under severe conditions, but the presence of an electron-releasing amino or oxo substituent enables more facile 3/6-nitration to occur. Even so, if a phenyl group is also present, extranuclear nitration occurs first.

1,8-Naphthyridine gave 3-nitro-1,8-naphthyridine (**1**) (fuming HNO_3, fuming H_2SO_4, 120°C, 20 h: 2% with 92% recovery of substrate).[682]

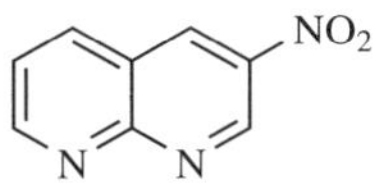

(**1**)

The Naphthyridines: The Chemistry of Heterocyclic Compounds, Volume 63, by D.J. Brown
Copyright © 2008 John Wiley & Sons, Inc.

1,8-Naphthyridin-2(1H)-one under somewhat similar conditions gave 3,6-dinitro-1,8-naphthyridin-2(1H)-one (**2**) (100°C: 56%).[860,1128]

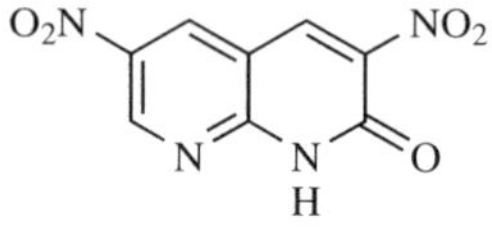

(2)

7-Amino-4-phenyl-1,8-naphthyridin-2(1H)-one (**3**, Q = R = H) gave a 3:2 mixture of 7-amino-4-*m*-nitrophenyl- (**3**, Q = NO$_2$, R = H) and 7-amino-4-*p*-nitrophenyl-1,8-naphthyridin-2(1H)-one (**3**, Q = H, R = NO$_2$)[HNO$_3$ (1 mol), H$_2$SO$_4$, 100°C: 84%]; an excess of nitric acid under similar conditions produced a 3:2 mixture of 7-amino-3,7-dinitro-4-*m*-nitrophenyl-1,8-naphthyridin-2(1H)-one (**4**, Q = NO$_2$, R = H) and its *p*-nitro isomer (**4**, Q = H, R = NO$_2$) in 64% yield.[749]

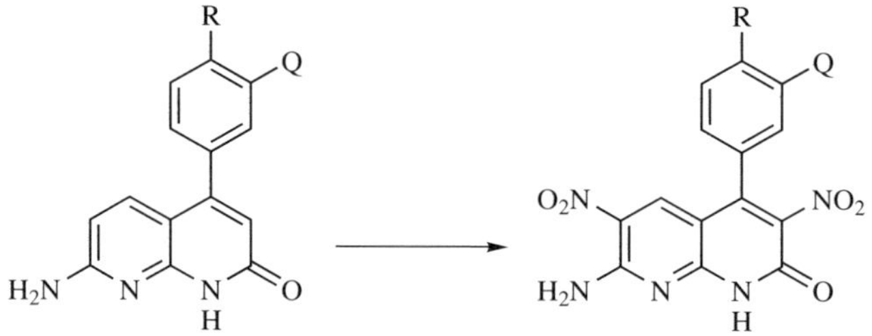

(3)   (4)

Also other examples.[69,160,364,387,721,1029,1112,1226]

By Oxidation of Nitroso-1,8-naphthyridines

Note: This good procedure is seldom used for lack of nitroso substrates.

4-Hydroxy-3-nitroso- (**5**, *n* = 1) gave 4-hydroxy-3-nitro-1,8-naphthyridin-2(1H)-one (**5**, *n* = 2) (H$_2$O$_2$, F$_3$CCO$_2$H, 5°C: 53%).[892]

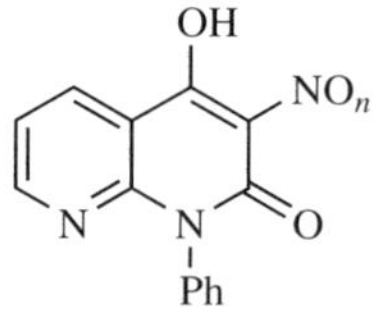

(5)

From 1,8-Naphthyridinamines

Note: This indirect route via a dimethylsulfimido intermediate offers a pathway to 2-, 4-, 5-, or 7-nitro derivatives.

1,8-Naphthyridin-2-amine (**6**, R = NH$_2$) gave 2-dimethylsulfimido-(**6**, R = N:SMe$_2$) [Me$_2$SO, (F$_3$CSO$_2$)O, −70°C: 33%] and thence by oxidation, 2-nitro-1,8-naphthyridine (**6**, R = NO$_2$) (*m*-ClC$_6$H$_4$CO$_3$H, CH$_2$Cl$_2$, 0°C: 47%).[692]

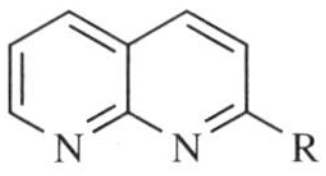

(**6**)

27.1.2. Reactions of Nitro-1,8-naphthyridines

Although nitro groups have a marked effect on the reactivity of other substituents on 1,8-naphthyridine,[1273] the only reported reaction of the nitro group itself is reduction to an amino group, as illustrated by the following examples.

Using Catalytic Hydrogenation

3-Nitro- (**7**, R = NO$_2$) gave 3-amino-6-phenyl-7-piperidino-1,8-naphthyridin-2(1*H*)-one (**7**, R = NH$_2$) (Pd/C, H$_2$, AcOH, 20°C: 85%);[1304] in a similar way, 7-methyl-2-*p*-nitrophenyl- (**8**, R = Me) gave 2-*p*-aminophenyl-7-methyl-1,8-naphthyridin-4(1*H*)-one (**8**, R = NH$_2$) (60%).[263]

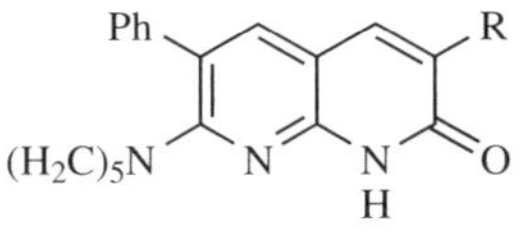

(**7**) 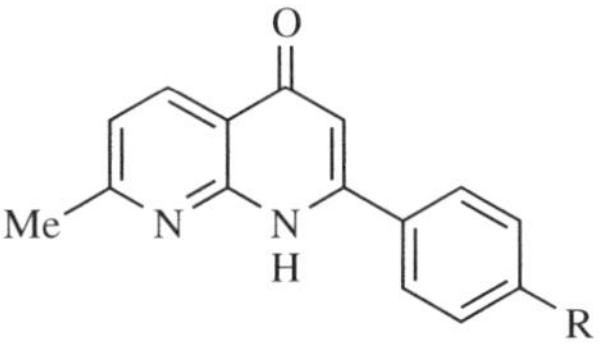   (**8**)

Using Iron/Acetic Acid or the Like

Ethyl 7-ethoxy-1-ethyl-6-nitro- (**9**, R = NO$_2$) gave ethyl 6-amino-7-ethoxy-1-ethyl-4-oxo-1,4-dihydro-1,8-naphthyridine-3-carboxylate (**9**, R = NH$_2$) (Fe powder, AcOH 85°C: 94%).[1211]

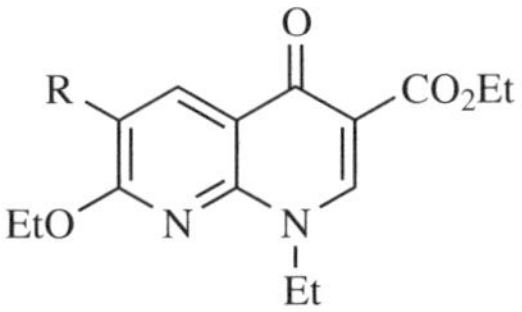

(**9**)

Also other examples.[364,387,474,1029]

Using Tin/Hydrochloric Acid

7-Amino-6-nitro- gave 6,7-diamino-1,8-naphthyridin-2(1H)-one (Sn, HCl, trace Cu^{2+}, 110°C: ~85%).[721]

Using Sodium Dithionite

6-Nitro- gave 6-amino-5-oxo-5,8-dihydro-1,8-naphthyridine-2-carboxylic acid (Na$_2$S$_2$O$_4$, NaOH, H$_2$O, 100°C: ~45%).[726]

27.2. NITROSO-1,8-NAPHTHYRIDINES

Unlike nitration, C-nitrosation cannot be forced at elevated temperatures because of the lability of nitrous acid. Thus nitrosation must be done at or below room temperature, and this requires the substrate to have a free unactivated position as well as at least two electron-releasing substituents. These criteria are met by 4-hydroxy-1-phenyl-1,8-naphthyridin-2(1H)-one (**10**, R = H), which does yield its 3-nitroso derivative (**10**, R = NO) (NaNO$_2$, AcOH, H$_2$O, 20°C: 96%).[892]

The foregoing nitroso compound (**10**, R = MO) may be oxidized to the corresponding 3-nitro compound (**10**, R = NO$_2$) (see Section 27.1.1) or (after aminolysis of its hydroxy group) reduced to 3,4-diamino-1,8-naphthyridin-2(1H)-one (**11**) (Na$_2$S$_2$O$_4$, H$_2$O, 20°C: ~90%).[892] N-Nitrosoperhydro-1,8-naphthyridines are known.[437]

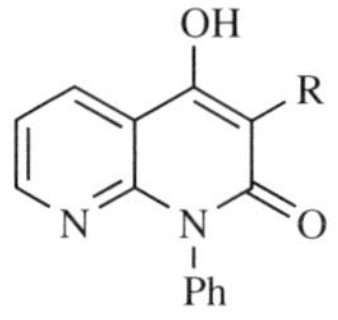

(10) 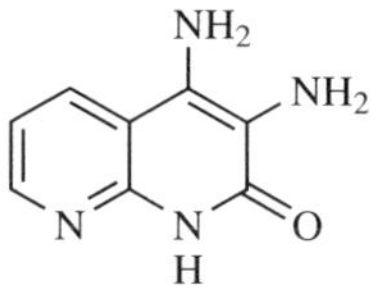   (11)

27.3. AMINO-1,8-NAPHTHYRIDINES

Because of a perceived importance of amino substituents for biological activities in 1,8-naphthyridines, e.g.,[359,422,478,526,1395,1410] such derivatives have been widely investigated, for example, with respect to their ionization,[1043] NMR spectra,[174] mass spectra,[1227] fluorescence properties,[1311] tautomerism,[82] and conversion into cyanine-type dyes.[768]

27.3.1. Preparation of Amino-1,8-naphthyridines

Several methods of preparation have been covered already: by *primary synthesis* (Chapter 22) and by *aminolysis of halogeno-1,8-naphthyridines* (Section 34.2),

1,8-naphthyridinones (Section 25.2), or *1,8-naphthyridine sulfones* (Section 26.3). Other preparative routes are illustrated by the following examples.

By Amination

Note. The *C*-aminations illustrated here proceed by addition of the amine across a double bond and subsequent dehydrogenation as indicated for the parent 1,8-naphthyridine (Section 21.1.4) and in two expert reviews.[54,1268] The *N*-aminations[655] and *C*-azo coupling (mentioned here) proceed differently.

3-Nitro-1,8-naphthyridine (**12**, R = H) gave 3-nitro-1,8-naphthyridin-4-amine (**13**, R = H) (liquid ammonia; then KMnO$_4$: 45%);[1320] 3,6-dinitro-1,8-naphthyridine (**12**, R = NO$_2$) likewise gave 3,6-dinitro-1,8-naphthyridin-4-amine (**13**, R = NO$_2$) (50%).[692]

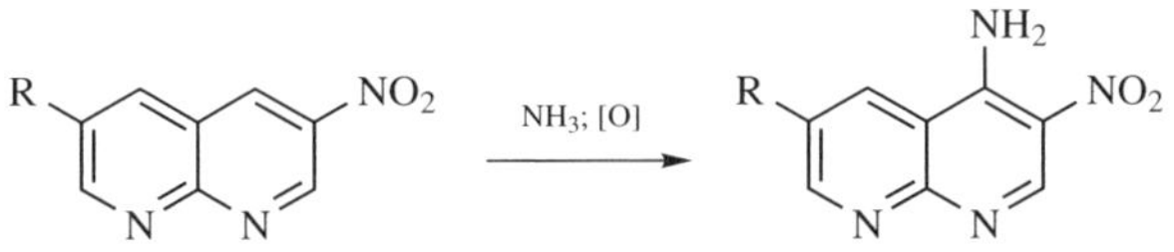

In a similar way, 3-nitro-1,8-naphthyridine gave 4-methylamino-3-nitro-1,8-naphthyridine (**14**, R = H) (neat MeNH$_2$; KMnO$_4$: 90%);[682] and 3-nitro-1,8-naphthyridin-2-amine gave 4-methylamino-3-nitro-1,8-naphthyridin-2-amine (**14**, R = NH$_2$ (86%).[682,785,860]

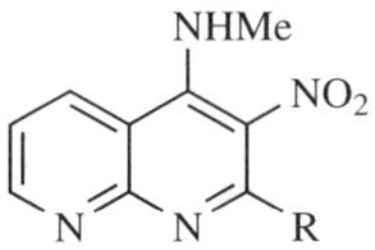

1,8-Naphthyridine methiodide gave 1-methyl-1,8-naphthyridin-2(1*H*)-imine (liquid NH$_3$; KMnO$_4$: 10%).[1093]

Also other such aminations.[830,1180]

4,5,7-Triamino-2-thioxo-1,2-dihydro-1,8-naphthyridine-3-carbonitrile (**15**, R = H) underwent azo coupling with benzenediazonium chloride to give its 6-phenylazo derivative (**15**, R = N:NPh) (NaOH, EtOH: 86%).[538]

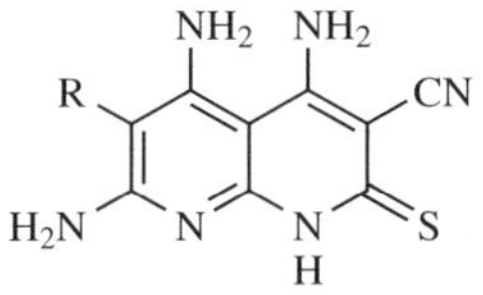

From Acylamino-1,8-naphthyridines

2-Acetamido-7-ethoxy-4-phenyl-1,8-naphthyridine (**16**, R = Ac) gave 7-ethoxy-4-phenyl-1,8-naphthyridin-2-amine (**16**, R = H) (EtONa/EtOH, reflux: ~90%).[718]

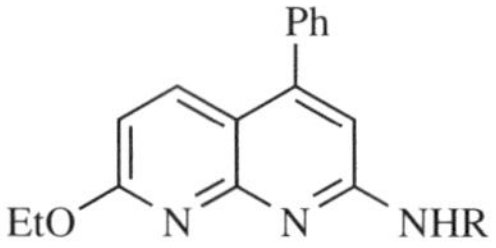

(**16**)

2-Acetamido-5,8-dimethyl-5,6,7,8-tetrahydro-1,8-naphthyridine (**17**, R = Ac) gave 5,8-dimethyl-5,6,7,8-tetrahydro-1,8-naphthyridin-2-amine (**17**, R = H) (KOH, EtOH, H_2O, N_2, reflux: 97%).[980]

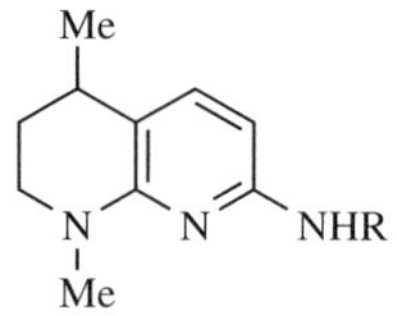

(**17**)

2-*p*-Acetamidophenyl- gave 2-*p*-aminophenyl-1,8-naphthyridine (5M HCl, reflux: 70%).

Also other examples.[158,472,650,720,741,743,764,972,1445]

By a Curtius Reaction

1-Ethyl-7-methyl-4-oxo-1,4-dihydro-1,8-naphthyridine-3-carbohydrazide (**18**, R = $CONHNH_2$) underwent the Curtius reaction to afford 3-amino-1-ethyl-7-methyl-1,8-naphthyridin-4(1*H*)-one (**18**, R = NH_2) (HNO_2, 8°C; then H^+: ~40% overall).[1116]

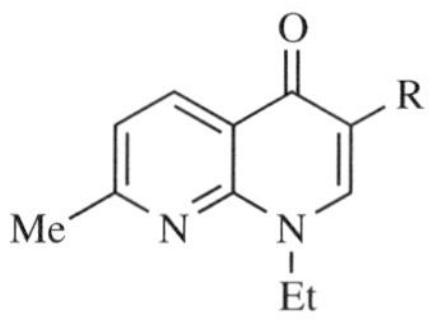

(**18**)

By Modification of Existing Amino-1,8-naphthyridines

Note: The examples presented here illustrate conversion of azido to amino derivatives, alkylamino to amino derivatives, amino to hydrazino derivatives, and hydrazino to azido derivatives.

4-Azido-7-methyl-2-phenyl-1,8-naphthyridine (**19**, R = N$_3$) gave 7-methyl-2-phenyl-1,8-naphthyridin-4-amine (**19**, R = NH$_2$) (Pd/C, MeOH, H$_2$, 20°C: 22%).[221] Also a better example.[1445]

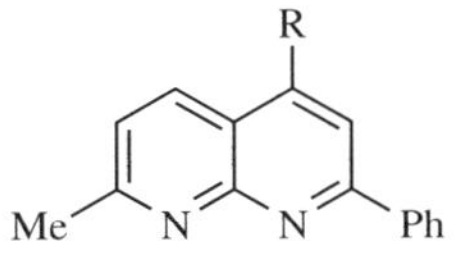

(**19**)

7-*p*-Methoxybenzylamino-1,8-naphthyridin-2-amine gave 1,8-naphthyridine-2,7-diamine (HCl, reflux: 95%).[1438]

3-*p*-Chlorophenyl-1,8-naphthyridin-2-amine (**20**, R = H) gave 3-*p*-chlorophenyl-2-hydrazino-1,8-naphthyridine (**20**, R = NH$_2$) (H$_2$NNH$_2$·HCl, HOCH$_2$CH$_2$OH, reflux: 82%; a true transamination).[653]

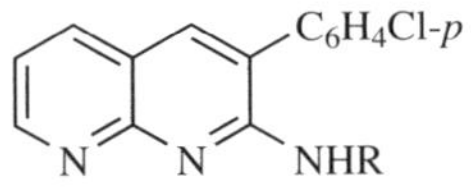

(**20**)

2-Hydrazino- (**21**, R = NHNH$_2$) gave 2-azido-5,7-dimethyl-1,8-naphthyridine (**21**, R = N$_3$ (HNO$_2$, 0°C: ?%);[763] analogs likewise.[701,771]

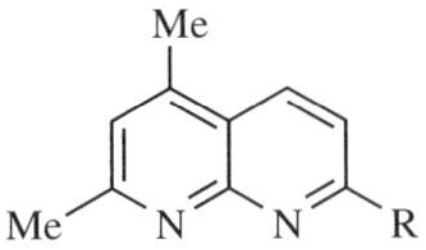

(**21**)

By Azo Coupling

2-*o*-Hydroxyphenyl- gave 2-(2-hydroxy-5-phenylazophenyl)-1,8-naphthyridine (PhN$_2$Cl, HO$^-$: 67%; analogs likewise).[1425]

27.3.2. Reactions of Amino-1,8-naphthyridines

The conversion of *amino- into halogeno-1,8-naphthyridines* has been covered in Section 24.1. Other reported reactions of amino-, hydrazino-, hydroxyamino-, and azido-1,8-naphthyridines are illustrated in the following groups of examples.

Acylation

5,7-Dimethyl-1,8-naphthyridin-2-amine (**22**, R = H) gave 2-acetamido-5,7-dimethyl-1,8-naphthyridine (**22**, R = Ac) (Ac$_2$O, PhH, reflux: ~75%);[1034]

the same substrate (**22**, R = H) with *p*-acetamidobenzenesulfonyl chloride gave 2-*p*-acetamidobenzenesulfonamido-5,7-dimethyl-1,8-naphthyridine (**22**, R = NHSO$_2$C$_6$H$_4$NHAc-*p*) (pyridine, 20°C: ∼60%).[84]

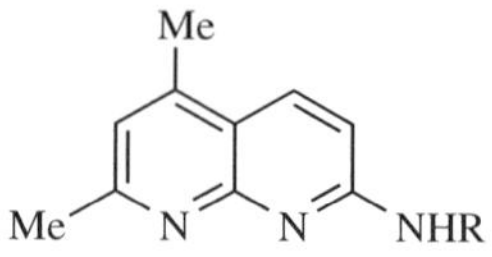

(**22**)

2-*p*-Aminophenyl- (**23**, R = H) gave 2-*p*-acetamidophenyl-7-methyl-1,8-naphthyridin-4(1*H*)-one (**23**, R = Ac) (Ac$_2$O, reflux: 33%).

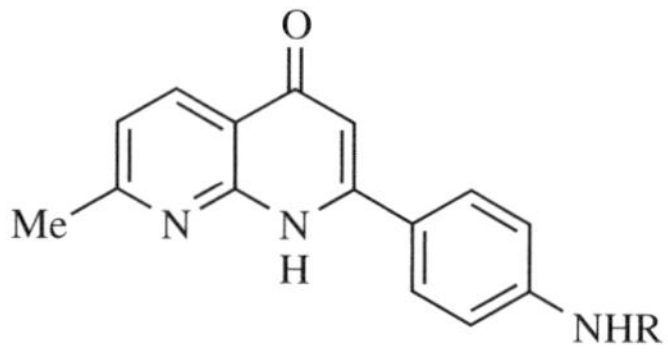

(**23**)

2-*p*-(Benzylideneamino)phenyl-1,8-naphthyridine (**24**, R = N:CHPh) underwent oxidation to give 2-*p*-benzamidophenyl-1,8-naphthyridine (**24**, R = NHBz) (*m*-ClC$_6$H$_4$CO$_3$H: 86%).[625]

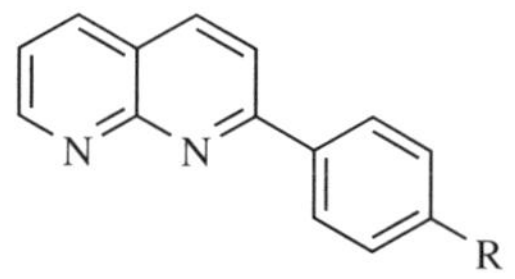

(**24**)

4-Hydroxyamino-7-methyl-1,2-dihydro-1,8-naphthyridine (**25**) with acetyl chloride gave 4-acetoxyamino-7-methyl-1,2-dihydro-1,8-naphthyridine (**26**, R = H) (20°C: 57%) but with acetic anhydride gave 4-acetoxyamino-1-acetyl-7-methyl-1,2-dihydro-1,8-naphthyridine (**26**, R = Ac) (80°C: 66%); analogs likewise.[1224]

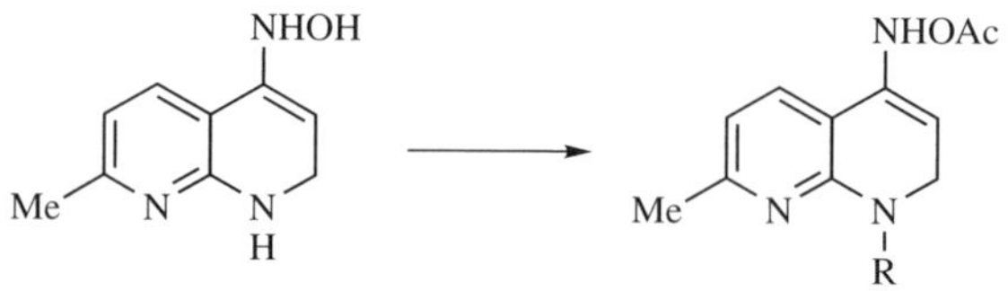

(**25**)   (**26**)

Also other examples.[73,79,98,205,207,239,533,594,649,718,761,956,972,1099,1269,1383,1415,1445]

Alkylation and Alkylidenation

1,2,3,4,4a,5,6,7-Octahydro-1,8-naphthyridine (**27**, R = H) gave its 1-methyl derivative (**27**, R = Me) (MeI, PhH, reflux; then CaO and K_2CO_3: 75%).[1080]

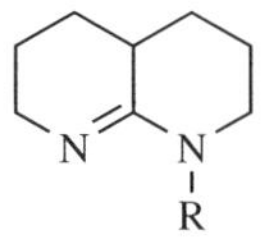

(**27**)

7-Amino-1,8-naphthyridin-4(1*H*)-one with diethyl ethoxymethylenemalonate gave 7-(2,2-diethoxycarbonylvinyl)amino-1,8-naphthyridin-4(1*H*)-one (**28**) (160°C: 73%).[720]

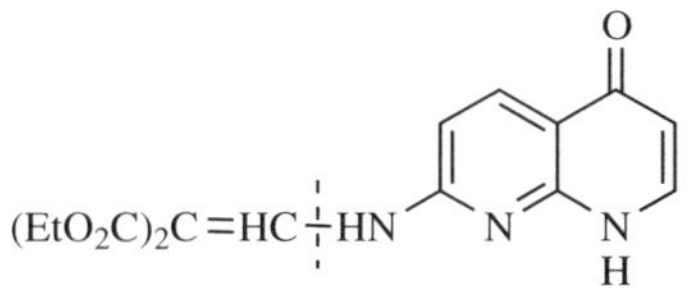

(**28**)

Ethyl 2-amino-6-cyano- with triethyl orthoformate gave ethyl 6-cyano-2-(ethoxymethyleneamino)-5,8-dimethyl-7-oxo-7,8-dihydro-1,8-naphthyridine-3-carboxylate (**29**) (neat reactants, reflux: 73%).[618]

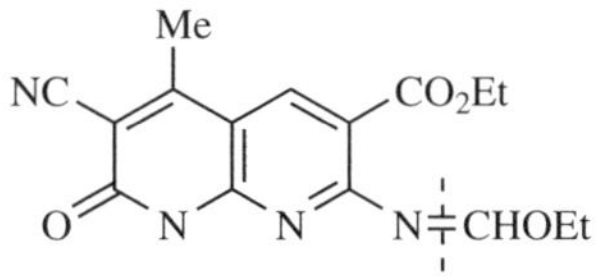

(**29**)

3-*o*-Chlorophenyl-2-hydrazino-1,8-naphthyridine with acetophenone gave 3-*o*-chlorophenyl-2-[(α-methylbenzylidene)hydrazino]-1,8-naphthyridine (**30**) (Me_2NCHO, microwave, 0.5 min: 94%); analogs likewise.[744]

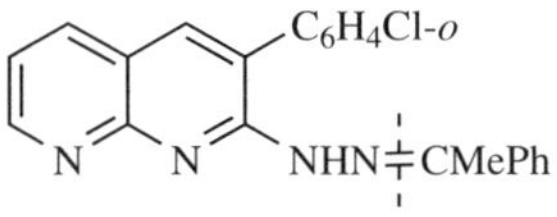

(**30**)

Also other examples.[74,77,201,355,478,529,532,552,573,637,650,653,729,755,856,861,870,1142,1265,1308,1328,1340,1343,1348,1364,1418]

To Azo-1,8-naphthyridines

1,8-Naphthyridine-2,7-diamine with nitrosobenzene gave 2,7-bisphenylazo-1,8-naphthyridine (31) (reactants, MeOH, KOH, H_2O: 33%).[494]

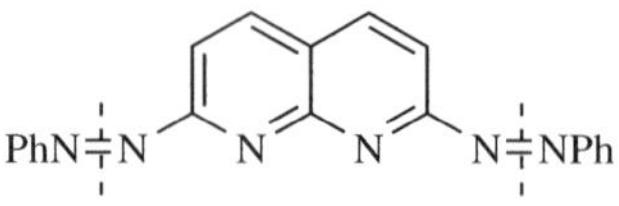

(31)

Complex Formation

Note: A review of some such metallic complexes has appeared.[899] Complexes with organic entities have been studied,[82,125,410,881,890] and some of the metallic complexes are referenced here.

Cu[329,382,494,891]

Re[375]

Rh[90,381,403,435,1305]

Ru[421,499,1230,1420]

Conversion into Cyano-1,8-naphthyridines

Ethyl 6-amino- (32, R = NH_2) gave ethyl 6-cyano-7-ethoxy-1-ethyl-4-oxo-1,4-dihydro-1,8-naphthyridine-3-carboxylate (32, R = CN) (diazotization; then $CuSO_4$, KCN: 82%).[1211]

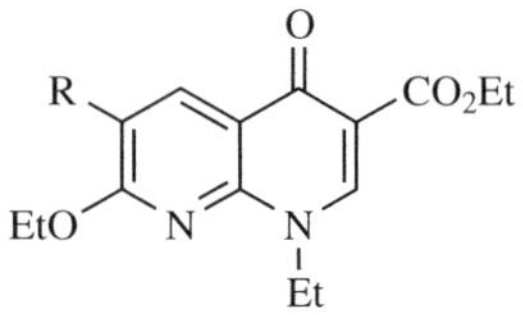

(32)

Also other examples.[387]

Conversion into Isocyanato- or Ureido-1,8-naphthyridines

5,7-Dimethyl-1,8-naphthyridin-2-amine (33, R = NH_2) gave 2-isocyanato-5,7-dimethyl-1,8-naphthyridine (33, R = NCO) as a minor product (7%) via an imidazo[1,2-*a*][1,8]naphthyridine intermediate.[1190] The same substrate (33, R = NH_2) with butyl isocyanate afforded 2-*N*-butylureido-5,7-dimethyl-1,8-naphthyridine (33, R = NHCONHBu) (PhMe, reflux: 72%).[123]

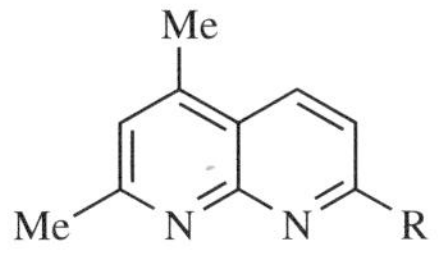

(33)

Also other examples.[559]

Cyclization Reactions

Note: Amino-1,8-naphthyridines and the like undergo a variety of cyclization reactions; a few typical examples are given here.

2,4-Diamino-1-phenyl-1,8-naphthyridin-2(1H)-one with triethyl orthoformate gave 5-phenyl[1H]imidazo[4,5-c][1,8]naphthyridin-4(5H)-one (**34**) (reactants, 120°C: 82%).[892]

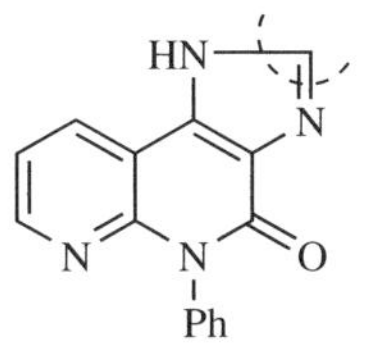

(34)

2-Amino-1,8-naphthyridine-3-carboxamide with benzaldehyde gave 2-phenyl-1,2-dihydropyrimido[4,5-b][1,8]naphthyridin-4(3H)-one (**35**) (AcOH, reflux: 65%).[875]

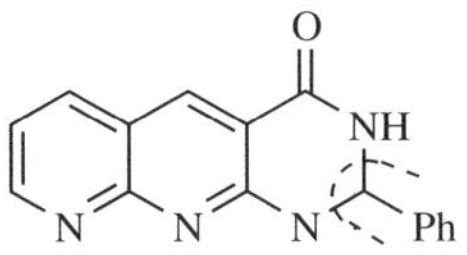

(35)

2-Hydrazino-3-*p*-methoxyphenyl-1,8-naphthyridine with α-benzoylstyrene gave 2-(3,5-diphenylpyrazolin-1-yl)-3-*p*-methoxyphenyl-1,8-naphthyridine (**36**) (AcOH, microwave: 86%).[735]

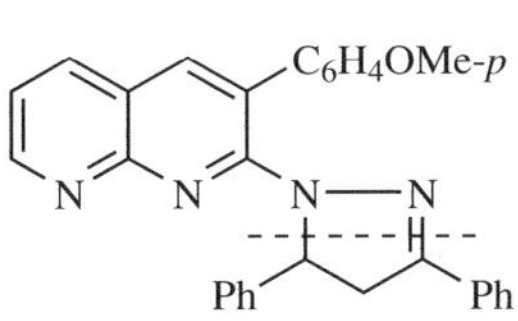

(36)

Also other examples.[282,358,525,601,692,730,836,850,866,894,1123,1191,1218,1293,1386]

CHAPTER 28

1,8-Naphthyridinecarboxylic Acids and Related Derivatives

This chapter deals with the carboxylic acids, carbonyl halides, carboxylic esters, carbonitriles, carbaldehydes, and ketone of 1,8-naphthyridine.

28.1. 1,8-NAPHTHYRIDINECARBOXYLIC ACIDS

The 1,8-naphthyridinecarboxylic acids are well represented in the literature, mainly because the powerful antimicrobial nalidixic acid (**1**) was adopted as a lead compound for a myriad of analogs,e.g.,[154,547,965]

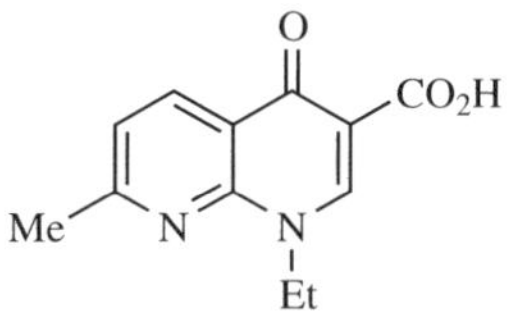

(**1**)

28.1.1. Preparation of 1,8-Naphthyridinecarboxylic Acids

These carboxylic acids have been made by *primary synthesis* (see Chapter 22), by *oxidation of alkyl-1,8-naphthyridines* (see Section 23.2.2), and by several procedures illustrated in the following examples.

By Hydrolysis of 1,8-Naphthyridinecarboxylic Esters

Note: This hydrolysis may be done under acidic or alkaline conditions.

The Naphthyridines: The Chemistry of Heterocyclic Compounds, Volume 63, by D.J. Brown

Ethyl 7-(3-aminopyrrolidin-1-yl)-4-oxo-1-(thiazol-2-yl)-1,4-dihydro-1,8-naphthyridine-3-carboxylate (**2**, R = Et) gave the corresponding 3-carboxylic acid (**2**, R = H) (10% HCl, reflux: 89%).[166]

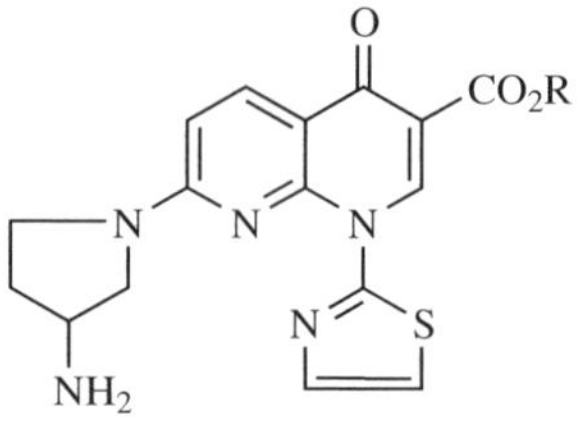

(**2**)

Ethyl 2,7-dimethyl-4-oxo-1,4-dihydro-1,8-naphthyridine-3-carboxylate (**3**, R = Et) gave the corresponding 3-carboxylic acid (**3**, R = H) (NaOH, EtOH, H$_2$O, reflux: 89%).[249]

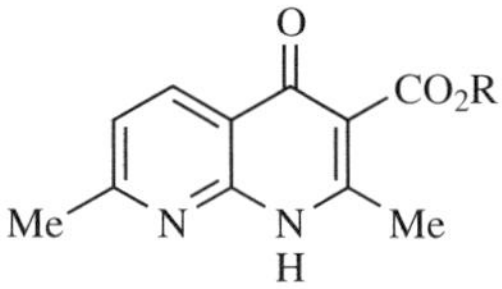

(**3**)

Also other examples.[418,611,681,687,690,702,765,963,1029,1041,1091,1205,1223,1236,1380]

By Hydrolysis of 1,8-Naphthyridinecarbonitriles

Note: Vigorous acidic or alkaline conditions may be used for these hydrolyses.

1-Ethyl-7-methyl-4-oxo-1,4-dihydro-1,8-naphthyridine-3-carbonitrile (**4**, R = CN) gave the corresponding 3-carboxylic acid (**4**, R = CO$_2$H) (9M H$_2$SO$_4$, 130°C: 86%).[1213]

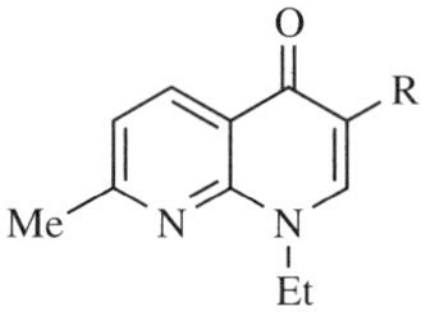

(**4**)

2-Phenyl-1,8-naphthyridine-3-carbonitrile (**5**, R = CN) gave the corresponding 3-carboxylic acid (**5**, R = CO$_2$H) (KOH, EtOH, H$_2$O, reflux, 14 h: 70–76%).[762,1100]

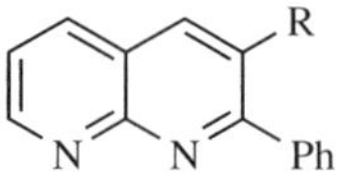

(**5**)

By Oxidation of 1,8-Naphthyridinecarbaldehydes

1,8-Naphthyridine-2,7-dicarbaldehyde (**6**, R = H) gave 1,8-naphthyridine-2,7-dicarboxylic acid (**7**, R = H) (80% HNO_3, reflux: 65%);[895,1133] similar oxidation of 4-octyloxy-1,8-naphthyridine-2,7-dicarbaldehyde (**6**, R = OC_8H_{17}) failed with nitric acid but succeeded with sodium chlorite in affording 4-octyloxy-1,8-naphthyridine-2,7-dicarboxylic acid (**7**, R = OC_8H_{17}) ($NaClO_2$, BuOH, $Me_2C=CHMe$, H_2O: 77%).[1376]

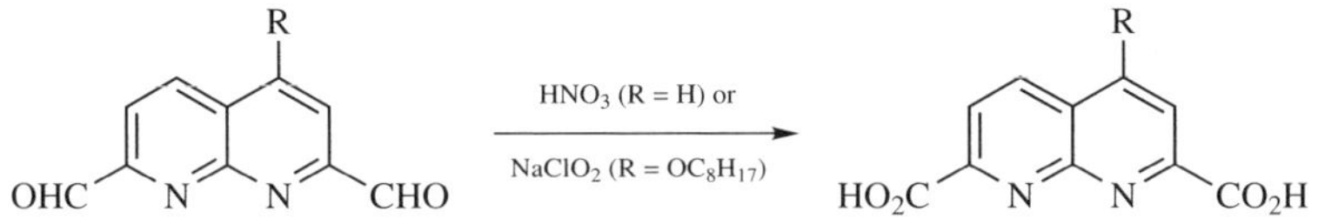

28.1.2. Reactions of 1,8-Naphthyridinecarboxylic Acids

1,8-Naphthyridinecarboxylic acids undergo the usual reactions, as illustrated in the following examples.

Decarboxylation

Note: Such decarboxylation has been done by heating dry, by heating in a high-boiling solvent, or by irradiation (in poor yield); the presence of cyanide ion (e.g., in Cu, KCN/Me_2SO) appears to facilitate decarboxylation.[518]

7-Amino-4-oxo-1,4-dihydro-1,8-naphthyridine-2-carboxylic acid (**8**, R = CO_2H) gave 2-phenyl-1,8-naphthyridine (**8**, R = H) (substrate, Cu bronze, 260°C: 78%).[1100]

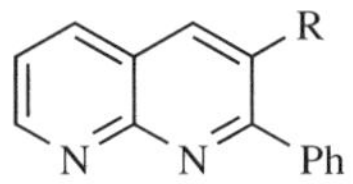

7-Amino-4-oxo-1,4-dihydro-1,8-naphthyridine-2-carboxylic acid (**9**, R = CO_2H) gave 7-amino-1,8-naphthyridin-4(1*H*)-one (**9**, R = H) Dowtherm A, trace copper chromite, reflux: 77%).[758]

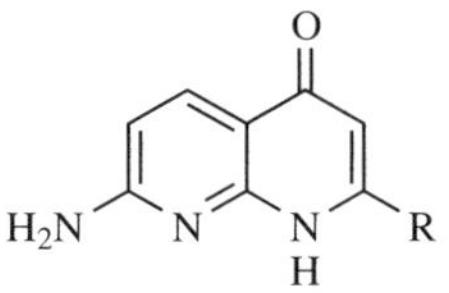

Also other examples.[153,251,344,681,687,702,1060,1257]

Esterification

Note: Esterification has been done directly or by the indirect route via an acid chloride.

Nalidixic acid (**10**, R = H) gave ethyl 1-ethyl-7-methyl-4-oxo-1,4-dihydro-1,8-naphthyridine-3-carboxylate (**10**, R = Et) (EtOH, H_2SO_4, reflux: 45%).[1276]

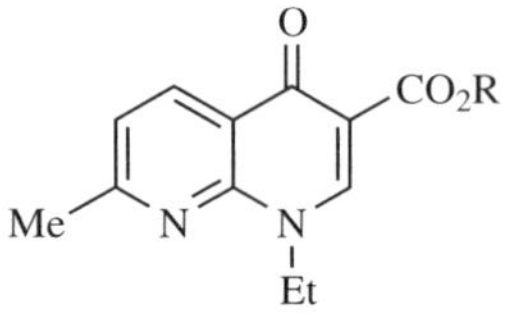

(**10**)

7-Chloro-1-ethyl-6-fluoro-4-oxo-1,4-dihydro-1,8-naphthyridine-3-carboxylic acid (**11**, R = H) gave its ethyl ester (**11**, R = Et) ($ClCO_2Et$, Et_3N, $CHCl_3$; then EtOH: 96%);[418] an analogous esterification used oxalyl chloride in the first stage.

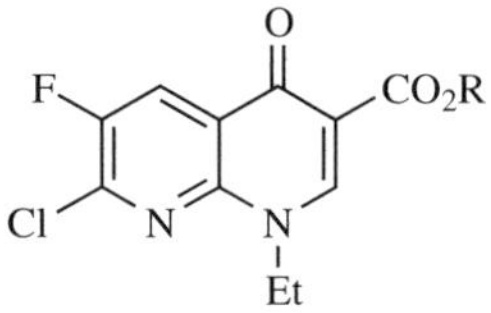

(**11**)

Also other examples.[1380,1450]

Conversion into Acid Halides

Note: Such acyl halides are only seldom characterized and rarely even isolated.

1,8-Naphthyridine-2,7-dicarboxylic acid gave 1,8-naphthyridine-2,7-dicarbonyl chloride (**12**, R = H) ($SOCl_2$, reflux: 57%);[1133] 4-octyloxy-1,8-naphthyridine-2,7-dicarbonyl chloride (**12**, R = OC_8H_{17}) was made similarly but not characterized.[1376]

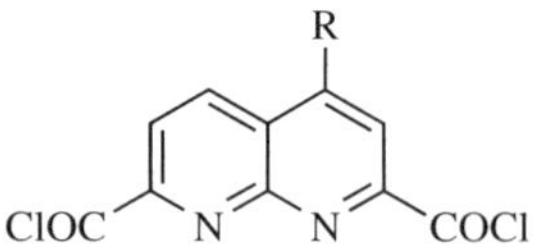

(**12**)

Also other examples.[738]

Conversion into Carboxamides

Note: Although this transformation can be done directly, all reported examples appear to have used the indirect route via an acid chloride.

5-(3-Aminopyrrolidin-1-yl)-6-fluoro-4-oxo-1-(thiazol-2-yl)-1,4-dihydro-1,8-naphthyridine-3-carboxylic acid (**13**, R = OH) gave the corresponding 3-carboxamide (**13**, R = NH$_2$) via the unisolated acid chloride (**13**, R = Cl) (ClCO$_2$Et, Et$_3$N; then NH$_3$/EtOH: 54%).[352]

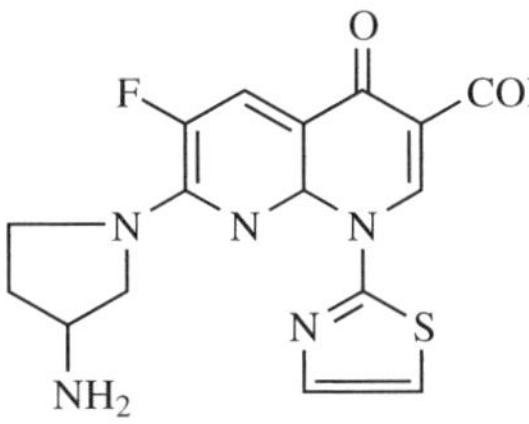

(**13**)

Also other examples.[212,235,966]

Complex Formation

Note: Typical metallic complexes are those of 1,8-naphthyridine-2,7-dicarboxylic acid with Ni,[368] Rh,[498] and Ru;[498] and of 2,7-bis[N, N-di(carboxymethyl)aminomethyl]-1,8-naphthyridine (**14**) with Eu and Tb (for luminescence analysis).[714]

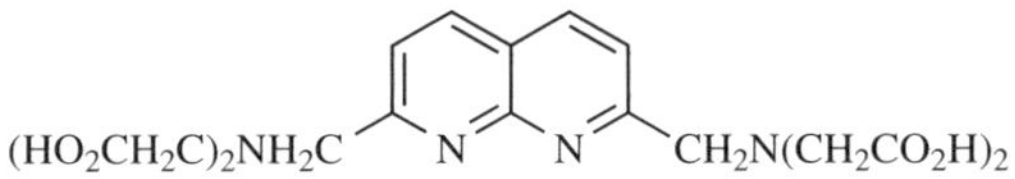

(**14**)

Degradation

Note: The oxidative degradation of a partly specified extranuclear carboxylated tetrahydro-1,8-naphthyridine has been studied by MS/NMR means.[1261]

28.2. 1,8-NAPHTHYRIDINECARBONYL HALIDES

The *preparation* of such chlorides from the corresponding carboxylic acids and their *conversion into esters or amides* have been covered briefly in Section 28.1.2. Other reported reactions are exemplified here.

Crude 2-methyl-1,8-naphthyridine-3-carbonyl chloride (**15**, R = Cl) gave 2-methyl-1,8-naphthyridine-2-carbohydrazide (**15**, R = NHNH$_2$) (H$_2$NNH$_2$·H$_2$O, EtOH, reflux: crude or 2-methyl-1,8-naphthyridine-3-carbonyl azide (**15**, R = N$_3$) (NaN$_3$, dioxane, 0°C: 43%).[738]

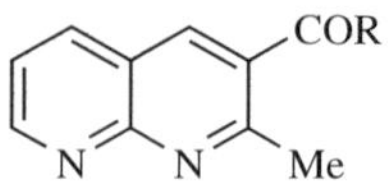

(**15**)

Although 1,8-naphthyridine-2,7-dicarbonyl chloride reacted with glycol to give di(2-hydroxyethyl)-1,8-naphthyridine-2,7-dicarboxylate (**16**) (excess HOCH$_2$-CH$_2$OH, 100°C: 83%),[1090] efforts to form a macrocyclic link between the carbonyl groups with a long-chain α,ω-dihydroxy synthon proved unsuccessful.[1090,1133]

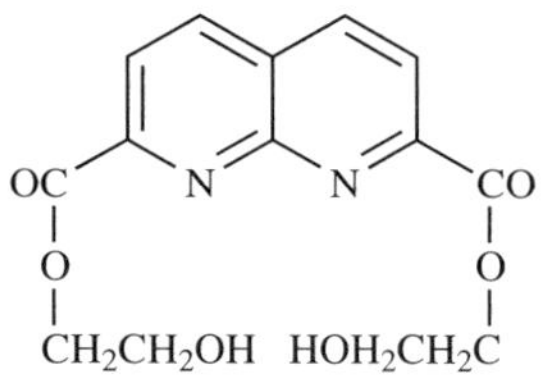

(**16**)

28.3. 1,8-NAPHTHYRIDINECARBOXYLIC ESTERS

28.3.1. Preparation of 1,8-Naphthyridinecarboxylic Esters

Most such esters have been made by *primary synthesis* (see Chapter 22) and some by *esterification* of corresponding carboxylic acids (see Section 28.1.2) or from *halogeno-1,8-naphthyridines* (see Section 24.2). A few extranuclear esters have been prepared by *passenger alkylations*. For example, by *S*-ethoxycarbonylation as in Chapter 26.

28.3.2. Reactions of 1,8-Naphthyridinecarboxylic Esters

These esters have been *reduced* to hydroxymethyl-1,8-naphthyridines (see Section 25.1.1) and *hydrolyzed* to 1,8-naphthyridinecarboxylic acids (see Section 28.1.1). Other reactions are illustrated by the following examples.

Dealkoxycarbonylation

Note: This has been done indirectly as a one-pot process by hydrolysis and decarboxylation; alternatively, the procedure may be performed in two distinct stages.

Methyl 6-bromo-2-oxo-1,2,3,4-tetrahydro-1,8-naphthyridine-3-carboxylate (**17**, R = CO$_2$Me) gave 6-bromo-3,4-dihydro-1,8-naphthyridin-2(1H)-one (**17**, R = H) (NaOH, MeOH, H$_2$O, reflux; then HCl to pH 7, reflux: 88%).[194]

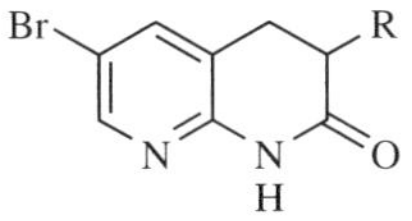

(**17**)

6-Ethoxycarbonyl-5-oxo-5,8-dihydro-1,8-naphthyridine-2-carboxylic acid (**18**, R = Et) underwent saponification to 4-oxo-1,4-dihydro-1,8-naphthyridine-5,7-dicarboxylic acid (**18**, R = H) (5M KOH, 100°C: crude solid), which was then decarboxylated to give 1,8-naphthyridin-4(1H)-one (**19**) (quinoline, reflux: ∼50% overall).[1186]

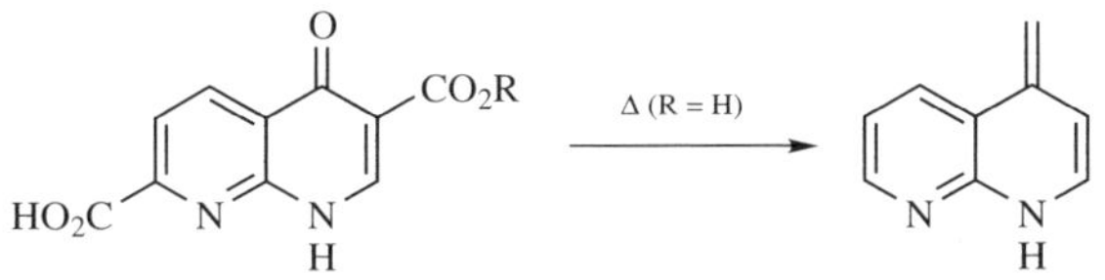

(**18**)   Δ (R = H)   (**19**)

Conversion into Carboxamides or the Like

Note: This partial aminolysis of esters has been used extensively.

Ethyl 1-ethyl-4-hydroxy-7-methyl-2-oxo-1,2-dihydro-1,8-naphthyridine-3-carboxylate (**20**, R = OEt) gave the corresponding 3-carboxamide (**20**, R = NH$_2$) (NH$_3$/EtOH, 100°C, sealed: 49%).[612]

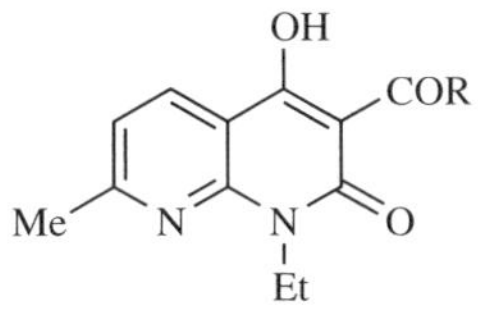

(**20**)

Ethyl 6-amino-7-chloro-1-ethyl-4-oxo-1,4-dihydro-1,8-naphthyridine-3-carboxylate (**21**, Q = Cl, R = Et) gave 6-amino-1-ethyl-N-methyl-7-methylamino-4-oxo-1,4-dihydro-1,8-naphthyridine-3-carboxamide (**21**, Q = R = NHMe) (MeNH$_2$/EtOH, 60°C: 95%; note the additional aminolysis of the chloro substituent).[1029]

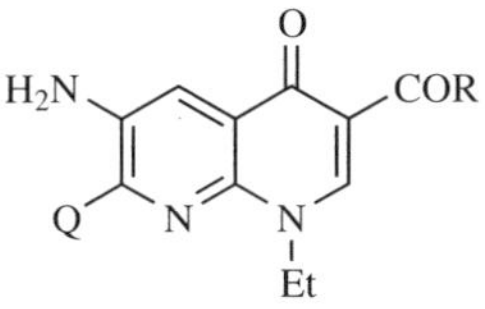

(**21**)

Ethyl 2-phenyl-1,8-naphthyridine-3-carboxylate (**22**, R = OEt) gave 2-phenyl-1,8-naphthyridine-3-carbohydrazide (**22**, R = NHNH$_2$) (H$_2$NNH$_2$·H$_2$O, EtOH, reflux: 95%).[859]

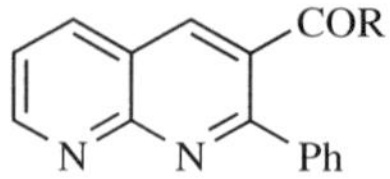

(**22**)

Ethyl 1-ethyl-7-methyl-4-oxo-1,4-dihydro-1,8-naphthyridine-3-carboxylate (**23**, R = OEt) gave N-amidino-1-ethyl-7-methyl-4-oxo-1,4-dihydro-1,8-naphthyridine-3-carboxamide [**23**, R = NHC(=NH)NH$_2$] [HN=C(NH$_2$)$_2$/EtOH, 20°C: 80% as hydrochloride].[635]

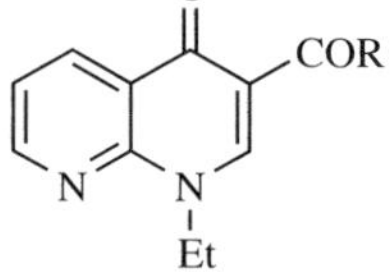

(**23**)

Also other examples.[76,305,362,618,622,632,651,728,740,1116,1416]

28.4. 1,8-NAPHTHYRIDINECARBOXAMIDES

Although 1,8-naphthyridinecarboxamides have shown several interesting bioactivities,[504,587] most of the described amides are in fact 1,8-naphthyridinecarbohydrazides.

28.4.1. Preparation of 1,8-Naphthyridinecarboxamides

Such carboxamides and the like have been made by *primary synthesis* (see Chapter 22), *from 1,8-naphthyridinecarbonyl halides* (see Sections 28.1.2 and 28.2), *from 1,8-naphthyridinecarboxylic esters* (see Section 28.3.2), or by the methods illustrated in the following examples.

By Controlled Hydrolysis of 1,8-Naphthyridinecarbonitriles

2-Phenyl-1,8-naphthyridine-3-carbonitrile (**24**, R = CN) gave 2-phenyl-1,8-naphthyridinecarboxamide (**24**, R = CONH$_2$) (KOH, H$_2$O, EtOH, reflux, 4 h: 51%; longer reflux gave the carboxylic acid).[1100]

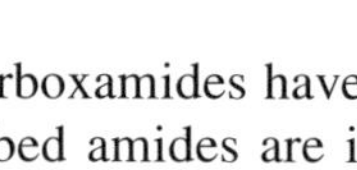

(**24**)

From Trichloromethyl-1,8-naphthyridines

4-Oxo-7-trichloromethyl-1,4-dihydro-1,8-naphthyridine-3-carboxylic acid (**25**) gave 7-(*N*-methylcarbamoyl)-4-oxo-1,4-dihydro-1,8-naphthyridine-3-carboxylic acid (**26**) (MeNH$_2$/EtOH, 90°C, sealed: 70%); also many analogs by slightly modified procedures.[1012]

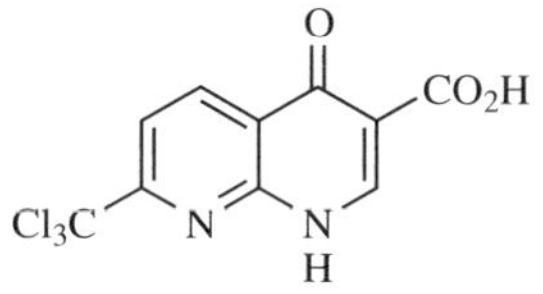

(**25**) 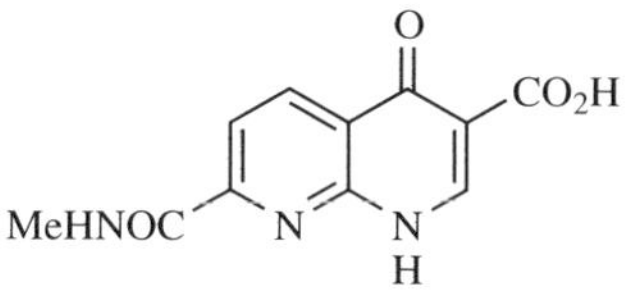   (**26**)

28.4.2. Reactions of 1,8-Naphthyridinecarboxamides

For the conversion of *hydrazides into amines* by the Curtius reaction, see Section 27.3; other reported reactions are illustrated in the following examples.

N-Alkylidenation

Note: All available examples of this procedure appear to use hydrazides rather than amides as substrates.

2-Oxo-1,2-dihydro-1,8-naphthyridine-3-carbohydrazide gave *N*-benzylidene-2-oxo-1,2-dihydro-1,8-naphthyridine-3-carbohydrazide (**27**) (PhCHO, trace AcOH, MeOH, reflux: 78%; analogs likewise).[632]

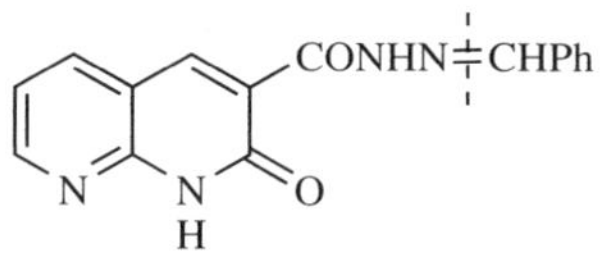

(**27**)

2-Methyl-1,8-naphthyridine-3-carbohydrazide gave 2-methyl-*N'*-methylbenzylidene-1,8-naphthyridine-3-carbohydrazide (**28**) (AcPh, trace AcOH, EtOH, reflux: 86%; analogs likewise).[871]

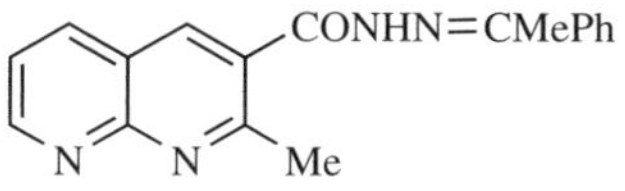

(**28**)

Also many other examples.[76,450,513,622,624,633,651,663,664,694,740,756,852,853,855,1341,1354]

Conversion into 1,8-Naphthyridinecarbonitriles

1-Ethyl-4-hydroxy-7-methyl-2-oxo-1,2-dihydro-1,8-naphthyridine-3-carboxamide (**29**) gave 4-chloro-1-ethyl-7-methyl-2-oxo-1,2-dihydro-1,8-naphthyridine-3-carbonitrile (**30**) ($POCl_3$, reflux: 55%; note chlorolysis as well as dehydration).[612]

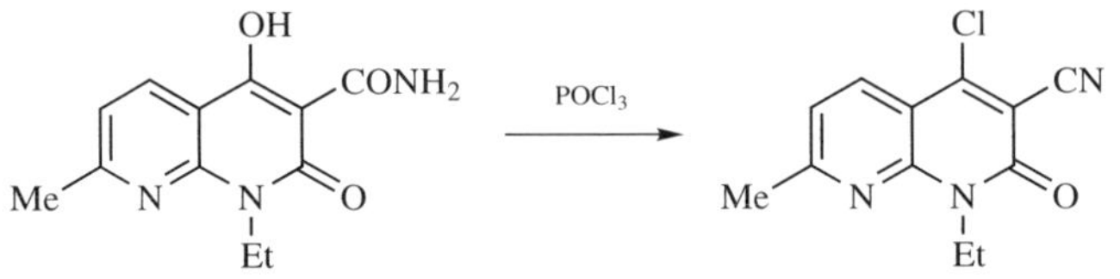

Conversion into N-Carbamoyl-1,8-naphthyridinecarbohydrazides

2-Phenyl-1,8-naphthyridine-3-carbohydrazide gave N'-[N-phenyl(thiocarbamoyl)]-2-phenyl-1,8-naphthyridine-3-carbohydrazide (**31**) (PhNCS, EtOH, reflux: 85%).[859]

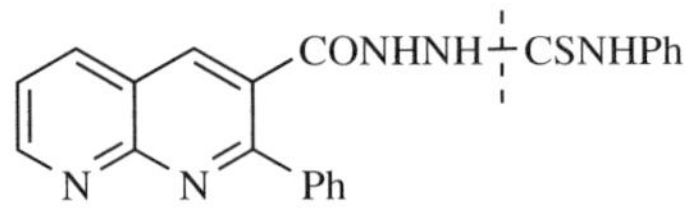

Also related examples.[475,631]

Cyclization Reactions

Note: Of the many cyclizations based on 1,8-naphthyridinecarbohydrazides, only a few random examples appear here.

2-Methyl-1,8-naphthyridine-3-carbohydrazide (**32**) gave 2-methyl-3-(5-thioxo-4,5-dihydro-1,3,4-oxadiazol-2-yl)-1,8-naphthyridine (**33**) (CS_2, KOH, H_2O, EtOH, reflux: 75%).[851]

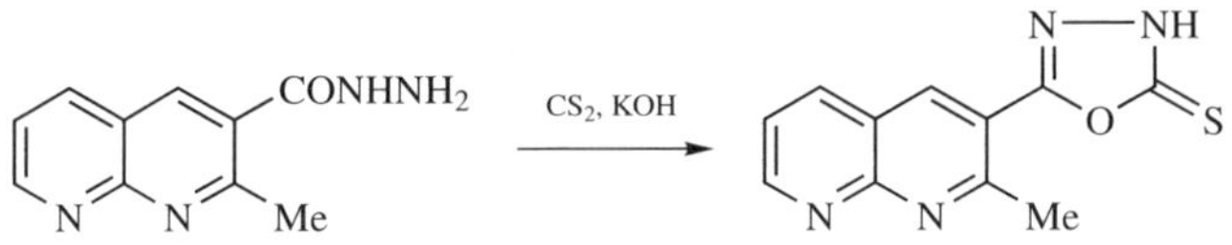

2-Trifluoromethyl-1,8-naphthyridine-3-carbohydrazide (**34**) with benzoylstyrene gave 3-(3,5-diphenyl-3-pyrazolin-1-yl)carbonyl-2-trifluoromethyl-1,8-naphthyridine (**35**) (BzCH=CHPh, trace TsOH, AcOH, reflux: 70%).[1356]

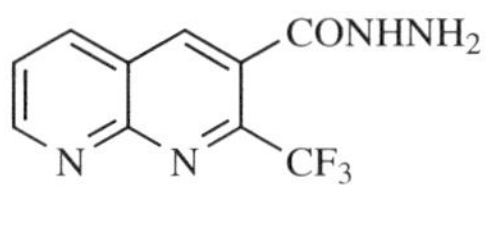

(34) structure with CONHNH₂ and CF₃ substituents on 1,8-naphthyridine

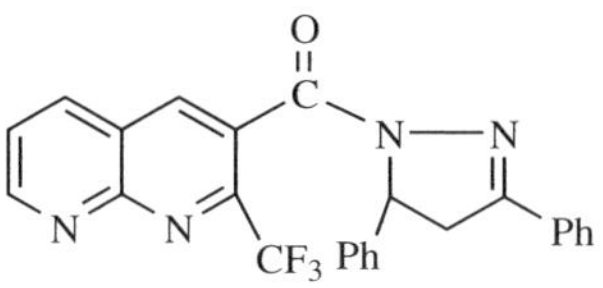

(35) structure with pyrazoline ring attached via carbonyl

Also a variety of other cyclizations.[1138,1239,1280,1337,1353,1421]

28.5. 1,8-NAPHTHYRIDINECARBONITRILES

Most of the information on these nitriles has been covered already

Preparation

Note: 1,8-Naphthyridinecarbonitriles have been made by *primary synthesis* (see Chapter 22), by *cyanolysis of halogeno-1,8-naphthyridines* (see Section 24.2), *from 1,8-naphthyridinamines* (see Section 27.3.2), or by *dehydration of 1,8-naphthyridinecarboxamides* (see Section 28.4.2).

Reactions

Note: 1,8-Naphthyridinecarbonitriles have been *hydrolyzed to the corresponding carboxylic acids* (see Section 28.1.1) or *carboxamides* (see Section 28.4.1). They may also be *converted into the corresponding carbaldehydes* as illustrated here.

Ethyl 7-chloro-6-cyano-1-ethyl- (**36**, R = CN) gave ethyl 7-chloro-1-ethyl-6-formyl-4-oxo-1,4-dihydro-1,8-naphthyridine-3-carboxylate (**36**, R = CHO) (HCl gas, SnCl₂, Et₂O; substrate in, 20°C, 48 h: 25%).[1211]

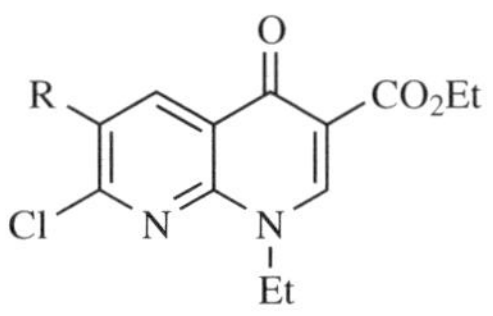

(36) structure

28.6. 1,8-NAPHTHYRIDINECARBALDEHYDES AND RELATED KETONES

28.6.1. Preparation of the Carbaldehydes and Ketones

One or both of these carbonyl entities has (have) been made by *primary synthesis* (see Chapter 22), by *oxidation of alkyl-1,8-naphthyridines* (see Section 23.2.2), or by the procedures illustrated in the following examples.

By C-Formylation

4-Hydroxy-1,8-naphthyridin-2(1H)-one (**37**, R = H) gave 4-hydroxy-2-oxo-1,2-dihydro-1,8-naphthyridine-3-carbaldehyde (**37**, R = CHO) (POCl$_3$, Me$_2$NCHO: ?%).[1281]

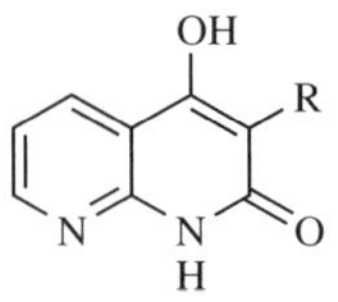

(**37**)

1,7-Dimethyl-2,3-dihydro-1,8-naphthyridin-4(1H)-one (**38**) gave its 3-hydroxymethylene derivative (**39**) (EtO$_2$CH, NaOEt, PhH, 20°C: 81%) that underwent oxidation in refluxing benzene to afford 1,7-dimethyl-4-oxo-1,4-dihydro-1,8-naphthyridine-3-carbaldehyde (**40**) as the main product (~45%).[1394]

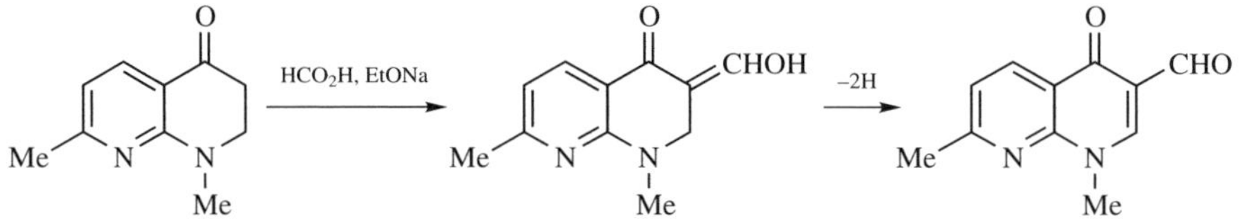

(**38**) →[HCO$_2$H, EtONa] (**39**) →[−2H] (**40**)

From Trimethylsilyl Analogs

Ethyl 7-chloro-1-cyclopropyl-6-fluoro-4-oxo-5-trimethylsilyl- (**41**, R=SiMe$_3$) gave ethyl 7-chloro-1-cyclopropyl-6-fluoro-5-formyl-4-oxo-1,4-dihydro-1,8-naphthyridine-3-carboxylate (**41**, R=CHO) (POCl$_3$, Me$_2$NCHO, o-Cl$_2$C$_6$H$_4$, 90°C: 24%).[1172]

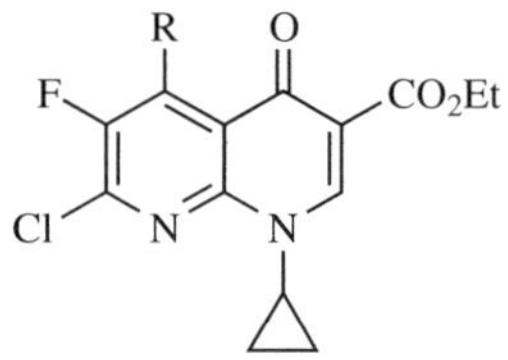

(**41**)

Ketones from Aldehydes

Methyl 1-ethyl-2-formyl-4-oxo-1,4-dihydro-1,8-naphthyridine-3-carboxylate (**42**) gave 7-acetyl-1-ethyl-4-oxo-1,4-dihydro-1,8-naphthyridine-3-carboxylic acid

(**43**) (CH$_2$N$_2$; then HCl, reflux: 74%).[1163]

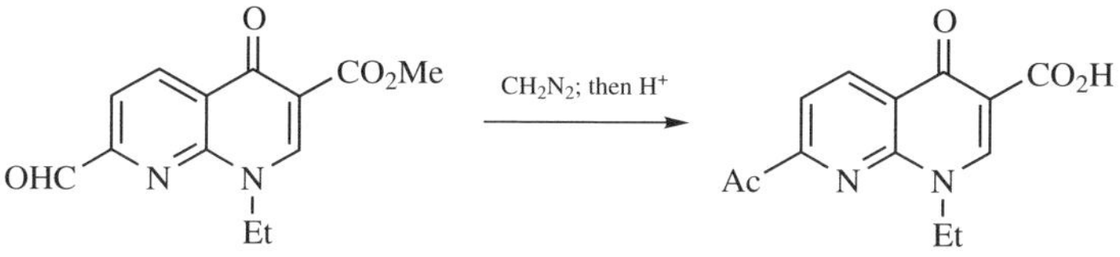

(**42**) →[CH$_2$N$_2$; then H$^+$] (**43**)

28.6.2. Reactions of the Carbaldehydes and Ketones

These carbonyl derivatives undergo the usual reactions associated with such compounds. The equilibrium of 1,8-naphthyridine-4-carbaldehyde with its hydrate has been studied.[953]

Anil, Oxime, and Hydrazone Formation

Ethyl 1-ethyl-7-formyl-4-oxo-1,4-dihydro- gave ethyl 1-ethyl-4-oxo-7-phenyliminomethyl-1,8-naphthyridine-3-carboxylate (**44**) (PhNH$_2$, EtOH, reflux: 34%; also many analogs).[106]

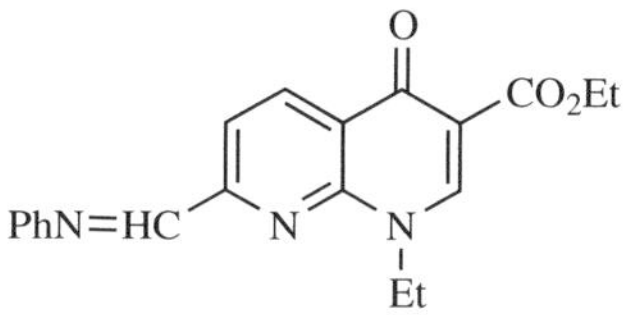

(**44**)

1-Ethyl-6-fluoro-7-formyl- gave 1-ethyl-6-fluoro-7-(hydroxyimino)methyl-4-oxo-1,4-dihydro-1,8-naphthyridine-3-carboxylic acid (**45**) (H$_2$NOH·HCl, pyridine, EtOH, reflux: 86%).[395]

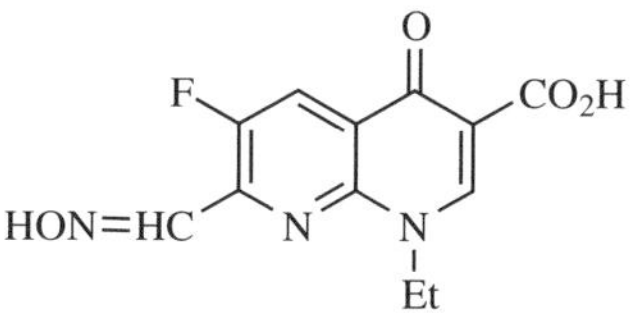

(**45**)

3-Acetyl-2-methyl-1,8-naphthyridine gave 3-[1-(benzoylhydrazono)ethyl]-1,8-naphthyridine (**46**) (BzHNNH$_2$: ?%).[516]

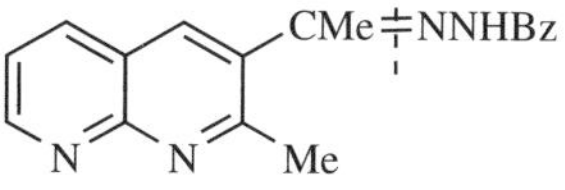

(**46**)

Alkylidenation (of Ketones)

3-Acetyl-2-methyl-1,8-naphthyridine gave 3-*p*-hydroxycinnamoyl-2-methyl-1,8-naphthyridine (**47**) (substrate, HOC_6H_4CHO-*p*, ground together, 10 min: 88%); many analogs likewise.[623]

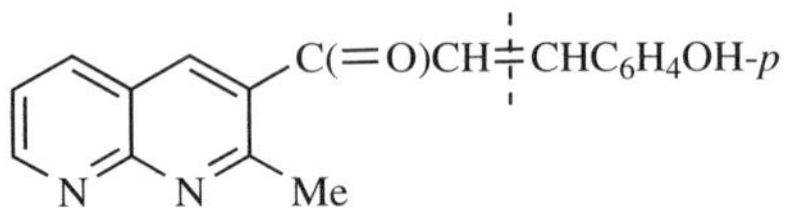

(**47**)

Also other examples.[657,1283]

Cyclization Reactions

2-Acetylstyryl-1,8-naphthyridine (**48**) gave 2-(3-methyl-1,5-diphenyl-2-pyrazolin-4-yl)-1,8-naphthyridine (**49**) ($PhNHNH_2$, AcOH, reflux: 65%).[753]

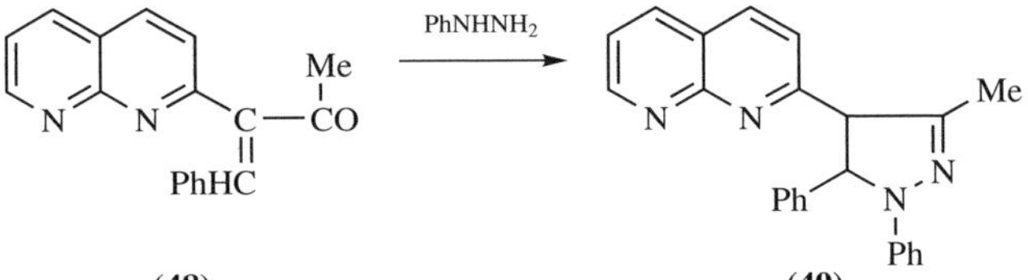

(**48**)   (**49**)

Reduction

Ethyl 7-chloro-1-cyclopropyl-6-fluoro-5-formyl- (**50**, R=CHO) underwent reduction to ethyl 7-chloro-1-cyclopropyl-6-fluoro-5-hydroxymethyl-4-oxo-1,4-dihydro-1,8-naphthyridine-3-carboxylate (**50**, R = CH_2OH) ($NaBH_4$, MeOH, 0°C: 47%).[1172]

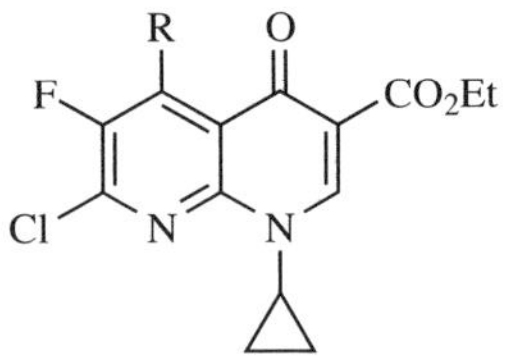

(**50**)

CHAPTER 29

The 2,6-Naphthyridines

Compared with the four naphthyridine systems already reviewed, there is so little information available on 2,6-naphthyridine (**1**) and its derivatives that a seven-chapter treatment is quite unwarranted.

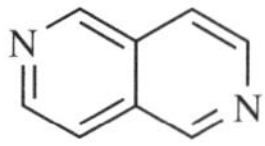

(**1**)

29.1. PRIMARY SYNTHESES OF 2,6-NAPHTHYRIDINES

Primary syntheses of 2,6-naphthyridines have been achieved by cyclization of appropriately substituted pyridine substrates, by cyclocondensation of pyridine substrates with synthons that provide one or more ring atoms, and from other heterocyclic substrates in several ways.

29.1.1. 2,6-Naphthyridines by Cyclization of Pyridine Derivatives

Such a synthesis is possible by completion of any one or five bonds of the final naphthyridine. All such possibilities have been attempted, as illustrated in the following examples.

By Completion of the 1,2-Bond

Ethyl 3-cyanomethyl-4-pyridinecarboxylate (**2**) gave 3-hydroxy-2,6-naphthyridin-1(2H)-one (**3**) (H$_2$SO$_4$, 80°C, 15 min: ~50%; presumably via an intermediate amide); also analogs.[710]

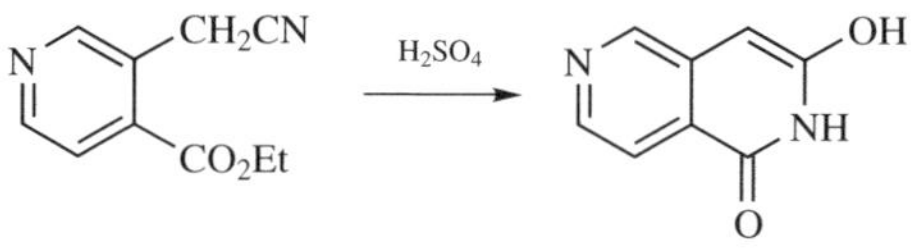

(**2**)   (**3**)

The Naphthyridines: The Chemistry of Heterocyclic Compounds, Volume 63, by D.J. Brown
Copyright © 2008 John Wiley & Sons, Inc.

By Completion of the 1,8a-Bond

Note: Only an unsuccessful attempt at this procedure has been reported.[46]

By Completion of the 2,3-Bond

2,6-Dimethyl-3-phenylethynyl-4-pyridinecarboxamide (**4**) gave 5,7-dimethyl-3-phenyl-2,6-naphthyridin-1(2*H*)-one (**5**) (EtONa/EtOH, reflux: 73%).[1055]

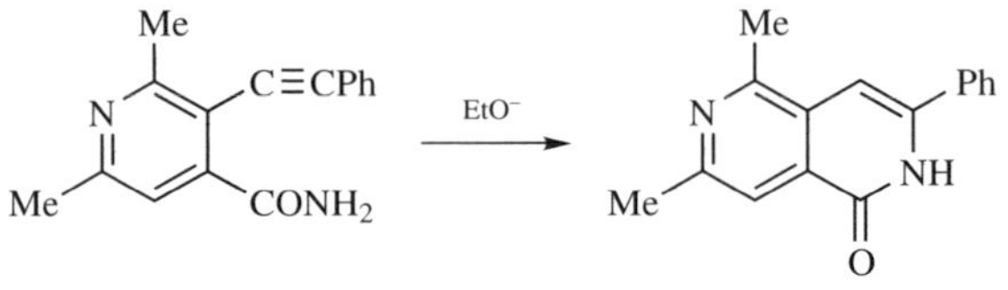

3-(2-Dimethylaminovinyl)-4-pyridinecarbonitrile (**6**) gave 2,6-naphthyridin-1(2*H*)-one (**7**) (HBr/AcOH, ~50°C: 66%; presumably via an amidic intermediate).[1115]

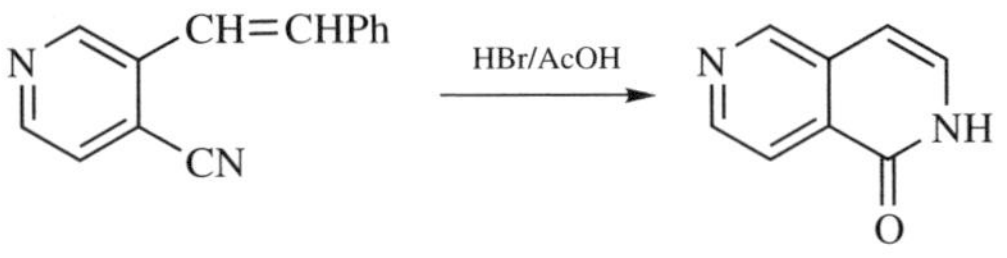

3-Cyanomethyl-2-pyridinecarbonitrile (**8**) gave 1-bromo-2,6-naphthyridin-3-amine (**9**) (HBr gas/Et$_2$O, −5°C: 80%).[186]

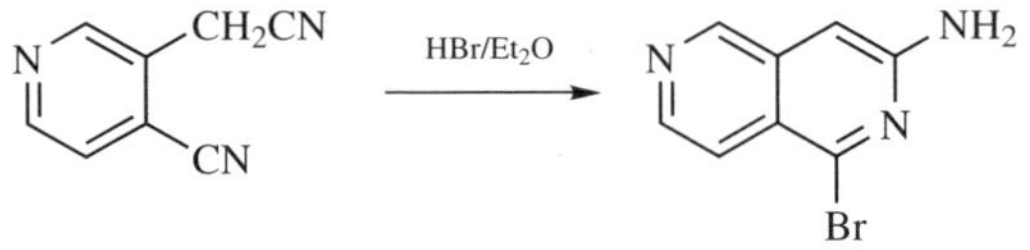

Also other examples.[133,147,441,621,669,677,710]

By Completion of the 3,4-Bond

Note: For an example of this synthesis, see Section 29.7.

By Completion of the 4,4a-Bond

4-[1-(Prop-1-enylimino)ethyl]pyridine (**10**) underwent pyrolysis to give a mixture of isomers in which 1,4-dimethyl-3,4-dihydro-2,6-naphthyridine (**11**) appeared to have been the major component.[400]

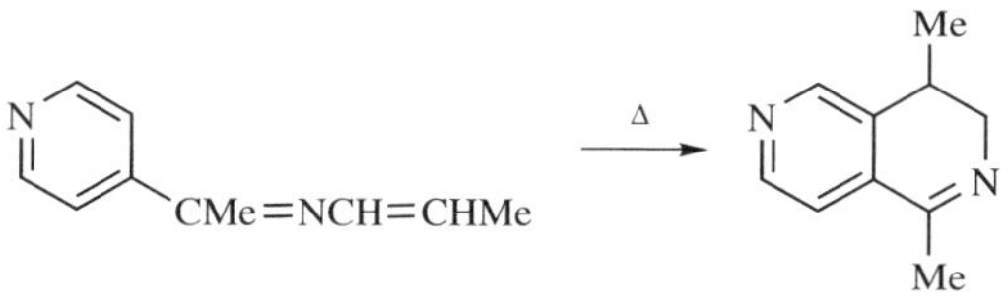

(10) → (11)

29.1.2. 2,6-Naphthyridines by Cyclocondensation of a Pyridine Derivative with a Synthon

Of the possiblities for such a cyclocondensation, only three types are represented in the 2,6-naphthyridine literature, although other types have been used e.g.,[308] to make fused analogs. Classified examples follow.

Where the Synthon Supplies C1

3-[2-Ethoxycarbonyl-2-(phosphoranylideneamino)vinyl]pyridine (12) with phenyl isocyanate gave a mixture of two isomers from which ethyl 1-amino-2,6-naphthyridine-3-carboxylate (13) was isolated (PhMe, 20°C; then 180°C, sealed: 22%); analogs likewise.[442]

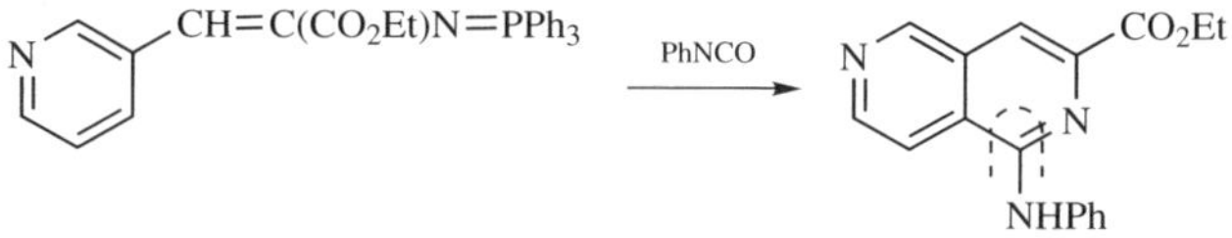

(12) → (13)

Where the Synthon Supplies N2

3-Phenylethynylpyridine-2-carbaldehyde (14) gave 3-phenyl-2,6-naphthyridine (15) (NH_3/EtOH, 80°C, sealed: 83%); analogs likewise.[621]

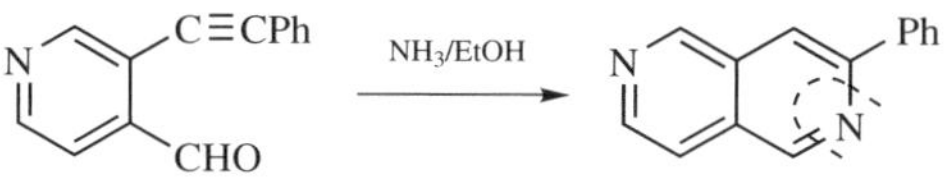

(14) → (15)

3-Acetonyl- (16, R=H) or 3-(diacetylmethyl)-2-pyridinecarbaldehyde (16, R=Ac) with aqueous ammonium acetate gave 3-methyl-2,6-naphthyridin-1(2H)-one

(17) (reflux: ~70%).[1072]

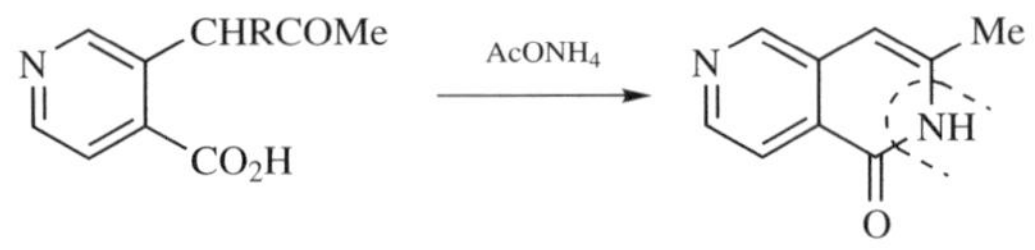

(16) → (17)

Also other examples.[229]

Where the Synthon Supplies C3

Treatment of 3-Acetoxy-5-acetoxymethyl-2-methyl-4-pyridinecarbonitrile (18) with ethyl magnesium bromide is reported to give 3-ethyl-8-hydroxy-3,7-dimethyl-3,4-dihydro-2,6-naphthyridin-1(2H)-one (19) (20%, after separation from another product).[240]

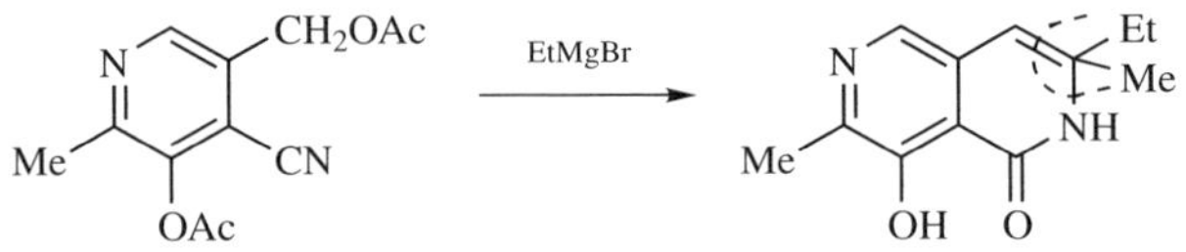

(18) → (19)

29.1.3. 2,6-Naphthyridines from Other Heterocyclic Substrates

Several heterocyclic systems other than pyridine have been used as substrates for the primary synthesis of 2,6-naphthyridines. The procedures are illustrated in the following examples.

1,6-Diazacyclodeca-3,8-diones as Substrates

1,6-Diisopropyl-1,6-diazacyclodeca-3,8-diyne (20) gave 2,6-diisopropyl-4-methoxy-1,2,3,5,6,7-hexahydro-2,6-naphthyridine (21) (MeOH, 140°C, sealed: 45%); also related syntheses.[684,cf. 580]

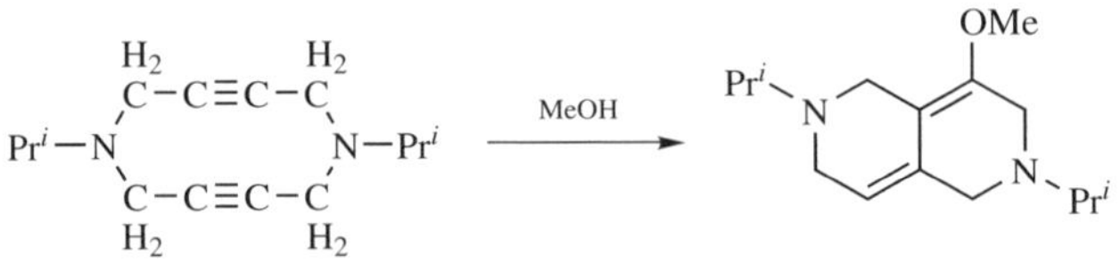

(20) → (21)

Pyrano [4,3-c]pyridines as Substrates

6-Methyl-3,7-diphenyl-1H-pyrano[4,3-c]pyridine-1,5(6H)-dione (**22**) gave 2,6-dimethyl-3,7-diphenyl-2,6-naphthyridine-1,5(2H,6H)-dione (**23**) (MeNH$_2$/MeOH, 20°C: 86%),[914] the structure of which was checked by X-ray analysis.[788]

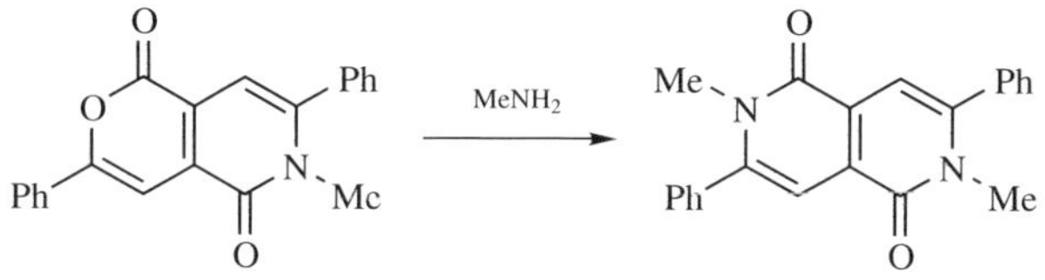

Pyrroles as Substrates

3,4-Dibenzoyl-1-methyl-2,4-diphenylpyrrole with benzylamine has been reported to afford a separable mixture of hexaphenyl-2,6- and hexaphenyl-2,7-naphthyridine (KOH, EtOH, reflux: 29% and 41%, respectively) by "an obscure pathway." The original paper should be consulted for details of this and other such syntheses.[1132]

1,2,4-Triazines as Substrates

Ethyl 1,2,4-triazine-3-carboxylate (**25**) underwent a Diels–Alder reaction with *tert*-butyl 4-(pyrrolidin-1-yl)-1,2,3,4-tetrahydropyridine-1-carboxylate (**24**) to give ethyl 6-*tert*-butoxycarbonyl-5,6,7,8-tetrahydro-2,6-naphthyridine-1-carboxylate (**26**) as a minor product (CHCl$_3$, 20°C: 8%).[941]

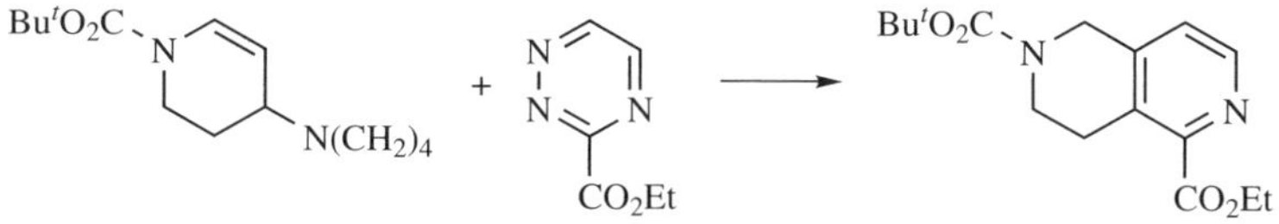

29.2. 2,6-NAPHTHYRIDINE AND ALKYL-2,6-NAPHTHYRIDINES

Despite a general paucity of information on 2,6-naphthyridines, the system has been included in several reviews:[50–52,57,58,61,265,436,1357,1430,1432] 4-methyl-2,6-naphthyridine has been found to occur naturally in several *Antirrhina* species,[226,241,267] and physical properties of the parent heterocycle (**1**) have been quite extensively studied, usually for comparison with those of the other unsubstituted naphthyridines (see Section 29.2.2).

29.2.1. Preparation of 2,6-Naphthyridine

2,6-Naphthyridine (**28**) is best made by hydrogenolysis of 1-chloro-2,6-naphthyridine (**27**) (Pd/C, QcONa, H$_2$, MeOH: 71%)[1115] or by oxidation of 1,3-dihydrazino-2,6-naphthyridine (**29**) (CuSO$_4$, AcOH, H$_2$O, 100°C: 53%);[186,677,1007] it has also been made from 3-hydroxy-2,6-naphthyridin-1(2H)-one (Zn dust, 170°C: poor yield).[677]

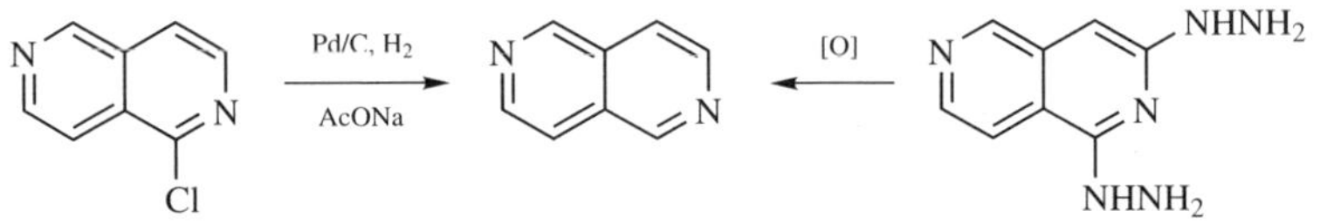

Efforts to prepare the parent naphthyridine by direct hydrogenolysis of 1,3-dichloro-2,6-naphthyridine (**30**) gave only 1,3-dichloro-5,6,7,8-tetrahydro-2,6-naphthyridine (**31**, R=Cl) (PtO$_2$, H$_2$: ?%),[677] and reductive deoxygenation of 1,4-dihydro-2,6-naphthyridin-3(2H)-one (**32**) gave 1,2,3,4-tetrahydro-2,6-naphthyridine (**31**, R=H) (LiAlH$_4$, dioxane, reflux: 15%).[669]

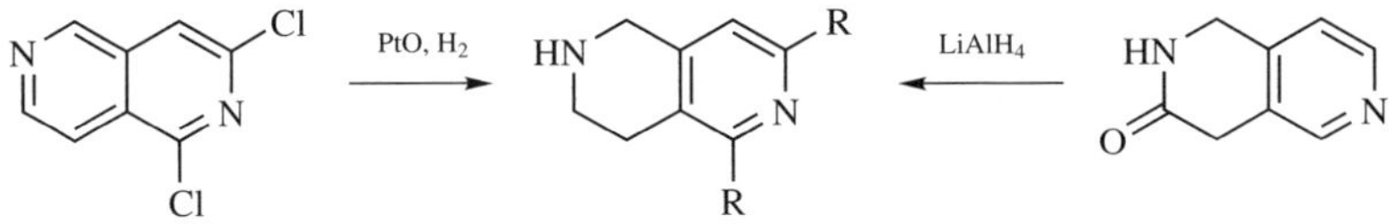

29.2.2. Properties of 2,6-Naphthyridine

Papers that report studies of the various physical properties of 2,6-naphthyridine are referenced here.

Aromaticity[676]
Electron density calculations[1126,1173,1174,1241]
Electron spin resonance spectra[1079,1083]
Ionization[813,1176,1192]
Nuclear magnetic resonance spectra[995,1174,1176,1192,1319]
Polarography[829,1329]
Ultraviolet spectra[820,840,878,1181,1250,1312,1426]
X-ray structure[193,798]

29.3. HALOGENO-2,6-NAPHTHYRIDINES

Halogeno-2,6-naphthyridines have good (but largely unused) potential as intermediates for metatheses, especially when the halogeno substituent occupies the activated 2-, 3-, 5-, or 7-position.

29.3.1. Preparation of Halogeno-2,6-naphthyridines

Some such derivatives have been made by *primary synthesis* (see Section 29.1). Other preparative routes are illustrated in the following examples.

By Halogenolysis of 2,6-Naphthyridinones

2,6-Naphthyridin-1(2H)-one (**33**) with a phosphoryl halide gave 1-bromo- (**34**, R = Br) (POBr$_3$, 135°C: 71%)[439] or 1-chloro-2,6-naphthyridine (**34**, R = Cl) (POCl$_3$, reflux: 77%).[1115]

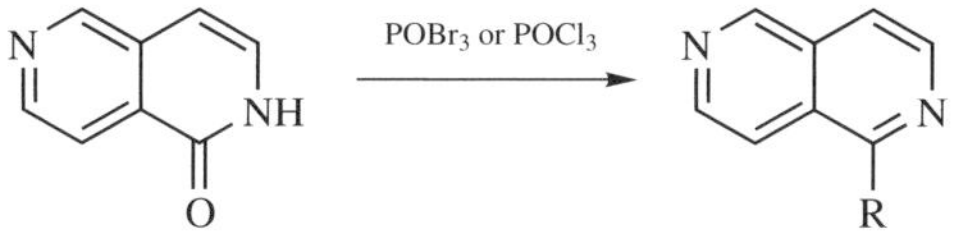

1,4-Dihydro-2,6-naphthyridin-3(2H)-one gave 3-chloro-1,4-dihydro-2,6-naphthyridine (**35**) (POCl$_3$, PCl$_5$, 210°C, sealed: 42%).[1389]

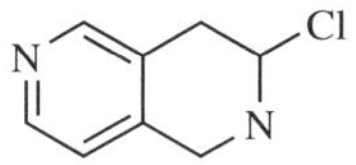

Also other examples.[677,1072]

From 2,6-Naphthyridinamines

2,6-Naphthyridin-3-amine (**36**, R = NH$_2$) gave 3-chloro-2,6-naphthyridine (**36**, R = Cl) (HNO$_2$, −5°C; then CuCl, 20°C: 17%).[1007]

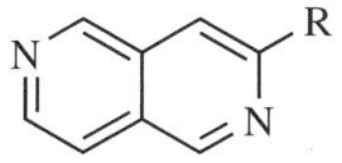

1-Bromo-4-methyl-2,6-naphthyridin-3-amine (**37**, R = Me) underwent diazotization in fuming hydrobromic acid to give 1,3-dibromo-4-methyl-2,6-naphthyridine (**38**, R = Me) (−5°C, then 20°C: 47%),[1007] but similar treatment of 1-bromo-2,6-naphthyridin-3-amine (**37**, R = H) afforded a separable mixture of 1,3-dibromo- (**38**, R = H) and 1,3,4-tribromo-2,6-naphthyridine (**38**, R = Br) (22% each).[186]

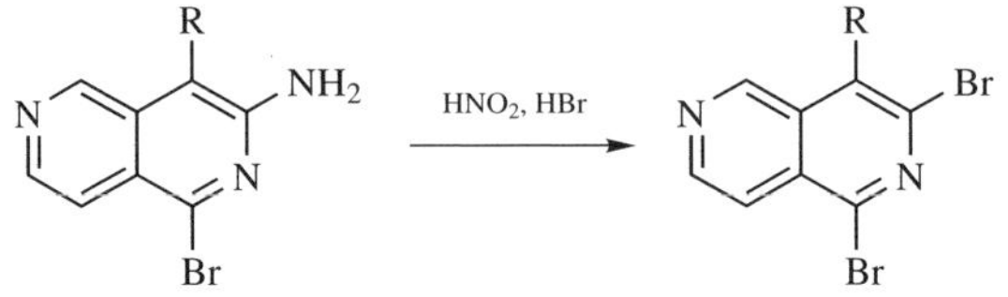

29.3.2. Reactions of Halogeno-2,6-naphthyridines

The relatively few reported reactions of halogeno-2,6-naphthyridines are illustrated by the following examples.

Aminolysis

1-Chloro-2,6-naphthyridine (**39**, R = Cl) gave 2,6-naphthyridin-1-amine (**39**, R = NH$_2$) (NH$_3$/PhOH, 175°C: 58%, identical to that obtained in 56% yield by amination of 2,6-naphthyridine);[1115] the analogous hydrazinolysis of 1-chloro-3-methyl-2,6-naphthyridine (**40**, R = Cl) gave 1-hydrazino-3-methyl-2,6-naphthyridine (**40**, R = NHNH$_2$) under quite mild conditions (H$_2$NNH$_2$·H$_2$O/EtOH, reflux, 10 min: 60%).[1072]

1,3-Dibromo- (**41**, R = Br) gave 1,3-dihydrazino-2,6-naphthyridine (**41**, R = NHNH$_2$) (H$_2$NNH$_2$ H$_2$O, dioxane, 20°C: 100%);[186,1007] 1,3-dichloro-2,6-naphthyridine reacted similarly (?%).[677]

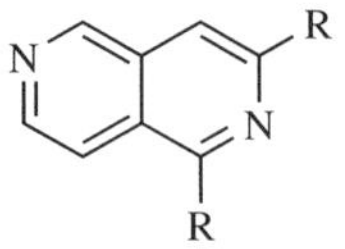

Also other examples.[439]

Alcoholysis

3-Chloro- (**42**, R = Cl) gave 3-ethoxy-1,4-dihydro-2,6-naphthyridine (**42**, R = OEt) (EtONa/EtOH, reflux: 85%).[1389]

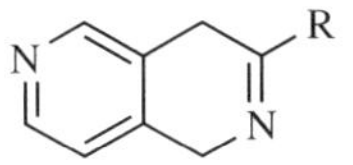

(**42**)

1,3-Dibromo- (**43**, R = Br) gave selectively 3-bromo-1-methoxy-2,6-naphthyridine (**43**, R = OMe) (NaOH, MeOH, 20°C: 80%).[1007]

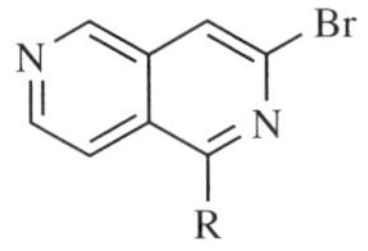

(**43**)

Hydrolysis

1,3-Dibromo-2,6-naphthyridine (**43**, R = Br) underwent direct hydrolysis to give selectively 3-bromo-2,6-naphthyridin-1(2H)-one (**44**) (3M H_2SO_4, reflux: 51%).[1007]

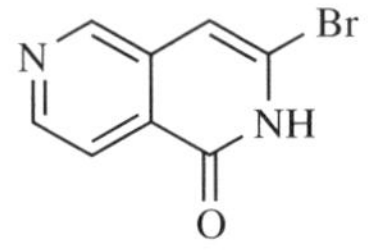

(**44**)

Dehalogenation

Note: Both direct and indirect dehalogenations have been exemplified in Section 29.2.1.

In addition, 2,6-naphthyridin-3-amine was obtained by direct hydrogenolysis of its 1-bromo derivative (Pd/C, KOH, H_2, EtOH: ?%).[186]

29.4. OXY-2,6-NAPHTHYRIDINES

Almost every reported aspect of 3,6-naphthyridinones, alkoxy-2,6-naphthyridines, and 2,6-naphthyridine *N*-oxides has been covered already.

2,6-Naphthyridinones—Preparation

These have been made either by *primary synthesis* (see Section 29.1) or by *hydrolysis of halogeno-2,6-naphthyridines* (see Section 29.3.2).

2,6-Naphthyridinones—Reactions

The halogenolysis of these naphthyridinones has been discussed in Section 29.3.1

Alkoxy-2,6-naphthyridines—Preparation

These ethers have been made by *primary synthesis* (see Section 29.1) or by *alcoholysis of halogeno-2,6-naphthyridines* (see Section 29.3.2).

Alkoxy-2,6-naphthyridines—Reactions

Only hydrogenolysis appears to have been reported; 3-ethoxy-1,4-dihydro- (**45**) gave 1,2,3,4-tetrahydro-2,6-naphthyridine (**46**) (NaBH$_4$, EtOH, 0°C, then 20°C: ?%).[1389]

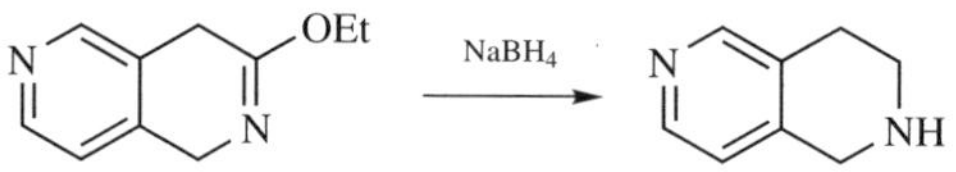

2,6-Naphthyridine *N*-Oxides: Preparation

Several such oxides have been made by primary synthesis; for example, 3-phenylethynyl-4-pyridinecarbaldehyde oxime (**47**) gave 3-phenyl-2,6-naphthyridine 2-oxide (**48**) (K$_2$CO$_3$, EtOH, reflux: 80%).[641]

29.5. THIO-2,6-NAPHTHYRIDINES

All types of thio-2,6-naphthyridine (see Chapter 5) seem to have escaped mention in the literature.

29.6. NITRO-, AMINO-, AND RELATED 2,6-NAPHTHYRIDINES

There appear to be no nitro- or nitroso-2,6-naphthyridines known,[cf. 434,1273] but amino- and hydrazino-2,6-naphthyridines have a modest representation in the literature.

29.6.1. Preparation of Amino-2,6-naphthyridines

Most such amines have been made by *primary synthesis* (see Section 29.1) or by *aminolysis of halogeno-2,6-naphthyridines* (see Section 29.3.2). In addition, some have been made by direct *C-amination*,[54] as illustrated in the following examples.

2,6-Naphthyridine (**49**, R = H) gave 2,6-naphthyridin-1-amine (**49**, R = NH$_2$) (KNH$_2$/NH$_3$: 54%).[1115,cf. 54]

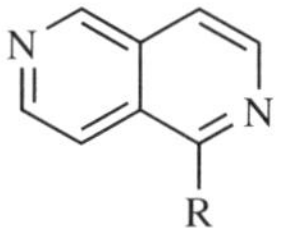

(**49**)

1-Chloro-2,6-naphthyridine (**50**) gave a separable mixture of 2,6-naphthyridin-1-amine (**51**, R=H) and 2,6-naphthyridine-1,5-diamine (**51**, R = NH$_2$) (KNH$_2$/NH$_3$, trace FeCl$_3$, −33°C: 20% and 17%, respectively).[439]

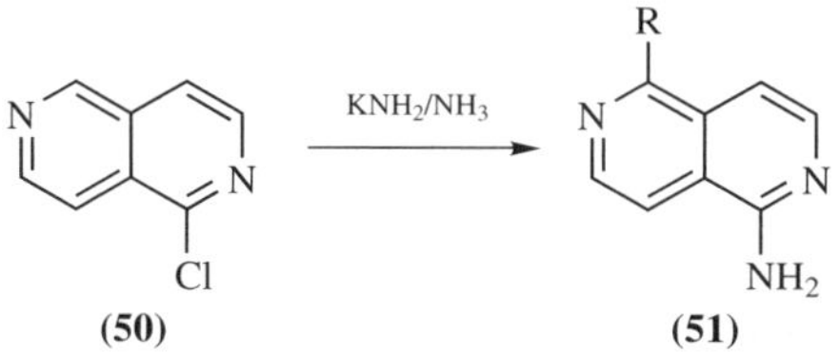

(**50**) → (**51**)

2,6-Naphthyridin-3-amine (**52**, R = H) gave 1-morpholino-2,6-naphthyridin-3-amine [**52**, R = N(CH$_2$CH$_2$)$_2$O] [(LiN(CH$_2$CH$_2$)$_2$O (made *in situ*), THF, −70°C; then −20°C: 91%]; analogs likewise.[133]

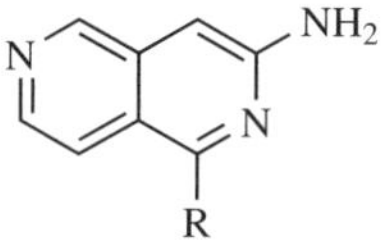

(**52**)

29.6.2. Reactions of Amino-2,6-naphthyridines

The conversion of *amino- into halogeno-2,6-naphthyridines* has been covered in Section 29.3.1. Other reported reactions are illustrated in the following examples.

Acylation

1-Bromo-4-methyl-2,6-naphthyridin-3-amine (**53**, R = H) gave 3-acetamido-1-bromo-4-methyl-2,6-naphthyridine (**53**, R = Ac) (Ac$_2$O, pyridine: ?%).[1007]

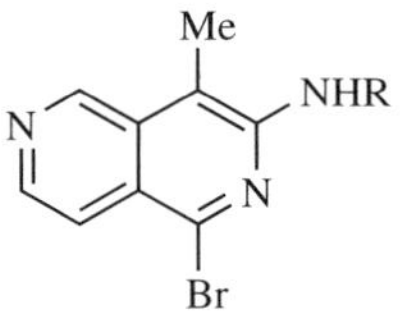

(53)

Also other examples.[686]

Dehydrazination

Note: This process has been used to make unsubstituted 2,6-naphthyridine (see Section 29.2.1). Other examples follow.

1-Hydrazino-3-methyl- (**54**, R = NHNH$_2$) gave 3-methyl-2,6-naphthyridine (**54**, R = H) (CuSO$_4$, AcOH, H$_2$O, 95°C: ~50%).[1072]

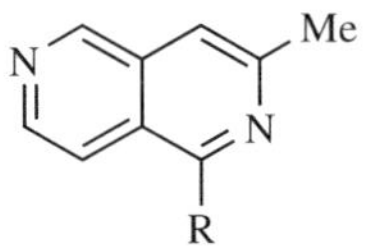

(54)

29.7. 2,6-NAPHTHYRIDINECARBOXYLIC ACIDS AND RELATED DERIVATIVES

A few such compounds have been made by *primary syntheses*, covered in Section 29.1; for example, *N*-(ethoxycarbonylmethyl)-3-(*p*-fluorobenzoyl)-*N*-methyl-4-pyridinecarboxamide (**55**) underwent cyclization to ethyl 4-*p*-fluorophenyl-2-methyl-1-oxo-1,2-dihydro-2,6-naphthyridine-3-carboxylate (**56**, R = Et) (substrate, diazabicycloundecene, PhMe, reflux: 38%).[1199] This ester underwent saponification to 4-*p*-fluorophenyl-2-methyl-1-oxo-1,2-dihydro-2,6-naphthyridine-3-carboxylic acid (**56**, R = H) (NaOH, EtOH, H$_2$O, THF, reflux: 95%) and subsequent conversion (via the unisolated carbonyl chloride) into *N*-[3,5-bis (trifluoromethyl)benzyl]-4-*p*-fluorophenyl-2,*N*-dimethyl-1-oxo-1,2-dihydro-2,6-naphthyridine-3-carboxamide (**57**) [SoCl$_2$, Me$_2$NCHO, PhH, reflux; then HNMeCH$_2$C$_6$H$_3$(CF$_3$)$_2$ −3,5 added, THF, 20°C: 54%].[1199]

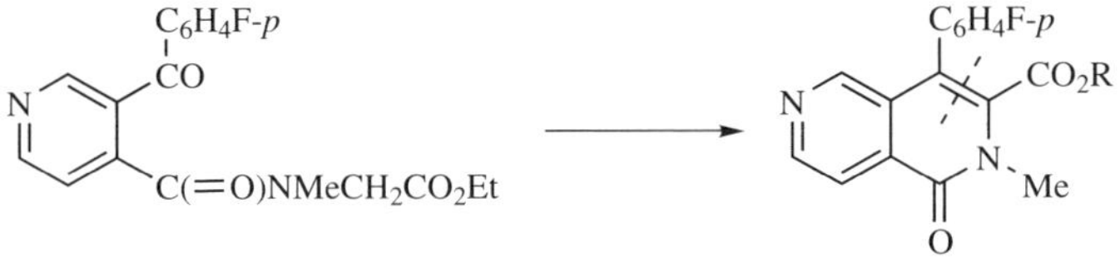

Other reactions of such carboxylic derivatives appear to be of marginal chemical interest. e.g.,[491]

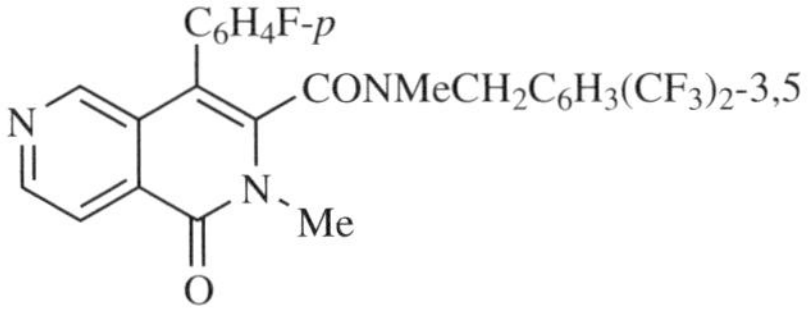

CHAPTER 30

The 2,7-Naphthyridines

Although a substantial number of papers are devoted, at least in part, to 2,7-naphthyridines, the spread of information is quite unbalanced in favor of primary syntheses. Accordingly, the 2,7-naphthyridine system will be covered in a single chapter. Several reviews of naphthyridine chemistry include material on the 2,7-system.[49,50,53,61,265,407,456,1357,1430,1432] The original system name, copyrin/copyrine,[13,cf. 39,40] is still used occasionally.

30.1. PRIMARY SYNTHESES OF 2,7-NAPHTHYRIDINES

Such syntheses may be classified into four main categories: from nonheterocyclic precursors; from single pyridine substrates, from pyridine substrates with ancillary synthons, and from other heterocyclic substrates.

30.1.1. 2,7-Naphthyridines from Nonheterocyclic Precursors

Of the many possibilities within this category, only a few distinct types have proved successful, as illustrated by the following examples.

Using Two Precursor Molecules

Heptane-2,4,6-trione (**1**) and malondiamidine (**2**) gave 3,6-dimethyl-2,7-naphthyridine-1,8-diamine (**3**) (NaOH, H_2O, 100°C: 9%).[736]

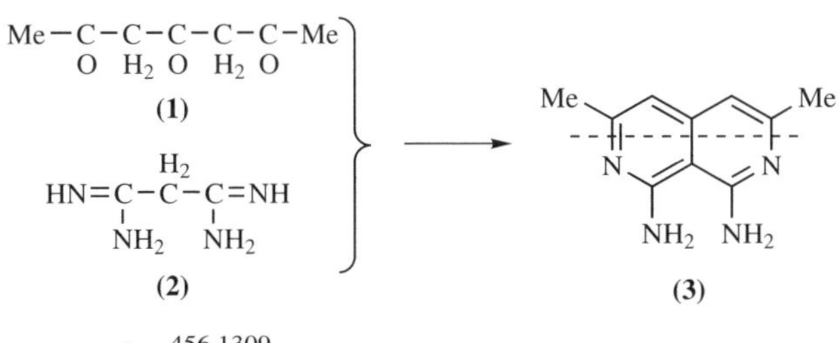

Also other examples.[456,1309]

The Naphthyridines: The Chemistry of Heterocyclic Compounds, Volume 63, by D.J. Brown
Copyright © 2008 John Wiley & Sons, Inc.

Using Three Precursor Molecules

2-Benzoyl-*N*-ethyl-1-methylthiovinylamine (**4**) with 2-cyano-*N*-methylacetamide (2 mol) gave 1-amino-2,7-dimethyl-3,8-dioxo-6-phenyl-2,3,7,8-tetrahydro-2,7-naphthyridine-4-carbonitrile (**5**) (reactants, Pr^{i}ONa, Pr^{i}OH, reflux: 75%; analogs likewise).[699]

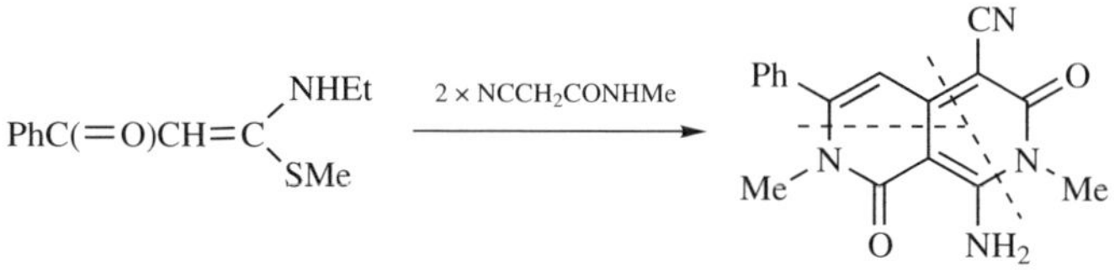

The (uncharacterized) intermediate (**6**) with ammonium ion (2 mol) gave 2,7-naphthyridine-4-carbaldehyde (**7**) (NH_4Cl, H_2O, reflux: 67%).[825]

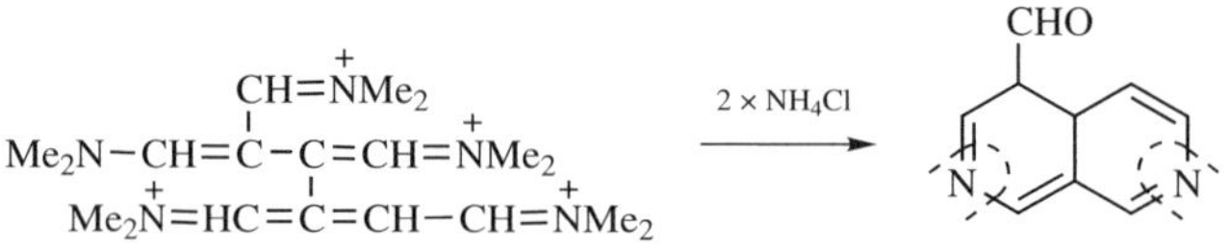

Using Seven Precursor Molecules

tert-Butanol (**8**) with a Vilsmeier salt (6 mol) and (subsequently) ammonium acetate (2 mol) gave 2,7-naphthyridine-4-carbaldehyde (**9**), for details and proposed mechanism, see original).[595]

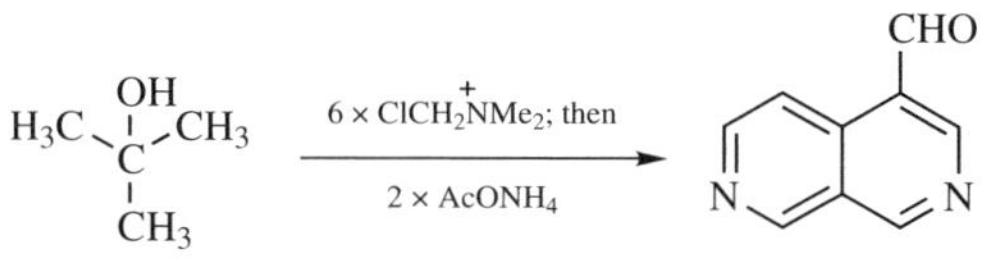

In a rather similar way, 2-methylpropene (**10**) (made *in situ* from *tert*-butanol), acetyl chloride, and aluminum chloride (followed by ammonia treatment) gave 1,3,6,8-tetramethyl-2,7-naphthyridine (**11**) (24%; homologs likewise).[388,1178]

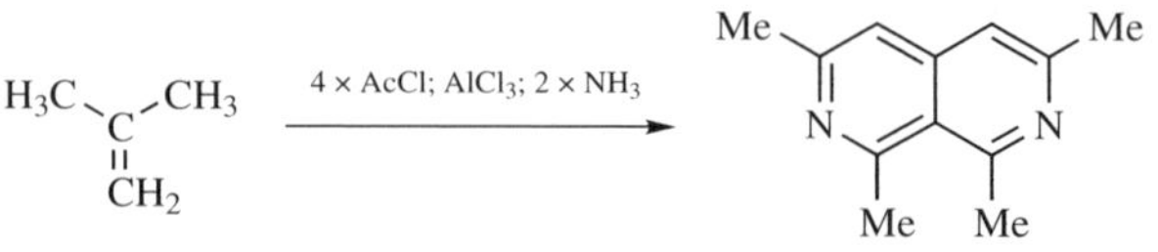

30.1.2. 2,7-Naphthyridines by Cyclization of Pyridine Substrates

This may be done by completion of any one of five bonds, but only two such cyclizations have been successful, as illustrated in the following examples.

By Completion of the 1,2-Bond

1-Amino-4-cyanomethyl-6-diethylamino-3,5-pyridinedicarbonitrile (12) gave 1,8-diamino-3-diethylamino-7-methoxy-2,7-naphthyridine-4-carbonitrile (14) via the unisolated iminoester (13) (MeONa/MeOH, reflux: 81%); analogs likewise.[668]

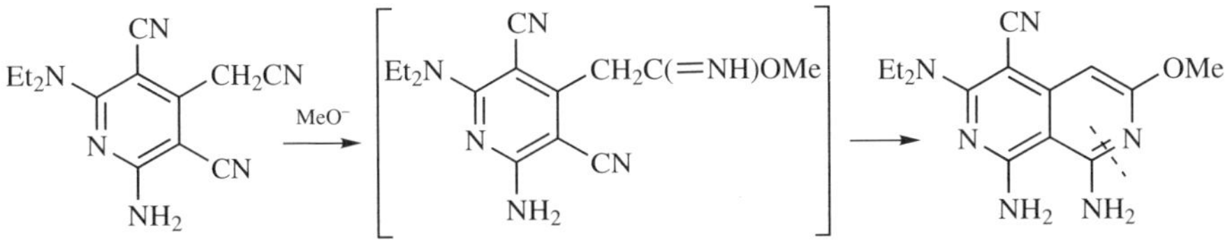

2-Amino-6-oxo-1-phenyl-4-(N-phenylcarbamoylmethyl)-1,6-dihydro-3-pyridinecarbonitrile (15) underwent thermal cyclization to 1-amino-6-hydroxy-8-imino-2,7-dipthenyl-2,3,7,8-tetrahydro-2,7-naphthyridin-3(2H)-one (16) (Et$_3$N, Me$_2$NCHO, reflux: 60%); also somewhat similar cyclizations.[802]

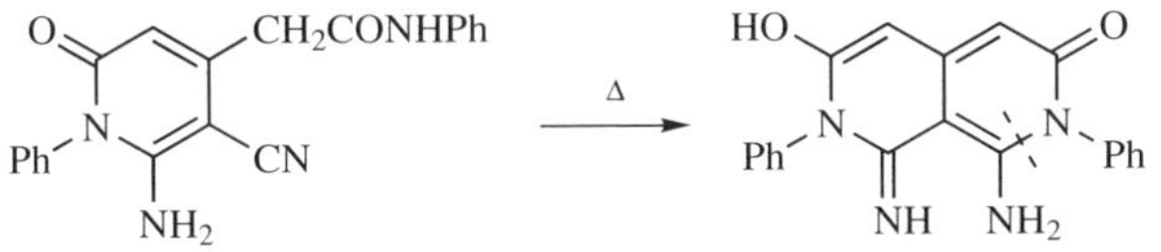

Also other examples, in both the unfused and fused systems.[802,960]

By Completion of the 2,3-Bond

4-Styrylpyridine-3-carboxamide (17) underwent cyclization to 3-phenyl-3,4-dihydro-2,7-naphthyridin-1(2H)-one (18) (P$_2$O$_5$/H$_3$PO$_4$, heat: 90%), easily dehydrogenated to 3-phenyl-2,7-naphthyridin-1(2H)-one (Pd/C, p-cymene, reflux: 91%).[188,cf. 999]

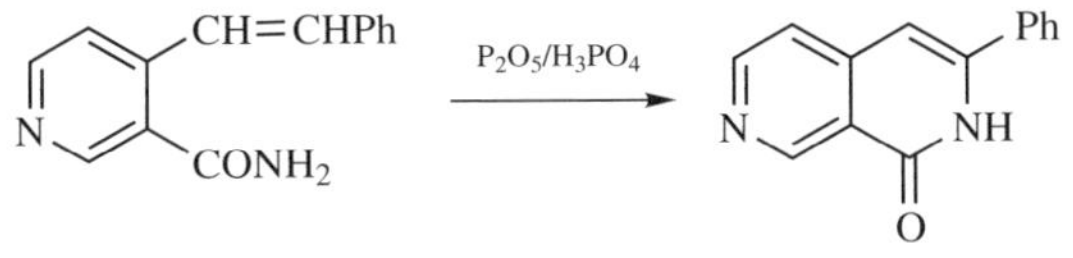

4-Ethoxalylmethylpyridine-3-carbonitrile (**19**) underwent cyclization to ethyl 1-oxo-1,2-dihydro-2,7-naphthyridine-3-carboxylate (**20**) (an Amberlite resin, H_2O, EtOH, reflux: 10%).[986]

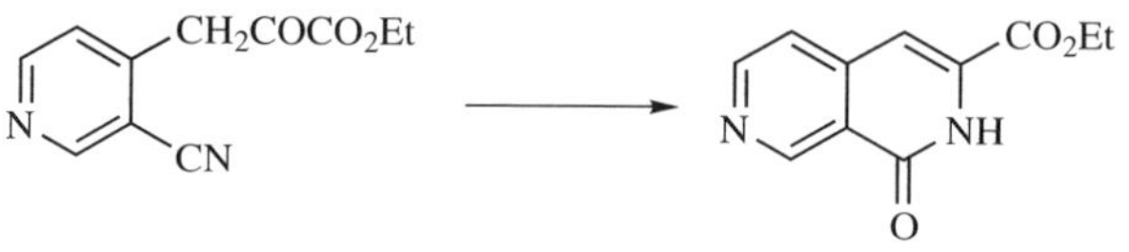

(19) → (20)

Also other examples including some fused systems.[203,211,253,410,598,621,786,1239,1270,1315]

By Completion of the 4,4a-Bond

Note: An attempt to use this route failed.[46]

30.1.3. 2,7-Naphthyridines by Cyclocondensation of Pyridine Substrates with Synthons

In this potentially wide category of primary syntheses, the few reported types are classified according to the ring atom(s) supplied to the naphthyridine by the synthon.

Where the Synthon Supplies C1

4-[2-Ethoxycarbonyl-2-(triphenylphosphoranylideneamino)vinyl]pyridine (**21**) and phenyl isocyanate gave ethyl 1-anilino-2,7-naphthyridine-3-carboxylate (**22**) (PhMe, 150°C: 59%) analogs likewise).[442]

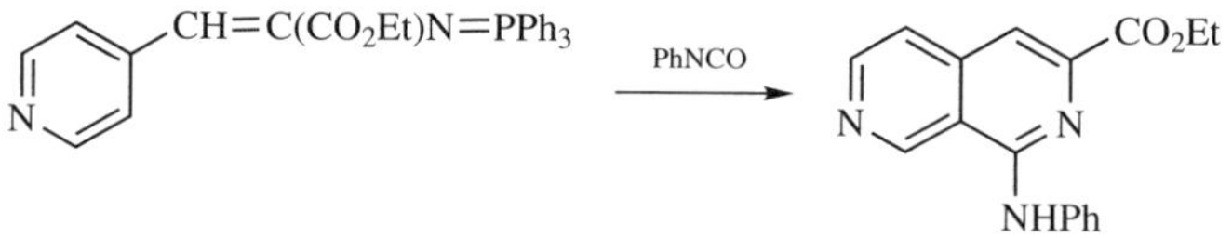

(21) PhNCO → (22)

Where the Synthon Supplies N2

Ethyl 2,6-dimethyl-4-phenylethynyl-3-pyridinecarboxylate (**23**) with ethanolic ammonia gave 6,8-dimethyl-3-phenyl-2,7-naphthyridin-1(2*H*)-one (**24**) (120°C, sealed: 85%).[1055]

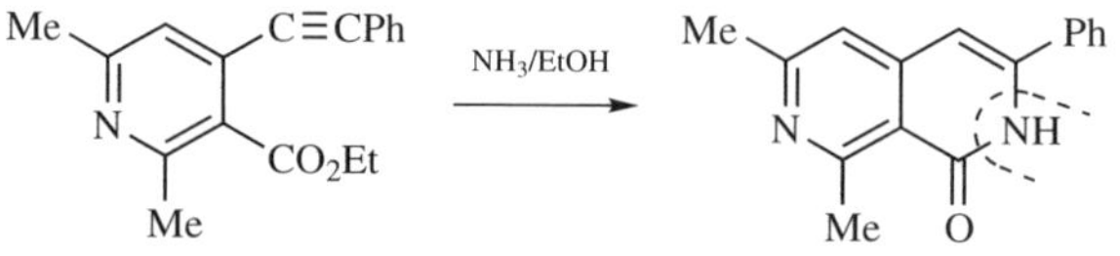

(23) NH$_3$/EtOH → (24)

Methyl 5-acetyl-4-methoxycarbonylmethyl-3-pyridinecarboxylate (**25**) gave methyl 8-methyl-6-oxo-6,7-dihydro-2,7-naphthyridine-4-carboxylate (**26**) (neat reactants, 150°C: 75%)[206] or the 5,6,7,8-tetrahydro derivative, jasminine [NH_4Cl, Et_3N, $Ti(OPr^i)_4$, MeOH, 20°C; then $NaBH_4$: 25%];[206] an alkaloid isolated[577] from several species of *Oleaceae.*

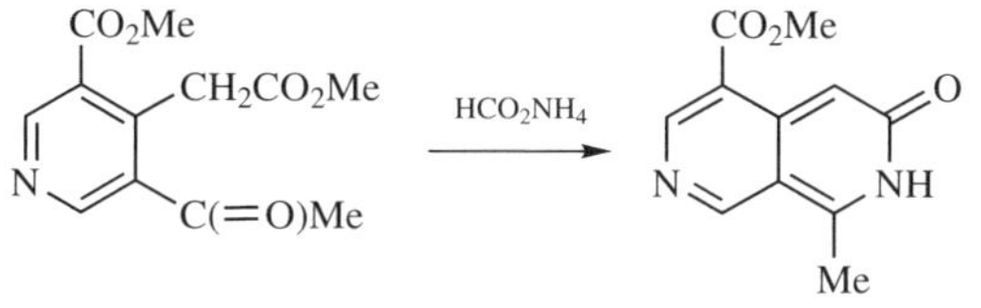

(25) **(26)**

Also other examples.[1110]

Where the Synthon Supplies N2 + C3

3,4-Dibenzoyl-2,5,6-triphenylpyridine (**27**) with benzylamine gave a mixture of hexaphenyl-2,7-naphthyridine (**28**) and the 2,6-isomer (KOH, EtOH, reflux: ~35% and ~25%, respectively, after separation); also analogous syntheses.[1132]

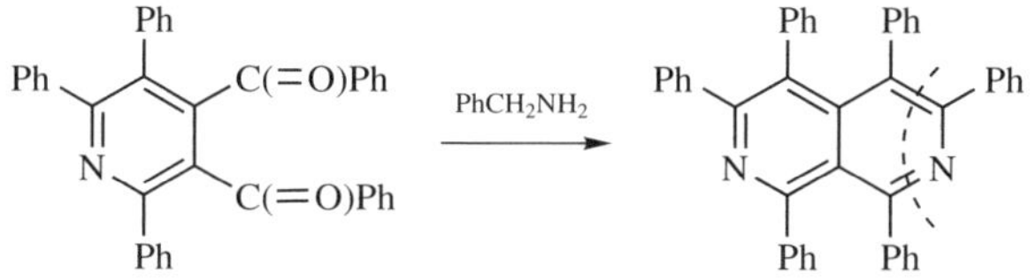

(27) **(28)**

Also other examples.[240,413]

Where the Synthon Supplies C3 + C4

N-Benzyl-4-chloropyridine-3-carboxamide (**29**) with malononitrile gave 3-amino-2-benzyl-1-oxo-1,2-dihydro-2,7-naphthyridine-4-carbonitrile (**30**) (K_2CO_3, Me_2NCHO, reflux: 75%).[1220]

Also other examples.[301,360,929,1001,1187,1433,cf. 120]

(29) **(30)**

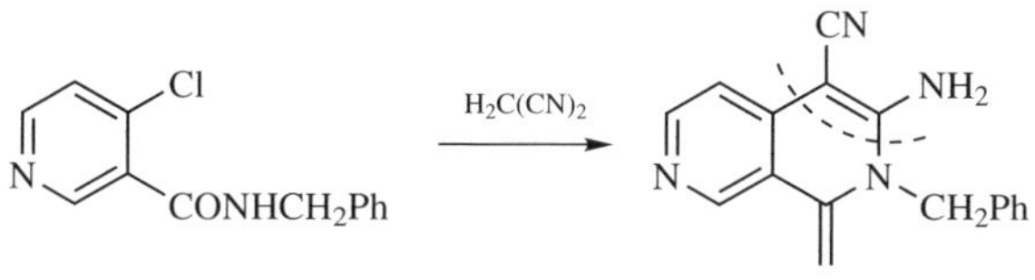

Where the Synthon Supplies N2 + C3 + C4

1-Methyl-4-methylthio-2-oxo-6-phenyl-1,2-dihydro-3-pyridinecarbonitrile (**31**) with 2-cyano-*N*-methylacetamide gave 1-amino-2,7-dimethyl-3,8-dioxo-6-phenyl-2,3,7,8-tetrahydro-2,7-naphthyridine-4-carbonitrile (**32**) (Pr^{i}ONa/Pr^{i}OH, reflux: 70%; analogs likewise).[1183]

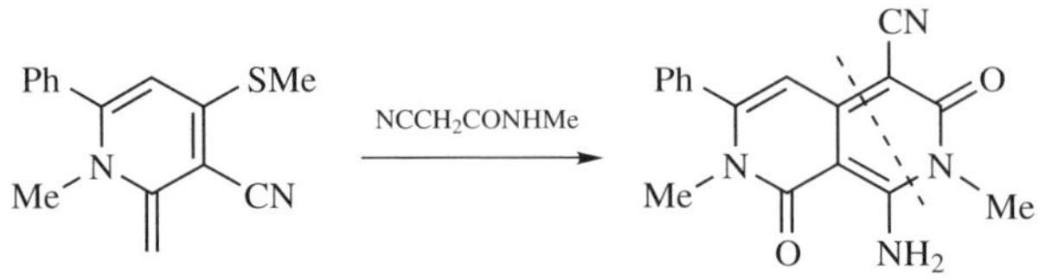

30.1.4. 2,7-Naphthyridines from Other Heterocyclic Substrates

Heterocyclic substrates other than pyridine derivatives have been used to prepare 2,7-naphthyridines (and fused analogs e.g.,[1229]). but many such syntheses are more of interest than utility.

From Isoquinolines

Several 2,7-naphthyridine derivatives have been isolated as byproducts during the alkaline treatment of polysubstituted isoquinolinecarbonitriles.[945]

From Isoquino[2,1-*b*][2,7]naphthyridines

2,3-Dimethoxy-8*H*-isoquino[2,1-*b*][1,7]naphthyridin-8-one (**33**) underwent reductive cleavage to 3-(3,4-dimethoxy-6-vinylphenyl)-2,7-naphthyridin-1(2*H*)-one (**34**) (NaOH, Me$_2$NCHO, 105°C: 55%).[476,cf. 1246]

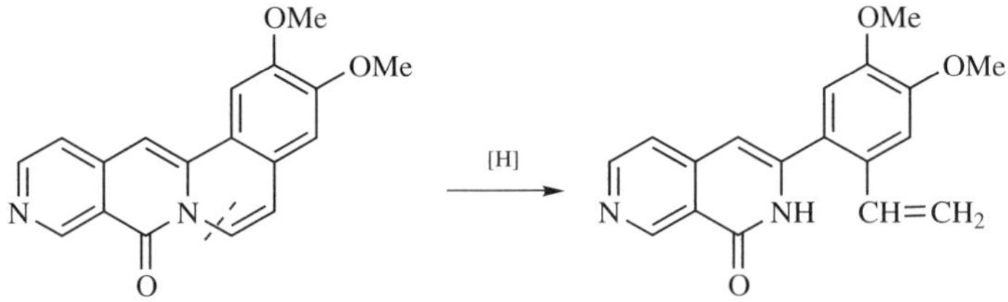

From Oxazoles

N-(But-3-enyl)-4-methyloxazole-5-carboxanilide (**35**) underwent thermal dehydration to give 8-methyl-2-phenyl-3,4-dihydro-2,7-naphthyridin-1(2*H*)-one

(**36**) (xylene, 140°C, 5 days: 32%).[1330]

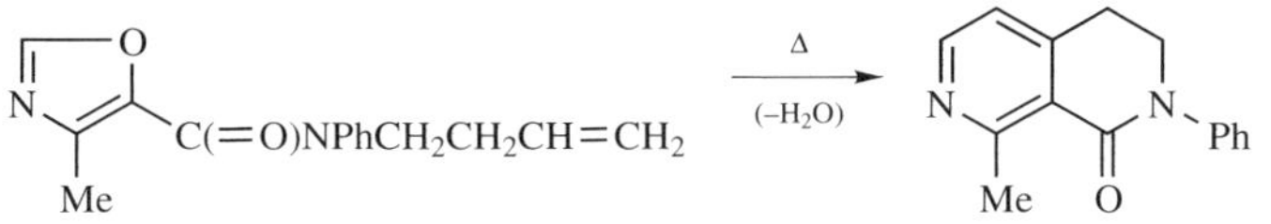

From Pyrans

4-[Bis(dimethyliminio)methylene]-2,6-diphenyl-4H-pyran diperchlorate (**37**) with methanolic ammonium hydroxide gave 3,6-diphenyl-2,7-naphthyridine (**38**) (reflux: 60%).[1065]

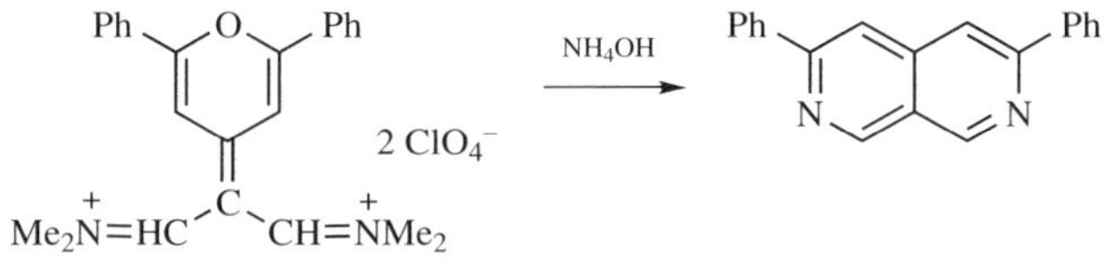

Also other essentially similar examples.[585,1239]

From Pyrano[3,4-c]pyridines

Methyl 1-methyl-3-oxo-3,4-dihydro-1H-pyrano[3,4-c]pyridine-5-carboxylate (**39**) with methanolic ammonia gave 5-(1-hydroxyethyl)-2,7-naphthyridine-1,3(2H,7H)-dione (**40**) (20°C: 78%).[206]

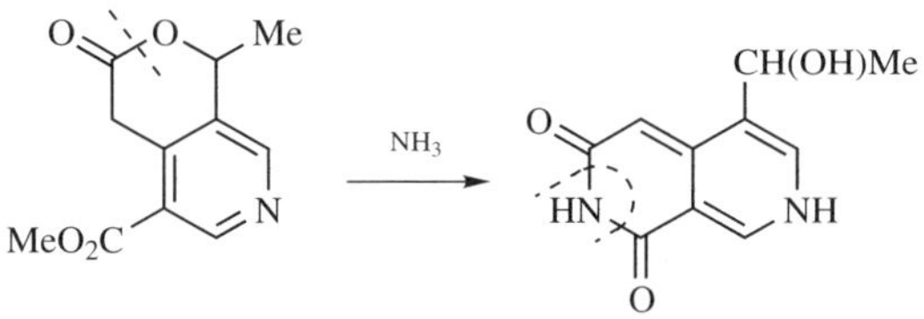

3-Phenyl-1H-pyrano[3,4-c]pyridin-1-one (**41**) gave 2-(2-hydroxyethyl)-3-phenyl-2,7-naphthyridin-1(2H)-one (**42**) ($H_2NCH_2CH_2OH$, MeOH, reflux: 82%).[902]

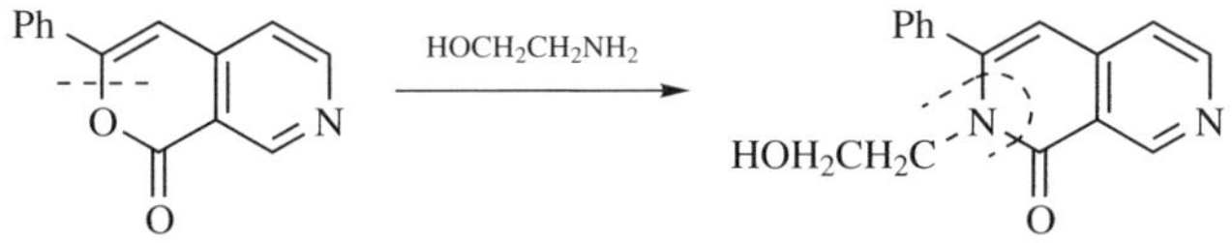

Also other examples.[291,713,916,989,1279]

From Pyrano[4,3-b]pyridines

4-(2-Dimethylaminovinyl)-5-oxo-2,7-diphenyl-5H-pyrano[4,3-b]pyridine-3-carbonitrile (**43**) with propylamine gave 8-oxo-1-phenacyl-3-phenyl-7-propyl-2,7-naphthyridine-4-carbonitrile (**44**) (PrNH$_2$, pyridine, reflux: 58%).[987]

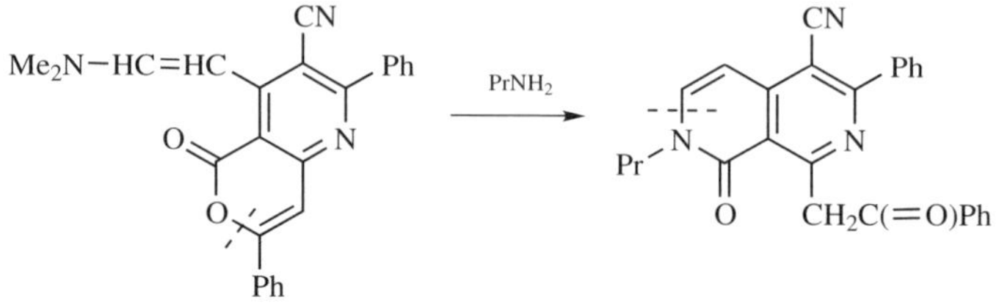

From Pyrazines

6-[N-Benzyl-N-(but-3-ynyl)aminomethyl]-5-chloro-3-methoxy-1-phenylpyrazin-2(1H)-one (**45**) underwent an intramolecular Diels–Alder reaction to afford 2-benzyl-8-chloro-6-methoxy-1,2,3,4-tetrahydro-2,7-naphthyridine (**46**) (PhBr, reflux: 78%).[602]

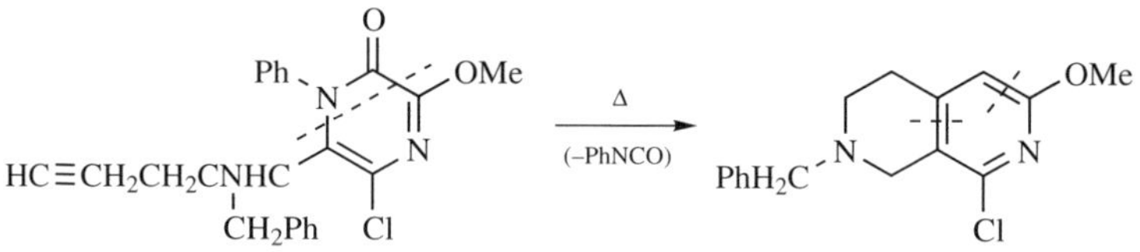

Also annelated 2,7-naphthyridine derivatives.[1324]

From Pyrrolo[3,4-c]pyridines

2-Methoxycarbonylmethyl-1H-pyrrolo[3,4-c]pyridine-1,3(2H)-dione (**47**) has been reported to undergo ring expansion to afford methyl 4-hydroxy-1-oxo-1,2-dihydro-2,7-naphthyridine-3-carboxylate (**48**) (MeONa/MeOH, 100°C: ~70%).[13,cf. 49]

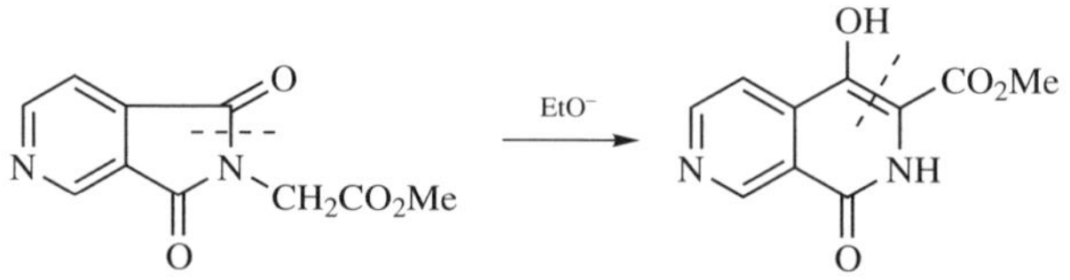

30.2. 2,7-NAPHTHYRIDINE AND ALKYL-2,7-NAPHTHYRIDINES

The parent 2,7-naphthyridine and some of its simple alkyl derivatives have found a natural place in many comparative studies on physical aspects of aza- and diazanaphthalenes (see Section 30.2.2). Moreover, the natural occurrence of the system as several alkaloids (see, e.g., the synthesis of jasminine, Section 30.1.3) and its use in analytical procedures e.g.,[1290] have stimulated interest in the system.

30.2.1. Preparation of 2,7-Naphthyridine

Unsubstituted 2,7-naphthyridine (**50**) is probably best made by oxidation of 2,7-naphthyridine-4-carbaldehyde to 2,7-naphthyridine-4-carboxylic acid (**49**) (KMnO$_4$, AcMe, 20°C: 88%) and subsequent thermal decarboxylation (Cu powder, gentle heat: 78%);[630] earlier syntheses involved direct hydrogenolysis of 1,3,6,8-tetrachloro-2,7 naphthyridine (**51**; R = Cl) (Pd/C, H$_2$, MeOH, AcOK: 71%)[1374,cf. 22] or indirect dechlorination of 1-chloro-2,7-haphthyridine (**51**, R = H) by hydrazinolysis and subsequent oxidation (H$_2$NNH$_2$·H$_2$O, EtOH, 100°C; then CuSO$_4$, H$_2$O: 76% overall);[1279] also homologs.[188]

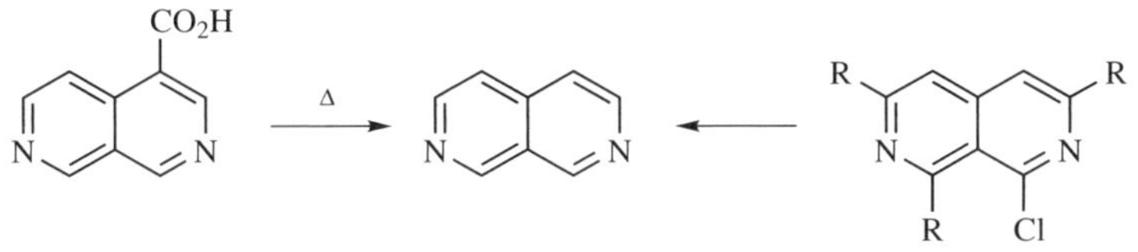

Simple alkyl-2,7-naphthyridines have been made by similar procedures; for example, 1-chloro-3-methyl-2,7-naphthyridine gave 3-methyl-2,7-naphthyridine (58%) via the 1-hydrazino intermediate.[1279]

30.2.2. Properties of 2,7-Naphthyridine

The following physical aspects of 2,7-naphthyridine and/or its simple alkyl derivatives have been included in comparative studies on naphthyridines and related azanaphthalenes.

Aromaticity[676]
Electron density calculations[840,1126,1174,1241]
Electron spin resonance spectra[1070,1083]
Infrared spectra[1251]

Ionization[813,1176]
Nuclear magnetic resonance spectra[1174,1176,1192,1288,1319]
Polarography[829,1229]
Ultraviolet spectra[820,878,1173,1181,1299,1310,1426]
X-ray structure[776]

30.2.3. Reactions of 2,7-Naphthyridine

Several reactions of 2,7-naphthyridine or its simple alkyl derivatives have been reported, as illustrated in the following examples.

Nuclear Reduction

2,7 Naphthyridine gave its 1,2,3,4 tetrahydro derivative (**52**) (Pd/CaCO$_3$, MeOH, H$_2$, 2 atm: 53% as its oxalate salt).[571]

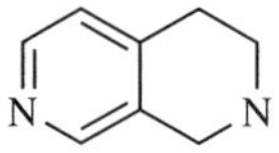

(52)

In contrast, 3-methyl-2,7-naphthyridine gave a 5:1 mixture of its 5,6,7,8- (**53**, R = H) and 1,2,3,4-tetrahydro (**54**, R = H) derivatives, separable after N-acylation to 2-benzoyl-6-methyl-1,2,3,4- tetrahydro- (**53**, R = Bz) and 2-benzoyl-3-methyl-1,2,3,4-tetrahydro-2,7-naphthyridine (**54**, R = Bz) (Pd/CaCO$_3$, MeOH, H$_2$: 95% of mixture; then BzCl/pyridine).[124]

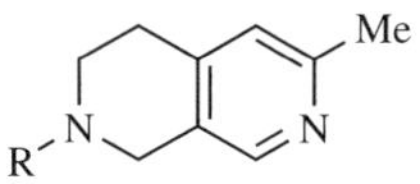

(53) 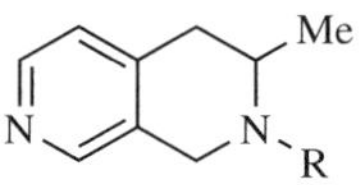   (54)

N-Oxidation

2,7-Naphthyridine gave its 2-oxide (**55**) (m-ClC$_6$H$_4$CO$_3$H, 1 mol, CHCl$_3$, 20°C: 72%) or its 2,7-dioxide (**56**) (m-ClC$_6$H$_4$CO$_3$H, 2 mol, CHCl$_3$, reflux: 20%).[1212]

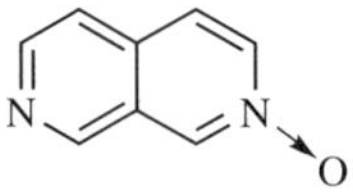

(55) 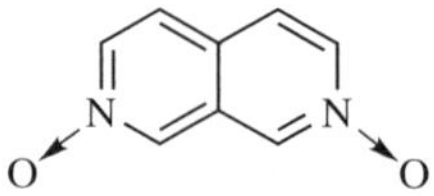   (56)

Amination

2,7-Naphthyridine underwent *C*-anination to 2,7-naphthyridin-1-amine (KNH$_2$/NH$_3$, 20°C, sealed: 22%).[1374,cf. 173]

Halogenation

2,7-Naphthyridine gave a separable mixture of 4-bromo- and 4,5-dibromo-2,7-naphthyridine (Br$_2$, CCl$_4$, pyridine, reflux: 48% and 8%, after separation).[1374,cf. 173]

Quaternization

2,7-Naphthyridine gave 2,7-dimethyl-2,7-naphthyridinediium bis(fluorosulfonate) (**57**) (neat FSO$_3$Me, exothermic: 92%);[1061] attempts to isolate the corresponding diiodide failed.[1002]

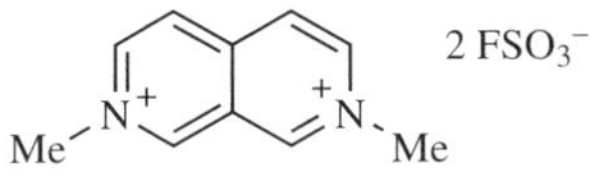

(**57**)

2,7-Naphthyridine underwent quaternization by 1-chloro-2,4-dinitrobenzene in methanol or acetone under reflux to afford 2-(2,4 dinitrophenyl)-2,7-naphthyridinium chloride (**58**) accompanied by either 2-(2,4-dinitrophenyl)-1-methoxy- (59, R=OMe) (10%) or 1-acetonyl-2-(2,4-dinitrophenyl)-1,2-dihydro-2,7-naphthyridine (**59**, R = CH$_2$Ac) (25%), according to the solvent used.[535]

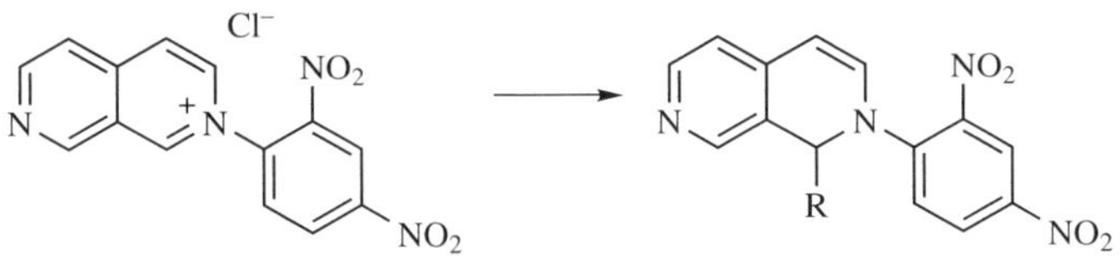

(**58**)   (**59**)

Also other examples.[200,535]

30.3. HALOGENO-2,7-NAPHTHYRIDINES

The literature contains some basic information on the preparation and reactions of halogeno-2,7-naphthyridines. In this system, halogeno substituents are active at the 1-, 3-, 6-, and 8-positions but inactive at the 4- and 5-positions. The effect of fluoro substituents on the ultraviolet spectra of 2,7-naphthyridines has been investigated.[1313] Reviews have appeared.[1266,1363]

30.3.1. Preparation of Halogeno-2,7-naphthyridines

Some halogeno-2,7-naphthyridines have been made by *primary synthesis* (see Section 30.1); other preparative routes are illustrated in the following examples.

By Direct Halogenation

Note: Halogenation of the parent heterocycle has been exemplified in Section 30.2.3, and a brief review is available.[1266]

1,3,6,8-Tetrachloro- (**60**, R = H) gave hexachloro-2,7-naphthyridine (**60**, R = Cl) (substrate, AlCl$_3$, Cl$_2$, 220°C, sealed: 81%).[918]

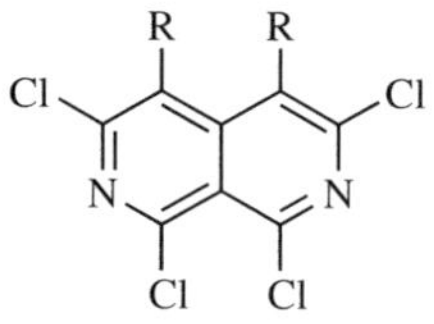

(**60**)

From 2,7-Naphthyridinones

2,7-Naphthyridin-1(2H)-one (**61**) gave 1-bromo- (**62**, X = Br) (POBr$_3$, 140°C: 60%)[838] or 1-chloro-2,7-naphthyridine (**62**, X = Cl) (POCl$_3$, 130°C, sealed: 80%).[1279]

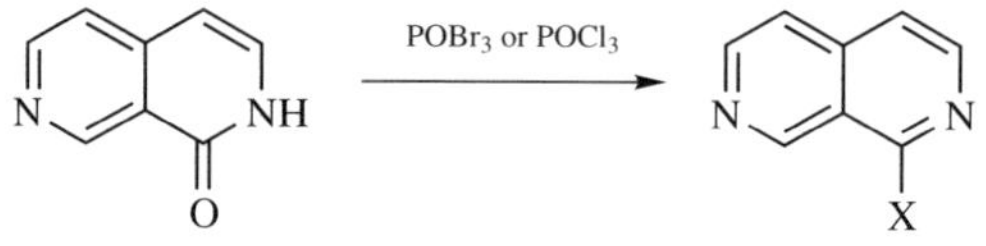

(**61**) → (**62**)

3,6-Dihydroxy-2,7-naphthyridine-1,8(2H,7H)-dione (**63**) gave 1,3,6,8-tetrachloro-2,7-naphthyridine (**64**) (POCl$_3$, 180°C, sealed: 82%).[1374,cf. 22]

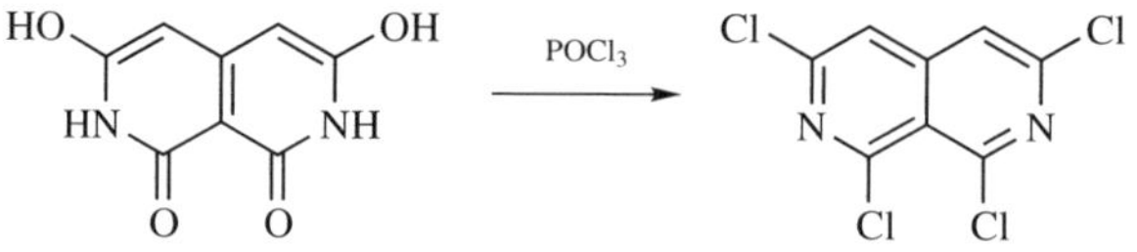

(**63**) → (**64**)

Also other examples.[188,1279]

By Halogen Interchange

1,3,6,8-Tetrachloro- (**65**, R = Cl) gave 1,3,6,8-tetrafluoro-2,7-naphthyridine (**65**, R = F) (KF, 240°C: 64%).[821,918]

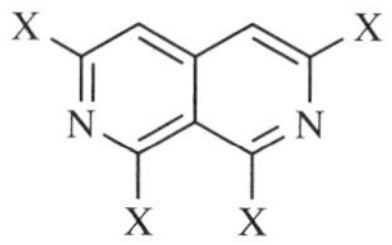

(**65**)

Hexachloro- (**66**) gave 4,5-dichloro-1,3,6,8-tetrafluoro- (**67**) (KF, 260°C: 65%) that underwent further interchange of its inactive chloro substituents to give a separable mixture of 4-chloro-1,3,5,6,8-pentafluoro- (**68**, R = Cl) and hexafluoro-2,7-naphthyridine (**68**, R = F) (KF, CsF, 200°C, 7 days: 35% and 33%, respectively).[918]

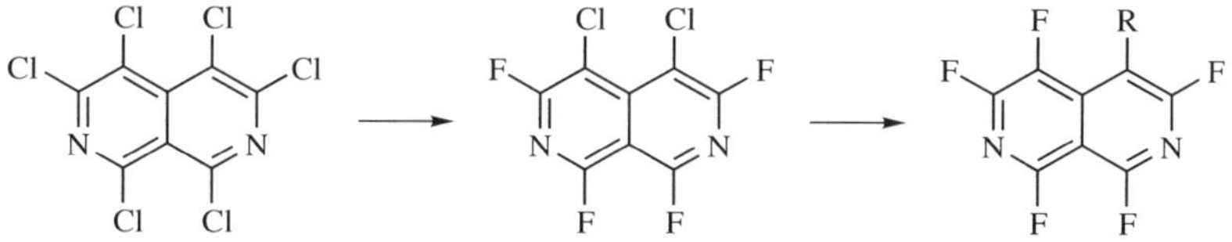

(**66**) → (**67**) → (**68**)

30.3.2. Reactions of Halogeno-2,7-naphthyridines

Of the many possible reactions of halogeno-2,7-naphthyridines, only alcoholysis, aminolysis, and dehalogenation have been reported.[cf. 60]

Alcoholysis

1-Chloro- (**69**, R = Cl) gave 1-methoxy-3-methyl-2,7-naphthyridine (**59**, R = OMe) (MeONa/MeOH, 100°C: ?%).[1431]

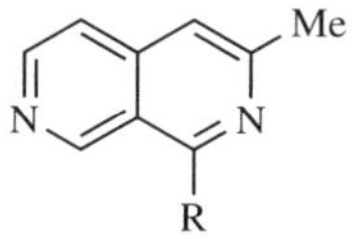

(**69**)

1,3,6,8-Tetrachloro- (**70**, R = Cl) underwent alcoholysis of its more active chloro substituents to give 3,6-dichloro-1,8-dimethoxy-2,7-naphthyridine

(**70**, R = OMe) (K$_2$CO$_3$, MeOH, 20°C; or K$_2$CO$_3$, MeOH, H$_2$O, reflux: ~80%).[22]

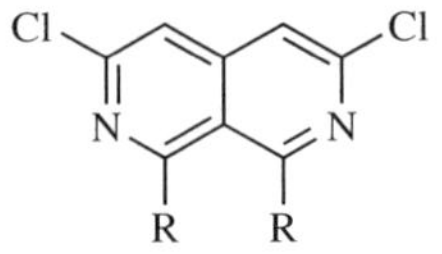

(**70**)

Aminolysis

Note: Examples of such hydrazinolyses have been given in Section 30.2.2; other aminolyses are exemplified here.

1-Chloro- gave 1-(2-diethylaminoethylamino)-3-methyl-2,7-naphthyridine (**71**) (neat Et$_2$NCH$_2$CH$_2$NH$_2$, 150°C: ?%).[1431]

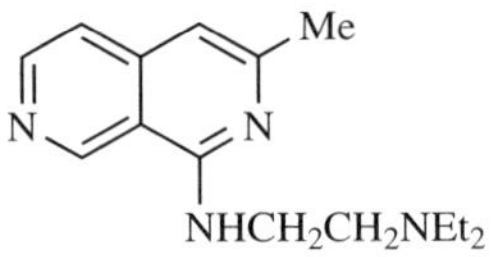

(**71**)

1-Chloro- (**72**, R = Cl) gave 1-hydrazino-3-phenyl-2,7-naphthyridine (**72**, R = NHNH$_2$) (H$_2$NNH$_2$·H$_2$O, EtOH, reflux: 90%).[188]

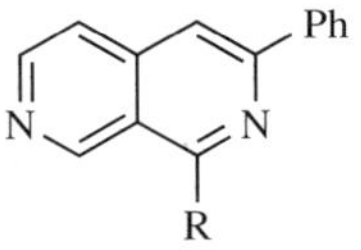

(**72**)

Dehalogenation

Note: Some examples have been given in Section 30.2.2; others are given here.

1-Chloro 3-methyl- (**73**, R = Cl) underwent hydrogenolysis to give 3-methyl-2,7-naphthyridine (**73**, R = H) (Pd/CaCO$_3$, MeOH, H$_2$: ~45%);[1431] see also Section 30.2,2 for a better indirect method.

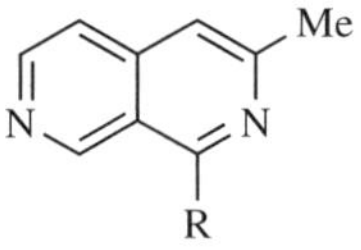

(**73**)

Hydrogenolysis of 1-chloro-3-phenyl-2,7-naphthyridine (**74**, R = Cl) proved unsatisfactory, but conversion into the 1-hydrazino analog (**74**, R = NHNH$_2$ and subsequent oxidation gave 3-phenyl-2,7-naphthyridine (**74**, R = H)

($H_2NNH_2 \cdot H_2O$/EtOH, reflux: 90%; then $CuSO_4$, H_2O, reflux: 55%).[188]

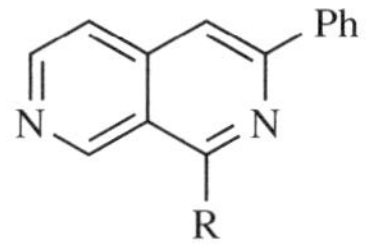

(74)

Also other examples.[1266]

30.4. OXY-2,7-NAPHTHYRIDINES

The few reported aspects of the various types of oxy-2,7-naphthyridine are summarized in the following paragraphs.

2,7-Naphthyridinones—Preparation

The only reported route to such naphthyridinones is by *primary synthesis* (see Section 30.1), but a few do occur naturally as alkaloids in several *Oleaceae* species: jasminine, methyl 8-methyl-6-oxo-5,6,7,8-tetrahydro-2,7-naphthyridine-4-carboxylate (**75**, R = CO_2Me);[577] jasminidine, 1-methyl-1,4-dihydro-2,7-naphthyridin-3(2H)-one (**75**, R = H);[919] dihydrojasminine, methyl 8-methyl-6-oxo-2,4a,5,6,7,8-hexahydro-2,7-naphthyridine-4-carboxylate (**76**);[988] and austrodimerine, formulated as methyl 2-[1-(5-methoxycarbonyl-4-methylpyridin-3-yl)ethyl]-8-methyl-6-oxo-1,2,4a,5,6,7,8,8a-octahydro-2,7-naphthyridine-4-carboxylate (**77**).[988] In addition, 4-phenyl-2,7-naphthyridin-1(2H)-one (lophocladine A) was isolated from a red alga, *Lophocladia* species.[1135] Also other products.[1339]

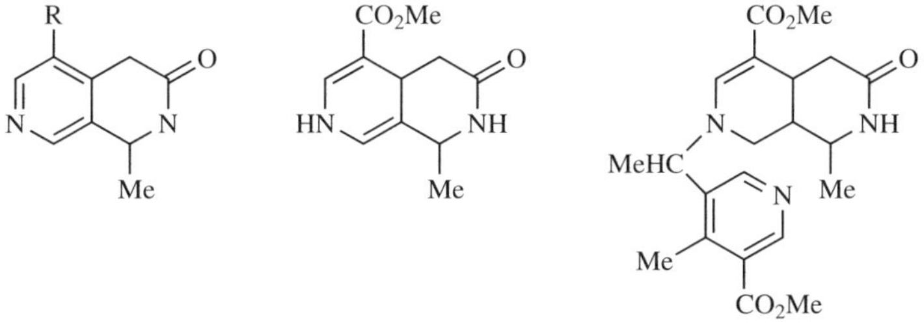

(75)   (76)   (77)

2,7-Naphthyridinones—Reactions

The *halogenolysis* of tautomeric 2,7-naphthyridinones has been covered in Section 30.3.1.

Alkylation is exemplified by the conversion of 3-methyl-2,7-naphthyridin-1(2H)-one (**28**, R = H) into a separable mixture of 2,3-dimethyl-2,7-naphthyridin-1(2H)-one (**78**, R=Me) and 1-methoxy-3-methyl-2,7-naphthyridin (**79**) (CH$_2$N$_2$, MeOH: ?%).[1431]

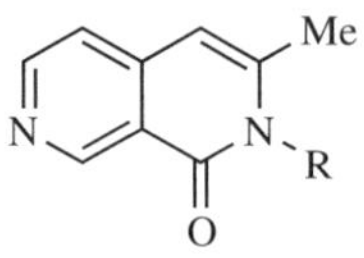

(78) 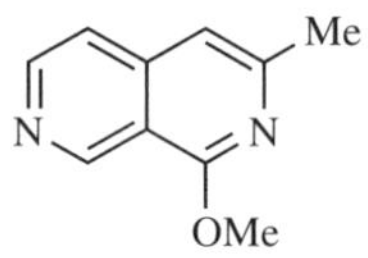   (79)

Intramolecular *cyclization*[540] and an *analytical use*[464] have also been reported.

Alkoxy-2,7-naphthyridines

These have been made by alcoholysis of halogeno substrates (see Section 30.3.2) and by alkylation of 2,7-naphthyridinones (see earlier in this section).

2,7-Naphthyridine *N*-Oxides

Simple examples have been made by *primary synthesis* (see Section 30.1.2; specifically, a 1999 paper by Numata, and colleagues[621]) and by *oxidation* with peroxyacids (see Section 30.2.3).

30.5. THIO-2,7-NAPHTHYRIDINES

These 2,7-naphthyridines appear to be completely missing from the literature as of 2005.

30.6. NITRO-, AMINO-, AND RELATED 2,7-NAPHTHYRIDINES

Most of the available information on these compounds has been covered already as indicated in the following paragraphs.[52,1277]

Nitro and Nitroso-2,7-naphthyridines

Apart from an extranuclear nitro compound made by a *primary synthesis*,[1230] no simple examples appear to have been reported.

Amino-2,7-naphthyridines: Preparation

All known examples of such amines have been made by *primary synthesis* (see Section 30.1), by *direct amination* (see Section 30.2.3), or by *aminolysis* of halogeno substrates (see Sections 30.2.1 and 30.3.2). 4-Phenyl-2,7-naphthyridin-1-amine (lophocladine B) was isolated from a red alga.[1435]

A possible example of *direct C-imination* has been described; a substrate, formulated as 1-amino-8-imino-2,7-diphenyl-5,8-dihydro-2,7-naphthyridine-3,6(2*H*,7*H*)-dione (**80**), was reported to condense with *N,N*-dimethyl-*p*-nitrosoaniline to give 1-amino-5-*p*-dimethylaminophenylimino-8-imino-2, 7-diphenyl-5,8-dihydro-2,7-naphthyridine-3,6(2*H*,7*H*)-dione (**81**) (reactants, trace piperidine, EtOH, reflux: 55%);[802] an isomeric azo compound might have been expected.

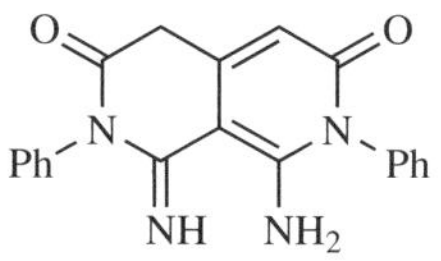

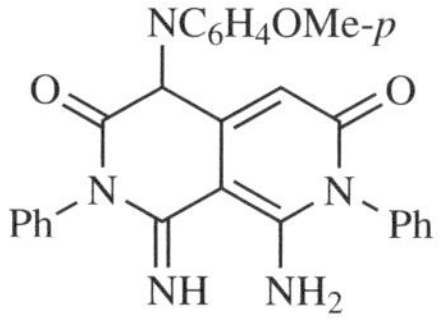

Amino-2,7-naphthyridines: Reactions

The *removal of hydrazino groups* by oxidation has been exemplified in Section 30.2.1. *Cyclizations* have been reported.[182]

30.7. 2,7-NAPHTHYRIDINECARBOXYLIC ACIDS AND RELATED DERIVATIVES

This potentially large family of 2,7-naphthyridines is not well represented in the literature despite the fact that several esters have been isolated from plant material.

30.7.1. 2,7-Naphthyridinecarboxylic Acids

Preparation

A few such acids have been made by *primary synthesis* (see Section 30.1) or by *oxidation of corresponding carbaldehydes* (see Section 30.2.1); other preparative routes are exemplified here.

Ethyl 4-*p*-fluorophenyl-2-methyl-1-oxo-1,2-dihydro-2,7-naphthyridine-3-carboxylate (**82**, R = Et) underwent *saponification* to afford the corresponding carboxylic acid (**82**, R = H) (NaPH, H$_2$O, THF, reflux: 63%);[1199] also other examples.[575,916]

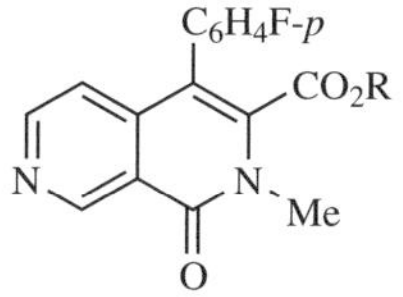

292   The 2,7-Naphthyridines

Reactions

The *decarboxylation* of such carboxylic acids has been exemplified in Section 30.2.1. Another, less simple, case is given here.

2-[2-(indol-3-yl)ethyl]-1-oxo-5-vinyl-1,2-dihydro-2,7-naphthyridine-3-carboxylic acid (**83**) suffered loss of CO_2 and intramolecular cyclization to give the natural product, angustine (**84**) (HCl, AcOH, reflux: 24%; presumably, aerial oxidation is involved).[916]

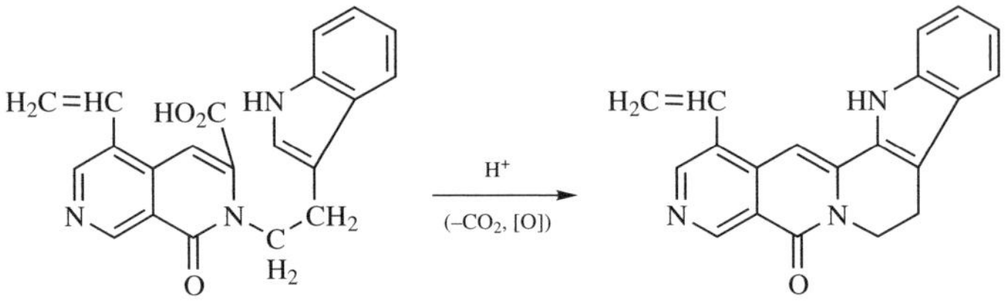

30.7.2. 2,7-Naphthyridinecarboxylic Esters

Preparation

A number of these esters have been prepared by *primary synthesis* (see Section 30.1). In addition, several such esters have been isolated from diverse plant species: neozeylanicine, methyl 1-methyl-2,7-naphthyridine-4-carboxylate (**85**), from *Strychnos cocculoides*[502] and *Neoneuclea zeylanica*;[597] methyl 5-methyl-2,7-naphthyridine 4-carboxylate (**86**), from *S. cocculoides*;[502] 4-methoxycarbonyl-7-methyl-2,4a,5,6,7,7a-hexahydro-1*H*-cyclopentapyridin-6-yl 1-methyl-1,2-dihydro-2,7-naphthyridine-4-carboxylate (**87**), from *Scaevola racemigera*;[931] and three others skin to the last and from the same source.[931]

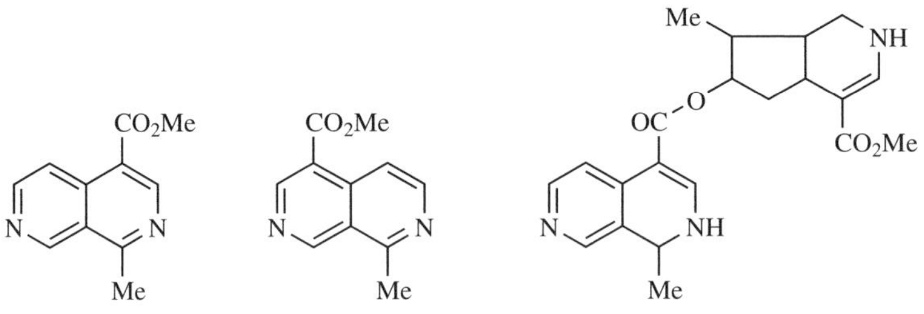

Reactions

The *saponification* of these esters has been covered in Section 30.7.1. No other reactions appear to have been reported.

30.7.3. 2,7-Naphthyridinecarboxamides and Carbonitriles

Such derivatives have been made by *primary synthesis* (see Section 30.1), but no other information has been reported.

30.7.4. 2,7-Naphthyridinecarbaldehydes and Ketones

Preparation

These carbonyl entities have been made by *primary synthesis* (see Section 30.1) or by *oxidation* (see Section 30.2.1).

Reactions

The *oxidation* of an aldehyde to a carboxylic acid has been exemplified in Section 30.2.1.

References

1. A. Reissert, *Ber. Dtsch. Chem. Ges.*, **1893**, *26*, 2137.
2. A. Reissert, *Ber. Dtsch. Chem. Ges.*, **1894**, *27*, 979.
3. A. Reissert, *Ber. Dtsch. Chem. Ges.*, **1895**, *28*, 119.
4. O. Rosenheim and J. Tafel, *Ber. Dtsch. Chem. Ges.*, **1893**, *26*, 1501.
5. A. E. Tschitschibabin, *Ber. Dtsch. Chem. Ges.*, **1925**, *58*, 1707.
6. E. Ochiai, K. Miyaki, and S. Sato, *Ber. Dtsch. Chem. Ges.*, **1937**, *70B*, 2018.
7. O. Seide, *Ber. Dtsch. Chem. Ges.*, **1925**, *58*, 352.
8. C. Räth, *Ber. Dtsch. Chem. Ges.*, **1925**, *58*, 346.
9. A. E. Tschitschibabin, *Ber. Dtsch. Chem. Ges.*, **1924**, *57*, 2092.
10. O. Seide, *Ber. Dtsch. Chem. Ges.*, **1924**, *57*, 1806.
11. A. E. Tschitschibabin, *Ber. Dtsch. Chem. Ges.*, **1924**, *57*, 1168.
12. B. Fels, *Ber. Dtsch. Chem. Ges.*, **1904**, *37*, 2129.
13. S. Gabriel and J. Colman, *Ber. Dtsch. Chem. Ges.*, **1902**, *35*, 1358.
14. B. Bobranski and E. Sucharda, *Rocz. Chem.*, **1927**, *7*, 241; *Chem. Abstr.*, **1928**, *22*, 777; also **1929**, *23*, 2712.
15. F. Knöhnke, K. Ellegasi, and E. Bertram, *Ann. Chem.*, **1956**, *600*, 198.
16. E. Sucharda, *Kosmos* (Warsaw), **1920**, 15 pp; *Chem. Abstr.*, **1928**, *22*, 2948.
17. L. Klisiecki and E. Sucharda, *Rocz. Chem.*, **1927**, *7*, 204; *Chem. Abstr.*, **1928**, *22*, 777.
18. F. C. Palazzo and A. Tamburini, *Atti Acad. Lincei*, **1911**, [5], *20*, I, 37; *Chem. Abstr.*, **1911**, *5*, 1586.
19. O. G. Backeberg, *J. Chem. Soc.*, **1933**, 390.
20. O. Seide, *Liebigs Ann. Chem.*, **1924**, *440*, 311.
21. W. Marchwald, *Liebigs Ann. Chem.*, **1894**, *279*, 1.
22. B. M. Ferrier and N. Campbell, *J. Chem. Soc.*, **1960**, 3513.
23. G. B. Crippa and E. Scevola, *Gazz. Chim. Ital.*, **1937**, *67*, 327.
24. S. von Niementowski and E. Sucharda, *J. Prakt. Chem.*, **1916**, [2], *94*, 193.
25. K. F. Schmidt, *Ber. Dtsch. Chem. Ges.*, **1924**, *57*, 704.
26. B. Fels, *Ber. Dtsch. Chem. Ges.*, **1904**, *37*, 2137.
27. S. von Niementowski and E. Sucharda, *Ber. Dtsch. Chem. Ges.*, **1919**, *52*, 484.
28. S. von Niementowski, *Ber. Dtsch. Chem. Ges.*, **1906**, *39*, 385.
29. S. von Niementowski, *Ber. Dtsch. Chem. Ges.*, **1896**, *29*, 76.
30. G. Koller, *Ber. Dtsch. Chem. Ges.*, **1927**, *60*, 1918.

31. G. Koller, *Ber. Dtsch. Chem. Ges.*, **1927**, *60*, 1572.
32. B. Bobranski and E. Sucharda, *Ber. Dtsch. Chem. Ges.*, **1927**, *60*, 1081.
33. G. Koller, *Ber. Dtsch. Chem. Ges.*, **1927**, *60*, 407.
34. O. Seide, *Ber. Dtsch. Chem. Ges.*, **1926**, *59*, 2465.
35. L. Schmid and B. Bangler, *Ber. Dtsch. Chem. Ges.*, **1926**, *59*, 1360.
36. K. F. Schmidt, *Ber. Dtsch. Chem. Ges.*, **1925**, *58*, 2413.
37. L. Schmid and B. Bangler, *Ber. Dtsch. Chem. Ges.*, **1925**, *58*, 1971.
38. P. K. Bose and D. C. Sen, *J. Chem. Soc.*, **1931**, 2840.
39. J. M. Gulland and R. Robinson, *J. Chem. Soc.*, **1925**, *127*, 1493.
40. W. Lawson, W. H. Perkin, and R. Robinson, *J. Chem. Soc.*, **1924**, *125*, 626.
41. W. Marckwald, L. Berndt, M. Busse, and C. Schmidt, *Liebigs Ann. Chem.*, **1893**, *274*, 331.
42. W. L. F. Armarego and T. J. Batterham, *J. Chem. Soc.* (B), **1966**, 750.
43. W. L. F. Armarego, T. J. Batterham, and J. R. Kershaw, *Org. Magn. Reson.*, **1971**, *3*, 575.
44. W. L. F. Armarego, G. B. Barlin, and E. Spinner, *Spectrochim. Acta*, **1966**, *22*, 117.
45. A. Albert, *J. Chem. Soc.*, **1960**, 1790.
46. E. P. Hart, *J. Chem. Soc.*, **1954**, 4030.
47. W. L. F. Armarego, *J. Chem. Soc.* (C), **1967**, 377.
48. A. Albert and W. L. F. Armarego, *J. Chem. Soc.*, **1963**, 4237.
49. C. F. H. Allen, *Chem. Rev.*, **1950**, *47*, 275.
50. W. W. Paudler and T. J. Kress, *Adv. Heterocycl. Chem.*, **1970**, *11*, 123.
51. W. W. Paudler and R. M. Sheets, *Adv. Heterocycl. Chem.*, **1983**, *33*, 147.
52. M. J. Weiss and C. R. Hauser, in *Heterocyclic Compounds*, R. C. Elderfield, ed., Wiley, New York, 1961, Vol. 7, p. 198.
53. M. Woźniak and H. C. van der Plas, *Heterocycles*, **1982**, *19*, 363.
54. H. C. van der Plas, M. Woźniak, and H. J. W. van den Haak, *Adv. Heterocycl. Chem.*, **1983**, *33*, 95.
55. W. Czuba, *Khim. Geteroysikl. Soedin.*, **1979**, 3.
56. N. Campbell, in *Rodd's Chemistry of Carbon Compounds*, 2nd ed., S. Coffey, ed., Elsevier, Amsterdam, 1978, Vol. 4H, p. 366.
57. D. T. Hurst, in *Rodd's Chemistry of Carbon Compounds*, 2nd ed., Suppl. I, M. F. Ansell, ed., Elsevier, Amsterdam, 1987, Vol. 4H, p. 113.
58. D. T. Hurst, in *Rodd's Chemistry of Carbon Compounds*, 2nd ed., Suppl. II, M. Sainsbury, ed., Elsevier, Amsterdam, 1998, Vol. 4H, p. 337.
59. G. F. Duffin, *Adv. Heterocycl. Chem.*, **1964**, *3*, 46.
60. R. G. Shepherd and J. L. Fedrick, *Adv. Heterocycl. Chem.*, **1965**, *4*, 377.
61. P. A. Lowe, in *Comprehensive Heterocyclic Chemistry*, A. J. Boulton and A. McKillop, eds., Pergamon, Oxford, 1984, Vol. 2, p. 581.
62. F. C. Palazzo and G. Marogna, *Atti R. Accad. Lincei*, **1912**, [5], *21*(II), 512; *J. Chem. Soc.* **1912**, *106*, Abstr., Part II, 1016.
63. A. Meyer and R. Vittenet, *Ann. Chim.* (Paris), **1932**, [10], *17*, 291.
64. G. Koller and E. Kandler, *Sitsungsber. Akad. Wiss. Wien, Abt. 2B*, **1931**, *140*, 213.
65. G. Koller and E. Strang, *Sitsungsber. Akad. Wiss. Wien, Abt. 2B*, **1928**, 620.
66. T. Miyaki and E. Kataoka, *Yakugaku Zasshi*, **1940**, *60*, 367; *Chem. Abstr.*, **1941**, *35*, 1404.
67. E. Ochiai and K. Miyaki, *Ber. Dtsch. Chem. Ges.*, **1941**, *74B*, 1115.
68. F. P. Mazza and C. Migliardi, *Atti Accad. Sci. Torino, Classe Sci. fis., Mat. Nat.*, **1940**, *75*, 438; *Chem. Abstr.*, **1942**, *36*, 5477.
69. A. Mangini, *Boll. Sci. Fac. Chim. Ind. Bologna*, **1940**, 165; *Chem. Abstr.*, **1942**, *36*, 5476.

70. A. Mangini and M. Colonna, *Boll. Sci. Fac. Chim. Ind. Bologna*, **1941**, 85; *Chem. Abstr.*, **1943**, *37*, 3096.
71. C. Migliardi, *Atti Accad. Sci. Torino, Classe Sci. Fis. Mat. Nat.*, **1940**, *75*(I), 548; *Chem. Abstr.*, **1944**, *38*, 1507.
72. I. N. Nazarov and A. K. Khomenko, *Bull. Acad. Sci. URSS, Classe Sci. Chim.*, **1942**, 137; *Chem. Abstr.*, **1945**, *39*, 1621.
73. A. Mangini and M. Colonna, *Gazz. Chim. Ital.*, **1942**, *72*, 183.
74. A. Mangini and M. Colonna, *Gazz. Chim. Ital.*, **1942**, *72*, 190.
75. O. A. Seide and A. I. Titow, *Ber. Dtsch. Chem. Ges.*, **1936**, *69*, 1884.
76. K. Mogilaiah and G. R. Reddy, *J. Chem. Res.*, **2004**, 477.
77. K. Mogilaiah and G. R. Reddy, *J. Chem. Res.*, **2004**, 145.
78. J. M. Quintela, C. Peinador, L. González, R. Islesias, A. Paramá, F. Alvarez, M. L. Sanmartin, and R. Riguero, *Eur. J. Med. Chem.*, **2003**, *38*, 265.
79. M. Badawneh, P. L. Ferrarini, V. Calderone, C. Marera, F. Martinotti, C. Mori, G. Saccomanni, and L. Testai, *Eur. J. Med. Chem.*, **2001**, *36*, 925.
80. G. Roma, M. Di Braccio, G. Grossi, F. Mattioli, and M. Ghia, *Eur. J. Med. Chem.*, **2000**, *35*, 1021.
81. P. L. Ferrarini, C. Mori, M. Badawneh, V. Calderone, R. Greco, C. Manera, A. Martinelli, P. Nieri, and G. Saccomanni, *Eur. J. Med. Chem.*, **2000**, *35*, 815.
82. C. Alvarez-Rua, S. Garcia-Granda, S. Goswami, R. Mukherjee, S. Dey, R. M. Claramunt, M. Dolores Santa Maria, I. Rozas, N. Jagerovic, I. Alkorta, and J. Elguero, *New J. Chem.*, **2004**, *28*, 700.
83. C.-M. Che, C.-W. Wan, K.-Y. Ho, and Z.-Y. Zhou, *New J. Chem.*, **2001**, *25*, 63.
84. R. Passerini, *Boll. Sci. Fac. Chim. Ind. Bologna*, **1950**, *8*, 138; *Chem. Abstr.*, **1951**, *45*, 7974.
85. E. Ochiai, T. Ishida, H. Nomura, and M. Hamana, *Yakugaku Zasshi*, **1945**, *65*, 69.
86. K. Fučik, Ž. Procházka, V. Hach, and J. Štrof, *Chem. Listy*, **1951**, *45*, 23; *Chem. Abstr.*, **1951**, *45*, 10245.
87. K. Miyaki, *Yakugaku Zasshi*, **1942**, *62*, 257.
88. K. Miyaki, *Yakugaku Zasshi*, **1942**, *62*, 26.
89. J. A. Al Jamal and M. Badawneh, *Arch. Pharm.*, **2003**, *336*, 285.
90. J.-L. Zuo, W.-F. Fu, C.-M. Che, and K.-K. Cheung, *Eur. J. Inorg. Chem.*, **2003**, 255.
91. B. Oskui and W. S. Sheldrick, *Eur. J. Inorg. Chem.*, **1999**, 1325.
92. A. El-Ghayoury and R. Ziessel, *J. Org. Chem.*, **2000**, *65*, 7757.
93. J. G. Murray and C. R. Hauser, *J. Org. Chem.*, **1954**, *19*, 2008.
94. T. Takahashi, T. Yatsuka, and S. Senda, *Yakugaku Zasshi*, **1945**, *65*(7/8A), 9.
95. Y. L. Gol'dfarb and M. S. Kondakova, *Izv. Akad. Nauk SSSR, Otd. Khim. Nauk*, **1951**, 610; *Chem. Abstr.*, **1952**, *46*, 8113.
96. M. Colonna and A. Risaliti, *Boll. Sci. Fac. Chim. Ind. Bologna*, **1951**, *9*, 82; *Chem. Abstr.*, **1952**, *46*, 7102.
97. K. A. Zhukova, M. S. Kondakova, and Y. L. Gol'dfarb, *Izv. Akad. Nauk SSSR, Otd. Khim. Nauk*, **1952**, 743; *Chem Abstr.*, **1953**, *47*, 401.
98. R. Passerini, *Rend. Ist. Super. Sanita (Rome)*, **1952**, *15*, 64; *Chem. Abstr.*, **1953**, *47*, 6948.
99. A. Albert and A. Hampton, *2nd Congr. Internatl. Biochim. Resumes Communs.* (*Paris, 1952*), 444; *Chem. Abstr.*, **1955**, *49*, 4648.
100. H. E. Baumgarten and A. L. Krieger, *J. Am. Chem. Soc.*, **1955**, *77*, 2438.
101. J. T. Adams, C. K. Bradsher, D. S. Brrslow, S. T. Amore, and C. R. Hauser, *J. Am. Chem. Soc.*, **1946**, *68*, 1317.
102. G. R. Lappin, *J. Am. Chem. Soc.*, **1948**, *70*, 3348.

103. E. Bacqué, M. El Qacemi, and S. Z. Zard, *Org. Lett.*, **2004**, *6*, 3671.
104. J. Bernstein, B. Stearns, E. Shaw, and W. A. Lott, *J. Am. Chem. Soc.*, **1947**, *69*, 1151.
105. L. J. Deady and S. M. Devine, *Tetrahedron*, **2006**, *62*, 2313.
106. I. N. Nazarov, G. A. Shvekhgeimer, and V. A. Rudenko, *Zh. Obshch. Khim.*, **1954**, *24*, 319; *Chem. Abstr.*, **1955**, *49*, 4651.
107. L. Toldy, T. Nógŕdi, L. Vargha, G. Ivánovics, and I. Koczka, *Acta Chim. Acad. Sci. Hung.*, **1954**, *4*, 303; *Chem. Abstr.*, **1956**, *50*, 362.
108. E. Ochiai and I. Arai, *Yakugaku Zasshi*, **1938**, *58*, 764; *Suppl.*, 207.
109. E. Ochiai and I. Arai, *Yakugaku Zasshi*, **1939**, *59*, 458; *Suppl.*, 152.
110. A. Albert and C. W. Rees, *Spec. Lectures Biochem.*, *Univ. Coll. London*, **1954–1955**, 96; *Chem. Abstr.*, **1958**, *52*, 8281.
111. D. Kluepfel, H. A. Baker, G. Piattoni, S. W. Sehgal, A. Sidorowicz, K. Singh, and C. Vézina, *J. Antibiot.*, **1975**, *28*, 497.
112. S. Gaswami, S. Dey, S. Jana, and A. K. Adak, *Chem. Lett.*, **2004**, *33*, 916.
113. C. R. Hauser and G. A. Reynolds, *J. Org. Chem.*, **1950**, *15*, 1224.
114. G. R. Lappin, Q. R. Petersen, and C. E. Wheeler, *J. Org. Chem.*, **1950**, *15*, 377.
115. C. R. Hauser and M. J. Weiss, *J. Org. Chem.*, **1949**, *14*, 453.
116. C. C. Price and R. M. Roberts, *J. Am. Chem. Soc.*, **1946**, *68*, 1204.
117. P. J. Manley and M. T. Bilodeau, *Org. Lett.*, **2004**, *6*, 2433.
118. J.-H. Liao, C.-T. Chen, H.-C. Chou, C.-C. Cheng, P.-T. Chou, J.-M. Fang, Z. Slanina, and T. J. Chow, *Org. Lett.*, **2002**, *4*, 3107.
119. Q. H. Huang, J. A. Hunter, and R. C. Larock, *Org. Lett.*, **2001**, *3*, 2973.
120. Y. Tominaga and K. Ueda, *J. Heterocycl. Chem.*, **2005**, *42*, 337.
121. Y. Hsiao, N. I. Rivera, N. Yasuda, D. L. Hughes, and P. J. Reider, *Org. Lett.*, **2001**, *3*, 1101.
122. J.-M. Fang, S. Selvi, J.-H. Liao, Z. Slanina, C.-T. Chen, and P.-T. Chou, *J. Am. Chem. Soc.*, **2004**, *126*, 3559.
123. U. Lüning, C. Kühl, and A. Uphoff, *Eur. J. Org. Chem.*, **2002**, 4063.
124. F. J. Rombouts, W. M. De Borggraeve, D. Delaere, M. Froeyen, S. M. Toppet, F. Compernolle, and G. J. Hoornaert, *Eur. J. Org. Chem.*, **2003**, 1868.
125. T. Witte, B. Decker, J. Mattay, and K. Huber, *J. Am. Chem. Soc.*, **2004**, *126*, 9276.
126. H. E. Baumgarten and K. C. Cook, *J. Org. Chem.*, **1957**, *22*, 138.
127. K. Tsuda, S. Saeki, S.-I. Imura, S. Okuda, Y. Sato, and H. Mishima, *J. Org. Chem.*, **1956**, *21*, 1481.
128. A. Bencini, E. Berti, A. Caneschi, D. Gatteschi, E. Giannasi, and I. Invernizzi, *Chem. Eur. J.*, **2002**, *8*, 3660.
129. S. Monti, S. Sortino, E. Fasani, and A. Albini, *Chem. Eur. J.*, **2001**, *7*, 2185.
130. A. B. Eldrup, B. B. Nielsen, G. Haaima, H. Rasmussen, J. S. Kastrup, C. Christensen, and P. E. Nielsen, *Eur. J. Org. Chem.*, **2001**, 1781.
131. E. M. Brun, S. Gil, and M. Parra, *Arkivoc*, **2002**(10), 80.
132. U. N. Rao, X. Han, and E. R. Biehl, *Arkivoc*, **2002**(10), 61.
133. X. Han and E. R. Biehl, *Arkivoc*, **2002**(10), 40.
134. B. M. Ferrier and N. Campbell, *Proc. R. Soc. Edinburgh, Sect. A*, **1959–1960**, *65*, 231.
135. S. Tamura, T. Kudo, and Y. Yanagihara, *Yakugaku Zasshi*, **1960**, *80*, 562.
136. J. H. Hutchinson, W. Halczenko, K. M. Brashear, M. J. Breslin, P. J. Coleman, L.-T. Duong, C. Fernandez-Metzler, M. A. Gentile, J. E. Fisher, G. D. Hartman, J. R. Huff, D. B. Kimmel, C.-T. Leu, R. S. Meissner, K. Merkle, R. Nagy, B. Pennypacker, J. J. Perkins, T. Prueksaritanont, G. A. Rodan, S. L. Varga, G. A. Wesolowski, A. E. Zartman, S. B. Rodan, and M. E. Duggan, *J. Med. Chem.*, **2003**, *46*, 4790.

137. B. R. Lahue, S.-M. Lo, Z.-K. Wan, G. H. C. Woo, and J. K. Snyder, *J. Org. Chem.*, **2004**, *69*, 7171.
138. H. Li, J. L. Petersen, and K. K. Wang, *J. Org. Chem.*, **2003**, *68*, 5512.
139. S. A. Springfield, K. Marcantonio, S. Ceglia, J. Albaneze-Walker, P. G. Dormer, T. D. Nelson, and J. A. Murry, *J. Org. Chem.*, **2003**, *68*, 4598.
140. C. A. Mitsos, A. L. Zografos, and O. Igglessi-Markopoulou, *J. Org. Chem.*, **2003**, *68*, 4567.
141. L. Zhuang, J. S. Wai, M. W. Embrey, T. E. Fisher, M. S. Egbertson, L. S. Payne, J. P. Guare, J. P. Vacca, D. J. Hazuda, P. J. Felock, A. L. Wolfe, K. A. Stillmock, M. V. Wilmer, G. Moyer, W. A. Schleif, L. J. Gabryelski, Y. M. Leonard, J. J. Lynch, S. R. Michelson, and S. D. Young, *J. Med. Chem.*, **2003**, *46*, 453.
142. E. J. Barreiro, C. A. Camara, H. Verli, L. Brasil-Más, N. G. Castro, W. M. Cintra, Y. Aracava, C. R. Rodrifues, and C. A. M. Fraga, *J. Med. Chem.*, **2003**, *46*, 1144.
143. K. Tomita, Y. Tsuzuki, K.-I. Shibamori, M. Tashima, F. Kajikawa, Y. Sato, S. Kashimoto, K. Chiba, and K. Hino, *J. Med. Chem.*, **2002**, *45*, 5564.
144. A. Hinschberger, S. Butt, V. Lelong, M. Boulouard, A. Dumuis, F. Dauphin, R. Bureau, B. Pfeiffer, P. Renard, and S. Rault, *J. Med. Chem.*, **2003**, *46*, 138.
145. P. G. Dormer, K. K. Eng, R. N. Farr, G. R. Humphrey, J. C. McWilliams, P. J. Reider, J. W. Sager, and R. P. Volante, *J. Org. Chem.*, **2003**, *68*, 467.
146. X. Lu, J. L. Petersen, and K. K. Wang, *J. Org. Chem.*, **2002**, *67*, 5412.
147. Q. Huang, J. A. Hunter, and R. C. Larock, *J. Org. Chem.*, **2002**, *67*, 3437.
148. W. Czuba, *Bull. Acad. Pol. Sci., Ser. Sci. Chim.*, **1963**, *11*, 423; *Chem. Abstr.*, **1964**, *60*, 2454.
149. W. Czuba, *Rocz. Chem.*, **1963**, *37*, 1589; *Chem. Abstr.*, **1964**, *60*, 8005.
150. W. Czuba, *Bull. Acad. Pol. Sci., Ser. Sci. Chim.*, **1963**, *11*, 375; *Chem. Abstr.*, **1964**, *60*, 2917.
151. E. W. McChlesney, E. J. Froelich, G. V. Lesher, A. V. R. Crain, and D. Rosi, *Toxicol. Appl. Pharmacol.*, **1964**, *6*, 292.
152. B. Frydman, M. E. Despuy, and H. Rapoport, *J. Am. Chem. Soc.*, **1965**, *87*, 3530.
153. E. V. Brown, *J. Org. Chem.*, **1965**, *30*, 1607.
154. G. Y. Lesher, E. J. Froelich, M. D. Gruett, J. H. Bailey, and R. P. Brujdage, *J. Med. Chem.*, **1962**, *5*, 1063.
155. H. Rapoport and A. D. Batcho, *J. Org. Chem.*, **1963**, *28*, 1753.
156. H. Mastalarz, R. Jasztold-Howorko, F. Rulko, A. Croisy, and D. Carrez, *Arch. Pharm. Pharm. Med. Chem.*, **2004**, *337*, 434.
157. H. I. El-Subbagh, A. H. Abadi, I. E. Al-Khawad, and K. A. Al-Rashood, *Arch. Pharm. Pharm. Med. Chem.*, **1999**, *332*, 19.
158. P. L. Ferrarini, C. Mori, V. Calderone, L. Calzolari, P. Nieri, G. Saccomanni, and E. Martinotti, *Eur. J. Med. Chem.*, **1999**, *34*, 505.
159. A. R. Kumar, J. S. Reddy, and B. V. Rao, *Tetrahedron Lett.*, **2003**, *44*, 5687.
160. H. Ban, M. Muraoka, and N. Ohashi, *Tetrahedron Lett.*, **2003**, *44*, 6021.
161. J. M. Chezal, E. Moreau, G. Delmas, A. Gueiffer, Y. Blache, G. Grassy, C. Lartigue, O. Chavignon, and J. C. Teulade, *J. Org. Chem.*, **2001**, *66*, 6576.
162. A. L. Zografos, C. A. Mitsos, and O. Igglessi-Markopoulou, *J. Org. Chem.*, **2001**, *66*, 4413.
163. H. E. Baumgarten, W. F. Murdock, and J. E. Dirks, *J. Org. Chem.*, **1961**, *26*, 803.
164. F. H. Case and J. A. Brennan, *J. Am. Chem. Soc.*, **1959**, *81*, 6397.
165. P.-W. Phuan and M. C. Kozlowski, *Tetrahedron Lett.*, **2001**, *42*, 3963.
166. Y. Tsuzuki, K. Tomita, K.-I. Shibamori, Y. Sato, S. Kashimoto, and K. Chiba, *J. Med. Chem.*, **2004**, *47*, 2097.
167. M. Palucki, D. L. Hughes, N. Yasuda, C. Yang, and P. J. Reider, *Tetrahedron Lett.*, **2001**, *42*, 6811.

168. W. Lu, L.-H. Zhang, K.-S. Ye, B. Su, and Z. Yu, *Tetrahedron*, **2005**, *62*, 1806.
169. J.-L. Vasse, V. Levacher, J. Bourguignon, and G. Dupas, *Tetrahedron*, **2003**, *59*, 4911.
170. V. Raghukumar, D. Thirumalai, V. T. Ramakrishnan, V. Karunakara, and P. Ramamurthy, *Tetrahedron*, **2003**, *59*, 3761.
171. G. Pagnini, *Antibiotica*, **1967**, *5*, 134; *Chem. Abstr.*, **1968**, *68*, 37757.
172. E. Wenkert, G. J. Daves, R. G. Lewis, and P. W. Sprague, *J. Am. Chem., Soc.*, **1967**, *89*, 6741.
173. W. W. Paudler and T. J. Kress, *J. Org. Chem.*, **1968**, *33*, 1384.
174. B. M. Lynch, B. C. Macdonald, and J. G. K. Webb, *Tetrahedron*, **1968**, *24*, 3595.
175. P.-L. Chien and C. C. Cheng, *J. Med. Chem.*, **1968**, *11*, 164.
176. W. D. Podmore, *J. Med. Chem.*, **1968**, *11*, 731.
177. W. W. Paudler and T. J. Kress, *J. Org. Chem.*, **1967**, *32*, 832.
178. S. A. Vartanyan and S. L. Shagbatyan, *Khim. Geterotsikl. Soedin., Akad. Nauk Latv. SSR*, **1966**, 427; *Chem. Abstr.*, **1967**, *66*, 2496.
179. K. D. Demidova, V. V. Kazakova, T. I. Kudryashova, M. B. Lis, B. A. Morozov, A. G. Pechenkin, O. E. Pylaeva, E. A. Rudzit, G. K. Khisamutdinov, and K. E. Chistyakov, *Med. Prom. SSSR*, **1966**, *20*, 18.
180. F. Haglid, *Ark. Kemi*, **1967**, *26*, 489.
181. W. Czuba, *Rocz. Chem.*, **1967**, *41*, 289; *Chem. Abstr.*, **1967**, *67*, 54056.
182. Y. Kitahara, M. Moshii, M. Mori, and A. Kubo, *Tetrahedron*, **2003**, *59*, 2885.
183. M. Valès, V. Lokshin, G. Pèpe, R. Guglielmetti, and A. Samat, *Tetrahedron*, **2002**, *58*, 8543.
184. G.-W. Wang, S. Cheng, J. Dong, and Y.-W. Dong, *Tetrahedron Lett.*, **2006**, *47*, 1059.
185. C. He and S. J. Lippard, *Tetrahedron*, **2000**, *56*, 8245.
186. R. Tan and A. Taurins, *Tetrahedron Lett.*, **1965**, 2737.
187. R. Tan and A. Taurins, *Tetrahedron Lett.*, **1966**, 1233.
188. J. M. Bobbitt and R. E. Doolittle, *J. Org. Chem.*, **1964**, *29*, 2298.
189. E. L. Little, W. J. Middleton, D. D. Coffman, V. A. Engelhardt, and G. N. Sausen, *J. Am. Chem. Soc.*, **1958**, *80*, 2832.
190. S. Boatman, T. M. Harris, and C. R. Hauser, *J. Am. Chem. Soc.*, **1965**, *87*, 5198.
191. G. Dondi and M. Di Marco, *Boll. Chim. Farm.*, **1966**, *105*, 491; *Chem. Abstr.*, **1966**, *65*, 20112.
192. G. Y. Lesher, *Internatl. Congr. Chemotherapy, Proc. 3rd, Stuttgart*, **1963**, *2*, 1367; *Chem. Abstr.*, **1966**, *65*, 15925.
193. M. Brufani, W. Fedeli, G. Giacomello, F. M. Riccieri, and A. Vaciago, *Atti Accad. Nazl. Lincei, Rend., Classe Sci. Fis., Mat. Nat.*, **1966**, *40*, 187; *Chem. Abstr.*, **1966**, *65*, 11464.
194. M. A. Seefeld, W. H. Miller, K. A. Newlander, W. J. Burgess, W. E. DeWolf, P. A. Elkins, M. S. Head, D. R. Jakas, C. A. Janson, P. M. Keller, P. J. Manley, T. D. Moore, D. J. Payne, S. Pearson, B. J. Polizzi, X. Qiu, S. F. Rittenhouse, I. N. Uzinskas, N. G. Wallis, and W. F. Huffman, *J. Med. Chem.*, **2003**, *46*, 1627.
195. F. J. R. Rombouts, W. De Borggraeve, S. M. Toppet, F. Compernolle, and G. J. Hoornaert, *Tetrahedron Lett.*, **2001**, *42*, 7397.
196. J. Pérard-Viret, G. van der Rest, and A. Rassat, *Tetrahedron Lett.*, **1999**, *40*, 7101.
197. V. I. Colandrea and E. M. Naylor, *Tetrahedron Lett.*, **2000**, *41*, 8053.
198. R. Hersperger, K. Bray-French, L. Mazzoni, and T. Müller, *J. Med. Chem.*, **2000**, *43*, 675.
199. M. E. Duggan, L. T. Duong, J. E. Fisher, T. G. Hamill, W. F. Hoffman, J. R. Huff, N. C. Ihle, C.-T. Liu, R. M. Nagy, J. J. Perkins, S. B. Rodan, G. Weselowski, D. B. Whitman, A. E. Zartman, G. A. Rodan, and G. D. Hartman, *J. Med. Chem.*, **2000**, *43*, 3736.
200. D. Urban, E. Duval, and Y. Langlois, *Tetrahedron Lett.*, **2000**, *41*, 9251.
201. K. Nakatani, S. Sando, K. Yoshida, and I. Saito, *Tetrahedron Lett.*, **1999**, *40*, 6029.

References

202. W. K. Easley and M. F. Meyer, *Proc. La. Acad. Sci.*, **1968**, *31*, 109; *Chem. Abstr.*, **1969**, *70*, 77833.
203. Y. Kitahara, M. Mochii, M. Mori, and A. Kubo, *Tetrahedron Lett.*, **2000**, *41*, 1481.
204. N. E. Austin, M. S. Hadley, J. D. Harling, F. P. Harrington, G. J. Macdonald, D. J. Mitchell, G. J. Riley, T. O. Stean, G. Stemp, S. C. Stratton, M. Thompson, and N. Upton, *Bioorg. Med. Chem. Lett.*, **2003**, *13*, 1627.
205. X.-Q. Li, D.-J. Feng, X.-K. Jiang, and Z.-T. Li, *Tetrahedron*, **2004**, *60*, 8275.
206. M.-L. Bennasar, T. Roca, E. Zulaica, and M. Monerris, *Tetrahedron*, **2004**, *60*, 6785.
207. X.-Q. Li, X.-K. Jiang, X.-Z. Wang, and Z.-T. Li, *Tetrahedron*, **2004**, *60*, 2063.
208. Y. D. Wang, D. H. Boschelli, S. Johnson, and E. Honores, *Tetrahedron*, **2004**, *60*, 2937.
209. M. J. S. Dewar, A. J. Harget, and N. Trinajstic, *J. Am. Chem. Soc.*, **1969**, *91*, 6321.
210. S. A. Vartanyan, V. N. Zhamagortsyan, S. L. Shagbatyan, and A. S. Norovyan, *Khim. Atsetilena*, **1968**, 239; *Chem. Abstr.*, **1969**, *71*, 61160.
211. A. V. Tverdokhlebov, E. V. Resnyanska, A. V. Zavada, A. A. Tolmachev, A. N. Kostyuk, and A. N. Chernega, *Tetrahedron*, **2004**, *60*, 5777.
212. R. Spano and R. Marri, *Boll. Chim. Farm.*, **1969**, 252; *Chem. Abstr.*, **1969**, *71*, 81231.
213. G. Abbiati, E. M. Beccalli, G. Broggini, and C. Zoni, *Tetrahedron*, **2003**, *59*, 9887.
214. L. Calcul, A. Longeon, A. Al Mourabit, M. Guyot, and M.-L. Bourguet-Kondracki, *Tetrahedron*, **2003**, *59*, 6539.
215. T. Takahashi, Y. Hamada, I. Tageuchi, and H. Uchiyama, *Yakugaku Zasshi*, **1969**, *89*, 1260.
216. Y. Kobayashi, T. Kumadaki, and H. Sato, *Tetrahedron Lett.*, **1970**, 2337.
217. H. Junek and A. R. O. Schmidt, *Tetrahedron Lett.*, **1969**, 2439.
218. T. L. Fisher and D. E. Metzler, *J. Am. Chem. Soc.*, **1969**, *91*, 5323.
219. A. Capek, A. Simek, E. Svatek, and M. Budesinsky, *Folia Microbiol.* (Prague), **1969**, *14*, 557; *Chem. Abstr.*, **1970**, *72*, 76019.
220. A. M. Thompson, C. J. C. Connolly, J. M. Hamby, S. Boushelle, B. G. Hartl, A. M. Amar, A. J. Kraker, D. L. Driscoll, R. W. Steinkampf, S. J. Patmore, P. W. Vincent, B. J. Roberts, W. L. Elliott, W. Klohs, W. R. Leopold, H. D. H. Showaltar, and W. A. Denny, *J. Med. Chem.*, **2000**, *43*, 4200.
221. P. L. Ferrarini, C. Mori, C. Manera, A. Martinelli, F. Mori, G. Saccomanni, P. L. Barili, L. Betti, G. Giannaccini, L. Trincavelli, and A. Lucacchini, *J. Med. Chem.*, **2000**, *43*, 2814.
222. E. E. Smissman and J. W. Ayres, *J. Org. Chem.*, **1972**, *37*, 1092.
223. F. W. Hartner, Y. Hsiao, K. K. Eng, N. I. Rivera, M. Paluki, L. Tan, N. Yasuda, D. L. Hughes, S. Weissman, D. Zewge, T. King, D. Tschaen, and R. P. Volante, *J. Org. Chem.*, **2004**, *69*, 8723.
224. Y. Kobayashi, I. Kumadaki, and H. Sato, *J. Org. Chem.*, **1972**, *37*, 3588.
225. D. J. Pokorny and W. W. Paudler, *J. Org. Chem.*, **1972**, *37*, 3101.
226. E. J. Harkiss, *Planta Med.*, **1971**, *20*, 108; *Chem. Abstr.*, **1972**, *76*, 23052.
227. R. Staroscik and J. Sulkowska, *Acta Pol. Pharm.*, **1971**, *28*, 601; *Chem. Abstr.*, **1972**, *76*, 158322.
228. H. Loewe, *Arzneim.: Entwickl. Wirkung, Darstell., Zweite Auflage*, **1972**, *4*, 231; *Chem. Abstr.*, **1972**, *77*, 92755.
229. E. W. Miles, H. M. Fales, and J. B. Gin, *Biochemistry*, **1972**, *11*, 4945.
230. L. C. Robertson and J. A. Merritt, *U.S. Natl. Tech. Inform. Serv. AD Rep.* **1971**, No. 742217, 18 pp; *Chem. Abstr.*, **1972**, *77*, 107209.
231. J. Pomorski, *Wiad. Chem.*, **1970**, *24*, 773; *Chem. Abstr.*, **1971**, *74*, 42289.
232. W. W. Paudler and D. J. Pokorny, *J. Org. Chem.*, **1971**, *36*, 1720.
233. E. V. Brown and A. C. Plasz, *J. Org. Chem.*, **1971**, *36*, 1331.
234. B. Frydman, M. Los, and H. Rapoport, *J. Org. Chem.*, **1971**, *36*, 450.
235. J. F. Magalhaes, Q. Mingoia, and M. G. Piros, *Rev. Farm. Bioquim. Univ. Sao Paulo*, **1970**, *8*, 125; *Chem. Abstr.*, **1970**, *74*, 125500.

236. W. W. Paudler, *U.S. Clearinghouse Fed. Sci. Tech. Inform.*, AD, **1970**, No. 711029, 16 pp.; *Chem. Abstr.*, **1971**, *75*, 35837.
237. J. Pomorski, *Arch. Immunol. Ther. Exp.*, **1971**, *19*, 261; *Chem. Abstr.*, **1971**, *75*, 48949.
238. S.-X. Zhang, J. Feng, S.-C. Kuo, A. Brossi, E. Hamel, A. Tropsha, and K.-H. Lee, *J. Med. Chem.*, **2000**, *43*, 167.
239. A. M. Thompson, G. W. Rewcastle, S. L. Boushelle, B. G. Hartl, A. J. Kracker, G. H. Lu, B. L. Batley, R. L. Panek, H. D. H. Showalter, and W. A. Denny, *J. Med. Chem.*, **2000**, *43*, 3134.
240. T. Kametani, S. Takano, H. Nemoto, and H. Takeda, *Yakugaku Zasshi*, **1971**, *91*, 966.
241. H. J. Harkiss and D. Swift, *Tetrahedron Lett.*, **1970**, 4773.
242. S.-X. Zhang, K. F. Bastow, Y. Tachibana, S.-C. Kuo, E. Hamel, A. Mauger, V. L. Narayanan, and K.-H. Lee, *J. Med. Chem.*, **1999**, *42*, 4081.
243. H. Natsugari, Y. Ikeura, I. Kamo, Tl Ishimaru, Y. Ishichi, A. Fujishima, T. Tanaka, F. Kasahara, M. Kawada, and T. Doi, *J. Med. Chem.*, **1999**, *42*, 3982.
244. L. Chan, H. Jin, T. Stefanac, J.-F. Lavallée, G. Falardeau, W. Wang, J. Bédard, S. May, and L. Yuen, *J. Med. Chem.*, **1999**, *42*, 3023.
245. B. Frydman, G. Buldain, and J. C. Repetto, *J. Org. Chem.*, **1973**, *38*, 1824.
246. J. Pomorski and H. J. den Hertog, *Rocz. Chem.*, **1973**, *47*, 549; *Chem. Abstr.*, **1973**, *79*, 66220.
247. E. M. Hawes, D. K. J. Gorecki, and D. D. Johnson, *J. Med. Chem.*, **1973**, *16*, 849.
248. J. J. Artús, J.-J. Bonet, and J. A. Jimenez, *Afinidad*, **1973**, *30*, 235; *Chem. Abstr.*, **1973**, *79*, 105108.
249. S. Nishigaki and F. Yoneda, *J. Med. Chem.*, **1971**, *14*, 638.
250. F. Voegtle and H. Foester, *Chem.-Ztg.*, **1973**, *97*, 386; *Chem. Abstr.*, **1973**, *79*, 386.
251. J. J. Artús, J.-J. Bonet, and A. E. Peña, *Tetrahedron Lett.*, **1973**, 3187.
252. A. S. Sharifkanov, T. M. Mukhametkaliev, and S. K. Alimzhanova, *Dokl. Vses, Konf. Khim. Atsetilena, 4th*, **1972**, *1*, 473; *Chem. Abstr.*, **1973**, *79*, 92059.
253. W. Trommer and H. Blume, *Tetrahedron Lett.*, **1973**, 1447.
254. H. Schmidbaur and K. C. Dash, *J. Am. Chem. Soc.*, **1973**, *95*, 4855.
255. R. L. Bodner and D. G. Hendricker, *Inorg. Chem.*, **1970**, *9*, 1255.
256. R. L. Bodner and D. G. Hendricker, *Inorg. Chem.*, **1973**, *12*, 33.
257. A. Emad and K. Emerson, *Inorg. Chem.*, **1972**, *11*, 2288.
258. L. Achremowicz and J. Miochowski, *Rocz. Chem.*, **1973**, *47*, 1383; *Chem. Abstr.*, **1974**, *80*, 70659.
259. J. Pomorski and H. J. den Hertog, *Rocz. Chem.*, **1973**, *47*, 2123; *Chem. Abstr.*, **1974**, *80*, 95783.
260. J. J. Artús, J.-J. Bonet, and M. Palau, *An. Quim.*, **1973**, *69*, 1203; *Chem. Abstr.*, **1974**, *80*, 82742.
261. D. G. Hendricker and R. J. Foster, *Inorg. Chem.*, **1973**, *12*, 349.
262. E. V. Brown and A. C. Plasz, *J. Org. Chem.*, **1967**, *32*, 241.
263. P. L. Ferrarini, L. Betti, T. Cavallini, G. Giannaccini, A. Lucacchini, C. Manera, A. Martinelli, G. Ortore, G. Saccomanni, and T. Tuccinardi, *J. Med. Chem.*, **2004**, *47*, 3019.
264. P. J. Coleman, K. M. Brashear, B. C. Askew, J. H. Hutchinson, C. A. M. Vean, L. T. Duong, B. P. Feuston, C. Ferandez-Metzler, M. A. Gentile, D. Hartman, D. B. Kimmel, C.-T. Leu, L. Lipfert, K. Merkle, B. Pennypacker, T. Prueksaritanont, G. A. Rodan, G. A. Wesolowski, S. B. Rodan, and M. E. Duggan, *J. Med. Chem.*, **2004**, *47*, 4829.
265. Y. Hamada and I. Takeuchi, *Yuki Gosei Kagaku Kyokai Shi*, **1974**, *32*, 602; *Chem. Abstr.*, **1975**, *82*, 57580.
266. W. Czuba and M. Wozniak, *Rocz. Chem.*, **1974**, *48*, 1815; *Chem. Abstr.*, **1975**, *82*, 111959.
267. S. E. Brooker and K. J. Harkiss, *Planta Med.*, **1974**, *26*, 305; *Chem. Abstr.*, **1975**, *82*, 95317.
268. F. Gellibert, J. Woolven, M.-H. Fouchet, N. Mathews, H. Goodland, V. Lovegrove, A. Laroze, V.-L. Nguyen, S. Sautet, R. Wang, C. Janson, W. Smith, G. Krysa, V. Boullay, A.-C. de Gouville, S. Huet, and D. Hartley, *J. Med. Chem.*, **2004**, *47*, 4494.

269. Y. Hamada, I. Takeuchi, and M. Sato, *Yakugaku Zasshi*, **1974**, *94*, 1328.
270. V. J. Catalano, H. M. Kar, and B. L. Bennett, *Inorg. Chem.*, **2000**, *39*, 121.
271. R. M. Titkova, A. S. Elina, E. N. Padeiskaya, and L. M. Polukhina, *Khim.-Farm. Zh.*, **1975**, *9*, 10; *Chem. Abstr.*, **1975**, *82*, 139983.
272. W. Czuba and M. Wozniak, *Rocz. Chem.*, **1973**, *47*, 2361; *Chem. Abstr.*, **1974**, *81*, 3791.
273. J. Pomorski, *Rocz. Chem.*, **1974**, *48*, 321; *Chem. Abstr.*, **1974**, *81*, 63508.
274. H. Abe, M. Ikeda, T. Takaishi, Y. Ito, and T. Okuda, *Tetrahedron Lett.*, **1977**, 735.
275. H. C. van der Plas, M. Woźniak, and A. van Veldhuizen, *Tetrahedron Lett.*, **1976**, 2087.
276. E. M. Hawes, D. K. J. Gorecki, and R. G. Gedir, *J. Med. Chem.*, **1977**, *20*, 838.
277. D. E. Beattie, R. Crossley, A. C. W. Curran, D. G. Hill, and A. E. Lawrence, *J. Med. Chem.*, **1977**, *20*, 718.
278. D. K. J. Gorecki and E. M. Hawes, *J. Med. Chem.*, **1977**, *20*, 124.
279. R. J. Staniewicz, R. F. Sympson, and D. G. Hendricker, *Inorg. Chem.*, **1977**, *16*, 2166.
280. K. R. Dixon, *Inorg. Chem.*, **1977**, *16*, 2618.
281. D. Kocjan, A. Azman, and D. Hadzi, *Vestn. Slov. Kem. Drus.*, **1976**, *23*, 37; *Chem. Abstr.*, **1977**, *87*, 5201.
282. A. Da Settimo, G. Primofiore, V. Santerini, G. Biagi, and L. d'Amico, *Chim. Ind.* (Milan), **1976**, *58*, 878; *Chem. Abstr.*, **1977**, *87*, 5542.
283. R. F. Zong, F. Naud, C. Segal, J. Burke, F. Wu, and R. Thummel, *Inorg. Chem.*, **2004**, *43*, 6195.
284. P. Caluwe and T. G. Majewicz, *J. Org. Chem.*, **1977**, *42*, 3410.
285. A. Da Settimo, G. Primofiore, V. Santerini, G. Biagi, and L. d'Amico, *J. Org. Chem.*, **1977**, *42*, 1725.
286. J. Tamas, G. Bujtas, J. Hollos, and M. Bihari, *Kem. Kozt.*, **1976**, *46*, 504; *Chem. Abstr.*, **1977**, *87*, 57472.
287. R. J. Staniewicz and D. G. Hendricker, *J. Am. Chem. Soc.*, **1977**, *99*, 6581.
288. H. Mastalarz and F. Rulko, *Pol. J. Chem.*, **1994**, *68*, 459; *Chem. Abstr.*, **1994**, *121*, 108481.
289. W. Czuba and M. Wozniak, *Uniw. Adama Mickiewicza Poznaniu, Wydz. Mat., Fiz. Chem.*, [Pr.], *Ser. Chem.*, **1975**, *18*, 295; *Chem. Abstr.*, **1976**, *85*, 46452.
290. A. S. Elina, R. M. Titkova, L. G. Tsyrul'nikova, and T. Y. Filipenko, *Khim.-Farm. Zh.* **1976**, *10*, 44; *Chem. Abstr.*, **1976**, *85*, 78081.
291. T. Kametani, M. Takeshita, M. Ihara, and K. Fukumoto, *J. Org. Chem.*, **1976**, *41*, 2542.
292. Z. Mészáros and I. Hermecz, *Tetrahedron Lett.*, **1975**, 1019.
293. J. M. Liesch, J. A. McMillan, R. C. Pandey, I. C. Paul, and K. L. Rinehart, *J. Am. Chem. Soc.*, **1976**, *98*, 299.
294. V. G. Zubenko and I. A. Shcherba, *Farm. Zh.* (Kiev), **1975**, *30*(3), 28; *Chem. Abstr.*, **1975**, *83*, 152437.
295. T. Perenyi, H. Graber, and E. K. Novak, *Acta Microbiol. Acad. Sci. Hung.*, **1975**, *22*, 433; *Chem. Abstr.*, **1976**, *84*, 85914.
296. W. Czuba and M. Wozniak, *Uniw. Adama Mickiewicza Poznaniu, Wydz. Mat., Fiz. Chem.*, [Pr.], *Ser. Chem.*, **1975**, *18*, 105; *Chem. Abstr.*, **1976**, *84*, 120688.
297. A. Clearfield, R. Gopal, and R. W. Olsen, *Inorg. Chem.*, **1977**, *16*, 911.
298. D. Gatteschi, C. Mealli, and L. Sacconi, *Inorg. Chem.*, **1976**, *15*, 2774.
299. E. V. Brown and S. R. Mitchell, *J. Org. Chem.*, **1975**, *40*, 660.
300. C. Reichardt and W. Scheibelein, *Tetrahedron Lett.*, **1977**, 2087.
301. S. B. Brown and M. J. S. Dewar, *J. Org. Chem.*, **1978**, *43*, 1331.
302. H. C. van der Plas, A. van Veldhuizen, M. Wozniak, and P. Smit, *J. Org. Chem.*, **1978**, *43*, 1673.

303. W. Czuba and M. Wozniak, *Zasz. Nauk. Uniw. Jagiellon., Pr. Chem.*, **1976**, *21*, 85; *Chem. Abstr.*, **1978**, *88*, 21573.

304. A. Da Settimo, G. Primofiore, V. Santerini, and G. Biagi, *Chim. Ind.* (Milan), **1977**, *59*, 454; *Chem. Abstr.*, **1978**, *88*, 22693.

305. P. L. Ferrarini, V. Calderone, T. Cavallini, C. Manera, G. Saccomanni, L. Pani, S. Ruiu, and G. L. Gessa, *Bioorg. Med. Chem.*, **2004**, *12*, 1921.

306. B. Sreenivasulu and K. V. Reddy, *Curr. Sci.*, **1977**, *46*, 597; *Chem. Abstr.*, **1977**, *88*, 37653.

307. E. Hayashi, T. Higashino, C. Iijima, E. Oishi, H. Makino, T. Irie, F. Yamamoto, Y. Yokoyama, Y. Iwai, H. Kato, N. Shimada, S. Suzuki, S. Sone, K. Morikawa, H. Mochizuki, M. Kohno, and D. Mizuno, *Yakugaku Zasshi*, **1977**, *97*, 1022.

308. A. L. Ruchelman, S. Zhu, N. Zhou, A. Liu, L. F. Liu, and E. J. La Voie, *Bioorg. Med. Chem. Lett.*, **2004**, *14*, 5585.

309. C. Christensen, A. B. Eldrup, G. Haaima, and P. E. Nielsen, *Bioorg. Med. Chem. Lett.*, **2002**, *12*, 3121.

310. L. Chan, T. Stefanac, J.-F. Lavallée, H. Jin, J. Bédard, S. May, and G. Falardeau, *Bioorg. Med. Chem. Lett.*, **2001**, *11*, 103.

311. A. A. Santilli, A. C. Scotese, and J. A. Yurchenco, *J. Med. Chem.*, **1975**, *18*, 1038.

312. D. R. Buckle, B. C. C. Cantello, H. Smith, and B. A. Spicer, *J. Med. Chem.*, **1975**, *18*, 726.

313. H. Graber, E. Ludwig, and M. Arr, *Gyogyszereink*, **1975**, *25*, 97; *Chem. Abstr.*, **1976**, *84*, 173528.

314. J. Kuzelka, S. Mukhopadhyay, B. Spingler, and S. J. Lippard, *Inorg. Chem.*, **2003**, *42*, 6447.

315. L. Sacconi, C. Mealli, and D. Gatteschi, *Inorg. Chem.*, **1974**, *13*, 1985.

316. R. A. Porter, W. N. Chan, S. Coulton, A. Johns, M. S. Hadley, K. Widdowson, J. C. Jerman, S. J. Brough, M. Coldwell, D. Smart, F. Jewitt, P. Jeffrey, and N. Austin, *Bioorg. Med. Chem. Lett.*, **2001**, *11*, 1907.

317. R. A. Van Dahm and W. W. Paudler, *J. Org. Chem.*, **1975**, *40*, 3068.

318. G. Falardeau, L. Chan, T. Stefanac, S. May, H. Jin, and J.-F. Lavallée, *Bioorg. Med. Chem. Lett.*, **2000**, *10*, 2769.

319. E. Goto, M. Usuki, H. Takenaka, K. Sakai, and T. Tanase, *Organometallics*, **2004**, *23*, 6042.

320. J. A. Hunt, F. Kallashi, R. D. Ruzek, P. J. Sinclair, I. Ita, S. X. McCormick, J. V. Pivnichny, C. E. C. A. Hop, S. Kumar, Z. Wang, S. J. O'Keefe, E. A. O'Neill, G. Porter, J. E. Thompson, A. Woods, D. M. Zaller, and J. B. Doherty, *Bioorg. Med. Chem. Lett.*, **2003**, *13*, 467.

321. D. J. Berg, J. M. Boncella, and R. A. Andersen, *Organometallics*, **2002**, *21*, 4622.

322. H. Yamanaka, T. Shiraishi, and T. Sakamoto, *Hukusakan Kagaku Toronkai Koen Yoshishu*, 8th, **1975**, 74; *Chem. Abstr.*, **1976**, *84*, 163797.

323. E. L. Stogryn, *U.S. NTIS, AD Rep.* **1977**, AD-A047160, 59 pp; *Chem. Abstr.*, **1978**, *89*, 24189.

324. S. Kitaura, M. Kato, T. Totani, T. Yamamoto, S. Kan, H. Nagasaka, and K. Nanjo, *Curr. Chemother., Proc. Internatl. Congr. Chemother.*, 10th, **1977**, *1*, 583; *Chem. Abstr.*, **1978**, *89*, 36409.

325. T. Koizumi and K. Tanaka, *Inorg. Chim. Acta*, **2004**, *357*, 3666.

326. B. Oskui, M. Mintert, and W. S. Sheldrick, *Inorg. Chim. Acta*, **1999**, *287*, 72.

327. T. Komatsu, A. Nagata, K. Iba, Y. Ohbe, K. Irie, and H. Miyawaki, *Curr. Chemother., Proc. Internatl. Congr. Chemother.*, 10th, **1977**, *1*, 641; *Chem. Abstr.*, **1978**, *89*, 36413.

328. W. Czuba, *Wiad. Chem.*, **1978**, *32*, 91; *Chem. Abstr.*, **1978**, *89*, 43162.

329. C. He and S. J. Lippard, *Inorg. Chem.*, **2000**, *39*, 5225.

330. R. M. Titkova, A. S. Elina, E. A. Trifonova, and T. A. Gus'kova, *Khim.-Farm. Zh.*, **1978**, *12*, 81; *Chem. Abstr.*, **1978**, *89*, 109176.

331. E. V. Brown, . R. Mitchell, and A. C. Plasz, *J. Org. Chem.*, **1975**, *40*, 2369.

332. M. Wozniak, W. Czuba, and H. C. van der Plas, *Rocz. Chem.*, **1976**, *50*, 451; *Chem. Abstr.*, **1976**, *85*, 94253.

References

333. J. F. Pilot and E. L. Stogryn, *U.S. NTIS, AD Rep.* **1975**, AD-A023974, 49 pp.; *Chem. Abstr.*, **1977**, *86*, 29679.
334. M. J. Breslin, M. E. Duggan, W. Halczenko, G. D. Hartman, L. T. Duong, C. Fernandez-Metzler, M. A. Gentile, D. B. Kimmel, C.-T. Liu, K. Merkle, T. Prueksaritanont, G. A. Rodan, S. B. Rodan, and J. H. Hutchinson, *Bioorg. Med. Chem. Lett.*, **2004**, *14*, 4515.
335. I. Takeuchi, I. Ozawa, I. Ogaki, Y. Hamada, and T. Ito, *Yakugaku Zasshi*, **1978**, *98*, 1279.
336. Y. Hamada, M. Sugiura, and M. Hirota, *Yakugaku Zasshi*, **1978**, *98*, 1361.
337. Y. Yamanaka, S. Kono, H. Tateishi, and H. Aratani, *Chemotherapy*, (Tokyo), **1978**, *26*(Suppl. 2), 111; *Chem. Abstr.*, **1979**, *90*, 48490.
338. W. A. Bolhofer, J. M. Hoffman, C. N. Habecker, A. M. Pietruszkiewicz, E. J. Craigoe, and M. L. Torchiana, *J. Med. Chem.*, **1979**, *22*, 301.
339. J. J. Baldwin, K. Mensler, and G. S. Ponticello, *J. Org. Chem.*, **1978**, *43*, 4878.
340. Y. Hamada and K. Shigemura, *Yakugaku Zasshi*, **1979**, *99*, 1225.
341. M. Singh, *J. Less-Common Met.*, **1979**, *65*, 279; *Chem. Abstr.*, **1980**, *92*, 14668.
342. A. M. Pedrini, *Antibiotics* (N.Y.), **1979**, *5*, 154; *Chem. Abstr.*, **1980**, *92*, 16253.
343. M. Wozniak and W. Roszkiewicz, *Zesz. Nauk. Uniw. Jagiellon.*, *Pr. Chem.*, **1979**, *24*, 31; *Chem. Abstr.*, **1980**, *92*, 21682.
344. N. Detzer and B. Huber, *Tetrahedron*, **1975**, *31*, 1937.
345. K. Kubo, N. Ito, Y. Isomura, I. Sozu, H. Homma, and M. Murakami, *Yakugaku Zasshi*, **1979**, *99*, 788.
346. M. Wozniak, *Pol. J. Chem.*, **1979**, *53*, 1665; *Chem. Abstr.*, **1980**, *92*, 76354.
347. J. Matsumoto, T. Miyamoto, A. Minamida, Y. Mishimura, H. Egawa, and H. Nishimura, *Curr. Chemother. Infect. Dis.*, *Proc. Internatl. Congr. Chemother.*, *11th*, **1979** *1*, 454; *Chem. Abstr.*, **1980**, *93*, 38346.
348. M. A. Cavanaugh, V. M. Cappo, C. J. Alexander, and M. L. Good, *Inorg. Chem.*, **1976**, *15*, 2615.
349. H. W. Richardson, J. R. Wasson, W. E. Hatfiled, E. V. Brown, and A. C. Plasz, *Inorg. Chem.*, **1976**, *15*, 2916.
350. L. F. Da Fonseca, *Rev. Port. Farm.*, **1978**, *28*, 7; *Chem. Abstr.*, **1978**, *89*, 163453.
351. H. Tobiki, N. Hamma, N. Tanno, and I. Umeda, *Yakugaku Zasshi*, **1980**, *100*, 206; *Chem. Abstr.*, **1980**, *93*, 53860.
352. Y. Tsuzuki, K. Tomita, Y. Sato, S. Kashimoto, and K. Chiba, *Bioorg. Med. Chem. Lett.*, **2004**, *14*, 3189.
353. M. Wozniak, *Zesz. Nauk. Politech. Krakow.*, *Chem.*, **1979**(13), 3; *Chem. Abstr.*, **1980**, *93*, 71598.
354. A. Tobiki, H. Yamada, T. Miyazaki, N. Tanno, H. Suzuki, K. Shimago, K. Okamura, and S. Ueda, *Yakugaku Zasshi*, **1980**, *100*, 49; *Chem. Abstr.*, **1980**, *93*, 95179.
355. A. Wissner, P. R. Hamann, R. Nilakantan, L. M. Greenberger, F. Ye, T. A. Rapuano, and F. Loganzo, *Bioorg. Med. Chem. Lett.*, **2004**, *14*, 1411.
356. H. Tobiki, H. Yamada, I. Nakatsuka, K. Shimago, Y. Eda, H. Naguchi, T. Komatsu, and T. Nakagame, *Yakugaku Zasshi*, **1980**, *100*, 38; *Chem. Abstr.*, **1980**, *93*, 95178.
357. H. Tobiki, H. Yamada, N. Tanno, K. Shimago, Y. Eda, H. Noguchi, T. Komatsu, and T. Nakegome, *Yakugaku Zasshi*, **1980**, *100*, 133; *Chem. Abstr.*, **1980**, *93*, 114423.
358. W. Czuba and J. A. Bajgrowicz, *Bull. Acad. Pol. Sci.*, *Ser. Sci. Chim.*, **1979**, *27*, 571; *Chem. Abstr.*, **1980**, *93*, 204497.
359. J. Trethewey, G. A. Stewart, E. M. Hawes, and D. K. J. Gorecki, *Can. J. Pharm. Sci.*, **1977**, *12*, 12; *Chem. Abstr.*, **1977**, *86*, 183234.
360. B. Stanovnik and M. Tišler, *Vestn. Slov. Kem. Drus.*, **1978**, *25*, 159; *Chem. Abstr.*, **1978**, *89*, 129364.

361. H. Takeda, M. Niwayama, M. Iwanaga, T. Kabasawa, Y. Tanaka, Y. Kinoshita, S. Kawashima, F. Yamasaku, and Y. Suzuki, *Chemotherapy* (Tokyo), **1978**, *26*(Suppl. 2), 287; *Chem. Abstr.*, **1978**, *89*, 157272.
362. R. Sarges, D. E. Kuhla, H. E. Wiedermann, and D. A. Mayhew, *J. Med. Chem.*, **1976**, *19*, 695.
363. M. Wozniak, *Zesz. Nauk. Uniw. Jagiellon., Pr. Chem.*, **1978**, *23*, 55; *Chem. Abstr.*, **1978**, *89*, 162657.
364. N. Suzuki, M. Kato, and R. Dohmori, *Yakugaku Zasshi*, **1979**, *99*, 155.
365. M. Witanowski, L. Stefaniak, and J. Wielgat, *Bull. Acad. Pol. Sci., Ser. Sci. Chim.*, **1978**, *26*, 865; *Chem. Abstr.*, **1979**, *91*, 38429.
366. M. Wozniak, *Zesz. Nauk. Uniw. Jagiellon., Pr. Chem.*, **1978**, *23*, 43; *Chem. Abstr.*, **1978**, *89*, 162656.
367. J. J. Baldwin, E. L. Engelhardt, R. Hirschmann, G. S. Ponticello, J. G. Atkinson, B. K. Wasson, C. S. Sweet, and A. Scriabine, *J. Med. Chem.*, **1980**, *23*, 65.
368. H. Aghabozorg, R. L. Palenik, and S. J. Palenik, *Inorg. Chem.*, **1985**, *24*, 4214.
369. N. Barba-Behrens, G. A. Müller-Carrera, D. M. L. Goodgame, A. S. Lawrence, and D. J. Williams, *Inorg. Chim. Acta*, **1985**, *102*, 173.
370. E. J. S. Reddy, S. M. Reddy, K. Mogilaiah, and B. Sreenivasulu, *Pesticides*, **1985**, *19*, 40; *Chem. Abstr.*, **1985**, *103*, 155711.
371. R. F. Borne, E. K. Fifer, and I. W. Waters, *J. Med. Chem.*, **1984**, *27*, 1271.
372. M. Wozniak, H. C. van der Plas, and S. Harkema, *J. Org. Chem.*, **1985**, *50*, 3435.
373. R. Cadorniga, L. Cadorniga, C. Saiz, and M. C. Saiz-Vadillo, *An. R. Acad. Farm.*, **1983**, *49*, 411; *Chem. Abstr.*, **1984**, *100*, 79426.
374. M. Wozniak and H. Jackiewicz, *Pol. J. Chem.*, **1983**, *57*, 587; *Chem. Abstr.*, **1984**, *101*, 90797.
375. D. Moya and A. Sergio, *Contrib. Cient. Tecnol.*, **1984**, *14*(65), 5; *Chem. Abstr.*, **1984**, *101*, 102866.
376. A. Kagemoto, T. Negoro, M. Nakao, Y. Sekine, and M. Nashimoto, *Chemotherapy* (Tokyo), **1984**, *32*(Suppl. 3), 147; *Chem. Abstr.*, **1984**, *101*, 230473.
377. K. Takegoshi, F. Imashiro, T. Terao, and A. Saika, *J. Org. Chem.*, **1985**, *50*, 2972.
378. E. C. Taylor, J. S. Skotnicki, and S. R. Fletcher, *J. Org. Chem.*, **1985**, *50*, 1005.
379. S. Löfås and P. Ahlberg, *J. Am. Chem. Soc.*, **1985**, *107*, 7534.
380. W. R. Vincent, S. G. Schulman, J. M. Midgley, W. J. van Oort, and R. H. A. Sorel, *Internatl. J. Pharm.*, **1981**, *9*, 191; *Chem. Abstr.*, **1982**, *96*, 34417.
381. A. T. Baker, W. R. Tikkanen, W. C. Kaska, and P. C. Ford, *Inorg. Chem.*, **1984**, *23*, 3254.
382. W. H. Tikkanen, C. Kruger, K. D. Bomben, W. L. Jolly, W. C. Kaska, and P. C. Ford, *Inorg. Chem.*, **1984**, *23*, 3633.
383. V. K. Jhalani and R. N. Usgaonkar, *Indian Drugs*, **1983**, *20*, 482; *Chem. Abstr.*, **1984**, *100*, 6367.
384. R. Mahesh, R. V. Perumal, and P. J. Pandi, *Bioorg. Med. Chem. Lett.*, **2004**, *14*, 5179.
385. T. Ukita, Y. Nakamura, A. Kubo, Y. Yamamoto, Y. Moratani, K. Saruta, T. Higashijima, J. Kotera, K. Fujishige, M. Takagi, K. Kikkawa, and K. Omori, *Bioorg. Med. Chem. Lett.*, **2003**, *13*, 2341.
386. W. Czuba and M. Wozniak, *Zesz. Nauk. Uniw. Jagiellon., Pr. Chem.*, **1975**, *20*, 61; *Chem. Abstr.*, **1976**, *85*, 21166.
387. J. Matsumoto, T. Miyamoto, A. Minamida, Y. Nishimura, H. Egawa, and H. Nishimura, *J. Med. Chem.*, **1984**, *27*, 292.
388. C. Erre, A. Pedra, M. Arnaud, and C. Roussel, *Tetrahedron Lett.*, **1984**, *25*, 515.
389. G. R. Newkome, S. J. Garbis, V. K. Majestic, F. R. Fronczek, and G. Chiara, *J. Org. Chem.*, **1981**, *46*, 833.
390. W. Czuba and T. Kowalska, *Zesz. Nauk. Uniw. Jagiellon., Pr. Chem.*, **1979**, *24*, 13; *Chem. Abstr.*, **1980**, *92*, 76357.

391. W. Czuba, T. Kowalska, and P. Kowalski, *Pol. J. Chem.*, **1978**, *52*, 2369; *Chem. Abstr.*, **1980**, *92*, 76355.
392. Y. Hamada and K. Shigemura, *Yakugaku Zasshi*, **1979**, *99*, 982.
393. J. Goliński, M. Mçkosza, and A. Rykowski, *Tetrahedron Lett.*, **1983**, *24*, 3279.
394. A. M. M. Lanfredi, A. Tiripicchio, R. Usón, L. A. Oro, M. A. Ciriano, and B. E. Villaroya, *Inorg. Chim. Acta*, **1984**, *88*, L9.
395. J. M. Domagala, L. D. Hanna, C. L. Heifetz, M. P. Hutt, T. F. Mich, J. P. Sanchez, and M. Solomon, *J. Med. Chem.*, **1986**, *29*, 394.
396. L. A. Oro, *Simp. Quim. Inorg. "Met. Transicion," 1st*, **1985**, 40; *Chem. Abstr.*, **1986**, *105*, 172674.
397. A. Da Settimo, G. Primofiore, P. L. Ferrarini, C. Mori, C. Martini, E. Pennacchi, and A. Lucacchini, *Farmaco, Ed. Sci.*, **1986**, *41*, 577; *Chem. Abstr.*, **1986**, *105*, 183403.
398. A. Tiripicchio, F. J. Lahoz, L. A. Oro, M. A. Ciriano, and B. E. Villarroya, *Inorg. Chim. Acta*, **1986**, *111*, L1.
399. R. P. Thummel and Y. Jahng, *Inorg. Chem.*, **1986**, *25*, 2527.
400. C. K. Govindan and G. Taylor, *J. Org. Chem.*, **1983**, *48*, 5348.
401. E. M. Kaiser, *Tetrahedron*, **1983**, *39*, 2055.
402. N. B. Marchenko, V. G. Granik, R. G. Glushkov, E. M. Peresleni, A. L. PolezHaeva, and M. D. Mashkovskii, *Khim.-Farm. Zh.*, **1976**, *10*, 49; *Chem. Abstr.*, **1977**, *86*, 155451.
403. J. L. Bear, L. K. Chau, M. Y. Chavan, F. Lefoulon, R. P. Thummel, and K. M. Kadish, *Inorg. Chem.*, **1986**, *25*, 1514.
404. I. Takeuchi, I. Ozawa, K. Shigemura, Y. Hamada, I. Yoshiki, T. Ito, and A. Ohyama, *Yakugaku Zasshi*, **1979**, *99*, 451.
405. M. P. Wentland and J. B. Cornetti, *Annu. Rep. Med. Chem.*, **1985**, *20*, 145; *Chem. Abstr.*, **1986**, *104*, 17384.
406. K. G. Naber, *J. Int. Biomed. Inform. Data*, **1985**, *6*(Suppl. 1), 5; *Chem. Abstr.*, **1986**, *104*, 141528.
407. A. S. Naravyan, E. G. Paronikyan, and S. A. Vartanyan, *Khim.-Farm. Zh.*, **1985**, *19*, 790; *Chem. Abstr.*, **1986**, *104*, 185324.
408. H. Nakamura, J. Kobayashi, and Y. Ohizumi, *Tetrahedron Lett.*, **1982**, *23*, 5555.
409. C. O. Dietrich-Buchecker, P. A. Marnot, and J. P. Sauvage, *Tetrahedron Lett.*, **1982**, *23*, 5291.
410. M. J. Wanner, G. J. Koomen, and U. K. Pandit, *Tetrahedron*, **1982**, *38*, 2741.
411. P. C. Keller, R. L. Marks, and J. V. Rund, *Polyhedron*, **1983**, *2*, 595.
412. W. Czuba, T. Kowalska, H. Poradowska, and P. Kowalski, *Pol. J. Chem.*, **1984**, *58*, 1221; *Chem. Abstr.*, **1986**, *104*, 186328.
413. O. Meth-Cohn and H. C. Taljaard, *Tetrahedron Lett.*, **1983**, *24*, 4607.
414. A. Sano, Y. Asabe, and S. Takatani, *Anal. Sci.*, **1985**, *1*, 441; *Chem. Abstr.*, **1986**, *105*, 17538.
415. M. J. Wanner, G. J. Koomen, and U. K. Pandit, *Tetrahedron*, **1983**, *39*, 3673.
416. W. Czuba, A. Kaniewska, and H. Poradowska, *Pol. J. Chem.*, **1980**, *54*, 591; *Chem. Abstr.*, **1981**, *94*, 3331.
417. W. Czuba, *Wiad. Chem.*, **1980**, *34*, 263; *Chem. Abstr.*, **1981**, *94*, 30592.
418. H. Egawa, T. Miyamoto, A. Minamida, Y. Nishimura, H. Okada, H. Uno, and J. Matsumoto, *J. Med. Chem.*, **1984**, *27*, 1543.
419. W. Czuba, *Wiad. Chem.*, **1980**, *34*, 593; *Chem. Abstr.*, **1981**, *94*, 24860.
420. T. Komatsu, H. Noguchi, H. Tobiki, and T. Nakagome, *Beta-Lactam Antibiot.*, **1981**, 87; *Chem. Abstr.*, **1981**, *95*, 214601.
421. W. Czuba, P. Kowalski, and M. Grzegozek, *Pol. J. Chem.*, **1980**, *54*, 1573; *Chem. Abstr.*, **1981**, *94*, 156858.

422. A. Da Settimo, P. L. Ferrarini, C. Mori, G. Primofiore, and A. Subissi, *Farmaco, Ed. Sci.*, **1986**, *41*, 827; *Chem. Abstr.*, **1987**, *106*, 95579.
423. H. Narita, Y. Konishi, J. Nitta, I. Kitayama, M. Miyazima, Y. Watanabe, A. Yotsugi, and I. Saikawa, *Yakugaku Zasshi*, **1986**, *106*, 802.
424. S. E. Denmark, *J. Org. Chem.*, **1981**, *46*, 3144.
425. H. J. W. Van den Haak, H. C. van der Plas, and B. van Veldhuizen, *J. Org. Chem.*, **1981**, *46*, 2134.
426. E. Binamira-Soriaga, S. D. Sprouse, R. J. Watts, and W. C. Kaska, *Inorg. Chim. Acta*, **1984**, *84*, 135.
427. M. Wozniak and M. Skiba, *Pol. J. Chem.*, **1981**, *55*, 2429; *Chem. Abstr.*, **1983**, *99*, 175616.
428. R. P. Thummel, F. Lefoulon, D. Cantu, and R. Mahadevan, *J. Org. Chem.*, **1984**, *49*, 2208.
429. M. Tan and X. Gan, *Huaxue Xuebao*, **1987**, *45*, 379; *Chem. Abstr.*, **1987**, *107*, 88471.
430. W. R. Tikkanen, E. Binamira-Soriaga, W. C. Kaska, and P. C. Ford, *Inorg. Chem.*, **1984**, *23*, 141.
431. Y. Noguchi, *Kagaku Ryoho no Ryolki*, **1987**, *3*, 531; *Chem. Abstr.*, **1987**, *107*, 112264.
432. C. Chen, X. Zheng, P. Zhu, and H. Guo, *Yaoxue Xuebao*, **1982**, *17*, 112; *Chem. Abstr.*, **1982**, *97*, 6191.
433. R. M. Titkova, A. S. Elina, E. A. Trifonova, O. A. Anisimova, and L. M. Polukhina, *Khim.-Farm. Zh.*, **1982**, *16*, 699; *Chem. Abstr.*, **1982**, *97*, 162862.
434. M. Wozniak, *Zesz. Nauk. Uniw. Jagiellon.*, *Pr. Chem.*, **1982**, *27*, 33; *Chem. Abstr.*, **1983**, *99*, 5013.
435. W. R. Tikkanen, E. Binamira-Soriaga, W. C. Kaska, and P. C. Ford, *Inorg. Chem.*, **1983**, *22*, 1147.
436. W. Czuba, *Wiad. Chem.*, **1981**, *35*, 441; *Chem. Abstr.*, **1982**, *96*, 52192.
437. R. L. Willer, C. K. Lowe-Ma, D. W. Moore, and L. F. Johnson, *J. Org. Chem.*, **1984**, *49*, 1481.
438. N. Kawamura, *Kagaku Ryoho no Ryoiki*, **1987**, *3*, 499; *Chem. Abstr.*, **1987**, *107*, 112261.
439. H. J. W. van den Haak and H. C. van der Plas, *J. Org. Chem.*, **1982**, *47*, 1673.
440. A. Cole, J. Goodfield, D. R. Williams, and J. M. Midgley, *Inorg. Chim. Acta*, **1984**, *92*, 91.
441. F. Alhaique, F. M. Riccieri, E. Santucci, and M. Marchetti, *Farmaco, Ed. Sci.*, **1983**, *99*, 53633.
442. P. Molina, A. Lorenzo, and E. Aller, *Tetrahedron*, **1992**, *48*, 4601.
443. Y. Combret, J.-J. Torché, N. Plé, J. Duflos, G. Dupas, J. Bourguignon, and G. Quéguiner, *Tetrahedron*, **1991**, *47*, 9369.
444. P. Remuzon, D. Bouzard, C. Guiol, and J.-P. Jacquet, *J. Med. Chem.*, **1992**, *35*, 2898.
445. P. Hradil, *Cesk. Farm.*, **1992**, *41*, 55; *Chem. Abstr.*, **1992**, *117*, 251247.
446. T. Fukuyama, *Adv. Heterocycl. Nat. Prod. Synth.*, **1992**, *2*, 189; *Chem. Abstr.*, **1992**, *117*, 130944.
447. J. P. Sanchez, J. M. Domagala, C. L. Heifetz, S. R. Priebe, J. A. Sesnie, and A. H. Trehan, *J. Med. Chem.*, **1992**, *35*, 1764.
448. W. Czuba and M. Grzegozek, *Univ. Iagellon. Acta Chim.*, **1991**, *35*, 63; *Chem. Abstr.*, **1992**, *117*, 48378.
449. C. S. Cooper, P. L. Klock, D. T. W. Chu, D. J. Hardy, R. N. Swanson, and J. P. Plattner, *J. Med. Chem.*, **1992**, *35*, 1392.
450. G. R. Rao, K. Mogilaiah, and B. Sreenivasulu, *Sulfur Lett.*, **1990**, *11*, 101; *Chem. Abstr.*, **1990**, *113*, 171937.
451. K. Shimada, *Jpn. J. Antibiot.*, **1990**, *43*, 583; *Chem. Abstr.*, **1990**, *113*, 90776.
452. M. Nakashima, T. Uematsu, M. Kanamaru, and H. Naganuma, *Chemotherapy* (Tokyo), **1990**, *38*, 533; *Chem. Abstr.*, **1990**, *113*, 144931.
453. T. Kuroda, F. Suzuki, T. Tamura, K. Ohmori, and H. Hosoe, *J. Med. Chem.*, **1992**, *35*, 1130.
454. D. Bouzard, P. Di Cesare, M. Essiz, J. P. Jacquet, B. Ledoussal, P. Remuzon, R. E. Kessler, and J. Fung-Tomc, *J. Med. Chem.*, **1992**, *35*, 518.
455. R. Leardini, D. Nanni, A. Tundo, G. Zanardi, and F. Buggieri, *J. Org. Chem.*, **1992**, *57*, 1842.
456. A. M. El-Reedy, S. M. Hussain, A. Essawy, and A. M. S. Youssef, *Egypt. J. Pharm. Sci.*, **1990**, *31*, 247; *Chem. Abstr.*, **1991**, *114*, 42601.

457. R. D. Toothaker, *Clin. Pharmacokinet.*, **1989**, *16*(Suppl. 1), 52; *Chem. Abstr.*, **1989**, *111*, 61.
458. Y. Mu, H. Guo, and Z. Zhang, *Yiyao Gongye*, **1988**, *19*, 433; *Chem. Abstr.*, **1989**, *111*, 39315.
459. M. Munakata, M. Maekawa, S. Kitagawa, M. Adachi, and H. Masuda, *Inorg. Chim. Acta*, **1990**, *167*, 181.
460. M. Wozniak, *Stud. Org. Chem.* (Amsterdam), **1988**, *35*, 575; *Chem. Abstr.*, **1989**, *110*, 95051.
461. M. Tai, H. Sakai, J. Nitta, H. Hayakawa, Y. Sugimoto, K. Hayashi, and T. Maeda, *Jpn. J. Antibiot.*, **1989**, *42*, 876; *Chem. Abstr.*, **1989**, *111*, 108423.
462. M. Tan, X. Gan, Y. Meng, X. Yin, and X. Wang, *Sci. Chin., Ser. B*, **1989**, *32*, 410; *Chem. Abstr.*, **1990**, *112*, 68599.
463. S. G. Pilosyan, V. V. Dabayeva, B. D. Enokyan, E. A. Abgryan, and A. S. Noravyan, *Arm. Khim. Zh.*, **1988**, *41*, 689; *Chem. Abstr.*, **1989**, *111*, 232614.
464. J. N. Reed, J. Rotchford, and D. Strickland, *Tetrahedron Lett.*, **1988**, *29*, 5725.
465. E. Laborde, L. E. Lesheski, and J. S. Kiely, *Tetrahedron Lett.*, **1990**, *31*, 1837.
466. J. Suffert and R. Ziessel, *Tetrahedron Lett.*, **1991**, *32*, 757.
467. J.-P. Jacquet, D. Bouzard, J.-R. Kiechel, and P. Remuzon, *Tetrahedron Lett.*, **1991**, *32*, 1565.
468. M. Di Braccio, G. Roma, A. Balbi, E. Sottofattari, and M. Carazzone, *Farmaco*, **1989**, *44*, 863; *Chem. Abstr.*, **1990**, *112*, 171774.
469. J. M. Robinson, L. W. Brent, C. Chau, K. A. Floyd, S. L. Gillham, T. L. McMahan, D. J. Magda, T. J. Motycks, N. J. Pack, A. L. Roberts, L. A. Seally S. L. Simpson, R. R. Smith, and K. N. Zalesny, *J. Org. Chem.*, **1992**, *57*, 7352.
470. J. A. Turner, *ACS Synop. Ser.*, **1991**, *443*, 214; *Chem. Abstr.*, **1991**, *114*, 122100.
471. F. Suzuki, T. Kuroda, T. Kawakita, H. Manabe, S. Kitamura, K. Ohmori, M. Ichimura, H. Kase, and S. Ichikawa, *J. Med. Chem.*, **1992**, *35*, 4866.
472. T. R. Kelly, G. J. Bridger, and C. Zhao, *J. Am. Chem. Soc.*, **1990**, *112*, 8024.
473. G. R. Newkome, K. J. Theriot, V. K. Majestic, P. A. Spruell, and G. R. Baker, *J. Org. Chem.*, **1990**, *55*, 2838.
474. S. Radl and L. Bruna, *Cesk. Farm.*, **1990**, *39*, 177; *Chem. Abstr.*, **1991**, *114*, 6323.
475. K. R. Reddy, K. Mogilaiah, B. Swamy, and B. Sreenivasula, *Acta Chim. Hung.*, **1990**, *127*, 45; *Chem. Abstr.*, **1990**, *113*, 231287.
476. B. Lal, R. M. Goodwani, and N. J. de Souza, *J. Org. Chem.*, **1990**, *55*, 5117.
477. J. A. Turner, *J. Org. Chem.*, **1990**, *55*, 4744.
478. P. L. Ferrarini, C. Mori, and N. Tellini, *Farmaco*, **1990**, *45*, 385; *Chem. Abstr.*, **1990**, *113*, 224120.
479. D. Bouzard, P. Di Cesare, M. Essiz, J. P. Jacquet, J. R. Kiechel, P. Remuzon, T. Oki, M. Masuyoshi, R. E. Kessler, J. Fung-Tomc, and J. Desiderio, *J. Med. Chem.*, **1990**, *33*, 1344.
480. C. Chen, Y. Zheng, and H. Z. Guo, *Yaoxue Xuebao*, **1993**, *28*, 594; *Chem. Abstr.*, **1994**, *120*, 134329.
481. K. V. Sadek, M. A. Selim, M. A. Elmaghraby, and M. H. Elnagdi, *Pharmazie*, **1993**, *48*, 419; *Chem. Abstr.*, **1994**, *120*, 163909.
482. E. E. Fenlon, T. J. Nurray, M. H. Baloga, and S. C. Zimmerman, *J. Org. Chem.*, **1993**, *58*, 6625.
483. G. Ulrich and R. Ziessel, *Tetrahedron Lett.*, **1994**, *35*, 1215.
484. Y. Yamashiro, M. Shimakura, S. Minami, Y. Fukuoka, T. Yasuda, Y. Watanabe, H. Narita, and M. Akama, *Jpn. J. Antibiot.*, **1994**, *47*, 245; *Chem. Abstr.*, **1994**, *120*, 319240.
485. M. Soufyane, C. Mirand, and J. Lévy, *Tetrahedron Lett.*, **1993**, *34*, 7737.
486. R. A. Bucsh, J. M. Domagala, E. Laborde, and J. C. Sesnie, *J. Med. Chem.*, **1993**, *36*, 4139.
487. J. M. Domagala, S. E. Hagen, T. Joannides, J. S. Kiely, E. Laborde, M. C. Schroeder, J. A. Sesnie, M. A. Shapiro, M. J. Suto, and S. Vanderroest, *J. Med. Chem.*, **1993**, *36*, 871.
488. J. H. Markgraf, J. F. Skinner, and G. T. Marshall, *J. Chem. Eng. Data*, **1988**, *33*, 9.

489. R. N. Gruenberg and J. D. Williams, *Internatl. Congr. Symp. Ser.—R. Soc. Med.*, **1990**, No. 169, 112 pp; *Chem. Abstr.*, **1991**, *114*, 35970.

490. B. Singh, G. Y. Lesher, K. C. Pluncket, E. D. Pagani, D. C. Bode, R. G. Bentley, M. J. Connell, L. T. Hamel, and P. J. Silver, *J. Med. Chem.*, **1992**, *35*, 4858.

491. R. F. Brown, M. D. Kinnick, J. M. Morin, R. T. Vasileff, F. T. Counter, E. O. Davidson, P. W. Ensminger, J. A. Eudaly, J. S. Kasher, A. S. Katner, R. E. Koehler, K. D. Kurz, T. D. Lindstrom, W. H. W. Lunn, D. A. Preston, J. L. Ott, J. F. Quay, J. K. Shadle, M. I. Steinberg, J. F. Stucky, J. K. Swartzendruber, J. R. Turner, J. A. Webber, W. E. Wright, and K. M. Zimmerman, *J. Med. Chem.*, **1990**, *33*, 2114.

492. M. Wozniak and M. Tomula, *Zesz. Nauk. Uniw. Jagiellon.*, *Pr. Chem.*, **1989**, *32*, 83; *Chem. Abstr.*, **1991**, *114*, 41862.

493. D. Bouzard, P. Di Cesare, M. Essiz, J. P. Jacquet, P. Remuzon, A. Weber, T. Oki, and M. Masuyoshi, *J. Med. Chem.*, **1989**, *32*, 537.

494. J.-P. Collin and M.-T. Youinou, *Inorg. Chim. Acta*, **1992**, *201*, 29.

495. F. A. E. M. Abd El-Aal, A. M. Negm, Y. Mahfouz, and H. A. El-Fahham, *Egypt. J. Pharm. Sci.*, **1991**, *32*, 861; *Chem. Abstr.*, **1993**, *118*, 80887.

496. X. Gan, N. Tang, X. Wang, Y. Zhu, and T. Minyu, *Polyhedron*, **1989**, *8*, 933.

497. H. Hayashi, Y. Miwa, S. Ichikawa, N. Yoda, I. Miki, A. Ishii, M. Kono, T. Yasuzawa, and F. Suzuki, *J. Med. Chem.*, **1993**, *36*, 617.

498. J.-P. Collin, A. Jouaiti, J.-P. Sauvage, W. C. Kaska, M. A. McLoughlin, N. L. Keder, W. T. A. Harrisom, and G. D. Stucky, *Inorg. Chem.*, **1990**, *29*, 2238.

499. E. Binamira-Soriaga, N. L. Keder, and W. C. Kaska, *Inorg. Chem.*, **1990**, *29*, 3167.

500. T. J. Murray, S. C. Zimmerman, and S. V. Kolotuchin, *Tetrahedron*, **1995**, *51*, 635.

501. D. Bouzard, P. Di Cesare, P. Hoffmann, J. Fung-Tomc, and R. Ressler, *Drugs Exp. Clin. Res.*, **1998**, *18*, 291; *Chem. Abstr.*, **1993**, *119*, 62461.

502. C. Delaude, P. Thepenier, M. J. Jacquier, G. Massiot, and L. Le Men-Olivier, *Bull. Soc. R. Sci. Liege*, **1992**, *61*, 429; *Chem. Abstr.*, **1993**, *119*, 91171.

503. P. Hradil, *Cesk. Farm.*, **1992**, *41*, 194; *Chem. Abstr.*, **1993**, *119*, 160155.

504. J. C. Zhou, T. H. Duan, and H. S. Zhou, *Yaoxue Xuebao*, **1989**, *24*, 512; *Chem. Abstr.*, **1990**, *112*, 178412.

505. M. Mintert and W. S. Sheldrick, *Inorg. Chim. Acta*, **1995**, *236*, 13.

506. S. C. Zimmerman and T. J. Murray, *Tetrahedron Lett.*, **1994**, *36*, 4077.

507. A. W. Erian, S. M. Sherif, A.-Z. A. Alassar, and Y. M. Elkholy, *Tetrahedron*, **1994**, *50*, 1877.

508. S. E. Hagen, J. M. Domagala, S. J. Gracheck, J. A. Sesnis, M. A. Stier, and M. J. Suto, *J. Med. Chem.*, **1994**, *37*, 733.

509. S. Atarashi, M. Imamura, Y. Kimura, A. Yoshida, and I. Hayakawa, *J. Med. Chem.*, **1993**, *36*, 3444.

510. O. G. Schramm, T. Oeser, and T. J. J. Müller, *J. Org. Chem.*, **2006**, *71*, 3494.

511. M. Woźniak, *Zesz. Nauk. Uniw. Jagiellon.*, *Pr. Chem.*, **1988**, *31*, 179; *Chem. Abstr.*, **1989**, *110*, 211824.

512. X. Gan, X. Wang, and M. Tan, *Huaxue Xuebao*, **1989**, *47*, 19; *Chem. Abstr.*, **1989**, *110*, 204598.

513. M. T. Chary, K. Mogilaiah, B. Swamy, and B. Sreenivasulu, *Sulfur Lett.*, **1988**, *8*, 79; *Chem. Abstr.*, **1989**, *110*, 231493.

514. M. Wozniak, *Politech. Krackow. in Tadeusza Kosciuszki*, **1987**, *52*, 25; *Chem. Abstr.*, **1989**, *110*, 135107.

515. E. G. Paronikyan, A. S. Noravyan, and S. A. Vartanyan, *Arm. Khim. Zh.*, **1987**, *40*, 587; *Chem. Abstr.*, **1989**, *110*, 23754.

516. C. R. Rao, K. Mogilaiah, M. T. Chary, B. Swamy, and B. Sreenivasulu, *Indian Drugs*, **1988**, *26*, 20; *Chem. Abstr.*, **1989**, *110*, 192682.

References

517. P. Garner, K. Sunitha, W.-B. Ho, W. J. Youngs, V. O. Kennedy, and A. Djebli, *J. Org. Chem.*, **1989**, *54*, 2041.
518. M. Reuman, M. A. Eissenstat, and J. D. Weaver, *Tetrahedron Lett.*, **1994**, *35*, 8303.
519. M. A. Convery, A. P. Davis, C. J. Dunne, and J. D. MacKinnon, *Tetrahedron Lett.*, **1995**, *36*, 4279.
520. W. S. Sheldrick and M. Mintert, *Inorg. Chim. Acta*, **1994**, *219*, 23.
521. Z. Wróbel, *Tetrahedron Lett.*, **1997**, *38*, 4913.
522. C. E. Neipp, P. E. Ranslow, Z. Wan, and J. S. Snyder, *Tetrahedron Lett.*, **1997**, *38*, 7499.
523. M. Mintert and W. S. Sheldrick, *Inorg. Chim. Acta*, **1997**, *254*, 93.
524. K. Chen, S.-C. Kuo, M.-C. Hsieh, A. Mauger, C. M. Lin, E. Hamel, and K.-H. Lee, *J. Med. Chem.*, **1997**, *40*, 2266.
525. H. S. Rani, K. Mogilaiah, and B. Sreenivasulu, *Indian J. Heterocycl. Chem.*, **1995**, *5*, 45; *Chem. Abstr.*, **1996**, *124*, 202094.
526. M. Refai, M. T. Omar, M. M. Kamel, Z. M. Nofal, and N. S. Ismail, *Egypt. J. Pharm. Sci.*, **1996**, *37*, 241; *Chem. Abstr.*, **1997**, *126*, 251088.
527. J. Kim, Y. H. Yoon, I. H. Cho, J. M. Lee, K. Lee, J. H. Kim, and K. H. Hong, *Korean J. Med. Chem.*, **1996**, *6*, 183; *Chem. Abstr.*, **1997**, *126*, 186005.
528. F. Abi Khalil, M. van Damme, M. Hanocq, P. Mauriac, J. Saint-Germain, D. Bouzard, and J. F. Dauphin, *Int. J. Pharm.*, **1996**, *140*, 1.
529. K. Mogilaiah, C. S. Reddy, and K. Vidya, *Indian J. Heterocycl. Chem.*, **2004**, *14*, 145.
530. M. E. Ernst, E. Ernst, and M. E. Klepser, *Am. J. Health-Syst.* **1997**, *54*, 2569; *Chem. Abstr.*, **1998**, *128*, 70299.
531. U. Lüning and C. Kühl, *Tetrahedron Lett.*, **1998**, *39*, 5735.
532. T. Seiner, U. Mueller, M. Jansen, and U. Holzgrabe, *Pharmazie*, **1998**, *53*, 442; *Chem. Abstr.*, **1998**, *129*, 189254.
533. P. L. Ferrarini, C. Mori, M. Bagawneh, V. Calderone, L. Calzolari, T. Loffredo, E. Martinotti, and G. Saccomanni, *Eur. J. Med. Chem.*, **1998**, *33*, 383.
534. H. Quast, A. Fuss, and W. Nüdling, *Eur. J. Org. Chem.*, **1998**, 317.
535. E. Magnier and Y. Langlois, *Tetrahedron Lett.*, **1998**, *39*, 837.
536. A. El-Ghayoury and R. Ziessel, *Tetrahedron Lett.*, **1998**, *39*, 4473.
537. S. V. Kolotuchin and S. C. Zimmerman, *J. Am. Chem. Soc.*, **1998**, *120*, 9092.
538. S. M. Fahmy and R. M. Mohareb, *Tetrahedron*, **1986**, *42*, 687.
539. Y. Combret, J. Duflos, G. Dupas, J. Bourguignon, and G. Quéguiner, *Tetrahedron Asym.*, **1993**, *4*, 1635.
540. R. Lavilla, T. Gotsens, F. Gullón, and J. Bosch, *Tetrahedron*, **1994**, *50*, 5233.
541. A. A. Alggasham and M. C. Nahata, *Ann. Pharmacother.*, **1999**, *33*, 48; *Chem. Abstr.*, **1999**, *130*, 332141.
542. H. Guo, *Yaoxue Xuebao*, **1998**, *33*, 629; *Chem. Abstr.*, **1999**, *130*, 81432.
543. P. L. Ferrarini, C. Manera, C. Mori, M. Badawneh, and G. Saccomanni, *Farmaco*, **1998**, *53*, 741.
544. K. K. Roesch and R. C. Larock, *Org. Lett.*, **1999**, *1*, 553.
545. E. L. Gaidarova, A. A. Borisenko, C. I. Chumakov, A. V. Mel'nikov, I. S. Orlov, and G. V. Grisina, *Tetrahedron Lett.*, **1998**, *39*, 7767.
546. M. Wozniak and B. Szpakiewicz, *Zesz. Nauk. Uniw. Jagiellon., Pr. Chem.*, **1987**, *30*, 87; *Chem. Abstr.*, **1988**, *108*, 140176.
547. H. Ichihashi, Y. Kawahito, and Y. Nakanishi, *Iyakuhin Kenkyu*, **1988**, *19*, 53; *Chem. Abstr.*, **1988**, *108*, 210072.
548. R. Singh, R. Fathi-Afshar, G. Thomas, M. P. Singh, E. Higashitani, A. Hyodo, N. Unemi, and R. G. Micetich, *Eur. J. Med. Chem.*, **1998**, *33*, 697.

549. Y. Wang, A. Decken, and G. Deslongchamps, *Tetrahedron*, **1998**, *54*, 9043.
550. J. Chylak, *Pol. Tyg. Lek.*, **1993**, *48*, 860; *Chem. Abstr.*, **1995**, *122*, 23076.
551. E. A. Mohamed, R. M. Abdel-Rahman, Z. El-Gendy, and M. M. Ismail, *J. Serb. Chem. Soc.*, **1993**, *58*, 1003; *Chem. Abstr.*, **1995**, *122*, 105711.
552. J. S. Rao, B. Sreenivasulu, and K. Mogilaiah, *Indian J. Heterocycl. Chem.*, **1995**, *4*, 211; *Chem. Abstr.*, **1995**, *123*, 83236.
553. J. Frigola, J. Parés, J. Corbera, D. Vañó, R. Mercè, A. Torrens, J. Más, and E. Valenti, *J. Med. Chem.*, **1993**, *36*, 801.
554. Z. Wróbel, *Tetrahedron*, **1998**, *54*, 2607.
555. X. Gan, Y. Zhu, N. Tang, and M. Tan, *Lanzhou Daxue Xuebao, Ziran Kexueban*, **1991**, *27*, 41; *Chem. Abstr.*, **1993**, *118*, 224254.
556. H. Natsugari, Y. Ikeura, I. Kamo, Y. Ishichi, A. Fujishima, T. Tanaka, F. Kasahara, M. Kawada, and T. Doi, *J. Med. Chem.*, **1999**, *42*, 3982.
557. A. Gil de Oliveiro Santos, W. Klute, J. Torode, V. P. W. Böhm, E. Cabrita, J. Runsink, and R. W. Hoffmann, *New J. Chem.*, **1998**, 993.
558. R. Sankaranarayanan, D. Velmurugan, S. S. S. Raj, and H.-K. Fun, *Acta Crystallogr., Sect. C*, **2001**, *57*, 726.
559. U. Lüning, C. Kühl, and M. Bolte, *Acta Crystallogr., Sect. C*, **2001**, *57*, 989.
560. L. Govindasamy, D. Velmurugan, V. Raghukumar, I.-H. Suh, and V. T. Ramakrishnan, *Acta Crystallogr., Sect. C*, **2000**, *56*, 80.
561. R. Sankaranarayanan, S. S. S. Raj, D. Velmurugan, H.-K. Fun, V. Raghukumar, and V. T. Ramakrishnan, *Acta Crystallogr., Sect. C*, **1999**, *55*, 1670.
562. R. Thirumurugan, S. S. S. Raj, G. Shanmugam, H.-K. Fun, V. Raghukumar, and V. T. Ramakrishnan, *Acta Crystallogr., Sect. C*, **1999**, *55*, 1522.
563. M. Parvez, S. Arayne, N. Sultana, and A. Z. Siddiqi, *Acta Crystallogr., Sect. C*, **2004**, *60*, 8281.
564. A. M. Thompson, H. D. H. Showalter, and W. A. Denny, *J. Chem. Soc., Perkin Trans. 1*, **2000**, 1843.
565. A. Fiksdahl, C. Plüg, and C. Wentrup, *J. Chem. Soc., Perkin Trans. 2*, **2000**, 1841.
566. P. Murugan, V. Raghukumar, and V. T. Ramakrishnan, *Synth. Commun.*, **1999**, *29*, 3881.
567. J.-H. Zhang, M.-X. Wang, and Z.-T. Huang, *J. Chem. Soc., Perkin Trans. 1*, **1999**, 2087.
568. E. Fasani, M. Rampi, and A. Albini, *J. Chem. Soc., Perkin Trans. 2*, **1999**, 1901.
569. K. Chen, S.-C. Kuo, M.-C. Hsieh, A. Mauger, C.-M. Lin, E. Hamel, and K.-H. Lee, *J. Med. Chem.*, **1997**, *40*, 3049.
570. C. Y. Hong, Y. K. Kim, J. H. Chang, S. H. Kim, H. Choi, D. H. Nam, Y. Z. Kim, and J. H. Kwak, *J. Med. Chem.*, **1997**, *40*, 3584.
571. M. Dukat, W. Fiedler, D. Dumas, I. Damaj, B. R. Martin, J. A. Rosecrans, J. R. James, and R. A. Glennon, *Eur. J. Med. Chem.*, **1996**, *31*, 875.
572. A. Belicova, M. Seman, V. Milata, D. Ilavsky, and L. Ebringer, *Folia Microbiol.* (Prague), **1997**, *42*, 193; *Chem. Abstr.*, **1997**, *127*, 63024.
573. P. L. Ferrarini, C. Mori, M. Badawneh, C. Manera, G. Saccomanni, V. Calderone, R. Scatizzi, and P. R. Barili, *Eur. J. Med. Chem.*, **1997**, *32*, 955.
574. A. Padwa, M. A. Brodneg, B. Liu, K. Satake, and T. Wu, *J. Org. Chem.*, **1999**, *64*, 3595.
575. L. Chan, H. Jin, T. Stefanac, W. Wang, J.-F. Lavallée, J. Bedard, and S. May, *Bioorg. Med. Chem. Lett.*, **1999**, *9*, 2583.
576. J. G. Montana, G. M. Buckley, N. Cooper, H. J. Dyke, L. Gowers, J. P. Gregory, P. G. Hellewell, H. J. Kendakk, C. Lowe, R. Maxey, J. Miotia, R. J. Naylor, K. A. Runcie, B. Tuladhar, and J. B. H. Warneck, *Bioorg. Med. Chem. Lett.*, **1998**, *8*, 2635.
577. N. K. Hart, S. R. Jones, and J. A. Lamberton, *Aust. J. Chem.*, **1968**, *21*, 1321; **1971**, *24*, 1739.

References

578. Y. Ikeura, Y. Ishichi, T. Tanaka, A. Fujishima, M. Murabayashi, M. Kawada, T. Ishimaru, I. Kamo, T. Doi, and H. Natsugari, *J. Med. Chem.*, **1998**, *41*, 4232.
579. F. Wu, J. Hardesty, and R. Thummel, *J. Org. Chem.*, **1998**, *63*, 4055.
580. R. Gleiter, H. Weigl, and G. Haberhauer, *Eur. J. Org. Chem.*, **1998**, 1447.
581. S. Matsuhashi, *Rev. Infect. Dis.*, **1988**, *10*(Suppl. 1), S27; *Chem. Abstr.*, **1988**, *109*, 3481.
582. A. A. A. Ebanany, *J. Chem. Soc. Pak.*, **1987**, *9*, 547; *Chem. Abstr.*, **1988**, *109*, 149460.
583. N. Barba-Behrens, G. Mendoza-Diaz, and D. M. L. Goodgane, *Inorg. Chim. Acta*, **1986**, *125*, 21.
584. M. A. Ciriano, B. E. Villarroya, and L. A. Oro, *Inorg. Chim. Acta*, **1986**, *120*, 43.
585. C. Roussel, A. Mercier, and M. Cartier, *J. Org. Chem.*, **1987**, *52*, 2935.
586. R. Clark and ?. Jahangir, *J. Org. Chem.*, **1987**, *52*, 5378.
587. M. Di Braccio, G. Roma, M. Ghia, and F. Mattiol, *Farmaco*, **1994**, *49*, 25; *Chem. Abstr.*, **1994**, *121*, 133994.
588. Z. Wróbel and M. Mąkosza, *Tetrahedron*, **1993**, *49*, 5315.
589. R. J. Friary, V. Seidl, J. H. Schwerdt, M. P. Cohen, D. Hou, and M. Nafissi, *Tetrahedron*, **1993**, *49*, 7169.
590. B. Singh, E. R. Bacon, G. Y. Lesher, S. Robinson, P. O. Pennock, D. C. Bode, E. D. Pagini, R. G. Bentley, M. J. Connell, L. T. Hamel, and P. J. Silver, *J. Med. Chem.*, **1995**, *38*, 2546.
591. M. Wozniak and M. Tomula, *Zesz. Nauk. Uniw. Jagiellon.*, *Pr. Chem.*, **1987**, *30*, 75; *Chem. Abstr.*, **1988**, *108*, 121288.
592. G. Klopman, O. T. Macina, M. E. Levinson, and H. S. Rosenkranz, *Antimicrob. Agents Chemother.*, **1987**, *31*, 1831.
593. H. Natsugari, Y. Ikeura, Y. Kiyota, Y. Ishichi, T. Ishimaru, O. Saga, H. Shirafuji, T. Tanaka, I. Kamo, T. Doi, and M. Otsuka, *J. Med. Chem.*, **1995**, *38*, 3106.
594. J. Frigola, D. Vañó, A. Torrens, A. Gómez-Gomar, E. Ortega, and S. Garcia-Granda, *J. Med. Chem.*, **1995**, *38*, 1203.
595. A. D. Thomas and C. V. Asokan, *J. Chem. Soc.*, *Perkin Trans. 1*, **2001**, 2583.
596. J. Frigola, A. Torrens, J. A. Castrillo, J. Mas, D. Vañó, J. M. Berrocal, C. Calvet, L. algado, J. Redondo, S. Garcia-Granda, E. Valenti, and J. R. Quintans, *J. Med. Chem.*, **1994**, *37*, 4195.
597. ?. Atta-ur-Rahman, I. I. Vohra, M. I. Choudhary, L. B. De Silva, W. H. M. W. Herath, and K. M. Navaratne, *Planta Med.*, **1988**, *54*, 461; *Chem. Abstr.*, **1989**, *110*, 111706.
598. T. Hirayama, K. Ueda, H. Ono, T. Watanabe, and S. Fujui, *Eisei Kagaku*, **1988**, *34*, 152; *Chem. Abstr.*, **1989**, *110*, 114645.
599. K. E. Brighty and T. D. Gootz, *J. Antimicrob. Chemother.*, **1997**, *39*(Suppl. B), 1; *Chem. Abstr.*, **1997**, *127*, 156066.
600. T. J. Murray and S. C. Zimmerman, *Tetrahedron Lett.*, **1995**, *36*, 7627.
601. J. M. Mellor and H. Rataj, *Tetrahedron Lett.*, **1996**, *37*, 2619.
602. K. J. Buysens, D. M. Vandenberghe, and G. J. Hoornaerts, *Tetrahedron*, **1996**, *52*, 9161.
603. H. Senboku, M. Takashima, M. Suzuki, K. Kobayashi, and H. Suginome, *Tetrahedron*, **1996**, *52*, 6125.
604. A. Boumendjel, J. C. Roberts, E. Hu, and P. V. Pallai, *J. Org. Chem.*, **1996**, *61*, 4434.
605. R. Ziessel, J. Suffert, and M. T. Youinou, *J. Org. Chem.*, **1996**, *61*, 6535.
606. S. Hanessian and M. J. Rozema, *J. Am. Chem. Soc.*, **1996**, *118*, 9884.
607. M. H. Sherlock, J. J. Kaminski, W. C. Tom, J. F. Lee, S.-C. Wong, W. Kreutner, R. W. Bryant, and A. T. McPhail, *J. Med. Chem.*, **1988**, *31*, 2108.
608. T. Rosen, D. T. W. Chu, I. M. Lico, P. B. Fernandes, K. Marsh, L. Shen, V. G. Cepa, and A. G. Pernet, *J. Med. Chem.*, **1988**, *31*, 1598.

609. T. Rosen, D. T. W. Chu, I. M. Lico, P. B. Fernandes, L. Shen, S. Borodkin, and A. G. Pernet, *J. Med. Chem.*, **1988**, *31*, 1586.
610. J. P. Sanchez, J. M. Domagala, S. E. Hagen, C. L. Heifetz, M. P. Hutt, J. R. Nichols, and A. K. Trehan, *J. Med. Chem.*, **1988**, *31*, 983.
611. Y. Nishimura and J.-I. Matsumoto, *J. Med. Chem.*, **1987**, *30*, 1622.
612. A. A. Santilli, A. C. Scotese, R. P. Bauer, and S. C. Bell, *J. Med. Chem.*, **1987**, *30*, 2270.
613. D. T. W. Chu, P. B. Fernandes, A. K. Claiborne, E. H. Gracey, and A. G. Pernet, *J. Med. Chem.*, **1986**, *29*, 2363.
614. M. H. Norman, H. D. Smith, C. W. Andrews, F. L. M. Tang, C. L. Cowan, and R. P. Steffen, *J. Med. Chem.*, **1995**, *38*, 4670.
615. P. Di Cesare, D. Bouzard, M. Essiz, J. P. Jacquet, B. Ledoussal, J. R. Kiechel, P. Remuzon, R. E. Kessler, J. Fing-Tonc, and J. Desiderio, *J. Med. Chem.*, **1992**, *35*, 4205.
616. F. A. Luzzio and F. S. Guziec, *Org. Prep. Proc. Internatl.*, **1988**, *20*, 533.
617. P. Remuzon, D. Bouzard, P. Di Cesare, M. Essiz, J. P. Jacquet, J. R. Kiechek, B. Ledoissal, R. E. Kessler, and J. Fung-Tomc, *J. Med. Chem.*, **1991**, *34*, 29.
618. R. A. Mekheimer, *Synthesis*, **2001**, 103.
619. P. Zakizewski, M. Gowan, L. A. Trimble, and C. K. Lau, *Synthesis*, **1999**, 1893.
620. C. Janiak, S. Deblon, and L. Uehlin, *Synthesis*, **1999**, 959.
621. A. Numata, Y. Kondo, and T. Sakamoto, *Synthesis*, **1999**, 306.
622. K. Mogilaiah, R. B. Rao, and K. N. Reddy, *Indian J. Chem., Sect. B*, **1999**, *38*, 818.
623. K. Mogilaiah and R. B. Rao, *Indian J. Chem., Sect. B*, **1999**, *38*, 869.
624. K. Mogilaiah, P. R. Reddy, and R. B. Rao, *Indian J. Chem., Sect. B*, **1999**, *38*, 1203.
625. K. Mogilaiah and R. B. Rao, *Indian J. Chem., Sect. B*, **2000**, *39*, 145.
626. K. Mogilaiah, D. S. Chowdary, and R. B. Rao, *Indian J. Chem., Sect. B*, **2001**, *40*, 43.
627. K. Janné and P. Ahlberg, *Synthesis*, **1976**, 452.
628. W. Czuba and M. Woźniak, *Synthesis*, **1974**, 809.
629. G. Fischer and M. Puza, *Synthesis*, **1973**, 218.
630. R. Danieli and A. Ricci, *Synthesis*, **1973**, 46.
631. K. Mogilaiah and R. B. Rao, *Indian J. Chem., Sect. B*, **2001**, *40*, 235.
632. K. Morilaiah, R. B. Rao, and G. R. Sudhakar, *Indian J. Chem., Sect. B*, **2001**, *40*, 336.
633. K. Mogilaiah and P. R. Reddy, *Indian J. Chem., Sect. B*, **2001**, *40*, 619.
634. R. L. Dow and S. R. Schneider, *J. Heterocycl. Chem.*, **2001**, *38*, 535.
635. C. Plisson and J. Chenault, *J. Heterocycl. Chem.*, **2001**, *38*, 467.
636. T. Ohta, H. Fujisawa, M. Kawazome, Y. Nakai, and I. Furukawa, *J. Heterocycl. Chem.*, **2001**, *38*, 159.
637. C. Afloroaei, N. Dulamita, M. Vlassa, J. Barbe, and P. Brouant, *J. Heterocycl. Chem.*, **2000**, *37*, 1289.
638. B. G. Harvey, A. M. Arif, and R. D. Ernst, *Polyhedron*, **2004**, *23*, 2725.
639. M. M. Blanco, G. Buldain, C. B. Schapira, and I. Perillo, *J. Heterocycl. Chem.*, **2002**, *39*, 341.
640. Y. Tominaga, K. Nomoto, and N. Yoshioka, *J. Heterocycl. Chem.*, **2001**, *38*, 1135.
641. D. Heber and E. V. Stoyanov, *J. Heterocycl. Chem.*, **2000**, *37*, 871.
642. C. Vu, D.-D. Walker, J. Wells, and S. Fox, *J. Heterocycl. Chem.*, **2002**, *39*, 829.
643. Y. Hamada, M. Sato, and I. Takeuchi, *Yakugaku Zasshi*, **1975**, *95*, 1492.
644. F. Al-Omran, R. M. Mohareb, and A. A. El-Khair, *J. Heterocycl. Chem.*, **2002**, *39*, 877.
645. S. J. Swamy and S. R. Reddy, *Indian J. Chem, Sect. A*, **2001**, *40*, 1093.
646. N. Susuki, K.-I. Nuhami, K. Matsumoto, N. Yoneda, and M. Miyoshi, *Synthesis*, **1978**, 461.

647. W. Roszkiewicz and M. Woźniak, *Synthesis*, **1996**, 691.
648. S. Sabatini, V. Cecchetti, O. Tabarrini, and A. Fravolini, *J. Heterocycl. Chem.*, **1999**, *36*, 953.
649. K. Mogilaiah, R. H. Babu, R. B. Rao, and N. V. Reddy, *Indian J. Chem.*, *Sect. B*, **2002**, *41*, 218.
650. K. Mogilaiah, P. R. Reddy, and H. B. Rao, *Indian J. Chem.*, *Sect. B*, **1999**, *38*, 495.
651. M. Kidwai, R. Sharma, and P. Misra, *Indian J. Chem.*, *Sect. B*, **2002**, *41*, 427.
652. B. Hutchinson and A. Sunderland, *Inorg. Chem.*, **1972**, *11*, 1948.
653. K. Mogilaiah and D. S. Chowdary, *Indian J. Chem.*, *Sect. B*, **2002**, *42*, 1894.
654. M. M. Blanco, I. A. Perillo, and C. B. Schapira, *J. Heterocycl. Chem.*, **1999**, *36*, 979.
655. Y. Tamura, J. Minamikawa, Y. Miki, S. Matsugashita, and M. Ikeda, *Tetrahedron Lett.*, **1972**, 4133.
656. A. T. Marcelis and H. C. van der Plas, *Tetrahedron*, **1989**, *45*, 2693.
657. K. Mogilaiah and G. Kankaiah, *Indian J. Chem.*, *Sect. B*, **2002**, *41*, 2194.
658. K. Mogilaiah and R. B. Rao, *Indian J. Chem.*, *Sect. B*, **2001**, *40*, 713.
659. K. Mogilaiah, N. V. Reddy, and R. B. Rao, *Indian J. Chem.*, *Sect. B*, **2001**, *40*, 837.
660. E. Fasani, F. F. B. Negra, M. Mella, S. Monti, and A. Albini, *J. Org. Chem.*, **1999**, *64*, 5388.
661. T. E. Reed and D. G. Hendricker, *Inorg. Chim. Acta*, **1970**, *4*, 471.
662. R. J. Foster and D. G. Hendricker, *Inorg. Chim. Acta*, **1972**, *6*, 371.
663. K. Mogilaiah and P. R. Reddy, *Indian J. Chem.*, *Sect. B*, **2001**, *40*, 839.
664. K. Mogilaiah, H. R. Babu, and R. B. Rao, *Indian J. Chem.*, *Sect. B*, **2001**, *40*, 1270.
665. M. Bacci, A. Dei, and R. Morassi, *Inorg. Chim. Acta*, **1973**, *7*, 209.
666. K. V. Reddy, K. Mogilaiah, and B. Sreenivasulu, *J. Indian Chem. Soc.*, **1984**, *61*, 888.
667. D. R. Sinha and A. B. Lal, *J. Indian Chem. Soc.*, **1979**, *56*, 164.
668. A. V. Tverdokhlebov, E. V. Resnyanska, A. A. Tolmachev, and A. V. Zavada, *Monatsh. Chem.*, **2003**, *134*, 1045.
669. F. Alhaique, F. M. Riccieri, and L. Campanella, *Ann. Chim.* (Rome), **1972**, *62*, 239.
670. Y. Hamada and I. Takeuchi, *Yakugaku Zasshi*, **2000**, *120*, 206.
671. H. Bock, T. T. H. Van, and H. Schödel, *Monatsh. Chem.*, **1996**, *127*, 391.
672. G. Zigeuner, K. Schweiger, and D. Habernig, *Monatsh. Chem.*, **1982**, *113*, 573.
673. H. Junek, O. S. Wolfbeis, H. Sprintschnik, and H. Wolny, *Monatsh. Chem.*, **1977**, *108*, 689.
674. K. Mogilaiah and N. V. Reddy, *Indian J. Chem.*, *Sect. B*, **2002**, *41*, 215.
675. W. Tikkanen, W. C. Kaska, S. Moya, T. Layman, R. Kane, and C. Krüger, *Inorg. Chim. Acta*, **1983**, *76*, L29.
676. B. A. Hess, L. J. Schaad, and C. W. Holyoke, *Tetrahedron*, **1975**, *31*, 295.
677. G. Giacomello, F. Gualtieri, F. M. Riccieri, and M. L. Stein, *Tetrahedron*, **1965**, *6*, 1117.
678. D. G. Hendricker, *Inorg. Chem.*, **1969**, *8*, 2328.
679. G. Zigeuner, W.-B. Lintschinger, A. Fuchsgruber, and K. Kollmann, *Monatsh. Chem.*, **1976**, *107*, 155.
680. P. L. Ferrarini, *Ann. Chim.* (Rome), **1971**, *61*, 318.
681. S. Carboni, A. Da Settimo, and G. Pirisino, *Ann. Chim.* (Rome), **1964**, *54*, 883.
682. M. Woźniak, M. Grzegozek, and P. Surylo, *Liebigs Ann. /Recl.*, **1997**, 2601.
683. I. C. Ivanov, E. V. Stoyanov, P. S. Denkova, and V. S. Dimitrov, *Liebigs Ann. /Recl.*, **1997**, 1777.
684. J. Ritter and R. Gleiter, *Liebigs Ann. /Recl.*, **1997**, 1179.
685. A. F. A. Harb, A.-H. M. Hesien, S. A. Metwally, and M. H. Elnagdi, *Liebigs Ann. Chem.*, **1989**, 585.
686. S. Carboni, A. Da Settimo, and G. Pirisino, *Ann. Chim.* (Rome), **1964**, *54*, 677.
687. S. Carboni and G. Pirisino, *Ann. Chim.* (Rome), **1962**, *52*, 340.
688. M. Wenzel, F. Lehmann, R. Reckert, W. Günther, and G. Göris, *Monatsh. Chem.*, **1999**, *130*, 1373.

689. J. Schurz, A. Ullrich, and H. Bayzer, *Monatsh. Chem.*, **1959**, *90*, 29.
690. S. Carboni and G. Pirisino, *Ann. Chim. (Rome)*, **1962**, *52*, 279.
691. K. Schweiger, D. Habernig, H.-W. Schramm, and G. Zigeuner, *Monatsh. Chem.*, **1983**, *114*, 79.
692. M. Woźniak and M. Tomula, *Liebigs Ann. Chem.*, **1993**, 471.
693. E. A. Mohamed, R. M. Abdel-Rahman, Z. El-Gendy, and M. M. Ismail, *J. Indian Chem. Soc.*, **1994**, *71*, 765.
694. M. T. Chary, K. Mogilaiah, C. M. Rao, and B. Sreenivasulu, *J. Indian Chem. Soc.*, **1990**, *69*, 691.
695. C. J. Fahrni and A. Pfaltz, *Helv. Chim. Acta*, **1998**, *81*, 491.
696. H. Zondler and W. Pfleiderer, *Helv. Chim. Acta*, **1975**, *58*, 2247.
697. K. R. Reddy, K. Mogilaiah, and B. Sreenivasulu, *J. Indian Chem. Soc.*, **1987**, *64*, 709.
698. S. M. Fahmy, S. O. A. Allah, and R. M. Moharab, *Synthesis*, **1984**, 976.
699. V. Aggarwal, G. Singh, H. Ila, and H. Junjappa, *Synthesis*, **1982**, 214.
700. G. Abbiati, A. Arcadi, F. Martinelli, and E. Rossi, *Synthesis*, **2002**, 1912.
701. S. Carboni, A. Da Settimo, G. Pirisino, and D. Segini, *Gazz. Chim. Ital.*, **1966**, *96*, 103.
702. S. Carboni, A. Da Settimo, and P. L. Ferrarini, *Gazz. Chim. Ital.*, **1965**, *95*, 1492.
703. M. Brufani, D. Duranti, G. Giacomello, and L. Zambonelli, *Gazz. Chim. Ital.*, **1961**, *91*, 287.
704. K. Emerson, A. Emad, R. W. Brookes, and R. L. Martin, *Inorg. Chem.*, **1973**, *12*, 978.
705. R. W. Stotz, J. A. Walmsley, and F. Walmsley, *Inorg. Chem.*, **1969**, *8*, 807.
706. K. R. Reddy, K. Mogilaiah, and B. Sreenivasulu, *J. Indian Chem. Soc.*, **1987**, *64*, 193.
707. K. V. Reddy, K. Mogilaiah, and B. Sreenivasulu, *J. Indian Chem. Soc.*, **1986**, *63*, 443.
708. M. H. Elnagdi, A. F. A. Harb, A. H. H. Elghandour, A. H. M. Hussien, and S. A. M. Metwally, *Gazz. Chim. Ital.*, **1992**, *122*, 299.
709. K. Mogilaiah, C. J. Rao, and A. K. Murthy, *J. Indian Chem. Soc.*, **1986**, *63*, 345.
710. F. Alhaique, F. M. Riccieri, and E. Santucci, *Gazz. Chim. Ital.*, **1975**, *105*, 1001.
711. F. Heinzer, M. Soukup, and A. Eschenmoser, *Helv. Chim. Acta*, **1978**, *61*, 2851.
712. M. T. Chary, K. Mogilaiah, and B. Sreenivasulu, *J. Indian. Chem. Soc.*, **1987**, *64*, 488.
713. L. Merlini, G. Nasini, and M. Palamareva, *Gazz. Chim. Ital.*, **1975**, *105*, 339.
714. V.-M. Mukkalo, C. Sund, M. Kwiatkowski, P. Pasenen, M. Högberg, J. Kankara, and H. Takalo, *Helv. Chim. Acta*, **1992**, *75*, 1621.
715. R. Huff, F. Mutterer, and C. D. Weis, *Helv. Chim. Acta*, **1977**, *60*, 907.
716. B. Singh, G. Y. Lesher, and R. P. Brundage, *Synthesis*, **1991**, 894.
717. P. L. Nyce, D. Gala, and M. Steinman, *Synthesis*, **1991**, 571.
718. S. Carboni, A. Da Settimo, P. L. Ferrarini, and F. Trusendi, *Gazz. Chim. Ital.*, **1968**, *98*, 1174.
719. N. Nishiwaki, M. Komatsu, and Y. Ohshiro, *Synthesis*, **1991**, 41.
720. S. Carboni, A. Da Settimo, P. L. Ferrarini, and I. Yonetti, *Gazz. Chim. Ital.*, **1971**, *101*, 129.
721. S. Carbon-, A. Da Settimo, P. L. Ferrarini, and I. Tonetti, *Gazz. Chim. Ital.*, **1969**, *99*, 823.
722. O. Süss and K. Möller, *Liebigs Ann. Chem.*, **1956**, *593*, 91.
723. O. Süss and K. Möller, *Liebigs Ann. Chem.*, **1956**, *599*, 233.
724. M. Weissenfels and B. Ulrici, *Z. Chem.*, **1978**, *18*, 20.
725. A. D. Dunn, *Z. Chem.*, **1990**, *30*, 20.
726. K. Möller and O. Süss, *Liebigs Ann. Chem.*, **1957**, *612*, 153.
727. K. Mogilaiah and O. R. Sudhakar, *Indian J. Chem., Sect. B*, **2003**, *42*, 1170.
728. K. Mogilaiah, P. R. Reddy, R. B. Rao, and N. V. Reddy, *Indian J. Chem., Sect. B*, **2003**, *42*, 1746.
729. K. Mogilaiah, G. R. Sudhakar, and N. V. Reddy, *Indian J. Chem., Sect. B*, **2003**, *42*, 1753.
730. K. Mogilaiah and N. V. Reddy, *Indian J. Chem., Sect. B*, **2003**, *42*, 2124.

References

731. L. Capuano, W. Hell, and C. Wamprecht, *Liebigs Ann. Chem.*, **1986**, 132.
732. P. Messinger and H. Meyer-Barrientos, *Liebigs Ann. Chem.*, **1981**, 2087.
733. K. Mogilaiah, S. Kavitha, and H. R. Babu, *Indian J. Chem., Sect. B*, **2003**, *42*, 1750.
734. H. Zondler and W. Pfleiderer, *Liebigs Ann. Chem.*, **1972**, *759*, 84.
735. K. Mogilaiah, N. V. Reddy, and R. B. Rao, *Indian J. Chem., Sect. B*, **2003**, *43*, 2618.
736. H. Meyer and J. Kurz, *Liebigs Ann. Chem.*, **1978**, 1491.
737. M. Weissenfels and B. Ulrichi, *Z. Chem.*, **1978**, *18*, 382.
738. D. Ramesh and B. Sreenivasulu, *Indian J. Chem., Sect. B*, **2004**, *43*, 897.
739. D. Galteschi, C. Mealli, and L. Sacconi, *J. Am. Chem. Soc.*, **1973**, *95*, 2736.
740. K. Mogilaiah, K. Srinivas, and G. R. Sudhakar, *Indian J. Chem., Sect. B*, **2004**, *43*, 2014.
741. S. Carboni, A. Da Settimo, P. L. Ferrarini, and I. Tonetti, *Gazz. Chim. Ital.*, **1967**, *97*, 1262.
742. F. M. Assai, J. Becher, J. Møller, and K. S. Varma, *Synthesis*, **1987**, 301.
743. S. Carboni, A. Da Settimo, and P. L. Ferrarini, *Gazz. Chim. Ital.*, **1967**, *97*, 1061.
744. K. Mogilaiah and C. S. Reddy, *Indian J. Chem., Sect. B*, **2004**, *43*, 2010.
745. P. Messinger and H. Meyer, *Liebigs Ann. Chem.*, **1979**, 443.
746. H. Junek and G. Stolz, *Monatsh. Chem.*, **1970**, *101*, 1234.
747. H. Rey-Bellet and H. Erlenmeyer, *Helv. Chim. Acta*, **1956**, *39*, 2106.
748. S. Carboni, A. Da Settimo, D. Segnini, and I. Tonetti, *Gazz. Chim. Ital.*, **1966**, *96*, 1443.
749. S. Carboni, A. Da Settimo, D. Bertini, P. L. Ferrarini, O. Livi, and I. Tonetti, *Gazz. Chim. Ital.*, **1974**, *104*, 499.
750. L. Capuano and C. Wamprecht, *Liebigs Ann. Chem.*, **1986**, 938.
751. H. Junek, *Monatsh. Chem.*, **1965**, *96*, 2046.
752. K. Mogilaiah and G. Kankaiah, *Indian J. Chem., Sect. B*, **2003**, *42*, 192.
753. K. Mogilaiah and G. R. Sudhakar, *Indian J. Chem., Sect. B*, **2003**, *42*, 636.
754. J. Kleinschroth, K. Mannhardt, J. Hartenstein, and G. Satzinger, *Synthesis*, **1986**, 859.
755. S. Carboni, A. Da Settimo, D. Bertini, and G. Biagi, *Gazz. Chim. Ital.*, **1969**, *99*, 677.
756. K. Mogilaiah and G. Kankaiah, *Indian J. Chem., Sect. B*, **2003**, *42*, 658.
757. H. Junek and A. R. O. Schmidt, *Monatsh. Chem.*, **1969**, *100*, 570.
758. S. Carboni, A. Da Settimo, D. Bertini, P. L. Ferrarini, O. Livi, C. Mori, and I. Tonetti, *Gazz. Chim. Ital.*, **1972**, *102*, 253.
759. W. Skoda and H. Bayzer, *Monatsh. Chem.*, **1958**, *89*, 5.
760. E. Ziegler and E. Nölken, *Monatsh. Chem.*, **1961**, *92*, 1184.
761. S. Carboni, A. Da Settimo, and P. L. Ferrarini, *Gazz. Chim. Ital.*, **1967**, *97*, 42.
762. K. R. Reddy, G. R. Rao, K. Mogilaiah, and B. Sreenivasulu, *J. Indian Chem. Soc.*, **1987**, *64*, 443.
763. A. Mangini and M. Colonna, *Gazz. Chim. Ital.*, **1943**, *73*, 323.
764. S. Carboni, A. Da Settimo, P. L. Ferrarini, and G. Pirisino, *Gazz. Chim. Ital.*, **1966**, *96*, 1456.
765. E. Vilsmaier and T. Goerz, *Synthesis*, **1998**, 739.
766. M. Brufani, D. Duranti, and G. Giacomello, *Gazz. Chim. Ital.*, **1959**, *89*, 2328.
767. P. Hemmerich and S. Fallab, *Helv. Chim. Acta*, **1958**, *41*, 498.
768. M. Pailer and E. Kuhn, *Monatsh. Chem.*, **1953**, *84*, 85.
769. M. Colonna and C. Runti, *Gazz. Chim. Ital.*, **1952**, *82*, 513.
770. W. Ciusa and G. Nebbia, *Gazz. Chim. Ital.*, **1950**, *80*, 518.
771. A. Mangini and M. Colonna, *Gazz. Chim. Ital.*, **1943**, *73*, 330.
772. V. D. Dyachenko, S. V. Roman, E. B. Rusanov, and V. P. Litvinov, *Khim. Geterotsikl. Soedin.*, **2002**, 1432.

773. M. Santo, L. Giacomelli, R. Catana, J. Silber, M. M. Blanco, C. R. Schapira, and I. A. Perillo, *Spectrochim. Acta, Part A*, **2003**, *59*, 1399.
774. V. K. Indirapriyadharshini, P. Ramanmurthy, V. Raghukumar, and V. T. Ramakrishnan, *Spectrochim. Acta, Part A*, **2002**, *58*, 1535.
775. E. L. Enwall and K. Emerson, *Acta Crystallogr., Sect. B*, **1979**, *35*, 2562.
776. C. Huiszoon, G. J. van Hummel, and D. M. W. van den Ham, *Acta Crystallogr., Sect. B*, **1977**, *33*, 1867.
777. R. L. Harlow and S. H. Simonsen, *Acta Crystallogr., Sect. B*, **1977**, *33*, 2662.
778. D. M. W. van den Ham and G. J. van Hummel, *Acta Crystallogr., Sect. B*, **1977**, *33*, 3866.
779. D. Gomez de Anderez, J. R. Helliwell, E. J. Dodson, J. F. Piniella, G. Germain, A. Alvarez-Larena, J. Texido, and P. Victory, *Acta Crystallogr., Sect. B*, **1992**, *48*, 104.
780. L.-C. Yu, H. Liang, C.-S. Zhou, Z.-F. Cheng, and Y. Zhang, *Acta Crystallogr., Sect. E*, **2004**, *60*, o1051.
781. S. Bhaskaran, M. Yogavel, V. Rajakannan, R. Krishna, D. Velmurugan, and V. Raghukumar, *Acta Crystallogr., Sect. E*, **2003**, *59*, 8200.
782. R. L. Harlow and S. H. Simonsen, *Acta Crystallogr., Sect. B*, **1978**, *34*, 2180.
783. C. Maelli and L. Sacconi, *Acta Crystallogr., Sect. B*, **1977**, *33*, 710.
784. S. V. Roman, V. D. Dyachenko, and V. P. Litvinov, *Khim. Geterotsikl. Soedin.*, **1999**, 1435.
785. M. Wozniak, P. Suryio, and H. C. van der Plas, *Khim. Geterotsikl. Soedin.*, **1996**, 1652.
786. V. A. Artemov, A. M. Shestopalov, and V. P. Litvinov, *Khim. Geterotsikil, Soedin.*, **1996**, 512.
787. R. M. Titkova, A. S. Elina, and N. P. Kostyuchenko, *Khim. Geterotsikl. Soedin.*, **1972**, 1237.
788. H. Irikawa and K. Iijima, *Acta Crystallogr., Sect. C*, **1998**, *54*, 1318.
789. K. Chinnakali, H.-K. Fun, I. A. Razak, P. Murugan, and V. T. Ramakrishnan, *Acta Crystallogr., Sect. C*, **1998**, *54*, 781.
790. D. Patel, D. S. Brown, and J. R. Traynor, *Acta Crystallogr., Sect. C*, **1992**, *48*, 1614.
791. A. K. Sheinkman, T. N. Nezdiiminoga, T. S. Chmilenko, and N. A. Klyuev, *Khim. Geterotsikl. Soedin.*, **1986**, 1218.
792. N. I. Smetskaya, A. M. Zhidkova, N. A. Mukhina, and V. G. Granik, *Khim. Geterotsikl. Soedin.*, **1984**, 1287.
793. V. T. Sigova and M. E. Konshin, *Khim. Geterotsikl. Soedin.*, **1984**, 783.
794. E. Weber and H.-J. Köhler, *J. Prakt. Chem. /Chem. Ztg.*, **1995**, *337*, 451.
795. H. Möhrle and H. Dwuletzki, *Chem. Ber.*, **1986**, *119*, 3600.
796. H. Wamhoff and L. Lichtenthäler, *Chem. Ber.*, **1998**, *111*, 2813.
797. R. M. Titkova and A. S. Elena, *Khim. Geterotsikl. Soedin.*, **1973**, 1279.
798. D. M. W. van den Ham, G. J. van Hummel, and C. Huiszoon, *Acta Crystallogr., Sect. B*, **1978**, *34*, 3134.
799. A. Achari and S. Neidle, *Acta Crystallogr., Sect. B*, **1976**, *32*, 600.
800. Y. D. Shen, H. Wu, X. Bu, and L. Gu, *Acta Crystallogr., Sect. E*, **2004**, *60*, 81641.
801. H. Haber, V. Hager, and M. Schlender, *J. Prakt. Chem.*, **1991**, *333*, 637.
802. A. M. El-Reedy, M. K. A. Ibrahim, and S. M. Hussain, *J. Prakt. Chem.*, **1989**, *331*, 745.
803. M. Makosza, J. Goliński, S. Ostrowski, A. Rykowski, and A. B. Sahasrabudhe, *Chem. Ber.*, **1991**, *124*, 577.
804. A. K. Sheinkman, T. S. Chmilenko, and T. N. Nezdiiminoga, *Khim. Geterotsikl. Soedin.*, **1983**, 706.
805. V. G. Granik, A. M. Zhidkova, and R. A. Dubinskit, *Khim. Geterotsikil. Soedin.*, **1982**, 518.
806. H. J. S. Machado and A. Hinchliffe, *J. Mol. Struct. (Theochem.)*, **1995**, *339*, 255.
807. B. Tinland, *Theor. Chim. Acta*, **1967**, *8*, 361.
808. H. C. van der Plas, M. Woźniak, and A. van Veldhuizen, *Recl. Trav. Chim. Pays-Bas*, **1976**, *95*, 233.

809. P. Dapporto, C. A. Ghilardi, C. Mealli, A. Orlandini, and S. Pacinotti, *Acta Crystallogr., Sect. C*, **1984**, *40*, 891.
810. N. P. Buu-Hoi and M. Declercq, *Recl. Trav. Chim. Pays-Bas*, **1954**, 73, 376.
811. W.Czuba, *Recl. Trav. Chim. Pays-Bas*, **1963**, *82*, 988.
812. M. J. S. Dewar and N. Trinajstić, *Theor. Chim. Acta*, **1970**, *17*, 235.
813. P. van de Weijer, D. van der Meer, and J. L. Koster, *Theor. Chim. Acta*, **1975**, *38*, 223.
814. R. M. Titkova, A. S. Elina, E. A. Trifonova, I. V. Persianova, N. P. Solov'eva, E. M. Peresleni, T. A. Gus'kova, and Y. N. Sheinker, *Khim. Geterotsikl. Soedin.*, **1981**, 792.
815. V. G. Granik, E. M. Peresleni, O. Y. Belyaeva, T. D. Zotova, O. S. Anisimova, R. G. Glushkov, and Y. N. Sheinker, *Khim. Geterotsikl. Soedin.*, **1977**, 793.
816. Y.-F. Ming, N. Horlemann, and H. Wamhoff, *Chem. Ber.*, **1987**, *120*, 1427.
817. F. Vögtle and D. Brombach, *Chem. Ber.*, **1975**, *108*, 1682.
818. M. Woźniak, H. C. van der Plas, M. Tomula, and A. van Veldhuizen, *J. R. Neth. Chem. Soc.*, **1983**, *102*, 511.
819. M. Woźniak, H. C. van der Plas, M. Tomula, and A. van Veldhuizen, *J. R. Neth. Chem. Soc.*, **1983**, *102*, 359.
820. G. Favini, I. Vandoni, and M. Simonetta, *Theor. Chim. Acta*, **1965**, *3*, 418.
821. D. M. W. van den Ham, G. F. S. Harrison, A. Spaans, and D. van der Meer, *Recl. Trav. Chim. Pays-Bas*, **1975**, *94*, 168.
822. J. Pomorski, H. J. den Hertog, D. J. Buurman, and N. H. Bakker, *Recl. Trav. Chim. Pays-Bas*, **1973**, *92*, 970.
823. V. G. Granik, E. O. Sochneva, N. P. Solov'eva, E. F. Kuleshova, and O. S. Anisimova, *Khim. Geterotsikl. Soedin.*, **1980**, 1120.
824. R. Gompper and R. Kunz, *Chem. Ber.*, **1965**, *98*, 1391.
825. C. Jutz, W. Müller, and E. Müller, *Chem. Ber.*, **1966**, *99*, 2479.
826. V. G. Granik, E. M. Peresleni, T. D. Kurochkina, A. M. Zhidkova, N. B. Marchenko, R. Glushkov, and Y. N. Sheinker, *Khim. Geterotsikl. Soedin.*, **1980**, 349.
827. H. J. W. van den Haak and H. C. van der Plas, *J. R. Neth. Chem. Soc.*, **1980**, *99*, 83.
828. H. G. M. Walraven, G. G. Choudry, and U. K. Pandit, *Recl. Trav. Chim. Pays-Bas*, **1976**, *95*, 220.
829. J. W. Wolsink, A. Spaans, D. M. W. van den Ham, and D. Feil, *J. R. Neth. Chem. Soc.*, **1982**, *101*, 141.
830. H. C. van der Plas, M. Woźniak, and A. van Veldhuizen, *Recl. Trav. Chim. Pays-Bas*, **1978**, *97*, 130.
831. A. S. Elina, I. S. Musatova, R. M. Titkova, R. A. Dubinskii, and M. S. Goizman, *Khim. Geterotsikl. Soedin.*, **1980**, 1106.
832. A. Clearfield, M. J. Sims, and P. Singh, *Acta Crystallogr., Sect. B*, **1972**, *28*, 350.
833. D. van der Meer, *Recl. Trav. Chim. Pays-Bas*, **1970**, *89*, 51.
834. H. C. van der Plas, M. Woźniak, and A. van Veldhuizen, *Recl. Trav. Chim. Pays-Bas*, **1977**, *96*, 151.
835. V. G. Granik, N. B. Marchenko, E. O. Sochneva, R. G. Glushkov, T. F. Vlasova, and Y. N. Sheinker, *Khim. Geterotsikl. Soedin.*, **1976**, 805.
836. V. G. Granik, I. V. Persianova, E. O. Sochneva, O. S. Anisimova, and Y. N. Sheinker, *Khim. Geterotsikl. Soedin.*, **1979**, 1255.
837. W. Czuba, *Recl. Trav. Chim. Pays-Bas*, **1963**, *82*, 997.
838. H. J. W. van den Haak and H. C. van der Plas, *J. R. Neth. Chem. Soc.*, **1983**, *102*, 235.
839. V. G. Granik, E. O. Sochneva, and N. P. Solov'eva, *Khim. Geterotsikl. Soedin.*, **1980**, 416.
840. F. W. Birss and N. K. D. Gupta, *Indian J. Chem., Sect. B*, **1979**, *17*, 610.
841. E. O. Sochneva, N. P. Solov'eva, and V. G. Granik, *Khim. Geterotsikl. Soedin.*, **1978**, 1671.

842. A. K. Sheinkman, V. P. Marshtupa, N. M. Shneer, L. G. Sharanina, and V. V. Petrenko, *Khim. Geterotsikl. Soedin.*, **1978**, 132.
843. J. Metz and M. Hanack, *Chem. Ber.*, **1987**, *120*, 1307.
844. W. Czuba and M. Woźniak, *Recl. Trav. Chim. Pays-Bas*, **1974**, *93*, 144.
845. V. G. Gnanik, N. B. Marchenko, T. F. Vlasova, and R. G. Glushkov, *Khim. Geterotsikl. Soedin.*, **1976**, 1509.
846. A. T. Amos, *Theor. Chim. Acta*, **1966**, *6*, 333.
847. V. G. Granik, O. Y. Belyaeva, R. G. Glushkov, T. F. Vlasova, and O. S. Anisimova, *Khim. Geterotsikl. Soedin.*, **1977**, 1106.
848. D. van der Meer and D. Feil, *Recl. Trav. Chim. Pays-Bas*, **1968**, *87*, 746.
849. W. Czuba and M. Woźniak, *Recl. Trav. Chim. Pays-Bas*, **1974**, *93*, 143.
850. H. S. Rani, K. Mogilaiah, and B. Sreenivasulu, *Indian J. Chem.*, Sect. B, **1996**, *35*, 106.
851. H. S. Rani, K. Mogilaiah, J. S. Rao, and B. Sreenivasulu, *Indian J. Chem.*, Sect. B, **1996**, *35*, 745.
852. J. S. Rao, K. Mogilaiah, and B. Sreenivasulu, *Indian J. Chem.*, Sect. B, **1996**, *35*, 713.
853. K. Mogilaiah and B. Sakram, *Indian J. Chem.*, Sect. B, **2004**, *43*, 2724.
854. M. Grzegożek and B. Szpakiewicz, *Can. J. Chem.*, **2004**, *82*, 567.
855. K. Mogilaiah, S. Kavitha, and G. R. Sudhakar, *Indian J. Chem.*, Sect. B, **2004**, *43*, 2713.
856. K. Mogilaiah, M. Prashanthi, and K. Vidya, *Indian J. Chem.*, Sect. B, **2004**, *43*, 2641.
857. Y. Kobayashi, T. Kutsuma, K. Morinaga, M. Fujita, and Y. Nahzawa, *Chem. Pharm. Bull.*, **1970**, *18*, 2489.
858. K. Mogilaiah, P. R. Reddy, and R. B. Rao, *Indian J. Chem.*, Sect. B, **1999**, *38*, 495.
859. G. R. Rao, K. Mogilaiah, and B. Sreenivasulu, *Indian J. Chem.*, Sect. B, **1996**, *35*, 339.
860. M. Woźnik, M. Grzegożek, and P. Surylo, *Can. J. Chem.*, **2000**, *78*, 950.
861. K. Mogilaiah and R. B. Rao, *Indian J. Chem.*, Sect. B, **1998**, *37*, 894.
862. P. B. Talukdar, S. K. Sengupta, and A. K. Datta, *Indian J. Chem.*, Sect. B, **1978**, *16*, 678.
863. R. A. Pawar, A. L. Kohak, and V. N. Gogte, *Indian J. Chem.*, Sect. B, **1976**, *14*, 375.
864. S. Singh, R. S. Taneja, and K. S. Narang, *Indian J. Chem.*, **1968**, *6*, 11.
865. M. Ogata and H. Matsumoto, *Chem. Pharm. Bull.*, **1972**, *20*, 2264.
866. V. P. Arya, F. Fernandes, and V. Sudarsanam, *Indian J. Chem.*, **1972**, *10*, 598.
867. V. K. Jhalani, L. P. Ghalsasi, S. P. Acharya, and R. N. Usgaonkar, *Indian J. Chem.*, Sect. B, **1989**, *28*, 173.
868. S. Nagasaki, N. Nakazawa, Y, Osada, T. Hashizume, and Y. Ôshima, *Chem. Pharm. Bull.*, **1972**, *20*, 639.
869. J. B. Sainani, A. C. Shah, and V. P. Arya, *Indian J. Chem.*, Sect. B, **1995**, *34*, 17.
870. H. S. Rani, K. Mogilaiah, J. S. Rao, and B. Sreenivasulu, *Indian J. Chem.*, Sect. B, **1995**, *34*, 1035.
871. J. S. Rao, B. Sreenivasulu, and K. Mogilaiah, *Indian J. Chem.*, Sect. B, **1995**, *34*, 734.
872. G. R. Rao, K. Mogilaiah, K. R. Reddy, and B. Sreenivasulu, *Indian J. Chem.*, Sect. B, **1988**, *27*, 200.
873. Y. Hamada and I. Takeuchi, *Chem. Pharm. Bull.*, **1971**, *19*, 1857.
874. K. Mogilaiah, K. V. Reddy, and B. Sreenivasulu, *Indian J. Chem.*, Sect. B, **1983**, *22*, 178.
875. K. R. Reddy, K. Mogilaiah, and B. Sreenivasulu, *Indian J. Chem.*, Sect. B, **1987**, *26*, 1194.
876. V. K. Jhalani and R. N. Usgaonkar, *Indian J. Chem.*, Sect. B, **1983**, *22*, 916.
877. K. Mogilaiah and B. Sreenivasulu, *Indian J. Chem.*, Sect. B, **1982**, *21*, 479.
878. N. K. Dasgupta, A. Dasgupta, and F. W. Birss, *Indian J. Chem.*, Sect. B, **1982**, *21*, 334.
879. V. N. Gogte, S. V. Kelkar, and B. D. Tilak, *Indian J. Chem.*, Sect. B, **1980**, *19*, 1011.
880. Y. Hamada, I. Takeuchi, and M. Hirota, *Chem. Pharm. Bull.*, **1971**, *19*, 1751.
881. R. Sairam and N. K. Ray, *Indian J. Chem.*, Sect. B, **1980**, *19*, 989.

References

882. J. Murgich, Y. Aray, H. J. Soscun, and R. A. Marino, *J. Phys. Chem.*, **1992**, *96*, 9198.
883. S. Bucknall, J. Silverlight, H. Coldham, L. Thorne, and R. Jackman, *Food Additives Contaminants*, **2003**, *20*, 221.
884. J. W. Huff, *J. Biol. Chem.*, **1947**, *167*, 151.
885. I. Ilmet and M. Krasij, *J. Phys. Chem.*, **1966**, *70*, 3755.
886. S. C. Zimmerman and T. J. Murray, *Philos. Trans. R. Soc. London, Ser. A*, **1993**, *345*, 49.
887. D. N. Bailey, D. M. Hercules, and T. D. Eck, *Anal. Chem.*, **1967**, *39*, 877.
888. C. Bessenbacher and W. Kaim, *J. Organomet. Chem.*, **1989**, *369*, 83.
889. C. Bessenbacher and W. Kaim, *J. Organomet. Chem.*, **1989**, *362*, 37.
890. S. Nishizawa, K. Yoshimoto, T. Seino, C.-Y. Xu, M. Minagawa, H. Satake, A. Tong, and N. Teramae, *Talanta*, **2004**, *63*, 175.
891. R. Ziessel, A. Harriman, A. El-Ghayoury, L. Douce, E. Leize, H. Nierengarten, and A. van Dorsselaer, *New J. Chem.*, **2000**, *24*, 729.
892. Y. Kuge, N. Kato, T. Sugata, and S. Tomioka, *Synth. Commun.*, **1994**, *24*, 3289.
893. D. J. Milner, *Synth. Commun.*, **1992**, *22*, 73.
894. G. Delmas, P. Deplat, J. M. Chezal, O. Chavignon, A. Gueiffier, Y. Blanche, J. L. Chabard, G. Dauphin, and J. C. Teulade, *Heterocycles*, **1996**, *43*, 1229.
895. P. Victory, N. Busquets, J. I. Borrell, J. Teixidó, B. Serra, J. L. Metallana, H. Junek, and H. Sterk, *Heterocycles*, **1995**, *41*, 1013.
896. J. M. Quintela, J. Vilar, C. Peinador, C. Veiga, and V. Ojea, *Heterocycles*, **1995**, *41*, 1001.
897. M. Woźniak, *Heterocycles*, **1982**, *19*, 363.
898. I. Tanaka and T. Mizukawa, *Appl. Organomet. Chem.*, **2000**, *14*, 863.
899. R. Ziessel, *Coord. Chem. Rev.*, **2001**, *216*, 195.
900. J. Villar, J. M. Quintela, C. Peinador, C. Veiga, and V. Ojea, *Heterocycles*, **1993**, *36*, 2697.
901. I. Cardinaud, A. Gueiffier, F. Fauvelle, J.-C. Milhavet, and J.-P. Chapat, *Heterocycles*, **1993**, *36*, 1945.
902. J. Epsztajn, M. M. Piotka, and J. Scianowski, *Synth. Commun.*, **1992**, *22*, 1239.
903. A. Staub, *Synth. Commun.*, **1993**, *23*, 365.
904. A. Døssing, S. Larsen, A. van Lelieveld, and R. M. Bruun, *Acta Chem. Scand.*, **1999**, *53*, 230.
905. C. Plisson and J. Chenault, *Heterocycles*, **1999**, *51*, 2627.
906. J. E. Dickeson, I. F. Eckhard, R. Fielden, and L. A. Summers, *J. Chem. Soc., Perkin Trans. 1*, **1973**, 2885.
907. T. Naito, S. Aburaki, H. Kamachi, Y. Narita, J. Okumura, and H. Kawaguchi, *J. Antibiot*, **1986**, *39*, 1092.
908. S. Cang, S. Ohta, H. Chiba, O. Johdo, H. Nomura, Y. Nagamatsu, and A. Yoshimoto, *J. Antibiot.*, **2001**, *54*, 304.
909. M. Balogh, I. Hermecz, I. Szilágyi, and Z. Meśzáros, *Heterocycles*, **1983**, *20*, 1083.
910. M. Sundbom, *Acta. Chem. Scand.*, **1971**, *25*, 487.
911. H. Yamada, H. Tobiki, K. Jimpo, K. Gooda, Y. Takeuchi, S. Ueda, T. Komatsu, T. Okuda, H. Noguchi, K. Irie, and T. Nakagome, *J. Antibiot.*, **1983**, *36*, 532.
912. A. Zografos, C. Mitsos, and O. Igglesi-Markopoulou, *Heterocycles*, **1999**, *51*, 1609.
913. J. Teixidó, J. I. Borrel, B. Serra, J. L. Metsllana, C. Colominas, F. Carrión, R. Pascual, J. L. Falcó, and X. Batllori, *Heterocycles*, **1999**, *50*, 739.
914. H. Irikawa, N. Adachi, and H. Muraoka, *Heterocycles*, **1998**, *48*, 1415.
915. Q. Li, L. A. Mitscher, and L. L. Shen, *Med. Res. Rev.*, **2000**, *20*, 231.
916. T. Kametani, M. Takeshita, M. Ihara, and K. Fukumoto, *Heterocycles*, **1975**, *3*, 627.
917. H. Yamanaka, T. Shiraishi, and T. Sakamoto, *Heterocycles*, **1975**, *3*, 1069.

918. D. M. W. van den Ham, *J. Fluorine Chem.*, **1975**, *5*, 537.
919. H. Ripperger, *Phytochemistry*, **1978**, *17*, 1069.
920. L. S. Vasilley, F. E. Surzhikov, O. G. Azarevich, V. S. Bogdanov, and V. A. Dorokhov, *Izv. Akad. Nauk, Ser. Khim.*, **1994**, 1510.
921. V. N. Nesterov, V. D. Dyachenko, Y. A. Sharanin, and Y. T. Struchkov, *Izv. Akad. Nauk. Ser. Khim.*, **1996**, 437.
922. Y. Nagawa, M. Ono, M. Hirota, Y. Hamada, and I. Takeuchi, *Bull. Chem. Soc. Jpn.*, **1976**, *49*, 1322.
923. M. Hirota, H. Masuda, Y. Hamada, and I. Takeuchi, *Bull. Chem. Soc. Jpn.*, **1974**, *47*, 2083.
924. B.-M. Swahn, A. Claesson, B. Pelcman, Y. Bsidski, H. Molin, M. P. Sandberg, and O.-G. Berge, *Bioorg. Med. Chem. Lett.*, **1996**, *6*, 1635.
925. C.-Y. Hong, Y. K. Kim, Y. H. Lee, and J. H. Kwak, *Bioorg. Med. Chem. Lett.*, **1998**, *8*, 221.
926. I. Hirao, Y. Kato, Y. Fukano, and S. Yanai, *Bull. Chem. Soc. Jpn.*, **1973**, *46*, 1826.
927. S. Hanessian, M. J. Rozema, G. B. Reddy, and J. F. Braganza, *Bioorg. Med. Chem. Lett.*, **1995**, *5*, 2535.
928. B. Singh, P. O. Pennock, G. Y. Lesher, E. R. Bacon, and D. F. Page, *Heterocycles*, **1993**, *36*, 133.
929. O. Meth-Cohn, *Heterocycles*, **1993**, *35*, 539.
930. P. J. Victory, J. Teixidó, J. I. Borrell, and N. Busquets, *Heterocycles*, **1993**, *36*, 1.
931. A.-L. Skaltsounis, F. Tillequin, M. Koch, J. Pusset, and R. Chauvière, *Heterocycles*, **1987**, *26*, 599.
932. A. E. Zbarskii, L. D. Smirnov, V. F. Zakharov, V. P. Zvolinskii, and K. M. Dyumaev, *Izv. Akad. Nauk SSSR, Ser. Khim.*, **1979**, 2350.
933. E. S. Kagan, V. I. Mikhailov, V. D. Sholle, B. V. Rozynov, and E. G. Rozantsev, *Izv. Akad. Nauk SSSR, Ser. Khim.*, **1977**, 2822.
934. M.-J. Bermejo, J.-I. Ruiz, and X. Solans, *J. Organomet. Chem.*, **1986**, *304*, 207.
935. M.-J. Bermejo, B. Martinez, and J. Vinaixa, *J. Organomet. Chem.*, **1993**, *463*, 143.
936. H. Takahata, T. Suzuki, and T. Yamazaki, *Heterocycles*, **1986**, *24*, 1247.
937. H. B. Davis, R. M. Sheets, W. W. Paudler, and G. L. Gard, *Heterocycles*, **1984**, *22*, 2029.
938. E. M. Kaiser, W. R. Thomas, T. E. Synos, J. R. McClure, T. S. Mansour, J. R. Garlich, and J. E. Chastain, *J. Organomet. Chem.*, **1981**, *213*, 405.
939. T. Takata, *Bull. Chem. Soc. Jpn.*, **1962**, *35*, 1438.
940. T. A. Koizumi, T. Tomon, and K. Tanaka, *Bull. Chem. Soc. Jpn.*, **2003**, *76*, 1969.
941. J. E. Macor, W. Kuipers, and R. J. Lachicotte, *Chem. Commun. (Cambridge)*, **1998**, 983.
942. K. Nakamura, T. Hasegawa, Y. Fukunaga, and K. Ienaga, *J. Chem. Soc., Chem. Commun.*, **1992**, 992.
943. C. S. Giam and D. Ambrozich, *J. Chem. Soc., Chem. Commun.*, **1984**, 265.
944. Y. Ikeura, T. Ishimaru, T. Doi, M. Kawada, A. Fujishima, and H. Natsugari, *Chem. Commun. (Cambridge)*, **1998**, 2141.
945. A. R. Andresen and E. B. Pedersen, *Heterocycles*, **1982**, *19*, 1467.
946. A. Geies, A. A. Abdel-Hafez, J. C. Lancelot, and H. S. El-Kashef, *Bull. Chem. Soc. Jpn.*, **1993**, *66*, 3716.
947. Y. Zhu, X. Gan, N. Tang, and M. Tan, *Bull. Chem. Soc. Jpn.*, **1989**, *62*, 1279.
948. H. Suezawa, M. Hirota, Y. Shibata, I. Takeuchi, and Y. Hamada, *Bull. Chem. Soc. Jpn.*, **1986**, *59*, 2362.
949. J. J. Artus, J.-J. Bonet, and A. E. Peña, *J. Chem. Soc., Chem. Commun.*, **1973**, 579.
950. M. Yamato, K. Sato, K. Hashigaki, and M. Ninomiya, *Heterocycles*, **1982**, *19*, 1263.
951. H. A. Daboun, S. E. Abdou, and M. M. Khader, *Heterocycles*, **1982**, *19*, 1925.
952. T. Naito and I. Ninomiya, *Heterocycles*, **1981**, *15*, 735.
953. K. Abe, M. Hirota, I. Takeuchi, and Y. Hamada, *Bull. Chem. Soc. Jpn.*, **1977**, *50*, 2028.

References

954. R. A. Bowie, *J. Chem. Soc. D. Chem. Commun.*, **1970**, 565.
955. A. Clearfield, P. Singh, and I. Bernal, *J. Chem. Soc. D. Chem. Commun.*, **1970**, 389.
956. S. Oae and Y. Kadoma, *Can. J. Chem.*, **1986**, *64*, 1184.
957. J. B. Brandon, M. Collins, and K. R. Dixon, *Can. J. Chem.*, **1978**, *56*, 950.
958. G. W. Bushnell, K. R. Dixon, and M. A. Khan, *Can. J. Chem.*, **1978**, *56*, 450.
959. S. V. Roman, V. D. Dyachenko, and V. P. Livinov, *Izv. Akad. Nauk. Ser. Khim.*, **2001**, 118.
960. V. P. Krivopalov, V. P. Sedova, and O. P. Shkurko, *Izv. Akad. Nauk, Ser. Khim.*, **2003**, 2307.
961. R. A. Y. Jones and N. Wagstaff, *J. Chem. Soc., Chem. Commun.*, **1969**, 56.
962. T. Yamazaki, H. Takahata, T. Matsuura, and R. N. Castle, *J. Heterocycl. Chem.*, **1979**, *16*, 527.
963. J. I. Matsumoto, M. Shinsaku, and S. Minami, *J. Heterocycl. Chem.*, **1979**, *16*, 1169.
964. W. W. Paudler and T. J. Kress, *J. Heterocycl. Chem.*, **1967**, *4*, 284.
965. D. T. W. Chu, A. K. Claiborne, J. J. Clement, and J. J. Plattner, *Can. J. Chem.*, **1992**, *70*, 1328.
966. T. Kametani, K. Kigasawa, M. Hiiragi, K. Wakisaka, S. Haga, H. Sugi, K. Tanigawa, Y. Suzuki, K. Fukawa, O. Irino, O. Saita, and S. Yamabe, *Heterocycles*, **1981**, *16*, 1205.
967. C. Temple, A. G. Laseter, and J. A. Montgomery, *J. Heterocycl. Chem.*, **1970**, *7*, 1219.
968. T. J. Kress and W. W. Paudler, *J. Chem. Soc., Chem. Commun.*, **1967**, 3.
969. A. H. Gawer and B. P. Dailey, *J. Chem. Phys.*, **1965**, *42*, 2658.
970. J. S. Vincent, *J. Chem. Phys.*, **1971**, *54*, 2237.
971. G. R. Newkome and S. J. Garbis, *J. Heterocycl. Chem.*, **1978**, *15*, 685.
972. S. Carboni, A. Da Settimo, P. L. Ferrarini, and P. L. Ciantelli, *J. Heterocycl. Chem.*, **1970**, *7*, 1037.
973. J. C. M. Henning, *J. Chem. Phys.*, **1966**, *44*, 2139.
974. E. M. Hawes and D. K. J. Gorecki, *J. Heterocycl. Chem.*, **1972**, *9*, 703.
975. W. W. Paudler and T. J. Kress, *J. Heterocycl. Chem.*, **1967**, *4*, 547.
976. F. Fülöp, I. Hermecz, Z. Mészáros, G. Dombi, and G. Bernáth, *J. Heterocycl. Chem.*, **1979**, *16*, 457.
977. W. T. Flowers, R. N. Haszeldine, C. R. Owen, and A. Thomas, *J. Chem. Soc., Chem. Commun.*, **1974**, 134.
978. S. B. Brown and M. J. S. Dewar, *J. Chem. Soc., Chem. Commun.*, **1977**, 87.
979. A. Da Settimo, G. Primofiore, O. Livi, P. L. Ferrarini, and S. Spinelli, *J. Heterocycl. Chem.*, **1979**, *16*, 169.
980. R. A. Henry and P. R. Hammond, *J. Heterocycl. Chem.*, **1977**, *14*, 1109.
981. S. Patai, *Glossary of Organic Chemistry*, Wiley, New York, 1962, p. 61.
982. S. Löfås and P. Ahlberg, *J. Chem. Soc., Chem. Commun.*, **1981**, 998.
983. R. P. Thummel and D. K. Kohli, *J. Heterocycl. Chem.*, **1977**, *14*, 685.
984. C. Mealli and F. Zanobini, *J. Chem. Soc., Chem. Commun.*, **1982**, 97.
985. H. C. van der Plas and M. Woźniak, *J. Heterocycl. Chem.*, **1976**, *13*, 961.
986. D. J. Sheffield and K. R. H. Wooldridge, *J. Chem. Soc., Perkin Trans. 1* **1972**, 2506.
987. V. A. Dorokhov, A. Y. Yagodkin, V. S. Bogdanov, and S. V. Baranin, *Izv. Akad. Nauk, Ser. Khim.*, **1997**, 1079.
988. R. Benkrief, Y. Ranarivelo, A.-L. Skaltsounis, F. Tillequin, M. Koch, J. Pusset, and T. Sévenet, *Phytochemistry*, **1998**, *47*, 825.
989. T. Kametani, M. Takeshita, and M. Ihara, *Heterocycles*, **1976**, *4*, 247.
990. M. Shibuya, Y. Jinbo, and S. Kubota, *Heterocycles*, **1983**, *20*, 1531.
991. H. B. Davis, R. M. Sheets, J. M. Brannfors, W. W. Paudler, and G. L. Gard, *Heterocycles*, **1983**, *20*, 2029.
992. V. J. Sigova and M. E. Konshin, *Zh. Obshch. Khim.*, **1984**, *54*, 2083.
993. J. W. Bunting and W. G. Meathrel, *Can. J. Chem.*, **1972**, *50*, 917.

994. G. Koitz, B. Thierrichter, and H. Junek, *Heterocycles*, **1983**, *20*, 2405.
995. V. P. Lee and D. F. R. Gilson, *Can. J. Chem.*, **1976**, *54*, 2783.
996. H. Yamamoto, H. Kawamoto, S. Morosawa, and A. Yokoo, *Heterocycles*, **1978**, *11*, 267.
997. Z. N. Timofeeva, A. V. Lizogub, S. V. Nekrasov, and A. V. El'tsov, *Zh. Obshch. Khim.*, **1973**, *43*, 1187.
998. M. Hirota, H. Masuda, Y. Hamada, and I. Takeuchi, *Bull. Chem. Soc. Jpn*, **1979**, *52*, 1408.
999. P. J. Villiani, E. A. Wefer, T. A. Mann, J. Mayer, L. Peer, and A. S. Levy, *J. Heterocycl. Chem.*, **1972**, *9*, 1203.
1000. H. E. Baumgarten, H. C.-F. Su, and R. P. Barkley, *J. Heterocycl. Chem.*, **1966**, *3*, 357.
1001. H. Inoue, I. Sonoda, and E. Imoto, *Bull. Chem. Soc. Jpn.*, **1979**, *52*, 1237.
1002. J. W. Bunting and W. G. Meathrel, *Can. J. Chem.*, **1994**, *52*, 962.
1003. Y. Tamura, Y. Miki, J.-I. Minamikawa, and M. Ikeda, *J. Heterocycl. Chem.*, **1974**, *11*, 675.
1004. T. Tsuchiya, M. Enkaku, and H. Sawanishi, *Heterocycles*, **1978**, *9*, 621.
1005. P. G. Gassman and C. T. Huang, *J. Chem. Soc., Chem. Commun.*, **1974**, 685.
1006. I. Hermecz and Z. Mészáros, *Heterocycles*, **1979**, *12*, 1407.
1007. A. Taurins and R. T. Li, *Can. J. Chem.*, **1974**, *52*, 843.
1008. R. A. van Dahm, D. J. Pokorny, and W. W. Paudler, *J. Heterocycl. Chem.*, **1972**, *9*, 1001.
1009. E. Eichler, C. S. Rooney, and H. W. R. Williams, *J. Heterocycl. Chem.*, **1976**, *13*, 43.
1010. F. Vargas, C. Rivas, and R. Machado, *J. Photochem. Photobiol. B*, **1991**, *11*, 81.
1011. Y. Kobayashi, I. Kumadaki, H. Sato, Y. Sekine, and T. Hara, *Chem. Pharm. Bull.*, **1974**, *22*, 2097.
1012. S. Nishigaki, M. Ichiba, S. Fukazawa, M. Kanehori, K. Shinomura, F. Yoneda, and K. Senga, *Chem. Pharm. Bull.*, **1975**, *23*, 3170.
1013. T. Higashino, K. Suzuki, and E. Hayashi, *Chem. Pharm. Bull.*, **1975**, *23*, 2939.
1014. T. Kato and T. Sakamoto, *Chem. Pharm. Bull.*, **1975**, *23*, 2629.
1015. J. Beaudin, D. E. Bourassa, P. Bowles, M. J. Castaldi, R. Clay, M. A. Couturier, G. Karrick, T. W. Makowski, R. E. McDermott, C. L. Meltz, J. E. Phillips, J. A. Ragan, D. H. B. Ripin, R. A. Singer, J. L. Tucker, and L. Wei, *Org. Process Res. Devel.*, **2003**, *7*, 873.
1016. S. F. Mason, *J. Chem. Soc.*, **1962**, 493.
1017. T. K. Adler and A. Albert, *J. Chem. Soc.*, **1960**, 1794.
1018. T. L. Stuk, B. K. Assink, R. C. Bates, D. T. Erdman, V. Fedij, S. M. Jennings, J. A. Lassig, R. J. Smith, and T. L. Smith, *Org. Process Res. Devel.*, **2003**, *7*, 851.
1019. F. J. C. Rossotti and H. S. Rossotti, *J. Chem. Soc.*, **1958**, 1304.
1020. S. F. Mason, *J. Chem. Soc.*, **1958**, 674.
1021. B. Olenik, R. Boese, and R. Sustmann, *Cryst. Growth Design*, **2003**, *3*, 175.
1022. R. F. C. Brown, V. M. Clark, M. Lamchen, and A. R. Todd, *J. Chem. Soc.*, **1959**, 2116.
1023. V. Oakes and H. N. Rydon, *J. Chem. Soc.*, **1958**, 204.
1024. B. Olenik, T. Smolka, R. Boese, and R. Sustmann, *Cryst. Growth Design*, **2003**, *3*, 183.
1025. A. B. Deyanov, M. E. Konshin, and Z. N. Semenova, *Khim. Geterotsikl. Soedin.*, **2004**, 1809.
1026. S. F. Mason, *J. Chem. Soc.*, **1957**, 5010.
1027. A. Albert and J. N. Phillips, *J. Chem. Soc.*, **1956**, 1294.
1028. T. Tsuchiya, M. Enkaku, T. Kurita, and H. Sawanishi, *Chem. Pharm. Bull.*, **1979**, *27*, 2183.
1029. N. Suzuki, Y. Tanaka, and R. Dohmori, *Chem. Pharm. Bull.*, **1980**, *28*, 235.
1030. T. Higashino, K. Suzuki, and E. Hayashi, *Chem. Pharm. Bull.*, **1978**, *26*, 3242.
1031. N. Suzuki and R. Dohmori, *Chem. Pharm. Bull.*, **1979**, *27*, 410.
1032. H. Hirano, M. Takamatsu, K. Sugiyama, and T. Kurihara, *Chem. Pharm. Bull.*, **1979**, *27*, 374.
1033. I. Takeuchi and Y. Hamada, *Chem. Pharm. Bull.*, **1976**, *24*, 1813.

1034. N. P. Buu-Hoi, R. Royer, and N. Hubert-Habart, *J. Chem. Soc.*, **1956**, 2048.
1035. S. F. Mason, *J. Chem. Soc.*, **1957**, 4874.
1036. A. A. Goldberg, R. S. Theobald, and W. Williamson, *J. Chem. Soc.*, **1954**, 2357.
1037. E. P. Hart, *J. Chem. Soc.*, **1956**, 212.
1038. A. Arany, O. Meth-Cohn, I. Berhés, and M. Nyerges, *Org. Biomol. Chem.*, **2003**, *1*, 2164.
1039. E. Eichler, C. S. Rooney, and H. W. R. Williams, *J. Heterocycl. Chem.*, **1976**, *13*, 841.
1040. A. Albert and A. Hampton, *J. Chem. Soc.*, **1954**, 505.
1041. K. Kubo, K. Ukawa, S. Kuzuna, and A. Nohara, *Chem. Pharm. Bull.*, **1986**, *34*, 1108.
1042. T. Sakamoto, N. Miura, Y. Kondo, and H. Yamanaka, *Chem. Pharm. Bull.*, **1986**, *34*, 2018.
1043. K. A. Allen, J. Cymerman-Craig, and A. A. Diamantis, *J. Chem. Soc.*, **1954**, 234.
1044. T. Sakamoto, Y. Kondo, and H. Yamanaka, *Chem. Pharm. Bull.*, **1985**, *33*, 4764.
1045. H. Yanagisawa, H. Nakao, and A. Ando, *Chem. Pharm. Bull.*, **1973**, *21*, 1080.
1046. M. Woźniak and H. C. van der Plas, *J. Heterocycl. Chem.*, **1978**, *15*, 431.
1047. E. P. Hart, *J. Chem. Soc.*, **1954**, 1879.
1048. A. Albert and A. Hampton, *J. Chem. Soc.*, **1952**, 4985.
1049. S. Mishio, T. Hirose, A. Minamida, J.-I. Matsumoto, and S. Mijami, *Chem. Pharm. Bull.*, **1986**, *33*, 4402.
1050. K. Tabei, H. Ito, and T. Takada, *Heterocycles*, **1980**, *14*, 1779.
1051. V. Petrov and B. Sturgeon, *J. Chem. Soc.*, **1949**, 1157.
1052. V. Petrov, E. L. Rewald, and B. Sturgeon, *J. Chem. Soc.*, **1947**, 1407.
1053. Y. Kobayashi, I. Kumadaki, and H. Sato, *Chem. Pharm. Bull.*, **1974**, *22*, 2812.
1054. T. Higashino and E. Hayashi, *Chem. Pharm. Bull.*, **1973**, *21*, 2643.
1055. T. Sakamoto, Y. Kondo, and H. Yamanaka, *Chem. Pharm. Bull.*, **1985**, *33*, 626.
1056. G. R. Clemo and G. A. Swan, *J. Chem. Soc.*, **1945**, 603.
1057. P. D. Sullivan and W. W. Paudler, *Can. J. Chem.*, **1973**, *51*, 4095.
1058. A. Shiozawa, Y.-I. Ichikawa, C. Komura, S. Kurashige, H. Miyazaki, H. Yamanaka, and T. Sakamoto, *Chem. Pharm. Bull.*, **1984**, *32*, 2522.
1059. C. Tapiéro and J. L. Imbach, *J. Heterocycl. Chem.*, **1975**, *12*, 439.
1060. I. Hayakawa, N. Suzuki, K. Suzuki, and Y. Tanaka, *Chem. Pharm. Bull.*, **1984**, *32*, 4914.
1061. D. J. Pokorny and W. W. Paudler, *Can. J. Chem.*, **1973**, *51*, 476.
1062. A. Shiozawa, Y.-I. Ichikawa, C. Komuro, M. Ishikawa, Y. Furuta, S. Kurashige, H. Miyazaki, H. Yamanaka, and T. Sakamoto, *Chem. Pharm. Bull.*, **1984**, *32*, 3981.
1063. Y. Kobayashi, I. Kumadaki, and H. Hamana, *Chem. Pharm. Bull.*, **1976**, *24*, 1704.
1064. S. Nishigaki, N. Mizushima, and K. Senga, *Chem. Pharm. Bull.*, **1976**, *24*, 1658.
1065. J. A. van Allan, G. A. Reynolds, D. P. Maier, and S. C. Chang, *J. Heterocycl. Chem.*, **1972**, *9*, 1229.
1066. S. Nishigaki, N. Mizushima, and K. Senga, *Chem. Pharm. Bull.*, **1977**, *25*, 349.
1067. W. W. Paudler and T. J. Kress, *J. Heterocycl. Chem.*, **1965**, *2*, 393.
1068. E. M. Hawes and D. K. J. Gorecki, *J. Heterocycl. Chem.*, **1974**, *11*, 151.
1069. D. J. Pokorny and W. W. Paudler, *J. Heterocycl. Chem.*, **1972**, *9*, 1151.
1070. R. M. Mohareb and S. M. Fahmy, *Z. Naturforsch., B*, **1985**, *40*, 1537.
1071. J. F. Harper and D. G. Wibberley, *J. Chem. Soc., C*, **1971**, 2985.
1072. D. E. Ames and W. D. Dodds, *J. Chem. Soc., Perkin Trans. 1*, **1972**, 705.
1073. E. B. Mullock, R. Searby, and H. Suschitzky, *J. Chem. Soc., C*, **1970**, 829.
1074. J. D. Atkinson and M. C. Johnson, *J. Chem. Soc., C*, **1968**, 1252.

1075. R. M. Mohareb and S. M. Fahmy, *Z. Naturforsch.*, *B*, **1986**, *41*, 105.
1076. I. Hermecz, Z. Mészáros, L. Vasvári-Debreczy, Á. Horváth, G. Horváth, and M. Pongor-Csákvári, *J. Chem. Soc., Perkin Trans 1*, **1977**, 789.
1077. D. G. Wibberley, *J. Chem. Soc.*, **1962**, 4529.
1078. P.-L. Compagnon, G. Gasquez, and T. Kimny, *Bull. Soc. Chim. Belg.*, **1986**, *95*, 49.
1079. D. M. W. van den Ham, J. J. du Sart, and D. van der Meer, *Mol. Phys.*, **1971**, *21*, 989.
1080. S. Löfås and P. Ahlberg, *J. Heterocycl. Chem.*, **1984**, *21*, 583.
1081. P. L. Ferrarini, C. Mori, G. Biagi, O. Livi, and I. Tonetti, *J. Heterocycl. Chem.*, **1984**, *21*, 417.
1082. N. Katagiri, T. Kato, and R. Niwa, *J. Heterocycl. Chem.*, **1984**, *21*, 407.
1083. P. Cavalieri d'Oro, R. Danieli, G. Maccagnani, G. P. Pedulli, and P. Palmieri, *Mol. Phys.*, **1971**, *20*, 365.
1084. A. N. Sargin, *j. Heterocycl. Chem.*, **1983**, *20*, 1749.
1085. P. L. Ferrarini, C. Mori, O. Livi, G. Biagi, and A. M. Marini, *J. Heterocycl. Chem.*, **1983**, *20*, 1053.
1086. W. R. Carper and J. Stengl, *Mol. Phys.*, **1969**, *16*, 627.
1087. K. Bláha, A. M. Farag, D. van der Helm, M. B. Hossain, M. Buděšinský, P. Maloń, J. Smoliková, and M. Tichý. *Collect. Czech. Chem. Commun*, **1984**, *49*, 712.
1088. S. Löfås and P. Ahlberg, *J. Chem. Soc., Perkin Trans 2*, **1986**, 1223.
1089. J. R. L. Smith and L. A. V. Mead, *J. Chem. Soc., Perkin Trans. 2*, **1976**, 1172.
1090. C. J. Chandler, L. W. Deady, and J. A. Reiss, *J. Heterocycl. Chem.*, **1986**, *23*, 1327.
1091. J. P. Sanchez and J. W. Bogowski, *J. Heterocycl. Chem.*, **1987**, *24*, 215.
1092. J. R. L. Smith and D. Masheder, *J. Chem. Soc., Perkin Trans. 2*, **1977**, 1732.
1093. M. Woźniak, D. J. Buurman, and H. C. van der Plas, *J. Heterocycl. Chem.*, **1985**, *22*, 765.
1094. A. D. Jordan, G. Fischer, and I. G. Ross, *J. Mol. Spectrosc.*, **1981**, *87*, 345.
1095. D. Bhattacharjee and F. D. Popp, *J. Heterocycl. Chem.*, **1980**, *17*, 1211.
1096. M. H. Elnagdi, F. A. M. A. Aal, E. A. A. Hafez, and Y. M. Yassin, *Z. Naturforsch., B*, **1989**, *44*, 683.
1097. R. A. Bowie and B. Wright, *J. Chem. Soc., Perkin Trans. 1*, **1972**, 1109.
1098. R. A. Bowie, M. J. C. Mullan, and J. F. Unsworth, *J. Chem. Soc., Perkin Trans. 1*, **1972**, 1106.
1099. J. F. Harper and D. G. Wibberley, *J. Chem. Soc., C*, **1971**, 2991.
1100. E. M. Hawes and D. G. Wibberley, *J. Chem. Soc., C*, **1967**, 1564.
1101. G. Leroy, F. van Remoortere, and C. Aussems, *Bull. Soc. Chim. Belg.*, **1968**, *77*, 191.
1102. S. Carboni, A. Da Settimo, D. Bertini, F. L. Ferrarini, O. Livi, and I. Tonetti, *J. Heterocycl. Chem.*, **1975**, *12*, 743.
1103. A. Decormeille, F. Guignant, G. Queguiner, and P. Pastour, *J. Heterocycl. Chem.*, **1976**, *13*, 387.
1104. J. Jilik, M. Rajšner, V. Valenta, M. Borovička, J. Holubek, M. Ryska, E. Svátek, J. Metyš, and M. Protiva, *Collect. Czech. Chem. Commun.*, **1990**, *55*, 1828.
1105. M. Balogh, I. Hermecz, K. Simon, and L. Pusztay, *J. Heterocycl. Chem.*, **1989**, *26*, 1755.
1106. A. H. M. Al-Shaar, D. W. Gilmour, D. J. Lythgoe, I. McClenaghan, and C. A. Ramsden, *J. Heterocycl. Chem.*, **1989**, *26*, 1819.
1107. J. S. Kiely, S. Huang, and L. E. Lesheski, *J. Heterocycl. Chem.*, **1989**, *26*, 1675.
1108. G. B. Barlin and S. J. Ireland, *Aust. J. Chem.*, **1988**, *41*, 1727.
1109. G. B. Barlin and C. Jiravinyu, *Aust. J. Chem.*, **1990**, *43*, 1367.
1110. J. P. Chupp and J. M. Molyneaux, *J. Heterocycl. Chem.*, **1989**, *26*, 645.
1111. Y. Combret, J. J. Torche, P. Binay, G. Dupas, J. Bourguignon, and G. Queguiner, *Chem. Lett.*, **1991**, 125.
1112. E. M. Hawes and D. G. Wibberley, *J. Chem. Soc., C*, **1966**, 315.
1113. P. J. Chappell, G. Fischer, J. R. Reimers, and I. G. Ross, *J. Mol. Spectrosc.*, **1981**, *87*, 316.

1114. J. S. Vincent, *Chem. Phys. Lett.*, **1971**, *8*, 37.
1115. H. J. W. van den Haak, H. C. van der Plas, and B. van Veldhuizen, *J. Heterocycl. Chem.*, **1981**, *18*, 1349.
1116. F. I. Carrol, B. D. Berrang, and C. P. Linn, *J. Heterocycl. Chem.*, **1981**, *18*, 941.
1117. A. Da Settimo, G. Biagi, G. Primofiore, P. L. Ferrarini, O. Livi, and A. M. Marini, *J. Heterocycl. Chem.*, **1980**, *17*, 1225.
1118. N. Saito, T. Kanbara, T. Kushida, K. Kubota, and T. Yamamoto, *Chem. Lett.*, **1993**, 1775.
1119. G. Fischer and I. G. Ross, *J. Mol. Spectrosc.*, **1981**, *87*, 331.
1120. J. Matsumoto, T. Miyamoto, H. Minamida, Y. Nishimura, H. Egawa, and H. Nishimura, *J. Heterocycl. Chem.*, **1984**, *21*, 673.
1121. E. Eichler, C. S. Rooney, and H. W. K. Williams, *J. Heterocycl. Chem.*, **1976**, *13*, 41.
1122. E. M. Hawes and H. L. Davis, *J. Heterocycl. Chem.*, **1973**, *10*, 39.
1123. Y. Tamura, Y. Miki, and H. Ikeda, *J. Heterocycl. Chem.*, **1975**, *12*, 119.
1124. J. T. Carrano and S. C. Wait, *J. Mol. Spectrosc.*, **1973**, *46*, 401.
1125. T. Mizukawa, K. Tsuge, H. Nakjima, and K. Tanaka, *Angew. Chem.*, **1999**, *111*, 373.
1126. S. C. Wait and J. W. Wesley, *J. Mol. Spectrosc.*, **1966**, *19*, 25.
1127. G. M. Coppola, J. D. Fraser, G. E. Hardtmann, and M. J. Shapiro, *J. Heterocycl. Chem.*, **1985**, *22*, 193.
1128. M. Woźiak, H. C. van der Plas, M. Yomula, and A. van Veldhuizen, *J. Heterocycl. Chem.*, **1985**, *22*, 761.
1129. S. Nishigaki, N. Mizushima, H. Kanezawa, M. Ichiba, and K. Senga, *J. Heterocycl. Chem.*, **1985**, *22*, 1029.
1130. S. Rádl and P. Hradil, *Collect. Czech. Chem. Commun.*, **1991**, *56*, 2420.
1131. G. R. Rao, K. Mogilaiah, and B. Sreenivasulu, *Collect. Czech. Chem. Commun.*, **1989**, *54*, 1716.
1132. S. Mataka, K. Takahashi, and M. Tashiro, *J. Heterocycl. Chem.*, **1983**, *20*, 971.
1133. C. J. Chandler, L. W. Deady, J. A. Reiss, and V. Tzimos, *J. Heterocycl. Chem.*, **1982**, *19*, 1017.
1134. M. T. Chary, K. Mogilaiah, and B. Sreenivasulu, *Collect. Czech. Chem. Commun.*, **1988**, *53*, 1543.
1135. H. Singh and G. Y. Lesher, *J. Heterocycl. Chem.*, **1983**, *20*, 491.
1136. P. L. Ferrarini, C. Mori, G. Primofiore, and L. Calzolari, *J. Heterocycl. Chem.*, **1990**, *27*, 881.
1137. L. W. Deady, *Aust. J. Chem.*, **1981**, *34*, 163.
1138. K. Mogilaiah, K. R. Reddy, G. R. Rao, and B. Sreenivasulu, *Collect. Czech. Chem. Commun.*, **1988**, *53*, 1539.
1139. S. E. Zayed, F. I. A. Elmaged, S. A. Metwally, and M. H. Einagdi, *Collect. Czech. Chem. Commun.*, **1991**, *56*, 2175.
1140. W. W. Paudler and T. J. Kress, *Chem. Ind. (London)*, **1966**, 1557.
1141. E. M. Beccalli, E. Erba, M. L. Gelmi, and D. Pocar, *J. Chem. Soc., Perkin Trans. 1*, **1996**, 1359.
1142. A. R. Katritzky, S. Rachwal, and B. Rachwal, *J. Chem. Soc., Perlin Trans. 1*, **1987**, 799.
1143. H. Nakamura, J. Kobayashi, and Y. Ohizumi, *J. Chem. Soc., Perkin Trans. 1*, **1987**, 173.
1144. I. F. Eckhard, R. Fielden, and L. A. Summers, *Chem. Ind. (London)*, **1973**, 275.
1145. R. E. Busby, S. M. Hussain, M. Iqbal, M. A. Khan, J. Parrick, and C. J. G. Shaw, *J. Chem. Soc., Perkin Trans. 1*, **1979**, 2782.
1146. A. Couture, P. Grandclaudon, C. Simion, and P. Woisel, *J. Chem. Soc., Perkin Trans. 1*, **1995**, 2643.
1147. M. Balogh, I. Hermecz, G. Náray-Szabó, K. Simon, and Z. Mészáros, *J. Chem. Soc., Perkin Trans. 1*, **1986**, 753.
1148. E. Wenkert, E. B. Spitzner, and R. L. Webb, *Aust. J. Chem.*, **1972**, *25*, 433.
1149. D. L. Kepert and D. Taylor, *Aust. J. Chem.*, **1974**, *27*, 1199.

1150. A. R. Forrester, H. Irikawa, R. H. Thomson, S. O. Woo, and T. J. King, *J. Chem. Soc., Perkin Trans. 1*, **1981**, 1712.
1151. G. B. Barlin and W.-L. Tan, *Aust. J. Chem.*, **1984**, *37*, 2469.
1152. G. B. Barlin, T. M. T. Nguyen, B. Kotecka, and K. H. Rieckmann, *Aust. J. Chem.*, **1992**, *45*, 1651.
1153. M. Ata, O. Sedoyama, and H. Yamaguchi, *Chem. Phys. Lett.*, **1982**, *90*, 133.
1154. R. L. Coppel, L. W. Deady, P. M. Loria, D. C. Olden, and J. D. Turnidge, *Aust. J. Chem.*, **1996**, *49*, 255.
1155. A. Tiripicchio, M. Tiripicchio-Camellini, B. Usón, L. A. Oro, M. A. Ciriano, and F. Viguri, *J. Chem. Soc., Dalton Trans*, **1984**, 125.
1156. N. G. Connelly, A. C. Loyns, M. A. Ciriano, M. J. Fernandez, L. A. Oro, and B. E. Villarroya, *J. Chem. Soc., Dalton Trans.*, **1989**, 689.
1157. G. Tóth, A. Kovács, M. Balogh, and I. Hermecz, *J. Heterocycl. Chem.*, **1991**, *28*, 497.
1158. E. Laborde, J. S. Kiely, L. E. Lesheski, and M. C. Schroeder, *J. Heterocycl. Chem.*, **1991**, *28*, 191.
1159. J. S. Kiely, *J. Heterocycl. Chem.*, **1991**, *28*, 541.
1160. D. Sternbach, M. Shibuya, F. Jaisli, M. Bonetti, and A. Eschenmoser, *Angew. Chem.*, **1979**, *91*, 670.
1161. G. B. Barlin and C. Jirovinyu, *Aust. J. Chem.*, **1990**, *43*, 1175.
1162. T. Miyamoto, H. Egawa, K. Shibamori, and J. Matsumoto, *J. Heterocycl. Chem.*, **1987**, *24*, 1333.
1163. J. M. Domagala and P. Peterson, *J. Heterocycl. Chem.*, **1989**, *26*, 1147.
1164. M. Grzegozek, M. Woźniak, A. Barański, and H. C. van der Plas, *J. Heterocycl. Chem.*, **1991**, *28*, 1075.
1165. G. B. Barlin and W.-L. Tan, *Aust. J. Chem.*, **1985**, *38*, 905.
1166. B. Singh and G. Y. Lesher, *J. Heterocycl. Chem.*, **1990**, *27*, 2085.
1167. A. J. Bridges and J. P. Sanchez, *J. Heterocycl. Chem.*, **1990**, *27*, 1527.
1168. C. H. Erre and C. Roussel, *Bull. Soc. Chim. Fr.*, **1984**, II-454.
1169. A. V. El'tsov, S. V. Nekrasov, and E. V. Smirnov, *Zh. Org. Khim.*, **1972**, *8*, ; 309.
1170. P. van de Weijer and D. van der Meer, *Org. Magn. Reson.*, **1977**, *9*, 71.
1171. Y. M. Yutilov and N. N. Smolyar, *Zh. Org. Khim.*, **1986**, *22*, 1793.
1172. P. Remuzon, D. Bouzard, P. Di Cesare, C. Dussy, J.-P. Jacquet, and A. Jaegly, *J. Heterocycl. Chem.*, **1992**, *29*, 985.
1173. R. W. Wagner, P. Hochmann, and M. A. El-Bayoumi, *J. Mol. Spectrosc.*, **1975**, *54*, 167.
1174. A. C. Boicelli, R. Danieli, A. Mangini, L. Lunazzi, and G. Placucci, *J. Chem. Soc., Perkin Trans. 2*, **1973**, 1024.
1175. J. P. Sanchez and R. D. Gogliotti, *J. Heterocycl. Chem.*, **1993**, *30*, 855.
1176. P. van de Weijer, H. Thijsse, and D. van der Meer, *Org. Magn. Reson.*, **1976**, *8*, 187.
1177. M. Pätzel, A. Ushmajev, J. Liebscher, V. Granik, S. Grisik, and M. Polievktov, *J. Heterocycl. Chem.*, **1992**, *29*, 1067.
1178. C. H. Erre and C. Roussel, *Bull. Soc. Chim. Fr.*, **1984**, II-449.
1179. G. B. Barlin, S. J. Ireland, C. Jiravinyu, T. M. T. Nguyen, B. Kotecka, and K. H. Rieckmann, *Aust. J. Chem.*, **1993**, *46*, 1695.
1180. M. Woźniak and H. C. van der Plas, *J. Heterocycl. Chem.*, **1986**, *23*, 473.
1181. D. M. W. van den Ham and D. van der Meer, *Chem. Phys. Lett.*, **1972**, *12*, 447.
1182. R. Ziessel, L. Suffert, and M.-T. Youinou, *J. Org. Chem.*, **1996**, *61*, 6535.
1183. R. R. Rastogi, A. Kumar, H. Ila, and H. Junjappa, *J. Chem. Soc., Perkin Trans. 1*, **1978**, 554.
1184. G. B. Barlin, C. Jiravinyu, and J.-H. Yan, *Aust. J. Chem.*, **1991**, *44*, 677.
1185. L. Vasvári-Debreczy, I. Hermecz, Z. Mészáros, P. Dvortsák, and G. Tóth, *J. Chem. Soc., Perkin Trans. 1*, **1980**, 227.

1186. G. B. Barlin and W.-L. Tan, *Aust. J. Chem.*, **1984**, *37*, 1065.
1187. M. N. Palfreyman and K. R. H. Wooldridge, *J. Chem. Soc., Perkin Trans. 1*, **1974**, 57.
1188. J. W. Bunting, *J. Chem. Soc., Perkin Trans. 1*, **1974**, 1833.
1189. M. Witanowski, L. Stefaniak, and G. A. Webb, *Org. Magn. Reson.*, **1981**, *16*, 309.
1190. A. Gueiffier, Y. Blanche, H. Viols, J. P. Chapat, O. Chavignon, J. C. Teulade, G. Dauphin, J. C, Debouzy, and J. L. Chabard, *J. Heterocycl. Chem.*, **1992**, *29*, 283.
1191. T. Kuroda and F. Suzuki, *J. Heterocycl. Chem.*, **1991**, *28*, 2029.
1192. P. van der Weijer, D. M. W. van den Ham, and D. van der Meer, *Org. Magn. Reson.*, **1977**, *9*, 281.
1193. G. Fischer, *Chem. Phys. Lett.*, **1973**, *21*, 305.
1194. G. B. Barlin, S. J. Ireland, T. M. T. Nguyen, B. Kotecka, and K. H. Rieckmann, *Aust. J. Chem.*, **1994**, *47*, 1143.
1195. H. Matsuzaki, I. Takeuchi, Y. Hamada, and K. Hatano, *Chem. Pharm. Bull.*, **2000**, *48*, 755.
1196. K. Takayama, M. Iwata, H. Hisamichi, Y. Okamoto, M. Aoki, and A. Niwa, *Chem. Pharm. Bull.*, **2002**, *50*, 1050.
1197. M. Fujita, K. Chiba, Y. Tominaga, and K. Hino, *Chem. Pharm. Bull.*, **1998**, *46*, 787.
1198. G. Fischer, *J. Mol. Spectrosc.*, **1974**, *49*, 201.
1199. Y. Ikeura, T. Tanaka, Y. Kiyota, S. Morimoto, M. Ogino, T. Ishimaru, I. Kamo, T. Doi, and H. Natsugari, *Chem. Pharm. Bull.*, **1997**, *45*, 1642.
1200. M. Grassi, G. De Munno, F. Nicolo, and S. L. Schiavo, *J. Chem. Soc., Dalton Trans.*, **1992**, 2367.
1201. Y. Todo, J. Nitta, M. Miyajima, Y. Fukuoka, Y. Yamashiro, N. Nishida, I. Saikawa, and H. Narita, *Chem. Pharm. Bull.*, **1994**, *42*, 2063.
1202. I. Takeuchi, K. Masuda, and Y. Hamada, *Chem. Pharm. Bull.*, **1992**, *40*, 2602.
1203. A. I. Kiprianov and G. A. Lezenko, *Zh. Org. Khim.*, **1973**, *9*, 2587.
1204. Y. Nishikawa, T. Shindo, K. Ishii, H. Nakamura, T. Kon, H. Uno, and J.-I. Matsumoto, *Chem. Pharm. Bull.*, **1989**, *37*, 1256.
1205. Y. Nishimura, A. Minamida, and J.-I. Matsumoto, *Chem. Pharm. Bull.*, **1988**, *36*, 1223.
1206. J. M. Epstein, J. C. Dawan, D. L. Kepert, and A. H. White, *J. Chem. Soc., Dalton Trans.*, **1974**, 1949.
1207. H. Yamanaka, T. Shiraishi, and T. Sakamoto, *Chem. Pharm. Bull.*, **1981**, *29*, 1056.
1208. H. Yamanaka, T. Shiraishi, T. Sakamoto, and H. Matsuda, *Chem. Pharm. Bull.*, **1981**, *29*, 1049.
1209. J. C. Dewan, D. L. Kepert, and A. H. White, *J. Chem. Soc., Dalton Trans.*, **1975**, 490.
1210. J. Aritomi and H. Nishimura, *Chem. Pharm. Bull.*, **1981**, *29*, 1193.
1211. N. Suzuki, *Chem. Pharm. Bull.*, **1980**, *28*, 761.
1212. W. W. Paudler, D. J. Pokorny, and S. J. Cornrich, *J. Heterocycl. Chem.*, **1970**, *7*, 291.
1213. R. M. Carr and D. R. Sutherland, *J. Labelled Comp. Radiopharm.*, **1994**, *34*, 961.
1214. W. W. Paudler and T. J. Kress, *J. Heterocycl. Chem.*, **1968**, *5*, 561.
1215. T. Hussenether and R. Troschütz, *J. Heterocycl. Chem.*, **2004**, *41*, 857.
1216. Y.-S. Oh, C.-W. Lee, Y.-H. Chung, S.-J. Yoon, and S.-H. Cho, *J. Heterocycl. Chem.*, **1998**, *35*, 541.
1217. T. Duelfer and D. Gala, *J. Labelled Comp. Radiopharm.*, **1991**, *29*, 651.
1218. A. Gueiffier, H. Viols, Y. Blanche, J. P. Chapat, O. Chavignon, J. C. Teulade, F. Fauvelle, G. Grassy, and G. Dauphin, *J. Heterocycl. Chem.*, **1997**, *34*, 765.
1219. C. Rivalle and E. Bisagni, *J. Heterocycl. Chem.*, **1997**, *34*, 441.
1220. A. Nemazany and N. Haider, *J. Heterocycl. Chem.*, **1997**, *34*, 397.
1221. M. Blanco, M. G. Lorenzo, I. Perillo, and C. B. Shapira, *J. Heterocycl. Chem.*, **1996**, *33*, 361.
1222. A. E. M. Boelrijk, T. X. Neenan, and J. Reeddijk, *J. Chem. Soc., Dalton Trans.*, **1997**, 4561.
1223. Y.-S. Oh and S.-H. Cho, *J. Heterocycl. Chem.*, **1998**, *35*, 17.

1224. P. L. Ferrarini, C. Mori, and S. Belfiore, *J. Heterocycl. Chem.*, **1996**, *33*, 1185.
1225. V. Delieza, A. Detsi, V. Bardakos, and O. Igglessi-Markopoulou, *J. Chem. Soc., Perkin Trans. 1*, **1997**, 1487.
1226. P. L. Ferrarini, C. Mori, M. Badawneh, C. Manera, A. Martinelli, M. Miceli, F. Romagnoli, and G. Saccomanni, *J. Heterocycl. Chem.*, **1997**, *34*, 1501.
1227. E. V. Brown, A. C. Plasz, and S. R. Mitchell, *J. Heterocycl. Chem.*, **1970**, *7*, 661.
1228. L. Stefaniak, J. D. Roberts, M. Witanowski, and G. A. Webb, *Org. Magn. Reson.*, **1984**, *22*, 201.
1229. A. Nemazany and N. Haider, *J. Heterocycl. Chem.*, **1996**, *33*, 1147.
1230. M. Mintert and W. S. Sheldrick, *J. Chem. Soc., Dalton Trans.*, **1995**, 2663.
1231. B. J. Tabner and J. R. Yandle, *J. Chem. Soc., A*, **1968**, 381.
1232. A. Yoshitake, T. Kamada, W. Gomi, and I. Nakatsuka, *L. Labelled Comp. Radiopharm.*, **1981**, *18*, 1415.
1233. A. Dancsó, M. Kajtár-Peredy, and C. Szántay, *J. Heterocycl. Chem.*, **1997**, *34*, 1267.
1234. S. J. Yoon, Y. H. Chung, C. W. Lee, Y. S. Oh, D. R. Choi, N. D. Kim, J. K. Lim, Y. H. Jin, D. K. Lee, and W. Y. Lee, *J. Heterocycl. Chem.*, **1997**, *34*, 1021.
1235. J. Volford, Z. Mészáros, and G. Kovács, *J. Labelled Comp.*, **1973**, *9*, 231.
1236. T. Miyamoto, H. Egawa, and J.-I. Matsumoto, *Chem. Pharm. Bull.*, **1987**, *35*, 2280.
1237. M. Shibuya, Y. Jinbo, and S. Kubota, *Chem. Pharm. Bull.*, **1984**, *32*, 1303.
1238. T. E. Reed and J. G. Hendricker, *J. Coord. Chem.*, **1972**, *2*, 83.
1239. J. A. van Allan, G. A. Reynolds, C. C. Petropoulos, and D. P. Maier, *J. Heterocycl. Chem.*, **1970**, *7*, 495.
1240. E. F. Elslager, S. C. Perricone, and D. F. Worth, *J. Heterocycl. Chem.*, **1970**, *7*, 543.
1241. C. Párkányi and W. C. Herndon, *Phosphorus Sulfur*, **1978**, *4*, 1.
1242. P. Ahlberg, K. Janné, S. Löfås, F. Nettelblad, and L. Swahn, *J. Phys. Org. Chem.*, **1989**, *2*, 429.
1243. N. Ikekawa, *Chem. Pharm. Bull.*, **1958**, *6*, 408.
1244. T. Kato, F. Hamaguchi, and T. Oiwa, *Chem. Pharm. Bull.*, **1956**, *4*, 178.
1245. T. D. Eck, E. L. Wehry, and D. M. Hercules, *J. Inorg. Nucl. Chem.*, **1966**, *28*, 2439.
1246. T. Naito and I. Ninomiya, *Heterocycles*, **1980**, *14*, 959.
1247. O. Wormell and A. R. Lacey, *Chem. Phys.*, **1987**, *118*, 71.
1248. J. Tani, Y. Mushika, and T. Yamaguchi, *Chem. Pharm. Bull.*, **1982**, *30*, 3517.
1249. R. Dohmori, S. Kadoya, Y. Tanaka, I. Takamura, R. Yoshimura, and T. Naito, *Chem. Pharm. Bull.*, **1969**, *17*, 1832.
1250. G. Fischer, I. G. Ross, and M. Puza, *Spectrochim. Acta, Part A*, **1982**, *38*, 603.
1251. N. Ikekawa, *Chem. Pharm. Bull.*, **1958**, *6*, 404.
1252. Y. Sato, T. Iwashige, and T. Miyadera, *Chem. Pharm. Bull.*, **1960**, *8*, 427.
1253. W. Czuba, M. Woźniak, T. Kowalska, and M. Grezegozek, *Org. Mass Spectrom.*, **1976**, *11*, 231.
1254. G. B. Barlin and C. Jiravinyu, *Aust. J. Chem.*, **1991**, *44*, 151.
1255. S. Okuda, *Chem. Pharm. Bull.*, **1957**, *5*, 460.
1256. Y. Nishimura, A. Minamida, and S. L. Matsumoto, *J. Heterocycl. Chem.*, **1988**, *25*, 479.
1257. P. L. Ferrarini, C. Mori, and G. Primofiore, *J. Heterocycl. Chem.*, **1986**, *23*, 501.
1258. K. Mogilaiah and C. S. Reddy, *Synth. Commun.*, **2003**, *33*, 3131.
1259. K. Mogilaiah, D. S. Chowdary, P. R. Reddy, and N. V. Reddy, *Synth. Commun.*, **2003**, *33*, 127.
1260. V. P. Litinov, S. V. Roman, and V. D. Dyachenko, *Usp. Khim.*, **2001**, *70*, 345.
1261. Y. Wu, T.-L. Hwang, K. Algayer, W. Xu, H. Wang, A. Procopio, L. de Busi, C.-Y. Yang, and B. Matuszewska, *J. Pharm. Biomed. Anal.*, **2003**, *33*, 999.

1262. S. Anandan, J. Madhavan, P. Maruthamuthu, V. Raghukumar, and V. T. Ramakrishnan, *Sol. Energ. Mat. Sol. Cell.*, **2004**, *81*, 419.

1263. V. P. Litvinov, *Usp. Khim.*, **2004**, *73*, 692.

1264. S. M. E. Khalil, *Synth. React. Inorg. Met.-Org. Chem.*, **2000**, *30*, 19.

1265. K. Mogilaiah and N. V. Reddy, *Synth. Commun.*, **2003**, *33*, 1067.

1266. E. Barbu and F. Cuiban, *Heterocycl. Commun.*, **2000**, *6*, 259.

1267. H. R. Park, O. H. Park, H. Y. Lee, J. J. Seo, and K. M. Bark, *Bull. Korean Chem. Soc.*, **2003**, *24*, 1618.

1268. H. C. van der Plas, *Adv. Heterocycl. Chem.*, **2004**, *86*, 1.

1269. B. D. Schober and T. Kappe, *J. Heterocycl. Chem.*, **1988**, *25*, 1231.

1270. M. Tada and Y. Yokoi, *J. Heterocycl. Chem.*, **1989**, *26*, 45.

1271. Y. Tamura, L. C. Chen, M. Fujita, and Y. Kita, *Chem. Pharm. Bull.*, **1982**, *30*, 1257.

1272. Y. Tamura, L. C. Chen, M. Fujita, H. Kiyokawa, and Y. Kita, *Chem. Pharm. Bull.*, **1981**, *29*, 2460.

1273. M. Woźiak and H. C. van der Plas, *Adv. Heterocycl. Chem.*, **2000**, *77*, 285.

1274. S. C. Zimmerman and W. S. Kwan, *Angew. Chem.*, **1995**, *107*, 2589.

1275. A. Shiozawa, Y.-I. Ich-kawa, M. Ishikawa, Y. Kogo, S. Kurashige, H. Miyazaki, H. Yamanaka, and T. Sakamoto, *Chem. Pharm. Bull.*, **1984**, *32*, 995.

1276. S. Nishigaki, F. Yoneda, K. Ogiwara, T. Naito, R. Dohmori, S. Kadoya, Y. Tanaka, and I. Takamura, *Chem. Pharm. Bull.*, **1969**, *17*, 1827.

1277. H. Graber, T. Perenyi, E. Ludwig, and M. Arr, *Int. J. Clin. Pharmacol. Biopharm.*, **1976**, *13*, 76; *Chem. Abstr.*, **1976**, *85*, 13610.

1278. K. Mogilaiah and R. B. Rao, *Synth. Commun.*, **2003**, *32*, 747.

1279. N. Ikekawa, *Chem. Pharm. Bull.*, **1958**, *6*, 269.

1280. K. Mogilaiah and C. S. Reddy, *Heterocycl. Commun.*, **2004**, *10*, 363.

1281. S. M. E. Khalil, *Synth. React. Inorg. Met.-Org. Chem.*, **2001**, *31*, 417.

1282. Y. Kobayashi, I. Kumadaki, and H. Sato, *Chem. Pharm. Bull.*, **1969**, *17*, 1045.

1283. K. Mogilaiah and N. V. Reddy, *Synth. Commun.*, **2003**, *33*, 73.

1284. T. Hirose, S. Mishio, J.-I. Matsumoto, and S. Minami, *Chem. Pharm. Bull.*, **1982**, *30*, 2399.

1285. N. Ikekawa, *Chem. Pharm. Bull.*, **1958**, *6*, 263.

1286. E. Veverková, M. Nosková, and Š. Toma, *Synth. Commun.*, **2002**, *32*, 2903.

1287. M. NgSee, H. W. Latz, and D. G. Hendricker, *J. Inorg. Nucl. Chem.*, **1977**, *39*, 71.

1288. E. Barbu, D. Mihaiescu, and F. Cuiban, *Molecules*, **2000**, *5*, 956.

1289. S. Selvanayagam, V. Rajakannan, S. N. Rao, S. S. S. Raj, H. K. Fun, V. Raghukumar, and D. Velmurugan, *Cryst. Res. Technol.*, **2004**, *39*, 172.

1290. A. Sano and H. Nakamura, *Anal. Sci.*, **2001**, *17*, 375.

1291. H. Misbahi, P. Brouant, A. Hever, A. M. Molnar, K. Wolfard, G. Spengler, H. Mefetah, J. Molnar, and J. Barbe, *Anticancer Res.*, **2002**, *22*, 2097.

1292. S. B. Brown and M. J. S. Dewar, *J. Inorg. Nucl. Chem.*, **1980**, *42*, 140.

1293. K. Mogilaiah, D. S. Chowdary, and P. R. Reddy, *Synth. Commun.*, **2002**, *32*, 857.

1294. J. D. Harling, F. P. Harrington, and M. Thompson, *Synth. Commun.*, **2001**, *31*, 787.

1295. W. Fabian, *Z. Naturforsch., A*, **1980**, *35*, 1408; *Chem. Abstr.*, **1981**, *94*, 174082.

1296. S. E. Lopez, M. E. Rosales, J. Salazar, N. Urdaneta, R. Ferrer, J. E. Angel, and J. E. Charris, *Heterocycl. Commun.*, **2003**, *9*, 345.

1297. N. Ikekawa, *Chem. Pharm. Bull.*, **1958**, *6*, 401.

1298. A. Matsuura, N. Ashizawa, R. Asakura, T. Kumonaka, T. Aotsuka, T. Hase, C. Shimizu, T. Kurihara, and F. Kobayashi, *Biol. Pharm. Bull.*, **1994**, *17*, 498.

1299. P. Wormell and A. R. Lacey, *Chem. Phys.*, **1992**, *160*, 55.
1300. H. Möhrle, *Arch. Pharm. (Weinheim, Ger.)*, **1975**, *308*, 499.
1301. A. Dornow and J. von Loh, *Arch. Pharm. (Weinheim, Ger.)*, **1957**, *290*, 136.
1302. A. Decormeille, G. Queguiner, and P. Pasteur, *C. R. Acad. Sci., Ser. C*, **1975**, *280*, 381.
1303. E. Laviron and L. Roullier, *C. R. Acad. Sci., Ser. C*, **1972**, *274*, 1489.
1304. P. L. Ferrarini, C. Mori, M. Badawneh, F. Franconi, C. Manera, M. Miceli, and G. Saccomanni, *Farmaco*, **2000**, *55*, 603.
1305. A. Petitjean, J.-M. Lehn, R. G. Khoury, A. de Cian, and N. Kyritsakas, *C. R. Chim.*, **2002**, *5*, 337.
1306. K. J. Schaper, *Arch. Pharm. (Weinheim, Ger.)*, **1978**, *311*, 650.
1307. F. Zymalkowski and P. Messinger, *Arch. Pharm. (Weinheim, Ger.)*, **1967**, *300*, 91.
1308. P. L. Ferrarini, O. Livi, and V. M. Menichetti, *Farmaco, Ed. Sci.*, **1979**, *34*, 165; *Chem. Abstr.*, **1979**, *90*, 152047.
1309. E. Gudriniece and B. Rigerte, *Latv. PSR Zinat. Akad. Vestnis, Kim. Ser.*, **1974**, 239; *Chem. Abstr.*, **1974**, *81*, 37490.
1310. K.-H. Nietsch and R. Troschütz, *Arch. Pharm. (Weinheim, Ger.)*, **1985**, *318*, 175.
1311. C. Hoock, J. Reichert, and M. Schmidtke, *Molecules*, **1999**, *4*, 264.
1312. J. Spanget-Larsen, *J. Electron Spectrosc. Relat. Phenom.*, **1974**, *3*, 369.
1313. D. M. W. van den Ham, M. Beerlage, D. van der Meer, and D. Feil, *J. Electron Spectrosc. Relat. Phenom.*, **1975**, *7*, 33.
1314. K. W. Garey and G. W. Amsden, *Pharmacotherapy*, **1999**, *19*, 1.
1315. K.-A. Kovar and M. Weidemann, *Arch. Pharm. (Weinheim, Ger.)*, **1984**, *317*, 977.
1316. Z. K. Si, W. Jiang, L. Wang, G. Ma, and J. T. Hu, *Mikrochim. Acta*, **2002**, *138*, 19.
1317. Y. Kobayashi, I. Kumadaki, and H. Sato, *Chem. Pharm. Bull.*, **1970**, *18*, 861.
1318. Y. Kobatashi, I. Kumadaki, and H. Sato, *Chem. Pharm. Bull.*, **1969**, *17*, 2614.
1319. P. van de Weijer, C. Mohan, and D. M. W. van den Ham, *Org. Magn. Reson.*, **1977**, *10*, 165.
1320. M. Woźniak, H. C. van der Plas, and B. van Veldhuizen, *J. Heterocycl. Chem.*, **1983**, *20*, 9.
1321. J. A. Merritt and R. J. Pirkle, *J. Mol. Spectrosc.*, **1970**, *35*, 251.
1322. H. A. Daboun, S. E. Abdou, and M. M. Khader, *Arch. Pharm. (Weinheim, Ger.)*, **1983**, *316*, 564.
1323. I. Tonetti and G. Primofiore, *Farmaco, Ed. Sci.*, **1980**, *35*, 1052.
1324. T. Hirota, K.-I. Tomita, K. Sasaki, K. Okuda, M. Yoshida, and S. Kashino, *Heterocycles*, **2001**, *55*, 741.
1325. G. Metz, *Arch. Pharm. (Weinheim, Ger.)*, **1987**, *320*, 285.
1326. N. Rosas, P. Sharma, A. Cabrera, G. Pènieres, J. L. Garcia, and C. Maldonado, *Heterocycles*, **2003**, *60*, 2631.
1327. J. Kim, C. Park, M. S. Chang, J. K. Cha, J. Lee, and Y. L. Chung, *Drug Devel. Ind. Pharm.*, **1999**, *25*, 1283.
1328. K. Mogilaiah, H. R. Babu, and N. V. Reddy, *Synth. Commun.*, **2003**, *32*, 2377.
1329. L. Roullier and E. Laviron, *Electrochim. Acta*, **1978**, *23*, 773.
1330. S. Shimada and T. Tojo, *Chem. Pharm. Bull.*, **1983**, *31*, 4247.
1331. J. R. Wagner and D. G. Hendricker, *J. Inorg. Nucl. Chem.*, **1975**, *37*, 1375.
1332. G. Abbiati, A. Arcadi, V. Canevari, L. Capezzuto, and E. Rossi, *J. Org. Chem.*, **2005**, *70*, 6454.
1333. J. T. Leonard, R. Gangadhar, S. K. Gnanasam, S. Ramachandran, M. Saravanan, and S. K. Sridhar, *Biol. Pharm. Bull.*, **2002**, *25*, 798.
1334. A. D. Hamilton, A. Muehldorf, S.-K. Chang, N. Pant, S. Goswami, and D. van Engen, *J. Inclusion Phenom. Mol. Recognit. Chem.*, **1989**, *7*, 27.
1335. U. N. Rao and E. R. Biehl, *Heterocycles*, **2002**, *56*, 443; **2002**, *57*, 1397.

1336. I. Yamazaki, M. Fujita, and H. Baba, *Photochem. Photobiol.*, **1976**, *23*, 69.
1337. K. Mogilaiah, D. S. Chowdary, and R. B. Rao, *Indian J. Heterocycl. Chem.*, **2000**, *9*, 311.
1338. H. S. Rani, H. Shailaja, K. Mogilaiah, and B. Sreenivasulu, *Indian J. Heterocycl. Chem.*, **1995**, *5*, 33; *Chem. Abstr.*, **1996**, *124*, 202062.
1339. M. Willems, *Arch. Pharm. (Weinheim, Ger.)*, **1987**, *320*, 1245.
1340. K. Mogilaiah and G. Kankaiah, *Indian J. Heterocycl. Chem.*, **2002**, *11*, 283.
1341. K. Mogilaiah, H. R. Babu, and R. B. Rao, *Indian J. Heterocycl. Chem.*, **2000**, *10*, 109.
1342. L. Roullier and E. Laviron, *Electrochim. Acta*, **1976**, *21*, 421.
1343. K. Mogilaiah and H. R. Babu, *Indian J. Heterocycl. Chem.*, **2000**, *10*, 113.
1344. R. Müller and F. Dörr, *Z. Elektrochem.*, **1959**, *63*, 1150.
1345. K. Mogilaiah and G. R. Sudhakar, *Indian J. Heterocycl. Chem.*, **2001**, *11*, 163.
1346. K. Janné and P. Ahlberg, *J. Chem. Res.*, **1977**, *Synop.* 286.
1347. K. Mogilaiah and G. R. Sushakar, *Indian J. Heterocycl. Chem.*, **2001**, *11*, 173.
1348. K. Mogilaiah, N. V. Reddy, and P. R. Reddy, *Indian J. Heterocycl. Chem.*, **2001**, *10*, 267.
1349. M. W. Read and P. S. Ray, *J. Heterocycl. Chem.*, **1995**, *32*, 1595.
1350. K. Mogilaiah and S. Kavitha, *Indian J. Heterocycl. Chem.*, **2002**, *12*, 79.
1351. E. G. Rozantsev, M. Dagonneau, E. S. Kagan, V. I. Mikhailov, and V. D. Sholle, *J. Chem. Res.*, **1979**, *Synop.* 260, *Minipr.* 2901.
1352. K. Mogilaiah, G. R. Reddy, and C. S. Reddy, *J. Chem. Res.*, **2004**, 832.
1353. D. Ramesh and B. Sreenivasulu, *Indian J. Heterocycl. Chem.*, **2003**, *13*, 163.
1354. K. Mogilaiah, G. R. Sudhakar, and J. U. Rani, *Indian J. Heterocycl. Chem.*, **2003**, *12*, 339.
1355. R. J. Foster, R. L. Badner, and D. G. Hendricker, *J. Inorg. Nucl. Chem.*, **1972**, *34*, 3795.
1356. K. Mogilaiah and B. Sakram, *Indian J. Heterocycl. Chem.*, **2004**, *13*, 289.
1357. D. G. Wibberley, *Compr. Org. Chem.*, **1979**, *4*, 247.
1358. T. Tanase, T. Igoshi, K. Kobayashi, and Y. Yamamoto, *J. Chem. Res.*, **1998**, *Synop*, 538, *Minipr.* 2140.
1359. K. Timmers and R. Sternglanz, *Bioinorg. Chem.*, **1978**, *9*, 145; *Chem. Abstr.*, **1978**, *9*, 145.
1360. I. A. El-Sakka, *J. Chem. Res.*, **1996**, *Synop.* 434.
1361. E. Barbu and F. Cuiban, *Heterocycl. Commun.*, **2000**, *6*, 259.
1362. P. Singh, A. Clearfield, and I. Bernal, *J. Coord. Chem.*, **1971**, *1*, 29.
1363. E. Barbu and F. Cuiban, *Rev. Roum. Chim.*, **2001**, *46*, 1139.
1364. K. Mogilaiah, S. Kavatha, and G. R. Reddy, *Indian J. Heterocycl. Chem.*, **2002**, *12*, 113.
1365. E. Dittmar, C. J. Alexander, and M. L. Good, *J. Coord. Chem.*, **1972**, *2*, 69.
1366. H. F. Schaeffer, *Microchem. J.*, **1969**, *14*, 415.
1367. D. G. Hendricker and R. J. Foster, *J. Inorg. Nucl. Chem.*, **1972**, *34*, 1949.
1368. V. Hanuš, F. Tureček, A. M. Farag, and M. Tichý, *Org. Mass Spectrom.*, **1984**, *19*, 459.
1369. B. Hutchinson, A. Sunderland, M. Neal, and S. Olbricht, *Spectrochim. Acta*, **1973**, *29*, 2001.
1370. R. J. Staniewcz, D. G. Hendricker, and P. R. Griffiths, *Inorg. Nucl. Chem. Lett.*, **1977**, *13*, 467.
1371. E. V. Brown and A. C. Plasz, *J. Heterocycl. Chem.*, **1970**, *7*, 593.
1372. D. G. Hendricker and R. L. Bodner, *Inorg. Nucl. Chem. Lett.*, **1970**, *6*, 187.
1373. D. J. McCaustland and C. C. Cheng, *J. Heterocycl. Chem.*, **1970**, *7*, 467.
1374. W. W. Paudler and S. J. Cornrich, *J. Heterocycl. Chem.*, **1970**, *7*, 419.
1375. Y. Hamada, I. Takeuchi, and H. Matsuoka, *Chem. Pharm. Bull.*, **1970**, *18*, 1026.
1376. T. J. Cardwell, R. W. Cattrall, L. W. Deady, K. A. Murphy, and S. S. Tan, *Aust. J. Chem.*, **1992**, *45*, 983.

1377. G. B. Barlin and W.-L. Tan, *Aust. J. Chem.*, **1985**, *38*, 459.
1378. M. H. Elnagdi, S. A. S. Ghozlan, F. M. Abdelrazek, and M. A. Selim, *J. Chem. Res.*, **1991**, *Synop.* 116, *Minipr.* 1021.
1379. G. B. Barlin and W.-L. Tan, *Aust. J. Chem.*, **1986**, *39*, 51.
1380. S. Carboni, A. Da Settimo, D. Bertini, P. L. Ferrarini, O. Livi, and I. Tonetti, *Farmaco, Ed. Sci.*, **1973**, *28*, 722.
1381. S. Carboni, A. Da Settimo, D. Bertini, P. L. Ferrarini, O. Livi, and I. Tonetti, *Farmaco, Ed. Sci.*, **1975**, *30*, 185.
1382. G. A. Portmann, E. W. McChesney, H. Stander, and W. E. Moore, *J. Pharm. Sci.*, **1966**, *55*, 59.
1383. S. Carboni, A. Da Settimo, D. Bertini, P. L. Ferrarini, O. Livi, and I. Tonetti, *Farmaco, Ed. Sci.*, **1975**, *30*, 237.
1384. O. Livi, E. Amato, G. Biagi, P. L. Ferrarini, and G. P. Primofiore, *Farmaco, Ed. Sci.*, **1978**, *33*, 838.
1385. A. Da Settimo, G. Biagi, G. Primofiore, P. L. Ferrarini, and O. Livi, *Farmaco, Ed. Sci.*, **1978**, *33*, 770.
1386. P. L. Ferrarini and O. Livi, *Farmaco, Ed. Sci.*, **1978**, *33*, 543.
1387. G. A. Portmann, E. W. McChesney, H. Stander, and W. E. Moore, *J. Pharm. Sci.*, **1966**, *55*, 72.
1388. A. Da Settimo, G. Primofiore, G. Biagi, and V. Santerini, *Farmaco, Ed. Sci.*, **1976**, *31*, 587.
1389. F. Alhaique, R. G. E. Cozzani, and F. M. Riccieri, *Farmico, Ed. Sci.*, **1976**, *31*, 845.
1390. S. Carboni, A. Da Settimo, F. L. Ferrarini, and O. Livi, *Farmaco, Ed. Sci.*, **1978**, *33*, 315.
1391. O. Livi, P. L. Ferrarini, D. Bertini, and I. Tonetti, *Farmaco, Ed. Sci.*, **1975**, *30*, 1017.
1392. I. Tonetti, D. Bertini, P. L. Ferrarini, and O. Livi, *Farmaco, Ed. Sci.*, **1976**, *31*, 175.
1393. S. Carboni, A. Da Settimo, D. Bertini, P. L. Ferrarini, O. Livi, and I. Tonetti, *Farmaco, Ed. Sci.*, **1976**, *31*, 322.
1394. A. Da Settimo, G. Primofiore, F. Da Settimo, and F. Simorini, *Drug Des. Discovery*, **1994**, *11*, 307.
1395. C. Mart-ni, W. Marrucci, A. Lucacchini, G. Biagi, and O. Livi, *J. Pharm. Sci.*, **1988**, *77*, 977.
1396. R. Johm and G. Seitz, *Arch. Pharm. (Weinheim, Ger.)*, **1989**, *322*, 561.
1397. K. Unferferth, R. Dörre, B. Körner, H. Scheibe, and E. Morgenstern, *Arch. Pharm. (Weinheim, Ger.)*, **1991**, *324*, 809.
1398. J. Heindl, H.-W. Kelm, E. Dogs, A. S. Seeger, and C. Herrmann, *Eur. J. Med. Chem.*, **1977**, *12*, 549.
1399. M. H. Elnagdi and W. W. Erian, *Arch. Pharm. (Weinheim, Ger.)*, **1991**, *324*, 853.
1400. H. L. Davis, R. G. Gedir, E. M. Hawes, and D. G. Wibberley, *Eur. J. Med. Chem.*, **1985**, *20*, 381.
1401. P. L. Ferrarini, C. Mori, M. Miceli, and F. Franconi, *Eur. J. Med. Chem.*, **1994**, *29*, 735.
1402. R. Troschütz and T. Dennstedt, *Arch. Pharm. (Weinheim, Ger.)*, **1994**, *327*, 33.
1403. P. L. Ferrarini, C. Mori, G. Primofiore, A. Da Settimo, M. C. Breschi, E. Martinotti, P. Nieri, and M. A. Ciucci, *Eur. J. Med. Chem.*, **1990**, *25*, 489.
1404. F. Haenel, R. John, and G. Seitz, *Arch. Pharm. (Weinheim, Ger.)*, **1992**, *325*, 349.
1405. P. M. Gilis, A. Haemers, and W. Bollaert, *Eur. J. Med. Chem.*, **1980**, *15*, 499.
1406. S. Carboni, A. Da Settimo, P. L. Ferrarini, G. Primofiore, O. Livi, V. Menichetti, M. del Tacca, E. Martinotti, C. Bernardini, and A. Bertelli, *Eur. J. Med. Chem.*, **1982**, *17*, 159.
1407. C. Rufer, H.-J. Kessler, and K. Schwarz, *Eur. J. Med. Chem.*, **1977**, *12*, 27.
1408. P. Strehlke, *Eur. J. Med. Chem.*, **1977**, *12*, 541.
1409. M. A. Khalil, *Phosphorus, Sulfur Silicon Relat. Elem.*, **2005**, *180*, 85.
1410. J. B.-H. Tok, L. Bi, and M. Saenz, *Bioorg. Med. Chem. Lett.*, **2005**, *15*, 827.
1411. G. Falardeau, H. Lachance, A. St-Pierre, C. G. Yannopoulis, M. Drouin, J. Bétard, and L. Chan, *Bioorg. Med. Chem. Lett.*, **2005**, *15*, 1693.
1412. G. B. W. I. Ligthart, H. Okawa, R. F. Siiresma, and E. W. Müller, *J. Org. Chem.*, **2006**, *71*, 375.

1413. A. Madhan and B. V. Rao, *Tetrahedron Lett.*, **2005**, *46*, 323.
1414. S. T. Hazeldine, L. Polin, J. Kushner, K. White, T. H. Corbett, J. Biehl, and J. P. Horwitz, *Bioorg. Med. Chem.*, **2005**, *13*, 1069.
1415. T. Peng, T. Murase, Y. Goto, A. Kobori, and K. Nakatani, *Bioorg. Med. Chem. Lett.*, **2005**, *15*, 259.
1416. K. Mogilaiah and M. Prashanthi, *Indian J. Heterocycl. Chem.*, **2005**, *14*, 185.
1417. D. Abbanat, G. Webb, B. Foleno, Y. Li, M. Macielag, D. Montenegro, E. Wira, and K. Bush, *Antimicrob, Agents Chemother.*, **2005**, *49*, 309.
1418. S. Hikishima, N. Minakawa, K. Kuramoto, Y. Fujisawa, M. Ogawa, and A. Matsuda, *Angew. Chem., Int. Ed.*, **2005**, *44*, 596.
1419. L. W. Deady, M. L. Rogers, L. Zuang, B. C. Baguley, and W. A. Denny, *Bioorg. Med. Chem.*, **2005**, *13*, 1341.
1420. D. Burdinski, K. Cheng, and S. J. Lippard, *Tetrahedron*, **2005**, *61*, 1587.
1421. J. Vilar, C. Peinador, and J. Quintela, *Heterocycles*, **2005**, *65*, 329.
1422. J. K. Park, S. K. Im, H. Ju, J. M. Jeon, C. G. Kim, D. U. Kim, J. C. Yoo, S. S. Cho, and S. K. Cho, *Bull. Korean Chem. Soc.*, **2005**, *26*, 371.
1423. G. Grossi, M. Di Nraccio, G. Roma, V. Ballabeni, M. Tognolini, and E. Barocelli, *Eur. J. Med. Chem.*, **2005**, *40*, 155.
1424. P. J. Silver, L. T. Hamel, R. G. Bentley, K. Dillon, M. J. Connell, B. O'Connor, R. A. Ferrari, and E. D. Pagani, *Drug Devel. Res.*, **1990**, *21*, 93; *Chem. Abstr.*, **1991**, *114*, 421.
1425. G. R. Rao, K. Mogilaiah, K. R. Reddy, and B. Sreenivasulu, *J. Indian Chem. Soc.*, **1987**, *64*, 710.
1426. G. Favini, I. Vandoni, and M. Simonetta, *Theor. Chim. Acta*, **1965**, *3*, 45.
1427. P. L. Ferrarini, G. Biagi, O. Livi, G. Primofiore, and M. Carpenè, *J. Heterocycl. Chem.*, **1981**, *18*, 1007.
1428. T. Posner, *Liebigs Ann. Chem.*, **1912**, *389*, 1.
1429. A. A. Shestopalov, A. V. Gromova, L. A. Rodinovskaya, K. G. Nikishin, V. P. Litinov, and A. M. Shestopalov, *Russ. Chem. Bull.*, **2004**, *53*, 2353.
1430. S. Stantorth, in *Comprehensive Heterocyclic Chemistry II*, C. A. Ramsden, ed., Pergamon, Oxford, 1996, Vol. 7, p. 527.
1431. L. Birkofer and C. Kaiser, *Chem. Ber.*, **1957**, *90*, 2933.
1432. V. P. Litvinov, *Adv. Heterocycl. Chem.*, **2006**, *91*, 189.
1433. V. Y. Vvedensky, Y. V. Ivanov, V. Kysil, C. Williams, S. Tkachenko, A. Kiselyov, A. V. Khvat, and A. V. Ivachtchenko, *Tetrahedron Lett.*, **2005**, *46*, 3953.
1434. A. M. Thompson, A. M. Delaney, J. M. Hamby, M. C. Schroeder, T. A. Spoon, S. M. Crean, H. D. Hollis Showalter, and W. A. Denny, *J. Med. Chem.*, **2005**, *48*, 4628.
1435. H. Gross, D. E. Goeger, P. Hills, S. L. Moobery, D. L. Ballantine, T. F. Murray, F. A. Valeriote, and W. H. Gerwick, *J. Nat. Prod.*, **2006**, *69*, 640.
1436. M. Basato, A. Biffis, G. Martinau, C. Tubaro, C. Graiff, A. Tiripicchio, L. A. Aronica, and A. M. Capporusso, *J. Organomet. Chem.*,, **2006**, *691*, 3464.
1437. J. P. Guare, J. S. Wai, R. P. Gomez, N. J. Anthony, S. M. Jolly, A. R. Cortes, J. P. Vacca, P. J. Felock, K. A. Stillmock, W. A. Schleif, G. Moyer, L. J. Gabryelski, L. Jin, I.-W. Chen, D. J. Hazuda, and S. D. Young, *Bioorg. Med. Chem. Lett.*, **2006**, *16*, 2900.
1438. T. Park, M. F. Mayer, S. Nakashima, and S. C. Zimmerman, *Synlett*, **2005**, 1435.
1439. K. Mogilaiah, M. Prashanthi, and S. Kavitha, *Indian J. Chem., Sect. B*, **2006**, *45*, 302.
1440. K. Mogilaiah and J. U. Rani, *Indian J. Chem., Sect. B.* **2006**, *45*, 1051.
1441. P. Zhichkin, C. M. Cillo Beer, W. M. Rennells, and D. J. Fairfax, *Synlett*, **2006**, 379.
1442. E. A. El Rady and M. A. Barsy, *J. Heterocycl. Chem.*, **2006**, *43*, 243.
1443. M. Grzegozek and B. Szpakiewicz, *J. Heterocycl. Chem.*, **2006**, *43*, 425.

1444. M. M. Blanco, C. B. Schapira, G. Levin, and I. A. Perillo, *J. Heterocycl. Chem.*, **2005**, *44*, 493.
1445. S. Goswami, R. Mukherjee, R. Mukherjee, S. Jana, A. C. Maity, and A. K. Adak, *Molecules*, **2005**, *10*, 929.
1446. W. L. F. Armarego and C. L. L. Chai, *Purification of Laboratory Chemicals*, 5th ed., Elsevier, London, **2003**, p. 307.
1447. H. D. H. Showalter, *J. Heterocycl. Chem.*, **2006**, *43*, 1311.
1448. V. D. Dyachenko, S. V. Roman, and V. P. Litvinov, *Russ. Chem. Bull.*, **2000**, *49*, 125.
1449. R. J. Cvetovich, R. A. Reamer, L. DiMichele, J. V. L. Chung, and J. R. Chilenski, *J. Org. Chem.*, **2006**, *71*, 8610.
1450. A. Kukrek, D. Wang, Y. Hou, R. Zong, and R. Thummel, *Inorg. Chem.*, **2006**, *45*, 10131.
1451. X. Jiang, G.-P. Chen, K. Prasad, O. Renič, and T. J. Blacklock, *J. Heterocycl. Chem.*, **2006**, *43*, 1725.

APPENDIX

Tables of Simple Naphthyridines

These six tables are reasonably comprehensive lists of 1,5- (Table A1), 1,6- (Table A2), 1,7- (Table A3), 1,8- (Table A4), 2,6- (Table A5), and 2,7-naphthyridines (Table A6) described before 2007. For each compund are recorded (1) melting and/or boiling point(s); (2) an indication of any reported spectra or other physical properties; (3) any reported salts or simple derivatives, especially when the parent compund was characterized poorly or not at all; and (4) direct references to the original literature.

To keep the tables within manageable length, the following categories of naphthyridine derivatives have been *excluded* on the grounds that they are not simple:

All fused or nucleus-reduced derivatives

Those with a cyclic substituent other than an unsubstituted cycloalkyl, morpholino, phenyl, or piperidino group

Those bearing a substituent with more than six carbon atoms except for an unsubstituted benzoyl, benzyl, benzylidene, phenethyl, phenylethynyl, or styryl group

Those with two or more independent functional groups on any one substituent, such as a trifluoromethyl group

The following conventions and abbreviations have been used in the tables:

Melting Point. This term covers not only a regular melting point or melting range but also such variations as "decomposing at" or "melting with decomposition at." The symbol > before the melting point indicates that the substance melts or decomposes above that temperature or that it does not melt or decompose below that temperature. Where two different melting points/ranges are reported in the literature, they appear in the tables as, for example, "102–103 or 104–107;" when more than two melting points/ranges are reported, they appear in the tables as, for example, "208 → 231."

Boiling Point. Boiling points/ranges are distinguished from melting points/ranges by the presence of a pressure in millimeters of mercury (mmHg) after the temperature(s): for example, 83–85/2.5.

The Naphthyridines: The Chemistry of Heterocyclic Compounds, Volume 63, by D.J. Brown
Copyright © 2008 John Wiley & Sons, Inc.

Appendix

Abbreviations for Physical Data

anal	Analytical data (usually assumed)
biol	Only bioactivity reported
crude	Compound not purified
dip	Dipole moment
ESR	Electron spin resonance data
fl sp	Fluorescence spectral data
IR	Infrared spectral data
liq	Liquid at 20°C (no boiling point given)
MS	Mass spectral data
N_D	Refractive index
NMR	Nuclear magnetic resonance data (any nucleus)
NQR	Nuclear quadrupole resonance data
pK_a	Ionization data
pol	Polarographic data
solid	Solid at 20°C (no melting point given)
st	Fine structure (e.g., tautomerism) discussed
th	Theoretical calculations reported (any area)
UV	Ultraviolet/visible spectral data
xl st	Crystal structure (X-ray data) reported

Abbreviations for Salts, Associated Anions, and Solvates

AcOH	Acetate salt
EtOH	Ethanolate
HBr, etc.	Appropriate hydrohalide salt
H_2O	Hydrate
HSO_4^-	Sulfate anion
H_2SO_4	Sulfate salt
I^-, etc.	Appropriate halide anion
MeI	Quaternary methiodide
NH_4	Ammonium salt
Na, etc.	Appropriate alkali metal salt
pic	Picrate salt or anion
TsOH	*p*-Toluenesulfonate salt

Abbreviations for Derivatives

dnp	2,4-Dinitrophenylhydrazone
Et_2 acetal, etc	Appropriate dialkyl acetal
$H_2NN=$	Hydrazone
$HON=$	Oxime
$MeCH=$, etc	Appropriate alkylidene derivative
$PhNHN=$	Phenylhydrazone
$PhN=$	Anil (Schiff base)
sc	Semicarbazone
tsc	Thiosemicarbazone

Other Notes. The use of "cf" before a reference usually indicates apparently inconsistent or doubtfully relevant information therein. An added query mark (?) indicates some doubt associated with a datum or reference. A dash (—) in the data column indicates that no physical data were gleaned from the given reference(s).

TABLE A.1. ALPHABETICAL LIST OF SIMPLE 1,5-NAPHTHYRIDINES

1,5-Naphthyridine	Melting Point (°C) etc.	Reference(s)
2-Acetamido-1,5-naphthyridine	190–192 or 210, NMR	811, 1443
4-Acetamido-1,5-naphthyridine	119–120	814
2-Acetamido-1,5-naphthyridine 1,5-dioxide	230–231	787
4-Acetamido-1,5-naphthyridine 1,5-dioxide	253–254, IR, NMR	814
2-Acetoxymethyl-1,5-naphthyridine 1,5-dioxide	—	290
1-Allyl-7-ethoxy-4-oxo-1,4-dihydro-1,5-naphthyridine-3-carboxylic acid	199–200, NMR	1398
1-Allyl-4-oxo-1,4-dihydro-1,5-naphthyridine-3-carboxylic acid	205	1398
4-Allylthio-1,5-naphthyridine	40–42, pK_a	814
3-Amino-2-ethoxy-1,5-naphthyridine	114–117, IR, NMR	827
4-Amino-2-ethoxy-1,5-naphthyridine	137–138, IR, NMR	827
3-Amino-1-ethyl-1,5-naphthyridin-2(1H)-one	205–207, IR, NMR	827
4-Amino-1-ethyl-1,5-naphthyridin-2(1H)-one	196–198, IR, NMR	827
2-Amino-1,5-naphthyridine-3-carboxamide	NMR	1302
1-Amino-1,5-naphthyridinium mesitylenesulfonate	193–195	1028
2-Amino-1,5-naphthyridin-4(1H)-one	320	1023
3-Amino-1,5-naphthyridin-4(1H)-one	268; HCl: —	722
4-Amino-3-nitro-1,5-naphthyridin-2(1H)-one	340–341	818–1273
4-Anilino-2-chloro-1,5-naphthyridine	146	1023
2-Anilino-6-methyl-1,5-naphthyridine	202–203	1051
4-Anilino-1,5-naphthyridin-2-amine	152	1023
4-Anilino-1,5-naphthyridin-2(1H)-one	251	1023
3-Benzyl-4-hydroxy-1,8-naphthyridin-2(1H)-one	300; *O/N*-acetyl: 210	760
4-Benzylsulfonyl-1,5-naphthyridine	179–180	814
4-Benzylsulfonyl-1,5-naphthyridine 1-oxide	212–213, MS	814
4-Benzylthio-1,5-naphthyridine	127–128, pK_a	814
4-Benzylthio-1,5-naphthyridine 1-oxide	167–168	814
2,4-Bis(benzylamino)-1,5-naphthyridine	147; HBr: 216	1023
2,4-Bis(methylamino)-3-nitro-1,5-naphthyridine	182–183, NMR	1443
3-Bromo-8-chloro-1,5-naphthyridine	183, NMR	1377
7-Bromo-6-chloro-4-oxo-1,4-dihydro-1,5-naphthyridine-2-carboxylic acid	—	273
3-Bromo-2-ethoxy-1,5-naphthyridine	96–97 or 134–135, NMR	827, 837
3-Bromo-1-ethyl-1,5-naphthyridin-2(1H)-one	134–135, IR, NMR	827
7-Bromo-1-ethyl-4-oxo-1,4-dihydro-1,5-naphthyridine-3-carboxylic acid	303–304	1398
3-Bromo-8-methoxy-1,5-naphthyridine	167–169, NMR	1377
2-Bromomethyl-1,5-naphthyridine 1,5-dioxide	—	290
6-Bromo-7-methyl-1,4-oxo-1,4-dihydro-1,5-naphthyridine-2-carboxylic acid	—	273
2-Bromo-1,5-naphthyridin-4-amine	199–201, NMR, UV,	822
3-Bromo-1,5-naphthyridin-2-amine	184–186, NMR	822
3-Bromo-1,5-naphthyridin-4-amine	MS	374, 546
6-Bromo-1,5-naphthyridin-2-amine	—	246

TABLE A.1. (*Continued*)

1,5-Naphthyridine	Melting Point (°C) etc.	Reference(s)
2-Bromo-1,5-naphthyridine	115–116, IR, NMR	259, 439, 811
6-*d*-2-Bromo-1,5-naphthyridine	—	439
3-Bromo-1,5-naphthyridine	105 to 108, IR, NMR; pic: 107–108	149, 150, 173, 811
2,6-d_2-3-Bromo-1,5-naphthyridine	—	439
4-Bromo-1,5-naphthyridine	102–103, IR; pic; 168–169	811
3-Bromo-1,5-naphthyridine 1,5-dioxide	287	811
2-Bromo-1,5-naphthyridine 5-oxide	154–155	811
3-Bromo-1,5-naphthyridine 1-oxide	174–175, NMR	317
3-Bromo-1,5-naphthyridine 5-oxide	161–163, NMR	317
3-Bromo-1,5-naphthyridin-2(1*H*)-one	295–297 or 325	822, 1037
3-Bromo-1,5-naphthyridin-4(1*H*)-one	318, MS	546, 1037
7-Bromo-1,5-naphthyridin-4(1*H*)-one	>360, NMR	1377
6-Bromo-4-oxo-1,4-dihydro-1,5-naphthyridine-2-carboxylic acid	—	273
7-Bromo-4-oxo-1,4-dihydro-1,5-naphthyridine-3-carboxylic acid	>295, NMR	1377
2-Butoxy-8-chloro-6-methyl-1,5-naphthyridine	62	1036
2-Butoxy-8-chloro-1,5-naphthyridine	40	1036
2-Butoxy-1-ethyl-4-oxo-1,4-dihydro-1,5-naphthyridine-3-carboxylic acid	188–189	1398
6-Butoxy-2-methyl-1,5-naphthyridin-4-amine	149	1036
2-Butoxy-6-methyl-1,5-naphthyridine	37; H$_2$O: 56	1036, 1051
6-Butoxy-2-methyl-1,5-naphthyridin-4(1*H*)-one	246	1036
2-Butoxy-6-methyl-8-phenoxy-1,5-naphthyridine	66–68	1036
2-Butoxy-1,5-naphthyridine	296–298/360	1036
6-Butoxy-1,5-naphthyridin-4(1*H*)-one	170	1036
6-Butoxy-4-oxo-1,4-dihydro-1,5-naphthyridine-3-carboxylic acid	246	1036
2-Butoxy-8-phenoxy-1,5-naphthyridine	liq, anal	1036
4-Carboxymethyl-1,5-naphthyridine 1-oxide	127–128	814
2-Chloro-6,8-dimethyl-1,5-naphthyridine	129	87
6-Chloro-1-ethyl-4-oxo-1,4-dihydro-1,5-naphthyridine-3-carboxylic acid	261–263	1398
7-Chloro-1-ethyl-4-oxo-1,4-dihydro-1,5-naphthyridine-3-carboxylic acid	304–306	1398
2-Chloro-6-hydrazino-1,5-naphthyridine	178–180	234
4-Chloro-2-hydrazino-1,5-naphthyridine	162	1023
7-Chloro-4-hydroxy-1,5-naphthyridin-2-(1*H*)-one	337–338, UV	1373
7-Chloro-4-hydroxy-3-nitro-1,5-naphthyridin-2(1*H*)-one	252–254	1273
4-Chloro-6-methoxy-2-methyl-1,5-naphthyridine	108	1036
2-Chloro-6-methoxy-1,5-naphthyridine	130–134, NMR, UV	234
4-Chloro-2-methoxy-1,5-naphthyridine	113–114 or 114–115, NMR	1151, 1373
2-Chloro-6-methyl-1,5-naphthyridine	177–178 or 181–182; 130–135/2, 5(?), NMR; MeI: 176	87, 94, 268, 1051
4-Chloro-1-methyl-1,5-naphthyridin-2(1*H*)-one	152–153, NMR, UV	1373
6-Chloro-7-methyl-4-oxo-1,4-dihydro-1,5-naphthyridine-2-carboxylic acid	—	273
4-Chloro-1,5-naphthyridin-2-amine	184	1023

TABLE A.1. (Continued)

1,5-Naphthyridine	Melting Point (°C) etc.	Reference(s)
2-Chloro-1,5-naphthyridine	109 to 116, NMR, UV	87, 225, 232, 234, 262, 1047, 1371
3-Chloro-1,5-naphthyridine	91	225, 232
4-Chloro-1,5-naphthyridine	102–103, NMR	101, 225, 232, 262, 1371
2-Chloro-1,5-naphthyridine-3-carboxamide	NMR	1302
2-Chloro-1,5-naphthyridine 5-oxide	—	271
4-Chloro-1,5-naphthyridine 1-oxide	171–172, NMR	814
4-Chloro-1,5-naphthyridin-2(1H)-one	263, NMR	1023, 1151, 1373
7-Chloro-1,5-naphthyridin-4(1H)-one	>360	1373
2-Chloro-3-nitro-1,5-naphthyridin-4-amine	252–254, IR, MS, NMR	492, 818, 1273
2-Chloro-3-nitro-1,5-naphthyridine	205 or 208–210, NMR	818, 827, 1273
4-Chloro-3-nitro-1,5-naphthyridine	162–165	48, 1273
6-Chloro-4-oxo-1,4-dihydro-1,5-naphthyridine-2-carboxylic acid	—	273
7-Chloro-4-oxo-1,4-dihydro-1,5-naphthyridine-3-carboxylic acid	>320	1373
2-Chloro-6-phenyl-1,5-naphthyridine	131–132, IR, NMR	619
2,4-Dianilino-1,5-naphthyridine	153; HCl: 253	1023
2,6-Dianilino-1,5-naphthyridine	278	1047
3-Diazonio-1,5-naphthyridin-4(1H)-one chloride	196	722
2,3-Dibromo-1,5-naphthyridine	203, NMR	822
2,4-Dibromo-1,5-naphthyridine	126–127, NMR	822
2,6-Dibromo-1,5-naphthyridine	—	259
2,7-Dibromo-1,5-naphthyridine	192–193, NMR	317
3,7-Dibromo-1,5-naphthyridine	239–240 or 240–241, NMR	149, 150, 173
4,7-Dibromo-1,5-naphthyridine	190–192, NMR	1179
2,7-Dibromo-1,5-naphthyridine 5-oxide	175–176, NMR	317
3,7-Dibromo-1,5-naphthyridine 1-oxide	207–208, NMR	317
6,7-Dibromo-4-oxo-1,4-dihydro-1,5-naphthyridine-2-carboxylic acid	—	273
3,8-Dichloro-6-methoxy-1,5-naphthyridine	140–141, NMR, UV	1373
2,3-Dichloro-1,5-naphthyridine	193–195, MS, NMR	818
2,4-Dichloro-1,5-naphthyridine	130 to 140, NMR	233, 1023, 1151
2,6-Dichloro-1,5-naphthyridine	190 to 260, IR, NMR, UV	233, 234, 389, 787, 971, 1047
2,8-Dichloro-1,5-naphthyridine	153–156	233
3,7-Dichloro-1,5-naphthyridine	150–152	232
3,8-Dichloro-1,5-naphthyridine	176–177 or 177–178	233
4,8-Dichloro-1,5-naphthyridine	274–276 or 278–279	233, 301
4-Diethoxyphosphinyl-6-methoxy-2-phenyl-1,5-naphthyridine	—	521
4,8-Dimethoxy-1,5-naphthyridine	214–216, NMR, xl st	301, 777, 978
7-Dimethylamino-1-ethyl-4-oxo-1,4-dihydro-1,5-naphthyridine-3-carboxylic acid	325–327	1398
4-Dimethylaminomethyl-1,5-naphthyridin-3(5H)-one	3HCl: 170–171, NMR	932

TABLE A.1. (*Continued*)

1,5-Naphthyridine	Melting Point (°C) etc.	Reference(s)
2,3-Dimethyl-1,5-naphthyridine	—	522
2,4-Dimethyl-1,5-naphthyridine	40; $HClO_4$: 189; pic: 190	87, 1297
2,6-Dimethyl-1,5-naphthyridine	139–140, IR, NMR, UV; 1-MeI: 265–267, IR, NMR, UV; 1,5-$(MeClO_4)_2$: >300, IR, NMR, UV	1169
2,8-Dimethyl-1,5-naphthyridine	NMR	268
4,8-Dimethyl-1,5-naphthyridine	112–113, IR, MS, NMR	880
1,5-Dimethyl-1,5-naphthyridine-2,6(1H,5H)-dione	220–222, UV	155
1,5-Dimethyl-1,5-naphthyridine-4,8(1H,5H)-dione	273–275, NMR; F_3CCO_2H: xl st	301, 782, 978
3,7-Dimethyl-1,5-naphthyridine-4,8(1H,5H)-dione	>300, NMR	301, 978
1,6-Dimethyl-1,5-naphthyridin-2(1H)-one	136; pic: 189–190; 5-MeI: 213–215	1051
6,8-Dimethyl-1,5-naphthyridin-2(1H)-one	272–275, IR, NMR	1044
1,6-Dimethyl-1,5-naphthyridin-2(1H)-one 5-oxide	246	1051
2-Dimethylsulfimido-1,5-naphthyridine	109–110, NMR	692
4,6-Dioxo-1,4,5,6-tetrahydro-1,5-naphthyridine-3-carboxylic acid	>350	233
4,8-Dioxo-1,4,5,8-tetrahydro-1,5-naphthyridine-3-carboxylic acid	>320	233, 301
2,4-Diphenoxy-1,5-naphthyridine	170	1023
2,3-Diphenyl-1,5-naphthyridine	MS, NMR	137, 522
4-Ethoxycarbonylmethyl-1,5-naphthyridine 6-oxide	92–93	814
7-Ethoxy-1-ethyl-4-oxo-1,4-dihydro-1,5-naphthyridine-3-carboxylic acid	286–287	1398
2-Ethoxy-4-methylamino-3-nitro-1,5-naphthyridine	103–105, NMR	1443
2-Ethoxy-1,5-naphthyridin-3-amine	115–117 or 200–202	827, 837
2-Ethoxy-1,5-naphthyridine	44–45, NMR; pic: 169–173	827
4-Ethoxy-1,5-naphthyridine	39–41; HCl: 229; pic: 195	202
2-Ethoxy-3-nitro-1,5-naphthyridin-4-amine	137–138, MS, NMR	492, 818, 1273
2-Ethoxy-3-nitro-1,5-naphthyridine	134–136, IR, MS, NMR	492, 827, 837, 1273
6-Ethoxy-4-oxo-1,4-dihydro-1,5-naphthyridine-3-carboxylic acid	260–268, NMR	1015
4-Ethoxy-6-phenyl-8-phenylsulfonyl-1,5-naphthyridine	243–245, MS, NMR	554
Ethyl 4-amino-1,5-naphthyridine-3-carboxylate	—	433
Ethyl 4-amino-2-oxo-1,2-dihydro-1,5-naphthyridine-3-carboxylate	216–218, IR	725
Ethyl 4-anilino-1,5-naphthyridine-3-carboxylate	—	433
Ethyl 7-bromo-1-ethyl-4-oxo-1,4-dihydro-1,5-naphthyridine-3-carboxylate	222–224	1398
Ethyl 7-bromo-4-oxo-1,4-dihydro-1,5-naphthyridine-3-carboxylate	318–320	1377, 1398
Ethyl 6-butoxy-4-oxo-1,4-dihydro-1,5-naphthyridine-3-carboxylate	268	1036
Ethyl 4-butylamino-1,5-naphthyridine-3-carboxylate	biol	433

TABLE A.1. (*Continued*)

1,5-Naphthyridine	Melting Point (°C) etc.	Reference(s)
Ethyl 6-chloro-1-ethyl-4-oxo-1,4-dihydro-1,5-naphthyridine-3-carboxylate	239–241	1398
Ethyl 7-chloro-1-ethyl-4-oxo-1,4-dihydro-1,5-naphthyridine-3-carboxylate	189–190	1398
Ethyl 2-chloro-1,5-naphthyridine-3-carboxylate	NMR	1302
Ethyl 4-chloro-1,5-naphthyridin-3-carboxylate	—	433
Ethyl 6-chloro-4-oxo-1,4-dihydro-1,5-naphthyridine-3-carboxylate	306–307	1398
Ethyl 7-chloro-4-oxo-1,4-dihydro-1,5-naphthyridine-3-carboxylate	309–311	1373, 1398
Ethyl 1,5-dimethyl-4,8-dioxo-1,4,5,8-tetrahydro-1,5-naphthyridine-3-carboxylate	282–283, NMR	301
3-Ethyl-1,5-dimethyl-1,5-naphthyridine-2,6(1H,5H)-dione	261–262, UV	155
Ethyl 6,7-dimethyl-4-oxo-1,4-dihydro-1,5-naphthyridine-3-carboxylate	280–283	1398
1-Ethyl-6,7-dimethyl-4-oxo-1,4-dihydro-1,5-naphthyridine-3-carboxylic acid	321–322	1398
Ethyl 4,6-dioxo-1,4,5,6-tetrahydro-1,5-naphthyridine-3-carboxylate	293–295	233
Ethyl 4,8-dioxo-1,5,8-tetrahydro-1,5-naphthyridine-3-carboxylate	>300	233
Ethyl 7-ethoxy-1-ethyl-4-oxo-1,4-dihydro-1,5-naphthyridine-3-carboxylate	160–161	1398
Ethyl 7-ethoxy-4-oxo-1,4-dihydro-1,5-naphthyridine-3-carboxylate	334–335	1398
Ethyl 1-ethyl-6,7-dimethyl-4-oxo-1,4-dihydro-1,5-naphthyridine-3-carboxylate	161–162	1398
Ethyl 1-ethyl-7-methoxy-4-oxo-1,4-dihydro-1,5-naphthyridine-3-carboxylate	160–161	1398
Ethyl 1-ethyl-7-methyl-4-oxo-1,4-dihydro-1,5-naphthyridine-3-carboxylate	165–166, NMR	1398
1-Ethyl-7-ethylthio-4-oxo-1,4-dihydro-1,5-naphthyridine-3-carboxylate	238–239	1398
1-Ethyl-7-hydroxy-4-oxo-1,4-dihydro-1,5-naphthyridine-3-carboxylic acid	>340, NMR	1398
1-Ethyl-7-isobutoxy-4-oxo-1,4-dihydro-1,5-naphthyridine-3-carboxylic acid	204–205, NMR	1398
1-Ethyl-7-isopropylthio-4-oxo-1,4-dihydro-1,5-naphthyridine-3-carboxylic acid	199–200	1398
Ethyl 6-methoxy-1,5-naphthyridine-4-carboxylate 1-oxide	128–130, IR, MS, NMR	588
Ethyl 6-methoxy-4-oxo-1,4-dihydro-1,5-naphthyridine-2-carboxylate	224–226	1036
Ethyl 6-methoxy-4-oxo-1,4-dihydro-1,5-naphthyridine-3-carboxylate	268	1036
Ethyl 7-methoxy-4-oxo-1,4-dihydro-1,5-naphthyridine-3-carboxylate	294–297	1398
Ethyl 8-methoxy-4-oxo-1,4-dihydro-1,5-naphthyridine-3-carboxylate	215, NMR	301

TABLE A.1. (*Continued*)

1,5-Naphthyridine	Melting Point (°C) etc.	Reference(s)
1-Ethyl-6-methoxy-4-oxo-1,4-dihydro-1,5-naphthyridine-3-carboxylic acid	273, NMR	1398
1-Ethyl-7-methoxy-4-oxo-1,4-dihydro-1,5-naphthyridine-3-carboxylic acid	298, NMR	1398
Ethyl 5-methyl-4,8-dioxo-1,4,5,8-tetrahydro-1,5-naphthyridine-3-carboxylate	262–264, NMR	301
3-Ethyl-1-methyl-1,5-naphthyridin-2(1H)-one	107–108, IR, UV	155
Ethyl 7-methyl-4-oxo-1,4-dihydro-1,5-naphthyridine-3-carboxylate	300–303, NMR	1398
1-Ethyl-7-methyl-4-oxo-1,4-dihydro-1,5-naphthyridine-3-carboxylic acid	302–303, NMR	1398
1-Ethyl-7-methylthio-4-oxo-1,4-dihydro-1,5-naphthyridine-3-carboxylic acid	259–260, NMR	1398
3-Ethyl-1,5-naphthyridine	162–170/40 or 181–182/56; IR, UV	155
1-Ethyl-1,5-naphthyridin-4(1H)-one	151–152	814
Ethyl 4-oxo-1,4-dihydro-1,5-naphthyridine-3-carboxylate	251–253 or 268, NMR	101, 116, 1245
1-Ethyl-4-oxo-7-piperidino-1,4-dihydro-1,5-naphthyridine-3-carboxylic acid	296–297	1398
1-Ethyl-4-oxo-7-propoxy-1,4-dihydro-1,5-naphthyridine-3-carboxylic acid	212–213, NMR	1398
4-Hydrazino-1,5-naphthyridine	170–171, pK_a	814
4-Hydrazino-1,5-naphthyridine 1-oxide	173, MS	814
2-Hydroxymethyl-1,5-naphthyridine 1,5-dioxide	—	290
1-Hydroxy-1,5-naphthyridin-4(1H)-one	189–191	814
4-Hydroxy-1,5-naphthyridin-2(1H)-one	>360, MS	1023, 1151, 1253
1-Hydroxy-1,5-naphthyridin-2(1H)-one 5-oxide	286	787
4-Hydroxy-3-nitro-1,5-naphthyridin-2(1H)-one	>350, biol	312, 1273
2-Iodo-1,5-naphthyridine	101–103, MS, NMR	692
4-Iodo-1,5-naphthyridine	97–98, MS, NMR	692
4-Isopropylamino-1,5-naphthyridine 1-oxide	150–151	814
2-Isopropyl-6-methoxy-1,5-naphthyridine-4-carbonitrile 1-oxide	160–161, IR, MS, NMR	588
2-Methoxy-6,8-diphenyl-1,5-naphthyridine	155–156, MS, NMR	455
6-Methoxy-2-methyl-1,5-naphthyridin-4-amine	172	1036
2-Methoxy-6-methyl-1,5-naphthyridine	50 to 54, IR, MS, NMR; pic: 190	619, 1036, 1051
2-Methoxy-6-methyl-1,5-naphthyridine 1/5-oxide	154	1051
6-Methoxy-2-methyl-1,5-naphthyridin-4(1H)one	294–296	1036
2-Methoxy-6-methyl-8-phenoxy-1,5-naphthyridine	114	1036
2-Methoxy-1,5-naphthyridine	38, UV	234
4-Methoxy-1,5-naphthyridine	47–49 or 60–63, pK_a, UV; pic: 176–178; Cu(II): biol	202, 814
6-Methoxy-1,5-naphthyridine-4-carbonitrile	180–183, IR, MS, NMR	588
6-Methoxy-1,5-naphthyridine2,3(1H,5H)-dione	240, NMR, UV	234
4-Methoxy-1,5-naphthyridine 1-oxide	92–93	814
6-Methoxy-1,5-naphthyridine-2(1H)-thione	—	271
6-Methoxy-1,5-naphthyridin-2(1H)-one	244–246, NMR, UV	234
6-Methoxy-1,5-naphthyridin-4(1H)-one	270	1036

TABLE A.1. (*Continued*)

1,5-Naphthyridine	Melting Point (°C) etc.	Reference(s)
1-Methoxy-1,5-naphthyridin-2(1H)-one 5-oxide	239–240, NMR	787
6-Methoxy-4-oxo-1,4-dihydro-1,5-naphthyridine-2-carboxylic acid	270	1036
6-Methoxy-4-oxo-1,4-dihydro-1,5-naphthyridine-3-carboxylic acid	>310	1036
2-Methoxy-6-phenyl-1,5-naphthyridine	114–115, IR, NMR	619
2-Methoxy-6-phenyl-8-phenylsulfonyl-1,5-naphthyridine	245–246, MS, NMR	521, 554
4-Methylamino-3-nitro-1,5-naphthyridin-2-amine	222–223, NMR	1443
2-Methylamino-3-nitro-1,5-naphthyridine	165–167, IR, MS, NMR	854
4-Methylamino-3-nitro-1,5-naphthyridine	208–210, NMR	1443
4-Methylamino-3-nitro-1,5-naphthyridin-2(1H)-one	275–277, NMR	1443
4-Methyl-6,7-diphenyl-1,5-naphthyridine	—	522
2-Methyl-1,5-naphthyridine	60 to 65, IR, NMR, UV; HCl: 240–245; pic: 161	87, 155, 268, 269, 828, 964, 1297
3-Methyl-1,5-naphthyridine	73–75, IR, NMR, UV	155, 269, 964
4-Methyl-1,5-naphthyridine	30–32, IR, NMR, UV	155, 269, 964
2-Methyl-1,5-naphthyridine 1,5-dioxide	180–181	797
4-Methyl-1,5-naphthyridine 1-oxide	128–129	814
1-Methyl-1,5-naphthyridin-2(1H)-imine	HI: 270–287, IR, NMR; pic: 244–246	1093
1-Methyl-1,5-naphthyridin-2(1H)-one	104–105, IR, UV; 5-MeI: 216	155
2-Methyl-1,5-naphthyridin-4(1H)-one	254–256, IR, NMR	163, 1000
3-Methyl-1,5-naphthyridin-2(1H)-one	261–262, UV	155, 865
4-Methyl-1,5-naphthyridin-2(1H)-one	261–262, IR, NMR	828
6-Methyl-1,5-naphthyridin-2(1H)-one	261–262, NMR	268, 1051
2-Methyl-6-phenoxy-1,5-naphthyridin	101–103	1051
3-Methyl-4-phenyl-1,5-naphthyridine	184–185	879
4-Methylsulfonyl-1,5-naphthyridine	152–153	814
1-(Methylthiomethyl)-1,5-naphthyridin-2(1H)-one	—	412
4-Methylthio-1,5-naphthyridine	87–88	814
4-Morpholinomethyl-1,5-naphthyridin-3(5H)-one	3HCl: 181–182, NMR	932
1,5-Naphthyridin-2-amine	196 to 205, IR, MS, NMR, pK_a, UV; HClO$_4$: 315; pic: 270 to 277	87, 148, 173, 174, 331, 425, 692, 787, 811, 837, 1047, 1227, 1371
1,5-Naphthyridin-3-amine	143–144 or 144–145, IR, MS, pK_a, UV	181, 331, 811, 837, 1227
1,5-Naphthyridin-4-amine	196 to 204, IR, MS, NMR, pK_a, UV; pic: 238–239	164, 331, 643, 811, 837, 1227, 1371
1,5-Naphthyridin-2-amine 1,5-dioxide	280–282, NMR; HCl:—	787
1,8-Naphthyridin-4-amine 1-oxide	178–180, IR	814
1,5-Naphthyridine	55(?), 70 to 75, ESR, IR, MS, NMR, NQR, pol, pK_a, Raman, st, th, UV, xl st; HClO$_4$: 217; 2HCl: 200; (CO$_2$H)$_2$:	xl st; 1-MeI: 254–255, NMR, pK_a;

TABLE A.1. (*Continued*)

1,5-Naphthyridine	Melting Point (°C) etc.	Reference(s)
	pic: 207–208; 2MeHSO$_4$: 200	14, 17, 32, 42–45, 47, 87, 113, 155, 183, 209, 230, 302, 331, 365, 425, 522, 676, 703, 705, 766, 791, 798, 806, 807, 812, 813, 929, 837, 840, 846, 873, 882, 906, 910, 923, 961, 969, 970, 973, 975, 993, 995, 997, 998, 1002, 1016, 1024, 1027, 1047, 1079, 1083, 1093, 1094, 1113, 1114, 1119, 1126, 1140, 1231, 1241, 1297, 1303, 1306, 1312, 1319, 1321, 1333, 1344, 1371, 1426, 1446
2-*d*-1,5-Naphthyridine	—	629
2,6-*d*$_2$-1,5-Naphthyridine	MS, NMR	302
1,5-Naphthyridine-4-carbaldehyde	IR, UV	953
1,5-Naphthyridine-2-carbaldehyde 1,5-dioxide	H$_2$O: 230–231, IR; PhNHH=: 243; sc: 283, tsc: 263	797
1,5-Naphthyridine-4-carbaldehyde 1-oxide	200–201, NMR	814
1,5-Naphthyridine-3-carboxamide	257	155
1,5-Naphthyridine-2-carboxamide 5-oxide	—	330
1,5-Naphthyridine-2-carboxylic acid	248–249	575, 797
1,5-Naphthyridine-3-carboxylic acid	279, UV	155
1,5-Naphthyridine-4-carboxylic acid	168–169, IR, NMR	814
1,5-Naphthyridine-2-carboxylic acid 1,5-dioxide	177–178	797
1,5-Naphthyridine-2,3-diamine	260–265, NMR	822, 827
1,5-Naphthyridine-2,4-diamine	225, NMR	822
1,5-Naphthyridine-2,6-diamine	—	246
1,5-Naphthyridine-4,8-diamine	255–257, NMR	301
1,5-Naphthyridine-2,6(1*H*,5*H*)-dione	>350 or 360, NMR	234, 1047
1,5-Naphthyridine-2,8(1*H*,5*H*)-dione	>360	233
1,5-Naphthyridine-4,8(1*H*,5*H*)-dione	>300	233
1,5-Naphthyridine 1,5-dioxide	298–301, NMR	233, 349, 971, 1047, 1212
1,5-Naphthyridine-2,6(1*H*,5*H*)-dithione	—	271
1,5-Naphthyridine-4,8(1*H*,5*H*)-dithione	>300	301
1,5-Naphthyridine 1-oxide	125–127 or 127–128, NMR	232, 1047, 1212
2-*d*-1,5-Naphthyridine 1-oxide	—	225

TABLE A.1. (Continued)

1,5-Naphthyridine	Melting Point (°C) etc.	Reference(s)
1,5-Naphthyridine-2(1H)-thione	—	271
1,5-Naphthyridine-4(1H)-thione	252–254	814
1,5-Naphthyridine-2(1H)-thione 5-oxide	—	271
1,5-Naphthyridinium-1-benzimidate	153–155, MS, UV	1003
1,5-Naphthyridin-2(1H)-one	254 to 261, IR, MS, NMR	163, 317, 787, 811, 1000, 1023, 1037, 1044, 1047, 1051, 1151, 1253
1,5-Naphthyridin-3(5H)-one	280, NMR	181, 932
1,5-Naphthyridin-4(1H)-one	>300 or 340, IR, MS, pK_a, UV	17, 101, 110, 116, 887, 1026, 1027, 1035, 1040, 1047, 1253, 1245
1,5-Naphthyridin-2(1H)-one 5-oxide	308	271, 787
2-Nitroamino-1,5-naphthyridine	204–206	811
3-Nitro-1,5-naphthyridin-2-amine	254-255, MS	818, 1273
3-Nitro-1,5-naphthyridin-4-amine	228–229	818, 1273
2-Nitro-1,5-naphthyridine	189–190, IR, MS, NMR	192, 1273
3-Nitro-1,5-naphthyridine	183–184, NMR, pK_a, UV	42, 48, 1273
2-d-3-Nitro-1,5-naphthyridine	183–184	818, 1273
3-Nitro-1,5-naphthyridine-2,4-diamine	267–269, IR, MS, NMR	492, 818, 1273
3-Nitro-1,5-naphthyridin-2(1H)-one	272–274; Aq:—	837, 1273, 1443
3-Nitro-1,5-naphthyridin-4(1H)-one	325 to >360	48, 722, 1037, 1073
2-Oxo-1,2-dihydro-1,5-naphthyridine-3-carbonitrile	NMR	1302
2-Oxo-1,2-dihydro-1,5-naphthyridine-3-carboxamide	NMR	1302
4-Oxo-1,4-dihydro-1,5-naphthyridine-2/3-carboxylic acid(?)	140, biol	17, 101, 107, 116
4-Oxo-1,4-dihydro-1,5-naphthyridine-3-carboxylic acid	>300 or >315, IR	116, 1232, 1245
4-Oxo-1,4-dihydro-1,5-naphthyridine-6,7-dicarboxylic acid	180	17
4-Phenoxy-1,5-naphthyridine	93–94, 140–150/0.005	101
2-Phenyl-1,5-naphthyridine	82–84, biol, NMR	1033
3-Phenyl-1,5-naphthyridine-2(1H)-one	solid, NMR	117
2-Piperidino-6-methyl-1,5-naphthyridine	62–64; pic: 199–200	1051
4-Piperidinomethyl-1,5-naphthyridin-3(5H)-one	3HCl: 171–172, NMR	932
2-Styryl-1,5-naphthyridine	120–121, UV	155
2,3,6-Tribromo-1,5-naphthyridine	—	259
2,4,6-Tribromo-1,5-naphthyridine	—	259
2,4,7-Trichloro-1,5-naphthyridine	194–195	1373

TABLE A.2. ALPHABETICAL LIST OF SIMPLE 1,6-NAPHTHYRIDINES

1,6-Naphthyridine	Melting Point (°C) etc.	Reference(s)
2-Acetamido-1,6-naphthyridine-3-carbonitrile	>300	247
2-Acetamido-1,6-naphthyridine-3-carboxamide	>300	247
5-(1-Acetoxyethyl)-1,6-naphthyridin-2(1*H*)-one	182–185, NMR	490
4-Acetoxy-1,6,7-trimethyl-1,6-naphthyridine-2,5(1*H*,6*H*)-dione	208–210, IR, NMR	950
8-Acetyl-2-chloro-3-fluoro-7-methyl-6-phenyl-1,6-naphthyridin-5(6*H*)-one	119–120, NMR	183
8-Acetyl-2-chloro-3-fluoro-7-methyl-6-propyl-1,6-naphthyridin-5(6*H*)-one	110–112, NMR	183
3-Acetyl-1,6-dibenzyl-4,7-dimethyl-1,6-naphthyridine-(2,5(1*H*,6*H*)-dione	117, IR, NMR, UV	1014
3-Acetyl-1,6-diisopropyl-4,7-dimethyl-1,6-naphthyridine-2,5(1*H*,6*H*)-dione	173–174, IR, NMR	636
8-Acetyl-7-(2-dimethylaminovinyl)-6-propyl-1,6-naphthyridin-5(6*H*)-one	129, NMR	183
3-Acetyl-4,7-dimethyl-1,6-dipropyl-1,6-naphthyridine-2,5(1*H*,6*H*)-dione	136–137, IR, NMR	636
3-Acetyl-7-ethyl-4,8-dimethyl-1,6-dipropyl-1,6-naphthyridine-2,5(1*H*,6*H*)-dione	liq(?), IR, NMR	636
3-Acetyl-4-hydroxy-1,6,7-trimethyl-1,6-naphthyridine-2,5(1*H*,6*H*)-dione	147–148, IR, NMR	950
3-Acetyl-2-methyl-1,6-naphthyridine	106–108, NMR	1068
8-Acetyl-7-methyl-1,6-naphthyridin-5(6*H*)-one	>250, NMR	183
8-Acetyl-7-methyl-6-phenyl-1,6-naphthyridin-5(6*H*)-one	91–92, NMR	183
8-Acetyl-7-methyl-6-propyl-1,6-naphthyridin-5(6*H*)-one	59–61, NMR	183
3-Acetyl-1,6-naphthyridin-2(1*H*)-one	>300, NMR	1068
5-Acetyl-1,6-naphthyridin-2(1*H*)-one	>300, NMR	490
3-Acetyl-1,4,6,7-tetramethyl-1,6-naphthyridin-2,5(1*H*,6*H*)-dione	239–240, IR, NMR	636
7-Amino-6-benzyl-5-oxo-5,6-dihydro-1,6-naphthyridine-8-carbonitrile	228, IR, NMR	1220
5-Amino-7-bromo-2,4-dimethyl-1,6-naphthyridine-8-carbonitrile	>300, IR, MS, NMR, UV	895
7-Amino-5-bromo-2,4-dimethyl-1,6-naphthyridine-8-carbonitrile	300, NMR	913, 994
5-Amino-7-bromo-2,4-diphenyl-1,6-naphthyridine-8-carbonitrile	>300, IR, MS, NMR, UV	895
7-Amino-5-bromo-2,4-diphenyl-1,6-naphthyridine-8-carbonitrile	crude, NMR	913
5-Amino-7-bromo-2-methyl-1,6-naphthyridine-8-carbonitrile	>300, IR, NMR	913
7-Amino-5-bromo-2-methyl-1,6-naphthyridine-8-carbonitrile	>300, IR, MS, NMR	
5-Amino-7-bromo-4-methyl-2-phenyl-1,6-naphthyridine-8-carbonitrile	>300, IR, NMR	913
7-Amino-5-bromo-4-methyl-2-phenyl-1,6-naphthyridine-8-carbonitrile	crude, NMR	913

TABLE A.2. (*Continued*)

1,6-Naphthyridine	Melting Point (°C) etc.	Reference(s)
5-Amino-7-bromo-2-phenyl-1,6-naphthyridine-8-carbonitrile	>300, IR, MS, NMR, UV	895
7-Amino-5-bromo-2-phenyl-1,6-naphthyridine-8-carbonitrile	>300, IR, NMR	913
5-Amino-7-chloro-2,4-dimethyl-1,6-naphthyridine-8-carbonitrile	>300, IR, MS, NMR, UV	895
5-Amino-7-chloro-2,4-diphenyl-1,6-naphthyridine-8-carbonitrile	>300, IR, MS, NMR, UV	895
5-Amino-7-chloro-2-phenyl-1,6-naphthyridine-8-carbonitrile	>300, IR, MS, NMR, UV	895
2-Amino-8-chloro-3-phenyl-1,6-naphthyridin-7(6H)-one	318–321, NMR	564
2-Amino-N,N-dimethyl-1,6-naphthyridine-3-carboxamide	207–210, IR, NMR	1068
5-Amino-2,4-dimethyl-7-oxo-6,7-dihydro-1,6-naphthyridine-8-carbonitrile	>300, IR, MS	673
5-Amino-3,4-dimethyl-7-oxo-6,7-dihydro-1,6-naphthyridine-8-carbonitrile	>300, IR	673
5-(2-Aminoethyl)-7-methyl-1,6-naphthyridin-5(6H)-one	—	383
2-Amino-N-ethyl-1,6-naphthyridine-3-carboxamide	210–212, NMR	974
5-Amino-7-hydrazino-2,4-dimethyl-1,6-naphthyridine-8-carbonitrile	>300, IR, MS, NMR, UV	895
5-Amino-7-hydrazino-2,4-diphenyl-1,6-naphthyridine-8-carbonitrile	265, IR, MS, UV	895
5-Amino-7-hydrazino-2-phenyl-1,6-naphthyridine-8-carbonitrile	>300, IR, MS, NMR, UV	895
5-Amino-7-methoxy-2,4-dimethyl-1,6-naphthyridine-8-carbonitrile	250 or 273, IR, MS, NMR, UV	895, 994
7-Amino-5-methoxy-2,4-dimethyl-1,6-naphthyridine-8-carbonitrile	250 or 258, IR, MS, NMR, UV	895, 994
5-Amino-7-methoxy-2-phenyl-1,6-naphthyridine-8-carbonitrile	280, IR, MS, NMR, UV	895
7-Amino-5-methoxy-2-phenyl-1,6-naphthyridine-8-carbonitrile	240–241, IR, MS, NMR, UV	895
2-Amino-N-methyl-1,6-naphthyridine-8-carboxamide	264–266, NMR	247, 974
3-Amino-5-methyl-1,6-naphthyridin-2(1H)-one	283–285, NMR	1166
6-Amino-7-methyl-1,6-naphthyridin-5(6H)-one	193–194, IR, NMR, UV; PhCH=:—	383, 1072
8-Amino-6-methyl-1,6-naphthyridin-5(6H)-one	crude, NMR	105, 1116
5-Amino-4-methyl-7-oxo-6,7-dihydro-1,6-naphthyridine-8-carbonitrile	>300, IR, MS	673
2-Amino-1,6-naphthyridine-3-carbohydrazide	268–269	247
2-Amino-1,6-naphthyridine-3-carbonitrile	>360, NMR	247, 974
2-Amino-1,6-naphthyridine-3-carbothioamide	294–297	247
2-Amino-1,6-naphthyridine-3-carboxamide	294–295, NMR	247, 974
2-Amino-1,6-naphthyridine-3-carboxamidrazone	240–242	247
2-Amino-1,6-naphthyridine-3-carboxylic acid	340–345, NMR	247, 974
3-Amino-1,6-naphthyridin-4(1H)-one	2HCl: anal	726

TABLE A.2. (*Continued*)

1,6-Naphthyridine	Melting Point (°C) etc.	Reference(s)
7-Amino-5-oxo-6-phenethyl-5,6-dihydro-1,6-naphthyridine-8-carbonitrile	244, IR, NMR	1220
7-Amino-5-oxo-6-phenyl-5,6-dihydro-1,6-naphthyridine-8-carbonitrile	285, IR, NMR	1220
2-Amino-3-phenyl-1,6-naphthyridin-7(6*H*)-one	270–276, NMR	564
6-Amino-7-phenyl-1,6-naphthyridin-5(6*H*)-one	188–191, UV	1072
7-Anilino-1-ethyl-4-oxo-1,4-dihydro-1,6-naphthyridine-3-carboxylic acid	276–279	1408
8-Benzoyl-7-benzyl-2-chloro-3-fluoro-6-phenyl-1,6-naphthyridin-5(6*H*)-one	157–158, NMR, xl st	183
8-Benzoyl-7-benzyl-5-chloro-1,6-naphthyridine	158–159, NMR	183
8-Benzoyl-6-benzyl-4-(2-dimethylaminovinyl)-5-oxo-2-phenyl-5,6-dihydro-1,6-naphthyridine-3-carbonitrile	238–239, MS, NMR	987
8-Benzoyl-7-benzyl-1,6-naphthyridin-5(6*H*)-one	>250, NMR	183
8-Benzoyl-7-benzyl-6-propyl-1,6-naphthyridin-5(6*H*)-one	167–168, NMR	183
8-Benzoyl-6,7-dibenzyl-2-chloro-3-fluoro-1,6-naphthyridin-5(6*H*)-one	160–161, NMR	183
8-Benzoyl-6,7-dibenzyl-3-fluoro-2-methoxy-1,6-naphthyridin-5(6*H*)-one	163–164, NMR	183
8-Benzoyl-6,7-dibenzyl-3-fluoro-2-piperidino-1,6-naphthyridin-5(6*H*)-one	100–101, NMR	183
8-Benzoyl-6,7-dibenzyl-1,6-naphthyridin-5(6*H*)-one	134–135, NMR	183
3-Benzoyl-1,6-diisopropyl-7-methyl-4-phenyl-1,6-naphthyridine-2,5(1*H*,6*H*)-dione	221–222, IR, NMR, xl st	636
8-Benzoyl-7-(2-dimethylaminovinyl)-6-propyl-1,6-naphthyridin-5(6*H*)-one	160, NMR	183
7-Benzoyl-8-hydroxy-1,6-naphthyridin-5(6*H*)-one	222, IR, NMR, UV	1221
8-Benzoyl-7-methyl-1,6-naphthyridin-5(6*H*)-one	>250	183
8-Benzoyl-7-methyl-6-propyl-1,6-naphthyridin-5(6*H*)-one	156–157, NMR	183
5-Benzylamino-7-methyl-1,6-naphthyridine	144–145, NMR	1072
Benzyl 1-benzyl-7-chloro-4-oxo-1,4-dihydro-1,6-naphthyridine-3-carboxylate	186–187, IR, NMR	1284
6-Benzyl-2-chloro-4,7-dimethyl-1,6-naphthyridin-5(6*H*)-one	126–127, IR, NMR, UV	1014
6-Benzyl-4,7-dimethyl-1,6-naphthyridine-2,5(1*H*,6*H*)-dione	283–285, IR, NMR, UV	1014
6-Benzyl-4,7-dimethyl-1,6-naphthyridin-5(6*H*)-one	111–112, IR, NMR, UV	1014
1-Benzyl-2-imino-6,7-dimethyl-5-oxo-1,2,5,6-tetrahydro-1,6-naphthyridine-3-carboxamide	>250, NMR	641
1-Benzyl-2-imino-7-methyl-5-oxo-6-phenethyl-1,2,5,6-tetrahydro-1,6-naphthyridine-3-carbonitrile	184–185, IR, NMR, UV	683
6-Benzyl-2-imino-7-methyl-5-oxo-1-phenethyl-1,2,5,6-tetrahydro-1,6-naphthyridine-3-carbonitrile	267–269, IR, NMR, UV	683
6-Benzyl-7-methyl-1,6-naphthyridin-5(6*H*)-one	—	383
5-Benzyl-1,6-naphthyridine	—	943
2-Benzyl-1,6-naphthyridine-4-carboxylic acid	245, NMR	1219
6-Benzyl-1,6-naphthyridin-5(6*H*)-one	122–123, IR, NMR, UV	1014

TABLE A.2. (*Continued*)

1,6-Naphthyridine	Melting Point (°C) etc.	Reference(s)
7-Bromo-1-(2-bromoethyl)-4-oxo-1,4-dihydro-1,6-naphthyridine-3-carboxylic acid	242–244	1408
3-Bromo-4-chloro-1,6-naphthyridine	123–125, NMR	844
3-Bromo-5-ethyl-1,6-naphthyridin-2-amine	194–195, NMR	590
3-Bromo-5-ethyl1,6-naphthyridin-2(1H)-one	215–216, NMR	590
7-Bromo-1-ethyl-4-oxo-1,4-dihydro-1,6-naphthyridine-3-carboxylic acid	256–258	1408
7-Bromo-1-(2-hydroxyethyl)-4-oxo-1,4-dihydro-1,6-naphthyridine-3-carboxylic acid	248–256	1408
3-Bromo-5-isobutyl-1,6-naphthyridin-2-amine	203–204, NMR	590
3-Bromo-5-isobutyl-1,6-naphthyridin-2(1H)-one	222–224, NMR	590
7-Bromo-1-(2-methoxyethyl)-4-oxo-1,4-dihydro-1,6-naphthyridine-3-carboxylic acid	236–239	1408
3-Bromo-5-methyl-1,6-naphthyridin-2-amine	208–210, NMR	928
3-Bromo-5-methyl-1,6-naphthyridin-2(1H)-one	245–247, NMR	928
5-Bromo-7-methyl-1,6-naphthyridin-2(1H)-one	276–278, IR, NMR	716
3-Bromo-1,6-naphthyridin-4-amine	MS	374, 546
8-Bromo-1,6-naphthyridin-2-amine	HBr: anal	310
2-Bromo-1,6-naphthyridine	—	390
3-Bromo-1,6-naphthyridine	124–125 or 125–126, NMR	173, 844
4-Bromo-1,6-naphthyridine	92–93, NMR	289, 390
5-Bromo-1,6-naphthyridine	112–113, NMR	339
8-Bromo-1,6-naphthyridine	82–84 or 84–86, NMR	173, 336, 390, 1042
8-Bromo-1,6-naphthyridine-2-carboxylic acid	—	244, 577, 1411
3-Bromo-1,6-naphthyridin-4(1H)-one	300–301, MS, NMR	546, 844
4-Bromo-1,6-naphthyridin-2(1H)-one	—	391
5-Bromo-1,6-naphthyridin-2(1H)-one	278–280, NMR	490
7-Bromo-4-oxo-1-vinyl-1,4-dihydro-1,6-naphthyridine-3-carboxylic acid	204–210	1408
3-Bromo-5-phenyl-1,6-naphthyridin-2-amine	218–221, NMR	928
3-Bromo-5-phenyl-1,6-naphthyridin-2(1H)-one	>300, NMR	928
3-Bromo-5-propyl-1,6-naphthyridin-2-amine	178–180, NMR	590
3-Bromo-5-propyl-1,6-naphthyridin-2(1H)-one	218–219, NMR	590
7-Butyl-1,6-naphthyridine	—	544
7-Butyl-8-phenylseleno-1,6-naphthyridine	—	119
7-Butyl-8-phenylthio-1,6-naphthyridine	—	119
7-Butylthio-1-ethyl-4-oxo-1,4-dihydro-1,6-naphthyridine-3-carboxylic acid	169–171	1408
7-*tert*-Butylthio-1-ethyl-4-oxo-1,4-dihydro-1,6-naphthyridine-3-carboxylic acid	210–212	1408
7-Chloro-1-cyclohexyl-8-fluoro-4-oxo-1,4-dihydro-1,6-naphthyridine-3-carboxylic acid	131–132, MS, NMR	1175
4-Chloro-5,7-dimethyl-1,6-naphthyridine	125	1255
5-Chloro-8-(2-ethoxycarbonylvinyl)-1,6-naphthyridine	129–131, IR, NMR	1042
7-Chloro-1-ethyl-4-oxo-1,4-dihydro-1,6-naphthyridine-3-carboxylic acid	260–262, IR	1284
5-Chloro-8-hydroxy-1,6-naphthyridine-7-carboxamide	288–289	108

TABLE A.2. (*Continued*)

1,6-Naphthyridine	Melting Point (°C) etc.	Reference(s)
5-Chloro-8-iodo-1,6-naphthyridine	127–129, NMR	1042
3-Chloro-4-methoxy-7-methyl-1,6-naphthyridine	158–159, NMR	950
8-Chloro-5-methoxy-1,6-naphthyridine	80–82, NMR	299
2-Chloro-5-methyl-1,6-naphthyridine	98–100, NMR	490
5-Chloro-7-methyl-1,6-naphthyridine	112–113, IR, NMR	1072, 1251, 1285
8-Chloro-6-methyl-1,6-nnaphthyridin-5(6H)-one	199–200	299
2-Chloro-1,6-naphthyridine	88–89, NMR	232, 1282
3-Chloro-1,6-naphthyridine	101–102 or 103, NMR	232, 844
4-Chloro-1,6-naphthyridine	90, NMR	45, 232, 289, 1067
5-Chloro-1,6-naphthyridine	104–105 or 106–107, NMR, UV	299, 1251, 1282, 1285
2-Chloro-1,6-naphthyridine-3-carbonitrile	>300	247
8-Chloro-1,6-naphthyridine-2-carboxylic acid	—	244, 577
4-Chloro-1,6-naphthyridine 6-oxide	crude, 152–154, NMR	1069
5-Chloro-1,6-naphthyridin-8-ol	215	108
3-Chloro-1,6-naphthyridin-4(1H)-one	299–301	844
2-Chloro-3-nitro-1,6-naphthyridin-4-amine	>330, IR, MS, NMR	492, 819, 1273
2-Chloro-3-nitro-1,6-naphthyridine	142–143, IR, MS, NMR	819, 1273
4-Chloro-3-nitro-1,6-naphthyridine	139–141	48, 1273
4-Chloro-8-nitro-1,6-naphthyridine	182–183	48, 1273
7-Chloro-3-phenyl-1,6-naphthyridin-2-amine	206–208, NMR	564
5-Chloro-1-phenyl-1,6-naphthyridin-2(1H)-one	NMR	1449
3-(Cyclohex-1-enyl)-1,6-naphthyridin-2-amine	235–237, NMR	974
1-Cyclopropyl-8-methyl-4-oxo-7-piperidino-1,4-dihydro-1,6-naphthyridine-3-carboxylic acid	182–184, NMR	648
2,4-Diamino-5,7-dioxo-1,5,6,7-tetrahydro-1,6-naphthyridine-3-carbonitrile	>270 or >300, IR, NMR	1070, 1096
2,5-Diamino-7-morpholino-4-phenyl-1,6-naphthyridine-3,8-dicarbonitrile	251–253, IR, NMR	784
2,5-Diamino-4-phenyl-7-piperidino-1,6-naphthyridine-3,8-dicarbonitrile	225–227, IR, NMR	784
3-Diazonio-1,6-naphthyridin-4-olate	150–151	726
1,6-Dibenzyl-4,7-dimethyl-1,6-naphthyridine-2,5(1H,6H)-dione	143–145, IR, NMR, UV	950, 1014
1,6-Dibenzyl-4-hydroxy-7-methyl-1,6-naphthyridine-2,5(1H,6H)-dione	165–166, IR, NMR	950
1,6-Dibenzyl-2-imino-7-methyl-5-oxo-1,3,5,6-tetrahydro-1,6-naphthyridine-3-carbonitrile	>200, IR, NMR, UV	683
1,6-Dibenzyl-2-imino-7-methyl-5-oxo-1,2,5,6-tetrahydro-1,6-naphthyridine-3-carboxylic acid	209–211, IR; K: 260–262, IR, NMR	683
2,4-Dibromo-1,6-naphthyridine	—	391
3,4-Dibromo-1,6-naphthyridine	140–141, NMR	844
3,8-Dibromo-1,6-naphthyridine	187–189, NMR	173
4,8-Dibromo-1,6-naphthyridine	—	390
2,3-Dichloro-4-methoxy-7-methyl-1,6-naphthyridine	181, IR, NMR	950
2,5-Dichloro-1,6-naphthyridine	173–174 or 175–176, NMR	1069, 1282
2,8-Dichloro-1,6-naphthyridine	93–94, NMR	1282
3,4-Dichloro-1,6-naphthyridine	122–123, NMR	844
3,5-Dichloro-1,6-naphthyridine	98–99, NMR	1069

TABLE A.2. (*Continued*)

1,6-Naphthyridine	Melting Point (°C) etc.	Reference(s)
4,5-Dichloro-1,6-naphthyridine	134–135, NMR	1069
5,8-Dichloro-1,6-naphthyridine	113–114, NMR	299
7,8-Dichloro-3-phenyl-1,6-naphthyridin-2-amine	271–274, NMR	564
3-Diethoxyphosphinylmethyl-1,6-naphthyridin-2(1*H*)-one	—	924
6-(2-Diethylaminoethyl)-7-methyl-1,6-naphthyridin-5(6*H*)-one	2HCl: 224, UV	1072
N,*N*-Diethyl-8-hydroxy-5-oxo-5,6-dihydro-1,6-naphthyridine-7-carboxamide	194	1444
4-Dimethylamino-1,6-naphthyridine	207–209, NMR	1067
2-Dimethylamino-1,6-naphthyridine-3-carbonitrile	164–168	247
2-Dimethylamino-1,6-naphthyridine-3-carboxamide	214–217	247
2,4-Dimethyl-5,7-dioxo-1,5,6,7-tetrahydro-1,6-naphthyridine-8-carboxamide	295 or 300	673, 757
3,4-Dimethyl-5,7-dioxo-1,5,6,7-tetrahydro-1,6-naphthyridine-8-carboxamide	315, IR	673
2,3-Dimethyl-1,6-naphthyridine	112–114, NMR	247, 1068
5,7-Dimethyl-1,6-naphthyridine	54, 190/10, IR	1251, 1255, 1297
1,6-Dimethyl-1,6-naphthyridine-2,5(1*H*,6*H*)-dione	205–207, NMR	1069
2,4-Dimethyl-1,6-naphthyridine-5,7(1*H*,6*H*)-dione	245, IR	673
5,7-Dimethyl-1,6-naphthyridine 6-oxide	127–132, UV; pic: 194–196	1244
1,5-Dimethyl-1,6-naphthyridin-2(1*H*)-one	203–205, NMR	1166
1,7-Dimethyl-1,6-naphthyridin-5(1*H*)-one(?)	HCl: anal, fl sp	884
5,7-Dimethyl-1,6-naphthyridin-2(1*H*)-one	263–264, IR, NMR	1044
5,7-Dimethyl-1,6-naphthyridin-4(1*H*)-one	240	1255
6,7-Dimethyl-1,6-naphthyridin-5(6*H*)-one	127–128, UV; HCl: 235	383, 1072
5,7-Dimethyl-4-oxo-1,4-dihydro-1,6-naphthyridine-3-carboxylic acid	247	1253
2,4-Dimethyl-7-phenyl-1,6-naphthyridin-5(6*H*)-one	244–245 or 250, IR, MS, NMR	1055, 1078
2-Dimethylsulfimido-1,6-naphthyridine	129–131	692
5,7-Dioxo-4-phenyl-1,5,6,7-tetrahydro-1,6-naphthyridine-8-carboxamide	250, IR	757
8-(2-Ethoxycarbonylvinyl)-5-methoxy-1,6-naphthyridine	101–104, IR, NMR	1042
8-(2-Ethoxycarbonylvinyl)-1,6-naphthyridine	90–92, IR, NMR	1042
7-Ethoxy-1-ethyl-4-oxo-1,4-dihydro-1,6-naphthyridine-3-carboxylic acid	248–250	1408
4-Ethoxy-1,6-naphthyridine	65–66, NMR	1067
2-Ethoxy-3-nitro-1,6-naphthyridin-4-amine	202–203, IR, MS, NMR	492, 819, 1273
2-Ethoxy-3-nitro-1,6-naphthyridine	157–159, MS, NMR	492, 819, 1273
Ethyl 5-amino-7-benzylseleno-8-cyano-2-methyl-1,6-naphthyridine-3-carboxylate	259–260, IR, NMR	959
Ethyl 6-amino-2-methyl-5-oxo-5,6-dihydro-1,6-naphthyridine-3-carboxylate	191, IR, NMR, UV	1105
2-Ethylamino-1,6-naphthyridine-3-carbonitrile	162–164	247
2-Ethylamino-1,6-naphthyridine-3-carboxamide	227–229	247

TABLE A.2. (*Continued*)

1,6-Naphthyridine	Melting Point (°C) etc.	Reference(s)
Ethyl 4-amino-2-oxo-1,2-dihydro-1,6-naphthyridin-3-carboxylate	H_2O: 270–271, IR	725
Ethyl 6-anilino-2-methyl-5-oxo-5,6-dihydro-1,6-naphthyridine-3-carboxylate	194–195, IR, NMR, UV	1105
Ethyl 1-benzyl-7-chloro-4-oxo-1,4-dihydro-1,6-naphthyridine-3-carboxylate	215–217, IR, NMR	1284
Ethyl 1-benzyl-2-imino-6,7-dimethyl-5-oxo-1,2,5,6-tetrahydro-1,6-naphthyridine-3-carboxylate	186–188, NMR	641
Ethyl 1-benzyl-2-imino-7-methyl-5-oxo-6-phenethyl-1,2,5,6-tetrahydro-1,6-naphthyridine-3-carboxylate	136–137, IR, NMR, UV	683
Ethyl 6-benzyl-2-imino-7-methyl-5-oxo-1-phenethyl-1,3,5,6-tetrahydro-1,6-naphthyridine-3-carboxylate	144–146, IR, NMR, UV	683
Ethyl 7-benzylseleno-8-cyano-2-methyl-5-oxo-5,6-dihydro-1,6-naphthyridine-3-carboxylate	207–209, IR, NMR	959
Ethyl 7-bromo-1-ethyl-4-oxo-1,4-dihydro-1,6-naphthyridine-3-carboxylate	178–179	1408
Ethyl 7-bromo-4-oxo-1-propyl-1,4-dihydro-1,6-naphthyridine-3-carboxylate	178–180	1408
Ethyl 7-chloro-1-cyclopropyl-8-fluoro-4-oxo-1,4-dihydro-1,6-naphthyridine-3-carboxylate	165–166, MS, NMR	1175
Ethyl 7-chloro-1-cyclopropyl-8-methyl-4-oxo-1,4-dihydro-1,6-naphthyridine-3-carboxylate	200–202, NMR	648
Ethyl 5-chloro-1-ethyl-4-oxo-1,4-dihydro-1,6-naphthyridine-3-carboxylate	195–197, IR, NMR	1284
Ethyl 7-chloro-1-ethyl-4-oxo-1,4-dihydro-1,6-naphthyridine-3-carboxylate	188–189, IR, NMR	1284
Ethyl 5-chloro-1(2-hydroxyethyl)-4-oxo-1,4-dihydro-1,6-naphthyridine-3-carboxylate	230–232, IR	1284
Ethyl 7-chloro-1-(2-hydroxyethyl)-4-oxo-1,4-dihydro-1,6-naphthyridine-3-carboxylate	188–189, IR	1284
Ethyl 3-cyano-7-hydroxy-4,5-dioxo-1,4,5,6-tetrahydro-1,6-naphthyridine-8-carboxylate	>300, IR, NMR	698
Ethyl-1-cyclopropyl-8-methyl-4-oxo-7-piperidino-1,4-dihydro-1,6-naphthyridine-3-carboxylate	170–172, NMR	648
Ethyl 1,6-dibenzyl-2-imino-7-methyl-5-oxo-1,2,5,6-tetrahydro-1,6-naphthyridine-3-carboxylate	193–195, IR, NMR, UV	683
Ethyl 6-(2-dimethylaminoethyl)-2-methyl-5-oxo-5,6-dihydro-1,6-naphthyridine-3-carboxylate	120–121, IR, NMR, UV	1105
Ethyl 6-(3-dimethylaminopropyl)-2-methyl-5-oxo-5,6-dihydro-1,6-naphthyridine-3-carboxylate	74–75, IR, NMR, UV	1105
Ethyl 1,2-dimethyl-4-oxo-1,4-dihydro-1,6-naphthyridine-3-carboxylate	IR, NMR	952
Ethyl 5,7-dimethyl-4-oxo-1,4-dihydro-1,6-naphthyridine-3-carboxylate	307–308	1255

TABLE A.2. (*Continued*)

1,6-Naphthyridine	Melting Point (°C) etc.	Reference(s)
Ethyl 2,4-dimethyl-5-oxo-7-phenyl-5,6-dihydro-1,6-naphthyridine-8-carboxylate	176, IR, MS, NMR	1078
1-Ethyl-4,7-dioxo-1,4,6,7-tetrahydro-1,6-naphthyridine-3-carboxylic acid	>310	1408
1-Ethyl-7-ethylthio-4-oxo-1,4-dihydro-1,6-naphthyridine-3-carboxylic acid	210–211	1408
Ethyl 6-(2-hydroxyethyl)-2-methyl-5-oxo-5,6-dihydro-1,6-naphthyridine-3-carboxylate	143–144, IR, NMR, UV	1105
Ethyl 8-hydroxy-6-methyl-5-oxo-5,6-dihydro-1,6-naphthyridine-7-carboxylate	95, IR, NMR, UV	639, 654
Ethyl 6-[1-(hydroxymethyl)propyl]-2-methyl-5-oxo-5,6-dihydro-1,6-naphthyridine-3-carboxylate	DL: 125–126, IR, NMR, UV; D: 110–112, IR, NMR, UV	1105
Ethyl 4-hydroxy-2-oxo-1,2-dihydro-1,6-naphthyridine-3-carboxylate	—	391
Ethyl 8-hydroxy-5-oxo-5,6-dihydro-1,6-naphthyridine-7-carboxylate	209, IR, NMR, UV	639, 1221
Ethyl 2-imino-6,7-dimethyl-5-oxo-1-phenethyl-1,2,5,6-tetrahydro-1,6-naphthyridine-3-carboxylate	168–170, NMR	641
Ethyl 2-imino-7-methyl-4-oxo-1,6-diphenethyl-1,2,5,6-tetrahydro-1,6-naphthyridine-3-carboxylate	137–140, IR, NMR, UV	683
1-Ethyl-7-isopropylthio-4-oxo-1,4-dihydro-1,6-naphthyridine-3-carboxylic acid	205–207	1408
1-Ethyl-7-methoxy-4-oxo-1,4-dihydro-1,6-naphthyridine-3-carboxylic acid	247–248	1408
Ethyl 2-methyl-6-methylamino-5-oxo-5,6-dihydro-1,6-naphthyridine-3-carboxylate	146–147, IR, NMR, UV	1105
Ethyl 2-methyl-1,6-naphthyridine-3-carboxylate	94–95, NMR	1068
Ethyl 2-methyl-5-oxo-5,6-dihydro-1,6-naphthyridine-3-carboxylate	260, IR, NMR, UV, xl st	1105, 1147
1-Ethyl-2-methyl-4-oxo-1,4-dihydro-1,6-naphthyridine-3-carboxylic acid	—	952
Ethyl 2-methyl-5-oxo-6-phenethyl-5,6-dihydro-1,6-naphthyridine-3-carboxylate	103–104, IR, NMR, UV	1105
Ethyl 2-methyl-5-oxo-4-phenyl-5,6-dihydro-1,6-naphthyridine-3-carboxylate	235–236, IR, NMR, UV	1105
Ethyl 2-methyl-5-oxo-6-phenyl-5,6-dihydro-1,6-naphthyridine-3-carboxylate	160–161, IR, NMR, UV	1105
1-Ethyl-7-methylthio-4-oxo-1,4-dihydro-1,6-naphthyridine-3-carboxylic acid	0.5H_2O: 230–231	1408
1-Ethyl-7-morpholino-4-oxo-1,4-dihydro-1,6-naphthyridine-3-carboxylic acid	>290	1408
3-Ethyl-1,6-naphthyridin-2-amine	233–234, NMR	974
2-Ethyl-1,6-naphthyridine	51–54, NMR	1068
5-Ethyl-1,6-naphthyridin-2(1H)-one	186–188	1166
7-Ethyl-1,6-naphthyridin-5(6H)-one	197–198, IR, MS, NMR	131
5-Ethyl-1,6-naphthyridin-2(1H)-one-6-oxide	262–264, NMR	490

TABLE A.2. (*Continued*)

1,6-Naphthyridine	Melting Point (°C) etc.	Reference(s)
Ethyl 8-nitro-4-oxo-1,4-dihydro-1,6-naphthyridine-3-carboxylate	273–274	48
Ethyl 2-oxo-1,2-dihydro-1,6-naphthyridine-3-carboxylate	212–213, NMR	1068
Ethyl 4-oxo-1,4-dihydro-1,6-naphthyridine-3-carboxylate	292–293, NMR	48, 113, 1067
1-Ethyl-4-oxo-7-phenylthio-1,4-dihydro-1,6-naphthyridine-3-carboxylic acid	235–238	1408
1-Ethyl-4-oxo-7-propoxy-1,4-dihydro-1,6-naphthyridine-3-carboxylic acid	161–163	1408
1-Ethyl-4-oxo-7-thioxo-1,4,6,7-tetrahydro-1,6-naphthyridine-3-carboxylic acid	>310	1408
Ethyl 2-phenyl-1,6-naphthyridine-3-carboxylate	122–124, NMR	1068
Ethyl 2,4,5-triamino-3-cyano-7-oxo-6,7-dihydro-1,6-naphthyridin-8-carboxylate	290–292, IR	1070
2-(N'-Ethylureido)-3-phenyl-1,6-naphthyridin-7-amine	181–182, NMR	
2-Hexyl-8-iodo-1,6-naphthyridine	liq, IR, NMR	119, 147
7-Hexyl-1,6-naphthyridine	85/3, NMR	147, 621
7-Hexyl-1,6-naphthyridine 6-oxide	97–98, NMR	621
7-Hexyl-8-phenylseleno-1,6-naphthyridine	liq, IR, NMR	119, 147
7-Hexyl-8-phenylthio-1,6-naphthyridine	liq, NMR	119
2-Hydrazino-5-methyl-1,6-naphthyridine	2HCl: 250–253, NMR	490
5-Hydrazino-7-methyl-1,6-naphthyridine	195	1285
4-Hydrazino-1,6-naphthyridine	270, NMR	45, 1067
5-Hydrazino-1,6-naphthyridine	210	1285
5-Hydroxy-2,7-dioxo-3-phenylhydrazono-4-styryl-1,2,3,7-tetrahydro-1,6-naphthyridine-8-carbonitrile (or tautomer)	210, IR, NMR	1360
6-(2-Hydroxyethyl)-7-methyl-1,6-naphthyridin-5(6H)-one	—	383
5-(1-Hydroxyethyl)-1,6-naphthyridin-2(1H)-one	213–215, NMR	490
6-(2-Hydroxyethyl)-7-phenyl-1,6-naphthyridin-5(6H)-one	113–115, IR, NMR	902
5-Hydroxy-4-methyl-2,7-dioxo-3-phenylhydrazono-1,2,3,7-tetrahydro-1,6-naphthyridin3-8-carbonitrile (or tautomer)	240, IR, NMR	1360
5-Hydroxymethyl-1,6-naphthyridin-2(1H)-one	292–295, NMR	490
6-Hydroxy-7-methyl-1,6-naphthyridin-5(6H)-one	161–162, IR, NMR, UV	383, 1072
1-Hydroxy-1,6-naphthyridin-2(1H)-one	350–352 or >350, IR, NMR(?)	215(?), 1282
4-Hydroxy-1,6-naphthyridin-2(1H)-one	MS	391, 1253
8-Hydroxy-1,6-naphthyridin-5(6H)-one	>310; pic: 230	12, 108
4-Hydroxy-3-nitro-1,6-naphthyridin-2(1H)-one	>370	312, 1273
6-Hydroxy-7-phenyl-1,6-naphthyridin-5(6H)-one	204–205	1072
4-Hydroxy-1,6,7-trimethyl-1,6-naphthyridine-2,5(1H,6H)-dione	191–192, IR, NMR	950
2-Imino-6,7-dimethyl-5-oxo-1-phenethyl-1,2,5,6-tetrahydro-1,6-naphthyridine-3-carboxamide	>250, NMR	641

TABLE A.2. (*Continued*)

1,6-Naphthyridine	Melting Point (°C) etc.	Reference(s)
2-Imino-5-oxo-1,6-diphenethyl-1,2,5,6-tetrahydro-1,6-naphthyridine-3-carbonitrile	183–185, IR, NMR, UV	683
8-Iodo-5-methoxy-1,6-naphthyridine	120–121, NMR	1042
8-Iodo-1,6-naphthyridin-5(6H)-one	274–277, IR, NMR	1042
8-Iodo-7-phenyl-1,6-naphthyridine	163–164, IR, NMR	119, 147
Isobutyl 8-hydroxy-6-methyl-5-oxo-5,6-dihydro-1,6-naphthyridine-7-carboxylate	liq, anal, IR, NMR, UV	639, 654
Isopropyl 1-benzyl-2-imino-6,7-dimethyl-5-oxo-1,2,5,6-tetrahydro-1,6-naphthyridine-3-carboxylate	185–186, NMR	641
Isopropyl 8-hydroxy-6-methyl-5-oxo-5,6-dihydro-1,6-naphthyridine-7-carboxylate	124, IR, NMR, UV	654
Isopropyl 8-hydroxy-5-oxo-5,6-dihydro-1,6-naphthyridine-7-carboxylate	180, IR, NMR, UV	639, 1221
Isopropyl 2-imino-6,7-dimethyl-5-oxo-1-phenethyl-1,2,5,6-tetrahydro-1,6-naphthyridine-3-carboxylate	178–179, NMR	641
7-Isopropyl-1,6-naphthyridin-5(6H)-one	190–192, IR, MS, NMR	131
2-Methoxy-5-methyl-1,6-naphthyridine	74–77, NMR	490
5-Methoxy-7-methyl-1,6-naphthyridine	$0.5H_2O$: 58–59, NMR	1072
2-Methoxy-1,6-naphthyridine	65–67, IR, NMR	1818
4-Methoxy-1,6-naphthyridine	115–116, NMR	1067
7-Methoxy-1,6-naphthyridine	IR	836
2-Methoxy-1,6-naphthyridine-3-carbonitrile	186–189	247
2-Methoxy-1,6-naphthyridine-5-carbonitrile	193, IR, NMR	1318
2-Methoxy-1,6-naphthyridine-5-carboxamide	236, IR, NMR	1318
2-Methoxy-1,6-naphthyridine-6-oxide	193, IR, NMR	1318
Methyl 8-acetoxy-5-chloro-1,6-naphthyridine-7-carboxylate	171	108
Methyl 8-acetoxy-1,6-naphthyridine-7-carboxylate	131–132	85
Methyl 8-acetoxy-5-oxo-5,6-dihydro-1,6-naphthyridine-7-carboxylate	224 or 226	85, 109
2-Methylamino-1,6-naphthyridine-3-carbonitrile	269–272	247
2-Methylamino-1,6-naphthyridine-3-carboxamide	229–231	247
Methyl 2-amino-1,6-naphthyridine-3-carboxylate	210–212	247
2-Methylamino-1,6-naphthyridine-3-carboxylic acid	>300	247
Methyl 1-benzyl-2-imino-6,7-dimethyl-5-oxo-1,2,5,6-tetrahydro-1,6-naphthyridine-3-carboxylate	207–209, NMR	641
Methyl 1-benzyl-2-imino-7-methyl-5-oxo-6-phenethyl-1,2,5,6-tetrahydro-1,6-naphthyridine-3-carboxylate	195–197, IR, NMR, UV	683
Methyl 6-benzyl-2-imino-7-methyl-5-oxo-1-phenethyl-1,2,5,6-tetrahydro-1,6-naphthyridine-3-carboxylate	162–164, IR, NMR, UV	683
Methyl 5-chloro-8-hydroxy-1,6-naphthyridine-7-carboxylate	226 or 227	6, 1048
Methyl 5-chloro-8-methoxy-1,6-naphthyridine-7-carboxylate	101	85

TABLE A.2. (*Continued*)

1,6-Naphthyridine	Melting Point (°C) etc.	Reference(s)
Methyl 1,6-dibenzyl-2-imino-7-methyl-5-oxo-1, 2,5,6-tetrahydro-1,6-naphthyridine-3-carboxylate	186–188, IR, NMR, UV	683
4-Methyl-5,7-dioxo-1,5,6,7-tetrahydro-1, 6-naphthyridine-8-carboxamide	198 or 310, IR	673, 757
Methyl 8-hydroxy-6-methyl-5-oxo-5, 6-dihydro-1,6-naphthyridine-7-carboxylate	120, IR, NMR, UV	639, 654
Methyl 8-hydroxy-5-oxo-5,6-dihydro-1, 6-naphthyridine-7-carboxylate	203 to 220, IR, NMR, UV; pic: 220; HCl: 163–165	6, 12, 85, 109, 639, 646, 773, 1048, 1221
Methyl 2-imino-6,7-dimethyl-5-oxo-1-phenethyl-1,2,5,6-tetrahydro-1, 6-naphthyridin-3-carboxylate	175–177, NMR	641
Methyl 2-imino-7-methyl-5-oxo-1, 6-diphenethyl-1,2,5,6-tetrahydro-1, 6-naphthyridine-3-carboxylate	183–184, IR, NMR, UV	683
3-Methyl-1,6-naphthyridin-2-amine	265–267, NMR	247, 974
5-Methyl-1,6-naphthyridin-2-amine	218–221, NMR	490
2-Methyl-1,6-naphthyridine	59–61, NMR	269, 964, 1068
3-Methyl-1,6-naphthyridine	59–61, NMR	269, 964
4-Methyl-1,6-naphthyridine	60–63 or 68–69, NMR, pic: 180–183	269, 880, 964
5-Methyl-1,6-naphthyridine	71–74 or 74–75, IR, UV; pic: 200–201	340, 943, 1251, 1285, 1297
7-Methyl-1,6-naphthyridine	86–87, IR, UV; pic: 180–181	1251, 1285, 1297
4-Methyl-1,6-naphthyridine-2-carboxylic acid	—	244, 577
4-Methyl-1,6-naphthyridine-5,7(1H,6H)-dione	250	673
5-Methyl-1,6-naphthyridine-2,3(1H,6H)-dione	>300, NMR	1166
5-Methyl-1,6-naphthyridine 6-oxide	145–147, MS, NMR	1202
7-Methyl-1,6-naphthyridine 6-oxide (?)	158–159, UV; pic: 216–219	1244
8-Methyl-1,6-naphthyridine 6-oxide	187–188, UV; pic: 170–172	1244
5-Methyl-1,6-naphthyridine-2(1H)-thione	225–228, NMR	490
6-Methyl-1,6-naphthyridin-5(6H)-imine	HI: 270–282, NMR; pic: 224–225	372
1-Methyl-1,6-naphthyridin-4(1H)-one	208–209, NMR	1067
3-Methyl-1,6-naphthyridin-2(1H)-one	>300 or 302–303, IR, NMR	716, 865
5-Methyl-1,6-naphthyridin-2(1H)-one	235–237 or 238–240, NMR	1166
6-Methyl-1,6-naphthyridin-5(6H)-one	95 to 99, IR, NMR; 2-MeI: 251–252	299, 372, 876, 1069, 1214
7-Methyl-1,6-naphthyridin-2(1H)-one	244–245, IR, NMR	716
7-Methyl-1,6-naphthyridin-5(6H)-one	244–246, IR, NMR, UV; HCl: 258–260	383, 1072, 1285, 1447
5-Methyl-1,6-naphthyridin-2(1H)-one 6-oxide	244–245, NMR	490
6-Methyl-8-nitro-1,6-naphthyridin-5(6H)-one	191–192 or 200–201, NMR	1116, 1171, 1273

TABLE A.2. (*Continued*)

1,6-Naphthyridine	Melting Point (°C) etc.	Reference(s)
6-Methyl-5-oxo-5,6-dihdyro-1,6-naphthyridine-8-carbaldehyde	217	412
5-Methyl-2-oxo-1,2-dihydro-1,6-naphthyridine-3-carbohydrazide	>300	1166
5-Methyl-2-oxo-1,2-dihydro-1,6-naphthyridine-3-carbonitrile	278–280, NMR	1166
5-Methyl-2-oxo-1,2-dihydro-1,6-naphthyridine-3-carboxamide	>300	1166
6-Methyl-5-oxo-5,6-dihdyro-1,6-naphthyridine-8-carboxamide	276–278, NMR	105
N-Methyl-2-oxo-1,2-dihydro-1,6-naphthyridine-3-carboxamide	>300, NMR	1068
5-Methyl-2-oxo-1,2-dihydro-1,6-naphthyridine-3-carboxylic acid	255–257	1166
6-Methyl-5-oxo-5,6-dihdyro-1,6-naphthyridine-8-carboxylic acid	240–241, IR, NMR	976
2-Methyl-3-phenyl-1,6-naphthyridine	112–114, NMR	1068
3-Methyl-2-phenyl-1,6-naphthyridine	140–142, NMR	1068
3-Methyl-4-phenyl-1,6-naphthyridine	196–197	879
1-Methyl-5-phenyl-1,6-naphthyridin-2(1*H*)-one	190–192, NMR	490
6-Methyl-2-phenyl-1,6-naphthyridin-5(6*H*)-one	154–155, IR, NMR	1033
1-(Methylthiomethyl)-1,6-naphthyridin-2(1*H*)-one	—	412
1,6-Naphthyridin-2-amine	236 to 245, MS, NMR, pK_a, th, UV	173, 174, 215, 247, 331, 425, 643, 692, 819, 974, 1227
1,6-Naphthyridin-3-amine	223–224, NMR, pK_a, UV	289, 331, 830, 849
1,6-Naphthyridin-4-amine	250 or 254, MS, NMR, pK_a, UV	289, 331, 830, 849, 1229
1,6-Naphthyridin-5-amine	204–206, NMR, pK_a, UV	331
1,6-Naphthyridin-8-amine	135–137, NMR, pK_a, UV	331
1,6-Naphthyridine	25 to 36, ESR, IR, MS, NMR, pK_a, pol, Raman, th, UV; 6-MeI: 154–156, NMR, UV; pic: 202–204 or 219–220	42–45, 134, 165, 173, 215, 299, 302, 331, 365, 372, 621, 676, 791, 806, 813, 829, 840, 873, 878, 880, 923, 943, 968, 975, 998, 1002, 1053, 1083, 1124, 1126, 1140, 1170, 1173, 1181, 1189, 1192, 1214, 1228, 1251, 1282, 1297, 1303, 1312, 1319, 1375, 1426, 1441
5-*d*-1,6-Naphthyridine	—	302
3,5-*d*$_2$-1,6-Naphthyridine	liq, NMR	1069
d$_6$-Naphthyridine	UV etc.	1198

TABLE A.2. (*Continued*)

1,6-Naphthyridine	Melting Point (°C) etc.	Reference(s)
1,6-Naphthyridine-2-carbonitrile	154–155, IR, NMR	1318
1,6-Naphthyridine-5-carbonitrile	142–143 or 143–145, IR, NMR, UV	1318, 1375
1,6-Naphthyridine-2-carbonitrile 1,6-dioxide	260, IR, NMR	224, 1317
1,6-Naphthyridine-5-carbonitrile 1,6-dioxide	257, IR, NMR	224, 1317
1,6-Naphthyridine-2-carbonitrile 1-oxide	235–236, IR, NMR	224, 1317
1,6-Naphthyridine-5-carbonitrile 6-oxide	95–96, IR, NMR	224, 1317
1,6-Naphthyridine-2-carboxylic acid	—	244, 577
1,6-Naphthyridine-3-carboxylic acid	—	575
1,6-Naphthyridine-5-carboxylic acid	204–205	340
1,6-Naphthyridin-2,4-diamine	MS	391, 492
1,6-Naphthyridine-2,5-dicarbonitrile	150–152, IR, NMR	1318
1,6-Naphthyridine-2,5-dicarbonitrile 1,6-dioxide	248, IR, NMR	224, 1317
1,6-Naphthyridine 1,6-dioxide	278 or 285, IR, NMR, UV	1011, 1053, 1212, 1282
1,6-Naphthyridine 1-oxide	158–160 or 235–237, IR, NMR, UV; 6-MeI: UV	215, 1053, 1212, 1282
1,6-Naphthyridine 6-oxide	150 to 158, IR, NMR, UV; pic: 185–186	135, 215, 621, 1011, 1053, 1212, 1244, 1282
1,6-Naphthyridin-8-ol	162, biol, IR, pK_a, UV; 6-MeCl: pK_2, UV	110, 1019, 1026, 1035, 1040, 1048
1,6-Naphthyridin-2(1*H*)-one	285 to 304, IR, MS, NMR, UV	215, 477, 819, 1044, 1166, 1253, 1282, 1318
1,6-Naphthyridin-4(1*H*)-one	297–305, MS, NMR	45, 726, 1067, 1253
1,6-Naphthyridin-5(6*H*)-one	241 to 244, IR, NMR	339, 1055, 1282, 1285, 1318, 1447
1,6-Naphthyridin-2(1*H*)-one 6-oxide	325 ot 350–352(?), NMR(?)	215(?), 216
3-Nitro-1,6-naphthyridin-2-amine	262, IR, MS, NMR	819, 1273
3-Nitro-1,6-naphthyridin-4-amine	298–299	514, 819, 1273
2-Nitro-1,6-naphthyridine	121–123, IR, MS, NMR	692, 1273
3-Nitro-1,6-naphthyridine	159 to 164, NMR, pK_a, UV	42, 48, 656, 1273
4-*d*-3-Nitro-1,6-naphthyridine	162–163 or 163–164, NMR	819, 1273
8-Nitro-1,6-naphthyridine	144–145, NMR, pK_a, UV	42, 48, 1273
3-Nitro-1,6-naphthyridine-2,4-diamine	330, IR, MS, NMR	819, 1273
3-Nitro-1,6-naphthyridin-2(1*H*)-one	>330, IR, MS, NMR	819, 1273
3-Nitro-1,6-naphthyridin-4(1*H*)-one	286	726, 1273
8-Nitro-1,6-naphthyridin-4(1*H*)-one	256–257	48, 1273
8-Nitro-4-oxo-1,4-dihydro-1,6-naphthyridine-3-carboxylic acid	283–284	48
2-Oxo-1,2-dihydro-1,6-naphthyridine-3-carbonitrile	>300, NMR	247, 1068, 1166
2-Oxo-1,2-dihydro-1,6-naphthyridine-3-carboxamide	>300, NMR	1068
4-Oxo-1,4-dihydro-1,6-naphthyridine-3-carboxylic acid	278, NMR	726, 1067
5-Oxo-5,6-dihydro-1,6-naphthyridine-2,8-dicarbonitrile	>300, IR, NMR	1378
4-Phenoxy-1,6-naphthyridine	86–87, NMR	1067

TABLE A.2. (*Continued*)

1,6-Naphthyridine	Melting Point (°C) etc.	Reference(s)
Phenyl 8-hydroxy-5-oxo-5,6-dihydro-1,6-naphthyridine-7-carboxylate	NMR	639
3-Phenyl-1,6-naphthyridin-2-amine	215–217, NMR	247, 974
2-Phenyl-1,6-naphthyridine	95–97 or 97–98, NMR, 6-MeI: 238–240	1033, 1068
3-Phenyl-1,6-naphthyridine	104–105, NMR	1068
7-Phenyl-1,6-naphthyridine	135–137 or 138–139, IR, NMR	147, 621
3-Phenyl-1,6-naphthyridine-5-carbaldehyde	160, NMR	903
2-Phenyl-1,6-naphthyridine-3-carbonitrile	202–203	247
2-Phenyl-1,6-naphthyridine-3-carboxamide	239–240	247
2-Phenyl-1,6-naphthyridine-4-carboxylic acid	>260, NMR	1219
3-Phenyl-1,6-naphthyridine-4-carboxylic acid	>260, NMR	1219
3-Phenyl-1,6-naphthyridine-2,7-diamine	200–201, NMR	564
4-Phenyl-1,6-naphthyridine-5,7(1H,6H)-dione	239, IR	757
7-Phenyl-1,6-naphthyridine 6-oxide	196–199, NMR	621
3-Phenyl-1,6-naphthyridin-2(1H)-one	>300, NMR	1068
5-Phenyl-1,6-naphthyridin-2(1H)-one	261–263	1166
7-Phenyl-1,6-naphthyridin-5(6H)-one	228–229 or 229–230, IR	1070, 1077
7-Phenyl-8-phenylseleno-1,6-naphthyridine	122–124, IR, NMR	147
4-Piperidino-1,6-naphthyridine	solid, anal, NMR; pic: 216–218	1067
2-Propylamino-1,6-naphthyridine-3-carbonitrile	182–184	247
2-Propylamino-1,6-naphthyridine-3-carboxamide	179–182	247
Propyl 8-hydroxy-6-methyl-oxo-5,6-dihydro-1,6-naphthyridine-7-carboxylate	90, IR, NMR, UV	639, 654
5-Propyl-1,6-naphthyridin-2(1H)-one	MeSO$_3$H: 201–203	1166
3,4,8-Tribromo-1,6-naphthyridine	177–179, NMR	1067
2,4,5-Trimethyl-7-oxo-1-phenyl-1,7-dihydro-1,6-naphthyridine-8-carbonitrile	296–297, IR, MS, NMR, UV	742

TABLE A.3. ALPHABETICAL LIST OF SIMPLE 1,7-NAPHTHYRIDINES

1,7-Naphthyridine	Melting Point (°C) etc.	Reference(s)
3-Acetyl-2-methyl-1,7-naphthyridine	112–113	126
5-Allyl-8-piperidino-1,7-naphthyridin-6-amine	solid, anal, IR, NMR	1335
2-Amino-1,7-naphthyridine-3-carbonitrile	NMR	1302
2-Amino-1,7-naphthyridine-3-carboxamide	NMR	1302
2-Amino-1,7-naphthyridin-4(1H)-one	2HCl: anal	723, 1017
6-Benzoyl-1,7-naphthyridine-4,8(1H,7H)-dione	170, IR, NMR, UV	639, 1221
8-Benzyl-1,7-naphthyridine	—	943
3-Bromo-4-chloro-1,7-naphthyridine	136–137, NMR	844
4-Bromo-3-chloro-1,7-naphthyridine	crude, NMR	844
3-Bromo-1,7-naphthyridine-4-amine	MS	374, 546
5-Bromo-1,7-naphthyridin-8-amine	220–221, IR, MS, NMR	546, 1046
8-Bromo-1,7-naphthyridin-6-amine	181 or 187, NMR	187, 198
2-Bromo-1,7-naphthyridine	—	266
3-Bromo-1,7-naphthyridine	92–93, IR, NMR, UV	985

TABLE A.3. (*Continued*)

1,7-Naphthyridine	Melting Point (°C) etc.	Reference(s)
5-Bromo-1,7-naphthyridine	69 to 76	173, 1046, 1374
8-Bromo-1,7-naphthyridine	87–88, NMR	834
3-Bromo-1,7-naphthyridin-4(1H)-one	298–300, MS, NMR	546, 844
4-Bromo-1,7-naphthyridir 2(1H)-one	MS	546
5-Bromo-1,7-naphthyridin-8(7H)-one	333–335, IR, NMR	546, 1046
5-(But-2-enyl)-8-morpholino-1,7-naphthyridin-6-amine	liq, IR, NMR	1335
8-Butyl-5-(3-methylbut-2-enyl)-1,7-naphthyridin-6-amine	liq, IR, NMR	1335
4-Chloro-6,8-dimethyl-1,7-naphthyridine	114–115	1098
6-Chloro-1-ethyl-8-morpholino-4-oxo-1,4-dihydro-1,7-naphthyridine-3-carboxylic acid	278–281	1130
3-Chloro-4-hydrazino-1,7-naphthyridine	200	225
2-Chloro-6-methoxy-1,7-naphthyridine	185–190, UV	234
4-Chloro-7-methyl-1,7-naphthyridin-8(7H)-one	172–173, NMR	1116
6-Chloro-5-methyl-3-phenyl-1,7-naphthyridine	136, MS, NMR	903
5-Chloro-1,7-naphthyridin-8-amine	216–217, IR, NMR	1046
2-Chloro-1,7-naphthyridine	NMR	225, 266
8-d-2-Chloro-1,7-naphthyridine	—	834
6,8-d_2-2-Chloro-1,7-naphthyridine	—	834
3-Chloro-1,7-naphthyridine	88–89, NMR	225, 808
4-Chloro-1,7-naphthyridine	121–122, IR, NMR	93, 225, 1251
5-Chloro-1,7-naphthyridine	75–76, NMR	1046
8-Chloro-1,7-naphthyridine	87–88 or 91–92, IR, UV	42, 155
8-d-5-Chloro-1,7-naphthyridine	crude	1046
3-Chloro-1,7-naphthyridin-4(1H)-one	303–305, NMR; HCl: 271–272, NMR	225, 844
4-Chloro-1,7-naphthyridin-2(1H)-one	—	266
5-Chloro-1,7-naphthyridin-8(7H)-one	345–347, IR, NMR	1046
6-Chloro-1,7-naphthyridin-2(1H)-one	304–306, IR, NMR	477
8-Chloro-5-nitro-1,7-naphthyridine	139–140, NMR	1046, 1273
6-Cyclohexyl-1,7-naphthyridin-8(7H)-one	>250, NMR	1451
3-Diazo-1,7-naphthyridin-4(1H)-one	124; HCl:—	723
5,8-Dibromo-1,7-naphthyridin-2-amine	222–224, NMR	819
2,4-Dibromo-1,7-naphthyridine	—	266
3,4-Dibromo-1,7-naphthyridine	134–135, NMR	844
3,5-Dibromo-1,7-naphthyridine	149–151 or 154–155, NMR	173, 985
5,8-Dibromo-1,7-naphthyridine	132–133, NMR	819, 1046
5,8-Dichloro-1,7-naphthyridin-2-amine	261–263, IR, NMR	819
2,4-Dichloro-1,7-naphthyridine	—	266
3,4-Dichloro-1,7-naphthyridine	122–123 or 125–127, NMR	225, 844
5,8-Dichloro-1,7-naphthyridine	136–137, NMR	819, 1046
2-Diethylamino-3-phenyl-1,7-naphthyridin-4-amine	134, IR, NMR	1141
N,N-Diethyl-5-hydroxy-8-oxo-7,8-dihydro-1,7-naphthyridine-6-carboxamide	189	1444
6,8-Dihydrazino-1,7-naphthyridine	147–149	187
2,3-Dimethyl-1,7-naphthyridine	103, NMR	1103
4,8-Dimethyl-1,7-naphthyridine	95–96, NMR	1033
6,8-Dimethyl-1,7-naphthyridin-2(1H)-one	245–251, IR, NMR	1044
6,8-Dimethyl-1,7-naphthyridin-4(1H)-one	>300	1098

TABLE A.3. (*Continued*)

1,7-Naphthyridine	Melting Point (°C) etc.	Reference(s)
2,4-Dimethyl-6-phenyl-1,7-naphthyridin-8(7H)-one	241–243, IR, NMR	1055
6,8-Diphenyl-1,7-naphthyridin-5-amine	170–171, IR, MS, NMR	1146
2,3-Diphenyl-1,7-naphthyridine	122, NMR	1103
6,8-Diphenyl-1,7-naphthyridin-5(1H)-one	188–189, IR, MS, NMR	1146
6-Ethoxy-1,7-naphthyridin-8-amine	123–124, NMR	710
8-Ethoxy-1,7-naphthyridin-6-amine	152–153, NMR	710
Ethyl 4-amino-1-(ethoxycarbonylmethyl)-2-oxo-1,2-dihydro-1,7-naphthyridine-3-carboxylate	228–230, IR	725
Ethyl 4-amino-2-oxo-1,2-dihydro-1,7-naphthyridine-3-carboxylate	270–271, IR	725
Ethyl 8-anilino-1,7-naphthyridine-6-carboxylate	104–105, IR, MS, NMR	442
Ethyl 6-chloro-1-ethyl-8-morpholino-4-oxo-1,4-dihydro-1,7-naphthyridine-3-carboxylate	191, IR, NMR, UV	1130
Ethyl 4-chloro-6-fluoro-8-morpholino-1,7-naphthyridine-3-carboxylate	134–136, IR, NMR	1130
Ethyl 6-chloro-8-morpholino-4-oxo-1,4-dihydro-1,7-naphthyridine-3-carboxylate	169–173, IR, NMR	1130
Ethyl 1,8-dichloro-4-oxo-1,4-dihydro-1,7-naphthyridine-3-carboxylate	298–302, IR, NMR	445, 1130
Ethyl 6,8-difluoro-4-oxo-1,4-dihydro-1,7-naphthyridine-3-carboxylate	274–275, IR, NMR	503, 1130
Ethyl 4-ethoxy-6-fluoro-8-morpholino-1,7-naphthyridine-3-carboxylate	110, IR, NMR, UV	1130
Ethyl 1-ethyl-6-fluoro-8-morpholino-4-oxo-1,4-dihydro-1,7-naphthyridine-3-carboxylate	135–138, NMR	1130
Ethyl 7-ethyl-5-hydroxy-8-oxo-7,8-dihydro-1,7-naphthyridine-6-carboxylate	NMR	639
1-Ethyl-6-fluoro-8-morpholino-4-oxo-1,4-dihydro-1,7-naphthyridine-3-carboxylic acid	257–265	1130
Ethyl 6-fluoro-8-morpholino-4-oxo-1,4-dihydro-1,7-naphthyridine-3-carboxylate	169–171, IR, NMR	1130
Ethyl 5-hydroxy-7-methyl-8-oxo-7,8-dihydro-1,7-naphthyridine-6-carboxylate	148, IR, NMR, UV	654
Ethyl 4-hydroxy-2-oxo-1,2-dihydro-1,7-naphthyridine-3-carboxylate	—	266
Ethyl 5-hydroxy-8-oxo-7,8-dihydro-1,7-naphthyridine-6-carboxylate	198, IR, NMR, UV	639, 1221
Ethyl 7-methyl-4,8-dioxo-1,4,7,8-tetrahydro-1,7-naphthyridine-3-carboxylate	283–285, NMR	1116
6-Ethyl-1,7-naphthyridin-8(7H)-one	188–189, NMR	1451
Ethyl 5-oxo-1,5-dihydro-1,7-naphthyridine-6-carboxylate	154–157	744
Ethyl 4-oxo-1,4-dihydro-1,7-naphthyridine-3-carboxylate 7-oxide	crude, 272	93
Ethyl 4-oxo-8-phenylthio-1,4-dihydro-1,7-naphthyridine-3-carboxylate	105–106, NMR	157
6-Hexyl-1,7-naphthyridine	78/3, NMR	621
6-Hexyl-1,7-naphthyridine 7-oxide	86–88, NMR	621
2-Hydrazino-6-methoxy-1,7-naphthyridine	170–172	234
4-Hydrazino-1,7-naphthyridine	200–215(?) or 234	45, 1292

TABLE A.3. (*Continued*)

1,7-Naphthyridine	Melting Point (°C) etc.	Reference(s)
8-Hydrazino-1,7-naphthyridine	98–99, UV	155
4-Hydrazino-1,7-naphthyridin-2(1H)-one	—	266
4-Hydroxy-1,7-naphthyridin-2(1H)-one	MS	1253
5-Hydroxy-1,7-naphthyridin-2(1H)-one	>300, IR, NMR	1271
8-Hydroxy-1,7-naphthyridin-6(7H)-one	229–231 or 275, biol	100, 110, 747
4-Hydroxy-3-nitro-1,7-naphthyridin-2(1H)-one	344	312, 1273
6-Isopropoxy-1,7-naphthyridin-8-amine	84–85, NMR	710
Isopropyl 5-hydroxy-8-oxo-7,8-dihydro-1,7-naphthyridine-6-carboxylate	190, IR, NMR, UV	639, 1221
6-Isopropyl-1,7-naphthyridin-8(7H)-one	245–246, NMR	1451
7-Isopropyl-8-oxo-5-phenyl-7,8-dihydro-1,7-naphthyridine-6-carbonitrile	241–243, IR	1397
7-Isopropyl-8-oxo-5-phenyl-7,8-dihydro-1,7-naphthyridine-6-carboxamide	0.5H_2O: 306–310, IR	1397
6-Methoxy-2,4-diphenyl-1,7-naphthyridine	165–166, MS, NMR	455
6-Methoxy-1,7-naphthyridin-8-amine	125–126, NMR	710
8-Methoxy-1,7-naphthyridin-6-amine	172–173, NMR	710
2-Methoxy-1,7-naphthyridine	52–54, NMR	225
4-Methoxy-1,7-naphthyridine	92–94, NMR	225
6-Methoxy-1,7-naphthyridine	45–46, NMR, UV	234
5-Methoxy-1,7-naphthyridin-2(1H)-one	274–276, IR, NMR	1271
6-Methoxy-1,7-naphthyridin-2(1H)-one	250–254, NMR, UV	234
7-Methyl-4,8-dioxo-1,4,7,8-tetrahydro-1,7-naphthyridine-3-carboxylic acid	crude, >259	1116
5-Methyl-6,8-diphenyl-1,7-naphthyridine	171–172, IR, MS, NMR	1146
Methyl 7-ethyl-5-hydroxy-8-oxo-7,8-dihydro-1,7-naphthyridine-6-carboxylate	NMR	639
Methyl 5-hydroxy-7-methyl-8-oxo-7,8-dihydro-1,7-naphthyridine-6-carboxylate	129, IR, NMR, UV	654
Methyl 5-hydroxy-8-oxo-7,8-dihydro-1,7-naphthyridine-6-carboxylate	190–205 or 204, IR, NMR, UV	85, 639, 773, 1221
2-Methyl-1,7-naphthyridine	52, NMR	1103
3-Methyl-1,7-naphthyridine	62, NMR	1103
8-Methyl-1,7-naphthyridine	—	943
7-Methyl-1,7-naphthyridine-4,8(1H,7H)-dione	>285, NMR	1116
7-Methyl-1,7-naphthyridin-8(7H)-imine	HI: 235–270, NMR; pic: 231–233	372
3-Methyl-1,7-naphthyridin-2(1H)-one	253–254	865
6-Methyl-1,7-naphthyridin-8(7H)-one	233–234, NMR	1451
7-Methyl-1,7-naphthyridin-8(7H)-one	119–121, NMR; pic: anal; 7-MeI: NMR	372, 1214
7-Methyl-8-oxo-5-phenyl-7,8-dihydro-1,7-naphthyridine-6-carbaldehyde	212–215, NMR	593
7-Methyl-8-oxo-5-phenyl-7,8-dihydro-1,7-naphthyridine-5-carbonitrile	274–276, NMR	593
7-Methyl-8-oxo-5-phenyl-7,8-dihydro-1,7-naphthyridin-6-carboxamide	>310, NMR	593
7-Methyl-8-oxo-5-phenyl-7,8-dihydro-1,7-naphthyridin-6-carboxylic acid	230–233, NMR; HCl: 180–185	593
5-Methyl-3-phenyl-1,7-naphthyridine	135–138, MS, NMR	903

TABLE A.3. (*Continued*)

1,7-Naphthyridine	Melting Point (°C) etc.	Reference(s)
1-(Methylthiomethyl)-1,7-naphthyridin-2(1*H*)-one	—	412
2-Morpholino-3-phenyl-1,7-naphthyridin-4-amine	243, IR, NMR	1141
1,7-Naphthyridin-2-amine	235–238	266, 275, 425, 834, 1046
1,7-Naphthyridin-3-amine	—	332, 808
1,7-Naphthyridin-4-amine	258–259	332, 425, 808, 834
1,7-Naphthyridin-6-amine	174–175	187
1,7-Naphthyridin-8-amine	165 to 169	173, 275, 425, 834, 1046, 1374
1,7-Naphthyridine	57 to 65, ESR, IR, MS, NMR, pK_a, pol, th, UV; 2-MeI: NMR; pic: 205–206	42–45, 155, 165, 173, 187, 225, 302, 372, 425, 621, 676, 806, 813, 829, 840, 878, 923, 943, 965, 998, 1083, 1126, 1140, 1173, 1181, 1192, 1214, 1251, 1297, 1312, 1319, 1329, 1426
8-*d*-1,7-Naphthyridine	—	302
1,7-Naphthyridine-2-carbaldehyde	HON=: 245–246	126
1,7-Naphthyridine-2-carboxylic acid	—	197, 575
1,7-Naphthyridine-3-carboxylic acid	—	575
1,7-Naphthyridine-2,6(1*H*,6*H*)-dione	350, NMR, UV	234
1,7-Naphthyridine 1,7-dioxide	273–275, NMR	225, 1212
1,7-Naphthyridine 1-oxide	190–192, NMR	225, 1212
1,7-Naphthyridine 7-oxide	142–145 or 150–151, NMR	621, 1212
1,7-Naphthyridin-2(1*H*)-one	291–292, IR, MS	100, 163, 266, 1253
1,7-Naphthyridin-4(1*H*)-one	297–298, MS	45, 93, 723, 1253
1,7-Naphthyridin-5(1*H*)-one	245	744
1,7-Naphthyridin-8(7*H*)-one	233 to 242, biol, IR, NMR, pK_a, UV	99, 155, 339, 834, 1026, 1035, 1040, 1046, 1049, 1055
5-Nitro-1,7-1,7-Naphthyridin-8-amine	247–248	1273
3-Nitro-1,7-naphthyridin-4(1*H*)-one	309–310 or 311–312	723, 1017, 1273
5-Nitro-1,7-naphthyridin-8(7*H*)-one	>350, IR, NMR	1046, 1273
3-Nitro-2-phenyl-1,7-naphthyridine	120–121	126, 1273
2-Oxa-1,2-dihydro-1,7-naphthyridine-3-carbonitrile	NMR	1302
4-Oxo-1,4-dihydro-1,7-naphthyridine-3-carboxylic acid	crude	93, 723
5-Oxo-1,5-dihydro-1,7-naphthyridine-6-carboxylic acid	234	744
4-Oxo-8-phenylthio-1,4-dihydro-1,7-naphthyridine-3-carboxylic acid	194–196, NMR	157
2-Phenyl-1,7-naphthyridine	113–115	100, 1033, 1103
3-Phenyl-1,7-naphthyridine	88, NMR	1103
6-Phenyl-1,7-naphthyridine	116–117, NMR	621
3-Phenyl-1,7-naphthyridine-5-carbaldehyde	148, MS, NMR	903

TABLE A.3. (*Continued*)

1,7-Naphthyridine	Melting Point (°C) etc.	Reference(s)
6-Phenyl-1,7-naphthyridine 7-oxide	212–214, NMR	621
3-Phenyl-1,7-naphthyridin-2(1*H*)-one	solid, NMR	117
6-Phenyl-1,7-naphthyridin-8(7*H*)-one	239–240, NMR	1451
Propyl 7-ethyl-5-hydroxy-8-oxo-7,8-dihydro-1,7-naphthyridine-6-carboxylate	NMR	639
Propyl 5-hydroxy-7-methyl-8-oxo-7,8-dihydro-1,7-naphthyridine-6-carboxylate	147, IR, NMR, UV	654

TABLE A.4. ALPHABETICAL LIST OF SIMPLE 1,8-NAPHTHYRIDINES

1,8-Naphthyridine	Melting Point (°C) etc.	Reference(s)
2-Acetamido-4-acetoxy-3-benzyl-1,8-naphthyridin-2(1*H*)-one	325, IR, NMR	1269
7-Acetamido-4-acetoxy-3-butyl-1,8-naphthyridin-2(1*H*)-one	292, IR, NMR	1269
7-Acetamido-4-acetoxy-3-ethyl-1,8-naphthyridin-2(1*H*)-one	301, IR, NMR	1269
7-Acetamido-4-acetoxy-3-methyl-1,8-naphthyridin-2(1*H*)-one	308, IR, NMR	1269
7-Acetamido-4-acetoxy-3-phenyl-1,8-naphthyridin-2(1*H*)-one	324, IR, NMR	1269
2-Acetamido-7-aminoethyl-1,8-naphthyridine	169–172, IR, NMR; HCl: 264–266	472
6-Acetamido-4-amino-1-ethyl-2-oxo-1,2-dihydro-1,8-naphthyridine-3-carboxylic acid	304–306	612
2-Acetamido-7-azido-3-methyl-1,8-naphthyridine	289–291	743
2-Acetamido-7-azido-4-methyl-1,8-naphthyridine	310, IR	761
2-Acetamido-7-azido-5-methyl-1,8-naphthyridine	>320, IR	761
2-Acetamido-7-azido-6-methyl-1,8-naphthyridine	>300, IR	764
2-Acetamido-5-azido-1,8-naphthyridine	210–215	720
2-Acetamido-7-azido-1,8-naphthyridine	304–305, IR	764
2-Acetamido-7-azido-3-phenyl-1,8-naphthyridine	230–233, IR	972
2-Acetamido-7-azido-4-phenyl-1,8-naphthyridine	286–288	718
2-Acetamido-7-azido-5-phenyl-1,8-naphthyridine	268–270	718
2-Acetamido-7-azido-6-phenyl-1,8-naphthyridine	269–271, IR	972
2-Acetamido-3-bromo-5-methoxy-7-phenyl-1,8-naphthyridine	209–210, IR, MS, NMR	472
2-Acetamido-7-bromomethyl-1,8-naphthyridine	178–180, IR, MS, NMR	112, 472
2-Acetamido-7-chloro-5-chloromethyl-1,8-naphthyridine	210–212	79
2-Acetamido-7-chloromethyl-1,8-naphthyridine	189–191, MS, NMR	472
2-Acetamido-7-chloro-3-methyl-1,8-naphthyridine	252–254	743
2-Acetamido-7-chloro-4-methyl-1,8-naphthyridine	260–261	761
2-Acetamido-7-chloro-5-methyl-1,8-naphthyridine	240 or 245–247	153, 1052
7-Acetamido-4-chloromethyl-1,8-naphthyridin-2(1*H*)-one	>320	79
2-Acetamido-7-chloro-5-morpholinomethyl-1,8-naphthyridine	254–256, NMR	89
2-Acetamido-5-chloro-1,8-naphthyridine	244–246 or 265–266	101, 720

TABLE A.4. (*Continued*)

1,8-Naphthyridine	Melting Point (°C) etc.	Reference(s)
2-Acetamido-7-chloro-1,8-naphthyridine	251–253	702
2-Acetamido-7-chloro-3-phenyl-1,8-naphthyridine	183–186	972
2-Acetamido-7-chloro-4-phenyl-1,8-naphthyridine	247–248	718
2-Acetamido-7-chloro-5-phenyl-1,8-naphthyridine	267–269	741
2-Acetamido-7-chloro-6-phenyl-1,8-naphthyridine	270–272	972
2-Acetamido-7-chloro-5-piperidinomethyl-1,8-naphthyridine	231–233	79
2-Acetamido-5,7-dichloro-1,8-naphthyridine	218–220 or 228–230, NMR	533, 1226
2-Acetamido-5,7-dimethyl-1,8-naphthyridine	297 or 300; HBr: 252–254; 8-PhCH$_2$Cl: 185–190	67, 980, 1034
2-Acetamido-5,7-dimethyl-1,8-naphthyridine-3-carboxamide	246, IR, NMR	1099
±-Acetamido-7-hydroxymethyl-1,8-naphthyridine	240	1445
6-Acetamido-4-hydroxy-1,8-naphthyridin-2(1*H*)-one	325	114
7-Acetamido-4-hydroxy-1,8-naphthyridin-2(1*H*)-one	>320	533
2-Acetamido-5-methyl-1,8-naphthyridine	197 to 262; TsOMe: 200–202	153, 680, 980
2-Acetamido-7-methyl-1,8-naphthyridine	280, IR, NMR	1445
7-Acetamido-3-methyl-1,8-naphthyridin-2(1*H*)-one	>320	764
7-Acetamido-3-methyl-1,8-naphthyridin-4(1*H*)-one	322–325	1076
7-Acetamido-4-methyl-1,8-naphthyridin-2(1*H*)-one	>285 or >310	73, 1052
7-Acetamido-5-methyl-1,8-naphthyridin-2(1*H*)-one	>320, IR	761
7-Acetamido-6-methyl-1,8-naphthyridin-2(1*H*)-one	282–285	743
2-Acetamido-5-methyl-7-phenoxy-1,8-naphthyridine	205	1052
2-Acetamido-7-methyl-5-phenyl-1,8-naphthyridine	207–208; 8-MeCl; 235–236; 8-MeI; 244–245	771, 1052
7-Acetamido-5-morpholinomethyl-1,8-naphthyridin-2(1*H*)-one	270–2272, NMR	89
7-Acetamido-1,8-naphthyridin-2-amine	261–263, NMR, xl st	82
7-Acetamido-1,8-naphthyridine-2-carbaldehyde	215, IR, NMR	1445
2-Acetamido-1,8-naphthyridine-3-carboxamide	>300 or >350, IR, NMR	278, 1099
7-Acetamido-1,8-naphthyridin-2(1*H*)-one	>320	702
7-Acetamido-1,8-naphthyridin-4(1*H*)-one	310–315 or 325–327	101, 720, 1076
7-Acetamido-3-nitro-1,8-naphthyridin-2(1*H*)-one	300–301, NMR	1401
2-Acetamido-7-phenyl-1,8-naphthyridine-3-carbonitrile	237–238, IR	1112
2-Acetamido-7-phenyl-1,8-naphthyridine-3-carboxamide	>350, IR, NMR	1099
7-Acetamido-2-phenyl-1,8-naphthyridin-4(1*H*)-one	267–269 or >320, NMR, UV	263, 1136
7-Acetamido-3-phenyl-1,8-naphthyridin-2(1*H*)-one	310, IR	972
7-Acetamido-4-phenyl-1,8-naphthyridin-2(1*H*)-one	>345	73
7-Acetamido-5-phenyl-1,8-naphthyridin-2(1*H*)-one	>340	718
7-Acetamido-6-phenyl-1,8-naphthyridin-2(1*H*)-one	280–282, IR	972
7-Acetamido-4-piperidinomethyl-1,8-naphthyridin-2(1*H*)-one	304–306	79
2-Acetamido-7-piperidino-5-piperidinomethyl-1,8-naphthyridine	315–317	79

TABLE A.4. (*Continued*)

1,8-Naphthyridine	Melting Point (°C) etc.	Reference(s)
4-Acetoxy-3-allyl-6-methoxy-1-phenyl-1,8-naphthyridin-2(1*H*)-one	164–166	607
4-Acetoxy-3-allyl-1-phenyl-1,8-naphthyridin-2(1*H*)-one	195–196	607
4-Acetoxy-1-benzyl-1,8-naphthyridin-2(1*H*)-one	133–134, IR, NMR	160
1-(3-Acetoxypropyl)-3-methyl-1,8-naphthyridin-2(1*H*)-one	solid (?), NMR	1298
6-Acetyl-4-amino-7-methyl-5-phenyl-2-thioxo-1,2-dihydro-1,8-naphthyridine-3-carbonitrile	310–312, IR, MS, NMR	1409
1-Acetyl-3-benzoyl-4-hydroxy-1,8-naphthyridin-2(1*H*)-one	206–211, IR, NMR	912
3-Acetyl-1-benzoyl-4-hydroxy-1,8-naphthyridin-2(1*H*)-one	211–214, IR, NMR	912
1-Acetyl-3-benzylidene-7-methyl-2,3-dihydro-1,8-naphthyridin-4(1*H*)-one (?)	112–115, NMR	1102
1-Acetyl-3-butyl-4-hydroxy-1,8-naphthyridin-2(1*H*)-one	237–238, IR, NMR	140
1-Acetyl-3-butyryl-4-hydroxy-1,8-naphthyridin-2(1*H*)-one	182–186, IR, NMR	912
6-Acetyl-2-ethoxy-7-methyl-4-phenyl-1,8-naphthyridine-3-carbonitrile	180–182, IR, MS, NMR	896
3-(2-Acetylethyl)-1-ethyl-7-methyl-1,8-naphthyridin-4(1*H*)-one	142, IR, MS, NMR	905
7-Acetyl-1-ethyl-6-fluoro-4-oxo-1,4-dihydro-1,8-naphthyridine-3-carboxylic acid	224–226, IR, NMR	1163
3-(2-Acetylethyl)-4-hydroxy-1-phenyl-1,8-naphthyridin-2(1*H*)-one	182–184	607
3-Acetyl-1-ethyl-7-methyl-1,8-naphthyridin-4(1*H*)-one	185–186, IR, MS, NMR	905
7-Acetyl-1-ethyl-4-oxo-1,4-dihydro-1,8-naphthyridine-3-carboxylic acid	283–285, IR, NMR	1163
1-Acetyl-3-hexanoyl-4-hydroxy-1,8-naphthyridin-2(1*H*)-one	179–182, IR, NMR	912
3-Acetyl-1-hexanoyl-4-hydroxy-1,8-naphthyridin-2(1*H*)-one	—	162
1-Acetyl-4-hydroxy-3-methyl-1,8-naphthyridin-2(1*H*)-one	209–211, IR, NMR	140
1-Acetyl-4-hydroxy-3-methylsulfonyl-1,8-naphthyridin-2(1*H*)-one	197–199, IR, NMR	912
1-Acetyl-4-hydroxy-2-oxo-1,2-dihydro-1,8-naphthyridine-3-carbonitrile	213–214, NMR	1225
1-Acetyl-4-hydroxy-3-phenyl-1,8-naphthyridin-2(1*H*)-one	220–222, IR, NMR	140
3-Acetyl-2-methyl-1,8-naphthyridine	146–147, IR, NMR	307, 658, 674, 707, 1112, 1440
3-Acetyl-7-methyl-1,8-naphthyridin-4(1*H*)-one	>300, UV	292, 1076
3-Acetyl-2-methyl-7-phenyl-1,8-naphthyridine	146–147, IR	1112
3-Acetyl-2-methyl-1-phenyl-1,8-naphthyridin-4(1*H*)-one	142–143, NMR	183

TABLE A.4. (*Continued*)

1,8-Naphthyridine	Melting Point (°C) etc.	Reference(s)
3-(2-Acetylvinyl)-1-ethyl-7-methyl-1,8-naphthyridin-4(1H)-one	184–186, IR, NMR	905
1-Allyl-7-allyloxy-1,8-naphthyridin-2(1H)-one	~235/1.25	1380
3-Allyl-4,6-dihydroxy-1-phenyl-1,8-naphthyridin-2(1H)-one	241–243	607
1-Allyl-4-hydroxy-3-nitro-1,8-naphthyridin-2(1H)-one	115–118	1127
3-Allyl-4-hydroxy-1-phenyl-1,8-naphthyridin-2(1H)-one	251–252	607
1-Allyl-7-methyl-1,8-naphthyridin-4(1H)-one	37–40, 245/1.25	1380
1-Allyl-7-methyl-4-oxo-1,4-dihydro-1,8-naphthyridine-3-carboxylic acid	207–208	154
4-Allyloxy-3-butyl-1-phenyl-1,8-naphthyridin-2(1H)-one	138–140	607
4-Allyloxy-6-methoxy-1-phenyl-1,8-naphthyridin-2(1H)-one	157–158	607
4-Allyloxy-7-methyl-1,8-naphthyridine	105–110	1308
2-Allyloxy-1,8-naphthyridine	—	427
4-Allyloxy-1-phenyl-1,8-naphthyridin-2(1H)-one	176–177	607
7-Amino-3-benzyl-4-hydroxy-1,8-naphthyridin-2(1H)-one	320, IR, NMR	1269
7-Amino-N-*tert*-butyl-2-[(*tert*-butylcarbamoyl)methyl]-1,8-naphthyridine-3-carboxamide	270–272, IR, NMR	482
2-Amino-N-butyl-5,7-dimethyl-1,8-naphthyridine-3-carboxamide	205–207	276
7-Amino-3-butyl-4-hydroxy-1,8-naphthyridin-2(1H)-one	301, IR, NMR	1269
2-Amino-N-butyl-1,8-naphthyridine-3-carboxamide	173–174	278
4-Amino-1-butyl-3-nitro-1,8-naphthyridin-2(1H)-one	221–225, NMR	471
2-Amino-N-butyl-7-phenyl-1,8-naphthyridine-3-carboxamide	280–282	1400
7-Amino-4-carboxymethyl-1,8-naphthyridin-2(1H)-one	>320, IR	681
6-Amino-1-(2-chloroethyl)-7-methoxy-4-oxo-1,4-dihydro-1,8-naphthyridine-3-carboxylic acid	>300, MS, NMR	1029
6-Amino-7-chloro-1-ethyl-4-oxo-1,4-dihydro-1,8-naphthyridine-3-carboxylic acid	>300	1029
7-Amino-4-chloromethyl-1,8-naphthyridin-2(1H)-one	>320, UV	1136
7-Amino-3-chloro-1,8-naphthyridin-2(1H)-one	300, IR	715
2-Amino-6-cyano-5,8-dimethyl-7-oxo-7,8-dihydro-1,8-naphthyridine-3-carbohydrazide	352, IR, NMR	618
7-(3-Aminocyclohex-1-enyl)-1-cyclopropyl-6-fluoro-4-oxo-1,4-dihydro-1,8-naphthyridine-3-carboxylic acid	HCl: IR, MS, NMR	1158
7-(3-Aminocyclopent-1-enyl)-1-cyclopropyl-6-fluoro-4-oxo-1,4-dihydro-1,8-naphthyridine-3-carboxylic acid	HCl: IR, MS, NMR	1158
7-(1-Aminocyclopropyl)-1-cyclopropyl-6-fluoro-4-oxo-1,4-dihydro-1,8-naphthyridine-3-carboxylic acid	220–222, IR, NMR	1201

TABLE A.4. (*Continued*)

1,8-Naphthyridine	Melting Point (°C) etc.	Reference(s)
4-Amino-1,*N*-diethyl-2-oxo-1,2-dihydro-1,8-naphthyridine-3-carboxamide	196–198	612
2-Amino-5,7-dimethyl-1,8-naphthyridine-3-carboguanide	298–300	276
2-Amino-5,7-dimethyl-1,8-naphthyridine-3-carbohydrazide	255–257	276
2-Amino-5,7-dimethyl-1,8-naphthyridine-3-carbonitrile	294–295	276
2-Amino-*N*,*N*-dimethyl-1,8-naphthyridine-3-carboxamide	228–230	278
2-Amino-5,7-dimethyl-1,8-naphthyridine-3-carboxamide	262 or 283–285, IR; HCl: 290–292	276, 1099
2-Amino-5,7-dimethyl-1,8-naphthyridine-3-carboxylic acid	258–260	276
3-Amino-1,7-dimethyl-1,8-naphthyridin-4(1*H*)-one	155–157, NMR	1116
6-Amino-5,7-dimethyl-1,8-naphthyridin-2(1*H*)-one	HCl: 225, IR, NMR	1039
7-Amino-1,4-dimethyl-1,8-naphthyridin-2(1*H*)-one	225–227	1392
7-Amino-2,3-dimethyl-1,8-naphthyridin-4(1*H*)-one	>320, UV	1136
2-Amino-5,7-dimethyl-4-oxo-1,4-dihydro-1,8-naphthyridine-3-carboxamide	420	1301
2-Amino-5,7-dimethyl-*N*-pentyl-1,8-naphthyridine-3-carboxamide	176–177	276
2-Amino-*N*,*N*-dimethyl-7-phenyl-1,8-naphthyridine-3-carboxamide	222–225	1400
2-Amino-5,7-dimethyl-*N*-propyl-1,8-naphthyridine-3-carboxamide	210–212	276
7-Amino-3,6-dinitro-1,8-naphthyridin-2(1*H*)-one	>320, IR	721
7-Amino-3,6-dinitro-4-phenyl-1,8-naphthyridin-2(1*H*)-one	>320, IR, NMR	749, 1273
7-Amino-4-ethoxycarbonylmethyl-1,8-naphthyridin-2(1*H*)-one	>320, IR	681
7-Amino-2-ethoxy-4-phenyl-1,8-naphthyridine-3-carbonitrile	268–270, IR, MS, NMR	78, 896
2-Amino-7-ethoxy-5-phenyl-1,8-naphthyridine-3,6-dicarbonitrile	293–295, IR, MS, NMR	896
2-Amino-*N*-ethyl-5,7-dimethyl-1,8-naphthyridine-3-carboxamide	221–222	276
1-(2-Aminoethyl)-5,7-dimethyl-1,8-naphthyridin-2(1*H*)-one	HCl: 253–254	338
4-Amino-1-ethyl-3,6-dinitro-1,8-naphthyridin-2(1*H*)-one	260–262, IR, MS, NMR	692, 1273
6-Amino-1-ethyl-4,7-dioxo-1,4,7,8-tetrahydro-1,8-naphthyridine-3-carboxylic acid	>300	1211
6-Amino-1-ethyl-7-ethylamino-4-oxo-1,4-dihydro-1,8-naphthyridine-3-carboxylic acid	>300	1029
7-Amino-1-ethyl-6-fluoro-4-oxo-1,4-dihydro-1,8-naphthyridine-3-carboxylic acid	>300, NMR	387, 568
7-Amino-3-ethyl-4-hydroxy-1,8-naphthyridin-2(1*H*)-one	308, NMR	1269

Appendix

TABLE A.4. (*Continued*)

1,8-Naphthyridine	Melting Point (°C) etc.	Reference(s)
6-Amino-1-ethyl-7-methylamino-4-oxo-1,4-dihydro-1,8-naphthyridine-3-carboxylic acid	>300	1029
6-Amino-1-ethyl-*N*-methyl-7-methylamino-4-oxo-1,4-dihydro-1,8-naphthyridine-3-carboxamide	300	1029
3-Amino-1-ethyl-7-methyl-1,8-naphthyridin-4(1*H*)-one	133–135, NMR	1116
7-Amino-1-ethyl-4-methyl-1,8-naphthyridin-2(1*H*)-one	146–148	1392
4-Amino-1-ethyl-7-methyl-2-oxo-1,2-dihydro-1,8-naphthyridine-3-carbonitrile	>300	612
4-Amino-1-ethyl-7-methyl-2-oxo-1,2-dihydro-1,8-naphthyridine-3-carboxamide	263–266	612
6-Amine-1-ethyl-7-methyl-4-oxo-1,4-dihydro-1,8-naphthyridine-3-carboxylic acid	298–300, IR, NMR	395
2-Amino-*N*-ethyl-1,8-naphthyridine-3-carboxamide	218–220	278
7-Amino-4-ethyl-1,8-naphthyridine-2(1*H*)-thione	272–274	1380
7-Amino-1-ethyl-1,8-naphthyridin-2(1*H*)-one	154–156, NMR	1392
4-Amino-1-ethyl-3-nitro-1,8-naphthyridin-2(1*H*)-one	285–286	692, 1273
4-Amino-1-ethyl-2-oxo-1,2-dihydro-1,8-naphthyridine-3-carbohydrazide	275–277	612
4-Amino-1-ethyl-2-oxo-1,2-dihydro-1,8-naphthyridine-3-carboxamide	231–232	612
4-Amino-1-ethyl-1-oxo-1,2-dihydro-1,8-naphthyridine-3-carboxylic acid	245–248	612
6-Amino-1-ethyl-4-oxo-7-thioxo-1,4,7,8-tetrahydro-1,8-naphthyridine-3-carboxylic acid	>300	1031
2-Amino-*N*-ethyl-7-phenyl-1,8-naphthyridine-3-carboxamide	302–305	1400
2-Amino-4-(4-hydroxybutyl)-1,8-naphthyridine-3-carbonitrile	245, NMR	500
7-Amino-1-(2-hydroxyethyl)-4-oxo-1,4-dihydro-1,8-naphthyridine-3-carboxylic acid	>300	1029
7-Amino-4-hydroxy-3-methyl-1,8-naphthyridin-2(1*H*)-one	305, IR, NMR	1269
2-Amino-8-hydroxy-5-methyl-7-oxo-7,8-dihydro-1,8-naphthyridine-3-carboxylic acid	>320	748
6-Amino-4-hydroxy-1,8-naphthyridin-2(1*H*)-one	325	114
7-Amino-4-hydroxy-1,8-naphthyridin-2(1*H*)-one	>320, NMR	533
2-Amino-8-hydroxy-7-oxo-7,8-dihdyro-1,8-naphthyridine-3-carboxylic acid	>320, IR	741
2-Amino-8-hydroxy-7-oxo-5-phenyl-7,8-dihydro-1,8-naphthyridine-3-carboxylic acid	300, IR	741
7-Amino-4-hydroxy-3-phenyl-1,8-naphthyridin-2(1*H*)-one	306, IR, NMR	1269
2-Amino-4-(4-iodobutyl)-1,8-naphthyridine-3-carbonitrile	235, NMR	500
7-Amino-2-methoxy-4,6-diphenyl-1,8-naphthyridine-3-carbonitrile	291–293, IR, MS, NMR	896
3-Amino-2-methoxy-7-morpholino-6-phenyl-1,8-naphthyridine	130–132, NMR	1304

TABLE A.4. (*Continued*)

1,8-Naphthyridine	Melting Point (°C) etc.	Reference(s)
3-Amino-2-methoxy-6-phenyl-7-piperidino-1,8-naphthyridine	182–184, NMR	1304
2-Amino-N'-methyl-1,8-naphthyridine-3-carbohydrazide	199–201	278
2-Amino-N-methyl-1,8-naphthyridine-3-carboxamide	222–223	278
7-Amino-4-methyl-1,8-naphthyridine-2(1H)-thione	275, NMR	1380
2-Amino-1-methyl-1,8-naphthyridin-2(1H)-one	269, NMR	1380, 1392
7-Amino-2-methyl-1,8-naphthyridin-4(1H)-one	268–270 or >320, IR, NMR, UV	701, 1136
7-Amino-3-methyl-1,8-naphthyridin-2(1H)-one	>340 or >350, IR	34, 681, 687
7-Amino-4-methyl-1,8-naphthyridin-2(1H)-one	301 to >360, IR, NMR, UV	73, 153, 681, 689, 690, 748, 759, 771, 1052, 1136, 1333
7-Amino-5-methyl-1,8-naphthyridin-2(1H)-one	>320, IR	761
7-Amino-6-methyl-1,8-naphthyridin-2(1H)-one	>320, IR	743
2-Amino-5-methyl-7-oxo-7,8-dihydro-1,8-naphthyridine-3-carboxylic acid	>320	748
7-Amino-3-methyl-2-oxo-1,2-dihydro-1,8-naphthyridine-4-carboxylic acid	>360	687
2-Amino-5-methyl-4-phenyl-7-thioxo-7,8-dihydro-1,8-naphthyridine-3,6-dicarbonitrile	237, IR, NMR	1399
3-Amino-7-morpholino-6-phenyl-1,8-naphthyridin-2(1H)-one	204–206, NMR	1304
7-Amino-1,8-naphthyridine-2-carbaldehyde	MS, NMR	1445
2-Amino-1,8-naphthyridine-3-carboguanide	>300	278
2-Amino-1,8-naphthyridine-3-carbohydrazide	>300	278
2-Amino-1,8-naphthyridine-3-carbonitrile	200 to 265, IR, NMR, UV	278, 307, 658, 674, 709, 1030, 1052, 1100
2-Amino-1,8-naphthyridine-3-carbothioamide	265–268	278
2-Amino-1,8-naphthyridine-3-carboxamide	287–288, IR, NMR; HCl: 289–290, biol	278, 359, 658, 674, 1100
2-Amino-1,8-naphthyridine-3-carboxamidrazone	218–220	278
2-Amino-1,8-naphthyridine-3-carboxanilide	>300	1347
2-Amino-1,8-naphthyridine-3-carboxylic acid	>300 or 336–339, IR	278, 1100
7-Amino-1,8-naphthyridin-2(1H)-one	>350 to 365, IR, NMR; H_2SO_4:—	389, 681, 687, 690, 702, 741, 1018
7-Amino-1,8-naphthyridin-4(1H)-one	300–302	101, 720, 758, 1388
7-Amino-3-nitro-1,8-naphthyridin-2(1H)-one	>320, IR	721, 1273, 1401
7-Amino-6-nitro-1,8-naphthyridin-2(1H)-one	>320, IR	721
7-Amino-6-nitro-1,8-naphthyridin-4(1H)-one	>320	758, 1273
7-Amino-6-nitro-4-oxo-1,4-dihydro-1,8-naphthyridine-3-carboxylic acid	284–285	758
4-Amino-3-nitro-1-phenyl-1,8-naphthyridin-2(1H)-one	>300, IR, NMR	1191, 1273

TABLE A.4. (*Continued*)

1,8-Naphthyridine	Melting Point (°C) etc.	Reference(s)
7-Amino-6-nitro-3-phenyl-1,8-naphthyridin-2(1H)-one	>320, NMR	1226
7-Amino-6-nitro-4-phenyl-1,8-naphthyridin-2(1H)-one	>320, IR, NMR	749
4-Amino-3-nitroso-1-phenyl-1,8-naphthyridin-2(1H)-one	287–290, IR, MS, NMR	892
4-Amino-2-oxo-1,2-dihydro-1,8-naphthyridine-3-carboxylic acid	297–300	612
6-Amino-5-oxo-5,8-dihydro-1,8-naphthyridine-2-carboxylic acid	>360	726
7-Amino-2-oxo-1,2-dihydro-1,8-naphthyridine-4-carboxylic acid	>360, IR	686, 690
7-Amino-4-oxo-1,4-dihydro-1,8-naphthyridine-2-carboxylic acid	290–291	758
7-Amino-4-oxo-1,4-dihydro-1,8-naphthyridin-3-carboxylic acid	307–310	720
4-Amino-2-oxo-5-phenyl-1,2-dihydro-1,8-naphthyridine-3-carbonitrile	>320, IR, NMR	746
7-Amino-4-oxo-1-vinyl-1,4-dihydro-1,8-naphthyridine-3-carboxylic acid	>300, NMR	1029
2-Amino-7-phenyl-1,8-naphthyridine-3-carboguanide	270–272	1400
2-Amino-1-phenyl-1,8-naphthyridine-3-carbohydrazide	255–257	1400
2-Amino-4-phenyl-1,8-naphthyridine-3-carbonitrile	293–295, IR	1400
2-Amino-7-phenyl-1,8-naphthyridine-3-carbonitrile	272–274, IR	1112
2-Amino-4-phenyl-1,8-naphthyridine-3-carboxamide	336–337, IR	1400
2-Amino-7-phenyl-1,8-naphthyridine-3-carboxamide	315, IR	1112
2-Amino-7-phenyl-1,8-naphthyridine-3-carboxylic acid	338–342, IR; HCl: 332–334	1100
3-Amino-7-phenyl-1,8-naphthyridin-2(1H)-one	177–178, IR	1112
4-Amino-1-phenyl-1,8-naphthyridin-2(1H)-one	246–250, IR, NMR	892
7-Amino-2-phenyl-1,8-naphthyridin-4(1H)-one	160–162, IR, MS, NMR, UV	221, 472, 1136
7-Amino-3-phenyl-1,8-naphthyridin-2(1H)-one	>310, IR	755, 972
7-Amino-4-phenyl-1,8-naphthyridin-2(1H)-one	>345 or >350, IR, UV; HCl: 350; pic: 245	73, 686, 718, 741, 791, 1136
7-Amino-5-phenyl-1,8-naphthyridin-2(1H)-one	>330, IR	741
7-Amino-6-phenyl-1,8-naphthyridin-2(1H)-one	306–308	972
3-Amino-6-phenyl-7-piperidino-1,8-naphthyridin-2(1H)-one	165–167, NMR	1304
2-Amino-7-phenyl-N-propyl-1,8-naphthyridine-3-carboxamide	282–284	1400
7-Amino-4-piperidinomethyl-1,8-naphthyridin-2(1H)-one	310	79
3-Amino-7-piperidino-1,8-naphthyridin-2(1H)-one	183–184, NMR	1401
2-Amino-N-propyl-1,8-naphthyridine-3-carboxamide	165–166	278
7-Amino-2-propyl-1,8-naphthyridin-4(1H)-one	270–271, UV	1136
7-Amino-3-propyl-1,8-naphthyridin-2(1H)-one	>320, UV	1136
2-Amino-5,7,N,N-tetramethyl-1,8-naphthyridine-3-carboxamide	308–310	276

TABLE A.4. (*Continued*)

1,8-Naphthyridine	Melting Point (°C) etc.	Reference(s)
2-Amino-5,7-*N*-trimethyl-1,8-naphthyridine-3-carboxamide	262–264	276
2-Anilino-3,6-dinitro-1,8-naphthyridine	300–303, IR, MS, NMR	860, 1273
2-Anilino-4-methylamino-3,6-dinitro-1,8-naphthyridine	254–256, IR, MS, NMR	860
2-Anilino-4-methylamino-3-nitro-1,8-naphthyridine	225–227, IR, NMR	682, 785, 1273
7-Anilino-5-methyl-1,8-naphthyridin-2-amine	269–270	1052
2-Anilino-7-methyl-5-phenyl-1,8-naphthyridine	286–287	1052
2-Anilino-1,8-naphthyridine	—	427
2-Anilino-3-nitro-1,8-naphthyridine	173–174, IR, NMR	682, 1273
2-Azido-4-chloro-5,7-dimethyl-3-phenyl-1,8-naphthyridine	255	1301
2-Azido-7-chloro-3-methyl-1,8-naphthyridine	248–249, IR	743
2-Azido-7-chloro-4-methyl-1,8-naphthyridine	242–245	761
2-Azido-7-chloro-5-methyl-1,8-naphthyridine	247–248	761
2-Azido-7-chloro-6-methyl-1,8-naphthyridine	246–248, IR	743
2-Azido-7-chloro-1,8-naphthyridine	260–265	764
2-Azido-7-chloro-3-phenyl-1,8-naphthyridine	220–222, IR	972
2-Azido-7-chloro-4-phenyl-1,8-naphthyridine	255–257	718
2-Azido-7-chloro-5-phenyl-1,8-naphthyridine	208–210	718
2-Azido-7-chloro-6-phenyl-1,8-naphthyridine	255–257	972
2-Azido-4-chloro-3,5,7-trimethyl-1,8-naphthyridine	248	1301
2-Azido-5,7-dimethyl-1,8-naphthyridine	240, NMR	680, 763
2-Azido-7-ethoxy-3-methyl-1,8-naphthyridine	188–189, IR	743
2-Azido-7-ethoxy-4-methyl-1,8-naphthyridine	182–183, IR	761
2-Azido-7-ethoxy-5-methyl-1,8-naphthyridine	221–222, IR	761
2-Azido-7-ethoxy-6-methyl-1,8-naphthyridine	235–237, IR	743
2-Azido-7-ethoxy-1,8-naphthyridine	170–171	764
2-Azido-7-ethoxy-3-phenyl-1,8-naphthyridine	206–208, IR	972
2-Azido-7-ethoxy-4-phenyl-1,8-naphthyridine	194–195	718
2-Azido-7-ethoxy-5-phenyl-1,8-naphthyridine	170–172	718
2-Azido-7-ethoxy-6-phenyl-1,8-naphthyridine	185–188, IR	972
2-Azido-7-methoxy-3-phenyl-1,8-naphthyridine	217–220, IR	285
7-Azido-3-methyl-1,8-naphthyridin-2-amine	>300, IR	743
7-Azido-4-methyl-1,8-naphthyridin-2-amine	>300, IR	761
7-Azido-5-methyl-1,8-naphthyridin-2-amine	320, IR	761
7-Azido-6-methyl-1,8-naphthyridin-2-amine	295–300, IR	743
2-Azido-5-methyl-1,8-naphthyridine	246–248, NMR	680
2-Azido-6-methyl-1,8-naphthyridine	233–235, NMR	680
4-Azido-7-methyl-1,8-naphthyridine	133–135, NMR	1391
7-Azido-1-methyl-1,8-naphthyridin-2(1*H*)-one	159–160, NMR	1384
7-Azido-4-methyl-1,8-naphthyridin-2(1*H*)-one	>240, IR	761
7-Azido-5-methyl-1,8-naphthyridin-2(1*H*)-one	240, IR	761
7-Azido-6-methyl-1,8-naphthyridin-2(1*H*)-one	250, IR	743
2-Azido-7-methyl-5-phenyl-1,8-naphthyridine	260–261	771
4-Azido-7-methyl-2-phenyl-1,8-naphthyridine	167–169, IR	1390
7-Azido-1,8-naphthyridin-2-amine	270 or 275, IR, NMR	764, 1445
2-Azido-1,8-naphthyridine	256–258, IR, NMR, UV	680, 1100
7-Azido-1,8-naphthyridin-2(1*H*)-one	>320, IR	764

TABLE A.4. (*Continued*)

1,8-Naphthyridine	Melting Point (°C) etc.	Reference(s)
7-Azido-3-phenyl-1,8-naphthyridin-2-amine	257–260, IR	972
7-Azido-4-phenyl-1,8-naphthyridin-2-amine	293–295	718
7-Azido-5-phenyl-1,8-naphthyridin-2-amine	285–288	718
7-Azido-6-phenyl-1,8-naphthyridin-2-amine	300–303, IR	972
2-Azido-3-phenyl-1,8-naphthyridine	208–210, IR, MS, NMR	680, 752
2-Azido-4-phenyl-1,8-naphthyridine	245–248	680
2-Azido-5-phenyl-1,8-naphthyridine	200–202, NMR	680
4-Azido-3-phenyl-1,8-naphthyridin-2(1*H*)-one	218–220, IR	972
7-Azido-4-phenyl-1,8-naphthyridin-2(1*H*)-one	212–214, IR	718
7-Azido-5-phenyl-1,8-naphthyridin-2(1*H*)-one	200, IR; K: IR	718
7-Azido-6-phenyl-1,8-naphthyridin-2(1*H*)-one	249–251, IR	
3-Benzoyl-2-benzyl-7-chloro-6-fluoro-1-phenyl-1,8-naphthyridin-4(1*H*)-one	132–133, NMR	183
1-Benzoyl-3-butyryl-4-hydroxy-1,8-naphthyridin-2(1*H*)-one	158–160, IR, NMR	912
3-Benzoyl-1,2-dibenzyl-7-chloro-6-fluoro-1,8-naphthyridin-4(1*H*)-one	153–154, NMR	183
1-Benzoyl-2-diethylamino-3-methyl-1,8-naphthyridin-4(1*H*)-one	141, NMR	816
1-Benzoyl-3-hexanoyl-4-hydroxy-1,8-naphthyridin-2(1*H*)-one	107–110, IR, NMR	912
3-Benzoyl-1-hexanoyl-4-hydroxy-1,8-naphthyridin-2(1*H*)-one	—	162
1-Benzoyl-4-hydroxy-3-methyl-1,8-naphthyridin-2(1*H*)-one	192–195, IR, NMR	140
1-Benzoyl-4-hydroxy-3-phenyl-1,8-naphthyridin-2(1*H*)-one	228–229, IR, NMR	
3-Benzoyl-2-methyl-1,8-naphthyridine	143, IR, NMR	658, 674, 872, 1258, 1439, 1440
3-Benzoyl-2-phenyl-1,8-naphthyridine	162, IR, NMR	658, 674, 872, 1258, 1439, 1440
3-Benzoyl-7-phenyl-1,8-naphthyridine	203–204, IR	1112
2-Benzylamino-5,7-dimethyl-1,8-naphthyridine	184–185	763
2-Benzylamino-7-methyl-1,8-naphthyridine	solid, IR, NMR, UV	1230
7-Benzylamino-1,8-naphthyridin-2-amine	300, IR, NMR	1445
2-Benzylamino-1,8-naphthyridine	—	427
4-Benzylamino-3-nitro-1-phenyl-1,8-naphthyridin-2(1*H*)-one	192–194	1191
1-Benzyl-7-benzyloxy-1,8-naphthyridin-2(1*H*)-one	114–115, NMR	1380
1-Benzyl-3-butyl-4-hydroxyl-1,8-naphthyridin-2(1*H*)-one	147–149	607
6-Benzyl-2-ethoxy-7-methyl-4-phenyl-1,8-naphthyridine-3-carbonitrile	184–185, IR, MS, NMR	896
3-Benzyl-1-ethyl-7-methyl-1,8-naphthyridin-4(1*H*)-one	147–149, IR, NMR	1102
1-Benzyl-3-hydroxy-7-methyl-1,8-naphthyridin-4(1*H*)-one	233–235, IR, NMR	1394
3-Benzyl-4-hydroxy-7-methyl-1,8-naphthyridin-2(1*H*)-one	265 or 266, IR, NMR	810, 1269, 1301

TABLE A.4. (*Continued*)

1,8-Naphthyridine	Melting Point (°C) etc.	Reference(s)
3-Benzylidene-7-methyl-2,3-dihydro-1,8-naphthyridin-4(1H)-one (?)	150, NMR	1102
3-Benzyl-2-methyl-1,8-naphthyridine	123–125, IR, NMR	213
1-Benzyl-7-methyl-1,8-naphthyridin-4(1H)-one	120–123 or 154–155, NMR	305, 1380
2-Benzyl-7-methyl-1,8-naphthyridin-4(1H)-one	246–250, NMR	263
3-Benzyl-7-methyl-1,8-naphthyridin-4(1H)-one	253, NMR	1102
1-Benzyl-7-methyl-4-oxo-1,4-dihydro-1,8-naphthyridine-3-carbaldehyde	170–173, IR, NMR	1394
1-Benzyl-4-methyl-2-oxo-1,2-dihydro-1,8-naphthyridine-3-carboxylic acid	crude	500
1-Benzyl-7-methyl-4-oxo-1,4-dihydro-1,8-naphthyridine-3-carboxylic acid	235–237, NMR	305
2-Benzyl-5,7-dimethyl-1,8-naphthyridine	235	763
6-Benzyloxy-1-ethyl-7-formylmethyl-4-oxo-1,4-dihydro-1,8-naphthyridine-3-carboxylic acid	Et$_2$-acetal: 201–202	1060
2,7-Bis(acetamido)-1,8-naphthyridine	238–240, NMR, xl st	82
2,7-Bis(benzylamino)-1,8-naphthyridine	solid, IR, NMR	83
2,7-Bis(benzylamino)-4-phenyl-1,8-naphthyridine	157–159, NMR	1381
2,7-Bis(bromomethyl)-1,8-naphthyridine	—	112
2,7-Bis[(*tert*-butylimino)methyl]-1,8-naphthyridine	127–128, IR, MS, NMR	642
2,7-Bis(chloromethyl)-1,8-naphthyridine	156–158, NMR, UV	473, 714, 794
2,7-Bis(diethylaminomethyl)-1,8-naphthyridine	liq, NMR	185
2,7-Bis(2,2-dimethylpropionamide)-1,8-naphthyridine-3-carbonitrile	181–183, IR, MS, NMR	123
2,7-Bis(2,2-dimethylpropionamido)-1,8-naphthyridine-3-carboxamide	204, IR, MS, NMR	123
2,7-Bis(hydroxymethyl)-1,8-naphthyridine	143–145, IR, MS, NMR	473
2,4-Bismethylamino-3,6-dinitro-1,8-naphthyridine	268–270, IR, MS, NMR	860, 1273
4,5-Bismethylamino-3,6-dinitro-1,8-naphthyridine	261–263, IR, MS, NMR	860
2,4-Bismethylamino-3-nitro-1,8-naphthyridine	191–192, IR, NMR	682, 785, 1273
2,7-Bis(trimethylsilylethynyl)-1,8-naphthyridine	278–279, IR, MS, NMR	466, 505
2-Bromo-3-bromomethyl-1,8-naphthyridine	—	924
6-Bromo-3-butyl-4-hydroxy-1-phenyl-1,8-naphthyridin-2(1H)-one	188–190	607
2-Bromo-3-chloro-1,8-naphthyridine	—	272
2-Bromo-5-ethoxy-7-phenyl-1,8-naphthyridine	182–184, NMR	263
2-Bromomethyl 1-benzyl-4-methyl-2-oxo-1,2-dihydro-1,8-naphthyridine-3-carboxylate	251–253, NMR	500
3-Bromo-4-hydroxy-1,8-naphthyridin-2(1H)-one	120, IR	693
6-Bromo-4-hydroxy-1-phenyl-1,8-naphthyridin-2(1H)-one	314–317	607
3-Bromo-5-methoxy-7-phenyl-1,8-naphthyridin-2-amine	241–242, IR, MS, NMR	472
3-Bromo-7-methyl-1,8-naphthyridine	213–215, then >350; UV	95, 97
4-Bromomethyl-1,8-naphthyridin-2(1H)-one	147, IR, NMR	1050
7-Bromomethyl-1,8-naphthyridin-2(1H)-one	199–200, IR, MS, NMR	472
3-Bromo-1,8-naphthyridin-4-amine	—	346
6-Bromo-1,8-naphthyridin-2-amine	210–212, MS	300, 546
2-Bromo-1,8-naphthyridine	152–153, IR, MS, NMR	303, 390, 647
2-*d*-2-Bromo-1,8-naphthyridine	—	830, 838

TABLE A.4. (Continued)

1,8-Naphthyridine	Melting Point (°C) etc.	Reference(s)
3-Bromo-1,8-naphthyridine	155 to 166, MS	173, 269, 272, 303, 985
2-d-Bromo-1,8-naphthyridine	—	830
4-Bromo-1,8-naphthyridine	72–73 or 78–79, IR, MS, NMR, UV	303, 628, 1186
3-Bromo-1,8-naphthyridine 1-oxide	174–176, NMR	317
3-Bromo-1,8-naphthyridin-2(1H)-one	304–306, MS, NMR	317, 546
3-Bromo-1,8-naphthyridin-4(1H)-one	MS	346, 546
6-Bromo-1,8-naphthyridin-2(1H)-one	—	390
7-Bromo-1,8-naphthyridin-4(1H)-one	—	304
7-Bromo-2-phenyl-1,8-naphthyridin-4(1H)-one	237–238, NMR	221
Butyl 4-amino-1-ethyl-2-oxo-1, 2-dihydro-1,8-naphthyridine-3-carboxylate	117–120	612
sec-Butyl 4-amino-1-ethyl-2-oxo-1, 2-dihydro-1,8-naphthyridine-3-carboxylate	157–159	612
7-sec-Butylcarbamoyl-1-ethyl-4-oxo-1, 4-dihydro-1,8-naphthyridine-3-carboxylic acid	267–269	1012
1-Butyl-4-carbamoylmethylene-7-methyl-1, 4-dihydro-1,8-naphthyridine	>170, UV	176
1-Butyl-4-chloro-3-nitro-1,8-naphthyridin-2(1H)-one	crude	471
3-Butyl-4,6-dihydroxy-1-phenyl-1, 8-naphthyridin-2(1H)-one	299–300	607
3-Butyl-1-(2-dimethylaminoethyl)-4-hydroxy-1, 8-naphthyridin-2(1H)-one	81–84	607
3-Butyl-2,7-dimethyl-1,8-naphthyridin-4(1H)-one	227–228	976
3-Butyl-4-ethoxy-1-phenyl-1,8-naphthyridin-2(1H)-one	171–172	607
5-Butyl-1-ethyl-6-hydroxy-7-methyl-1, 8-naphthyridin-4(1H)-one	170–172, UV	949
5-Butyl-7-ethyl-1,8-naphthyridine-3-carboxamide	>275, IR, NMR	1326
3-Butyl-1-hexyl-4-hydroxy-1,8-naphthyridin-2(1H)-one	125–126	607
7-(N-Butylhydrazino)-1-ethyl-4-oxo-1,4-dihydro-1, 8-naphthyridine-3-carboxylic acid	199–200	1129
3-Butyl-4-hydroxy-6,7-dimethyl-1-phenyl-1, 8-naphthyridin-2(1H)-one	239–240	607
3-Butyl-4-hydroxy-6-methoxy-1-phenyl-1, 8-naphthyridin-2(1H)-one	181–182	607
3-Butyl-4-hydroxy-7-methyl-1,8-naphthyridin-2(1H)-one	262, IR, NMR	1269
3-Butyl-4-hydroxy-7-methyl-1-phenyl-1, 8-naphthyridin-2(1H)-one	281–283	607
3-Butyl-4-hydroxy-1,8-naphthyridin-2(1H)-one	222–223	607
1-Butyl-4-hydroxy-3-nitro-1,8-naphthyridin-2(1H)-one	106–109, NMR	471
N-Butyl-4-hydroxy-2-oxo-1-phenyl-1, 2-dihydro-1,8-naphthyridin-3-carboxamide	181–184, IR, NMR	453
3-Butyl-4-hydroxy-1-phenyl-1,8-naphthyridin-2(1H)-one	236–237, NMR; Na: 245–262	607

TABLE A.4. (*Continued*)

1,8-Naphthyridine	Melting Point (°C) etc.	Reference(s)
1-Butyl-4-methylamino-3-nitro-1,8-naphthyridin-2(1*H*)-one	233–235, NMR	471
1-Butyl-7-(*N*-methylhydrazino)-4-oxo-1,4-dihydro-1,8-naphthyridine-3-carboxylic acid	201–203	1129
3-Butyl-2-methyl-1,8-naphthyridine	99–101, NMR	121
Butyl 7-methyl-4-oxo-1,4-dihydro-1,8-naphthyridine-3-carboxylate	234–237	1249
2-Butyl-1,8-naphthyridine	liq, IR, NMR	145
2-*sec*-Butyl-1,8-naphthyridine	liq, IR, NMR	145
2-*tert*-Butyl-1,8-naphthyridine	anal, NMR	1013
7-(*N*-Butyl-*N*-nitrosoamino)-1-ethyl-4-oxo-1,4-dihydro-1,8-naphthyridine-3-carboxylic acid	219–221	1129
1-Butyl-4-phenyl-1,8-naphthyridin-2(1*H*)-one	96–97, IR, NMR	160
3-Butyl-1-phenyl-1,8-naphthyridin-2(1*H*)-one	198–200	607
4-Carbamoylmethyl-7-methyl-1,8-naphthyridine	190–192, UV	176
1-(1-Carboxyethyl)-7-methyl-1,8-naphthyridin-4(1*H*)-one	239–240, NMR	1380
1-(2-Carboxyethyl)-7-methyl-1,8-naphthyridin-4(1*H*)-one	217–218	1380
1-Carboxymethyl-7-methyl-1,8-naphthyridin-4(1*H*)-one	289–290	1380
3-Carboxymethyl-7-methyl-1,8-naphthyridin-4(1*H*)-one	315, NMR	305
2-Carboxymethyl-1,8-naphthyridine-3-carboxylic acid	Na$_2$: >300, IR, NMR	1041
4-Carboxymethyl-1,8-naphthyridine-2,7(1*H*, 8*H*)-dione	>320, IR	681
3-Carboxymethyl-1,8-naphthyridin-2(1*H*)-one	crude	130
1-(1-Carboxypropyl)-7-methyl-1,8-naphthyridin-4(1*H*)-one	216–219	1380
1-(3-Carboxypropyl)-7-methyl-1,8-naphthyridin-4(1*H*)-one	190–193	1380
2-(4-Chlorobutyl)-1,8-naphthyridine	liq, NMR	145
7-Chloro-1-cyclopropyl-6-fluoro-4-oxo-1,4-dihydro-1,8-naphthyridine-3-carboxylic acid	210–212, IR, NMR	487
7-Chloro-1-(2-dimethylaminoethyl)-5-methyl-1,8-naphthyridin-2(1*H*)-one	110–111	338
3-Chloro-7-dimethylamino-1,8-naphthyridin-2(1*H*)-one	252–253, IR, NMR	715
2-Chloro-5,7-dimethyl-1,8-naphthyridine	146–147, UV	67, 69, 88, 759, 763
3-Chloro-2,4-dimethyl-1,8-naphthyridine	158–160, IR, NMR	1145
4-Chloro-2,7-dimethyl-1,8-naphthyridine	84 to 87, NMR	1085, 1133, 1186
4-Chloro-3,7-dimethyl-1,8-naphthyridine	134–135, NMR, UV	1076
4-Chloro-5,7-dimethyl-1,8-naphthyridine	155–156, NMR	1098
3-Chloro-2,7-dimethyl-1,8-naphthyridin-4(1*H*)-one	>300 or >320, IR, NMR, UV	909, 1085
7-Chloro-3,6-dinitro-1,8-naphthyridin-4-amine	>330, NMR	1180, 1273
2-Chloro-3,6-dinitro-1,8-naphthyridine	204–206, NMR	1128, 1273
7-*d*-2-Chloro-3,6-dinitro-1,8-naphthyridine	—	1128

TABLE A.4. (*Continued*)

1,8-Naphthyridine	Melting Point (°C) etc.	Reference(s)
2-Chloro-7-ethoxy-3-methyl-1,8-naphthyridine	132–134	743
2-Chloro-7-ethoxy-4-methyl-1,8-naphthyridine	122–124	761
2-Chloro-7-ethoxy-6-methyl-1,8-naphthyridine	119–120	743
2-Chloro-7-ethoxy-1,8-naphthyridine	153–154	764
2-Chloro-7-ethoxy-3-phenyl-1,8-naphthyridine	183–186	972
2-Chloro-7-ethoxy-5-phenyl-1,8-naphthyridine	129–130	718
2-Chloro-7-ethoxy-6-phenyl-1,8-naphthyridine	156–158	972
2-Chloro-7-ethoxy-5-phenyl-1,8-naphthyridine-3, 6-dicarbonitrile	238–239, IR, NMR	78
6-Chloro-1-ethyl-4,7-dioxo-1,4,7,8-tetrahydro-1, 8-naphthyridine-3-carboxylic acid	>300, IR	387
7-Chloro-1-ethyl-6-fluoro-4-oxo-1,4-dihydro-1, 8-naphthyridine-3-carboxylic acid	264–266 or 265–267, IR, NMR	387, 1091
6-Chloro-1-ethyl-7-methoxy-4-oxo-1,4-dihydro-1, 8-naphthyridine-3-carboxylic acid	271–275, IR	387
4-Chloro-1-ethyl-7-methyl-2-oxo-1,2-dihydro-1, 8-naphthyridine-3-carbonitrile	215–217	612
7-Chloro-1-ethyl-1,8-naphthyridin-2(1H)-one	NMR	1401
7-Chloro-1-ethyl-6-nitro-4-oxo-1,4-dihydro-1, 8-naphthyridine-3-carboxylic acid	235–240	364
7-Chloro-1-ethyl-4-oxo-1,4-dihydro-1, 8-naphthyridine-3-carboxylic acid	249–250	1284
4-Chloro-2-hydrazino-3,5-dimethyl-7-phenyl-1, 8-naphthyridine	204; PhCH=: 197; N'-Ac: 276	1301
4-Chloro-2-hydrazino-5,7-dimethyl-3-phenyl-1, 8-naphthyridine	191; PhCH=: 142; N'-Ac: 208	1301
4-Chloro-2-hydrazino-3,5,7-trimethyl-1, 8-naphthyridine	181; PhCH=: 173–174; N'-Ac: 255	1301
2-Chloro-4-isobutylamino-N,N-dimethyl-1, 8-naphthyridine-3-carboxamide	187–188, IR, NMR	80
4-Chloro-2-isobutylamino-N,N-dimethyl-1, 8-naphthyridine-3-carboxamide	95–96, IR, NMR	80
7-Chloro-3-methyl-1,8-naphthyridin-2-amine	245–246	743
7-Chloro-4-methyl-1,8-naphthyridin-2-amine	273–275	761
7-Chloro-5-methyl-1,8-naphthyridin-2-amine	257–259	761, 1052
7-Chloro-6-methyl-1,8-naphthyridin-2-amine	234–236	743
2-Chloro-5-methyl-1,8-naphthyridine	130–131	680
2-Chloro-6-methyl-1,8-naphthyridine	206–207	680
2-Chloro-7-methyl-1,8-naphthyridine	215–216	153
4-Chloro-7-methyl-1,8-naphthyridine	116–118 or 121–122, NMR	153, 1186
7-Chloro-1-methyl-1,8-naphthyridin-2(1H)-one	162–164, NMR	1384
7-Chloro-3-methyl-1,8-naphthyridin-2(1H)-one	256–257	743
7-Chloro-4-methyl-1,8-naphthyridin-2(1H)-one	226–229	761
7-Chloro-5-methyl-1,8-naphthyridin-2(1H)-one	268–270	761
7-Chloro-6-methyl-1,8-naphthyridin-2(1H)-one	254–256	743
2-Chloro-7-methyl-5-phenyl-1,8-naphthyridine	161 or 164	771, 1052
4-Chloro-7-methyl-2-phenyl-1,8-naphthyridine	102–104	1390
4-Chloro-7-methyl-2-propyl-1,8-naphthyridine	pic: 156–158	1085
5-Chloro-7-morpholino-1,8-naphthyridin-2-amine	208–210, NMR	1226
5-Chloro-7-morpholino-1,8-naphthyridin-2(1H)-one	265–267, NMR	1226

TABLE A.4. (*Continued*)

1,8-Naphthyridine	Melting Point (°C) etc.	Reference(s)
5-Chloro-7-morpholino-3-nitro-1,8-naphthyridin-2-amine	>320, NMR	1226
5-Chloro-7-morpholino-nitro-1,8-naphthyridin-2(1*H*)-one	>320, NMR	1226
5-Chloro-7-morpholino-6-nitro-1,8-naphthyridin-2(1*H*)-one	285–287, NMR	1226
2-Chloro-7-morpholino-3-nitro-6-phenyl-1,8-naphthyridine	158–160	1304
2-Chloro-7-morpholino-6-phenyl-1,8-naphthyridine	205–207, NMR	1304
2-Chloro-1,8-naphthyridin-3-amine	213–215, IR, NMR	647
3-Chloro-1,8-naphthyridin-4-amine	—	346
5-Chloro-1,8-naphthyridin-2-amine	173–175, NMR	158
6-Chloro-1,8-naphthyridin-2-amine	208–209	300
7-Chloro-1,8-naphthyridin-2-amine	170 or 183–184, xl st	764, 800, 1445
2-Chloro-1,8-naphthyridine	135 to 142, IR, NMR, UV	225, 647, 680, 1100
7-*d*-2-Chloro-1,8-naphthyridine	138–139	830
3-Chloro-1,8-naphthyridine	143 to 157, IR, NMR	225, 269, 272, 1145, 1441
2-*d*-3-Chloro-1,8-naphthyridine	—	830
4-Chloro-1,8-naphthyridine	49 to 64, NMR	225, 830, 1441
2-Chloro-1,8-naphthyridine-3-carbonitrile	273 or >300, IR, MS, NMR	278, 367, 384
3-Chloro-1,8-naphthyridin-2(1*H*)-one	—	272
3-Chloro-1,8-naphthyridin-4(1*H*)-one	—	346
4-Chloro-1,8-naphthyridin-2(1*H*)-one	185, IR	386, 693
5-Chloro-1,8-naphthyridin-2(1*H*)-one	234–235, NMR	158
6-Chloro-1,8-naphthyridin-2(1*H*)-one	307–309, IR, NMR	477
7-Chloro-1,8-naphthyridin-2(1*H*)-one	246–253 or 255–257, IR, NMR	477, 764
2-Chloro-3-nitro-1,8-naphthyridin-4-amine	>350, MS	492, 1273
2-Chloro-3-nitro-1,8-naphthyridine	267–268 or >280, IR, MS, NMR	647, 1273, 1320
4-Chloro-3-nitro-1,8-naphthyridine	169–170	346, 1273
7-Chloro-3-nitro-1,8-naphthyridin-2(1*H*)-one	290, NMR	1273, 1401
7-Chloro-6-nitro-1,8-naphthyridin-2(1*H*)-one	240–242, NMR	1273, 1401
4-Chloro-3-nitro-1-phenyl-1,8-naphthyridin-2(1*H*)-one	228–232, IR, NMR	1191, 1273
2-Chloro-3-nitro-6-phenyl-7-piperidino-1,8-naphthyridine	162–164, NMR	1304
2-Chloro-3-nitro-7-piperidino-1,8-naphthyridine	170–171, NMR	1401
2-Chloro-6-nitro-7-piperidino-1,8-naphthyridine	146–147, NMR	1401
7-Chloro-3-phenyl-1,8-naphthyridin-2-amine	265–267	972
7-Chloro-4-phenyl-1,8-naphthyridin-2-amine	214–215	718
7-Chloro-5-phenyl-1,8-naphthyridin-2-amine	263–265	741
7-Chloro-6-phenyl-1,8-naphthyridin-2-amine	237–239	972
2-Chloro-3-phenyl-1,8-naphthyridine	193 or 204–205, IR, MS, NMR	680, 850
2-Chloro-4-phenyl-1,8-naphthyridine	145–147	680

TABLE A.4. (*Continued*)

1,8-Naphthyridine	Melting Point (°C) etc.	Reference(s)
2-Chloro-5-phenyl-1,8-naphthyridine	178–180	680
7-Chloro-2-phenyl-1,8-naphthyridin-4(1*H*)-one	293–295, NMR	221
7-Chloro-3-phenyl-1,8-naphthyridin-2(1*H*)-one	208–210	972
7-Chloro-4-phenyl-1,8-naphthyridin-2(1*H*)-one	220–222	718
7-Chloro-5-phenyl-1,8-naphthyridin-2(1*H*)-one	245–247	741
7-Chloro-6-phenyl-1,8-naphthyridin-2(1*H*)-one	248–250 or 256–258, NMR	972, 1226
2-Chloro-6-phenyl-7-piperidino-1,8-naphthyridine	170–172, NMR	1304
2-Chloro-7-piperidino-5-piperidinomethyl-1,8-naphthyridine	158–160	79
4-Chloro-2,3,5,7-tetramethyl-1,8-naphthyridine	136–137	1085
4-Chloro-2,3,7-trimethyl-1,8-naphthyridine	177–179	1085
4-Chloro-2,5,7-trimethyl-1,8-naphthyridine	118–119	1085
6-Cyano-7-ethoxy-2-methyl-5-phenyl-1,8-naphthyridine-3-carboxamide	294–295, IR, MS, NMR	896
6-Cyano-1-ethyl-7-methoxy-4-oxo-1,4-dihydro-1,8-naphthyridine-3-carboxylic acid	281–284	387
3-(2-Cyanovinyl)-1-ethyl-7-methyl-1,8-naphthyridin-4(1*H*)-one	204, IR, NMR	905
4-Cyclohexylamino-7-methyl-2-phenyl-1,8-naphthyridine	235–237, NMR	221
2-Cyclohexylamino-1,8-naphthyridine	—	427
7-Cyclohexylamino-5-phenyl-1,8-naphthyridin-2-amine	119–122	1381
7-Cyclohexylamino-2-phenyl-1,8-naphthyridin-4(1*H*)-one	156–159, NMR	263
3-Cyclohexyl-4-hydroxy-1-phenyl-1,8-naphthyridin-2(1*H*)-one	292–294	607
7-(*N*′-Cyclopentylidene-*N*-methylhydrazino)-1-ethyl-4-oxo-1,4-dihydro-1,8-naphthyridine-3-carboxylic acid	249–251	1129
N-Cyclopentyl-7-methyl-1-(2-morpholinoethyl)-4-oxo-1,4-dihydro-1,8-naphthyridine-3-carboxamide	208–210	305
N-Cyclopentyl-7-methyl-4-oxo-1,4-dihydro-1,8-naphthyridine-3-carboxamide	280–283, NMR	305
1-Cyclopropyl-6-fluoro-4,7-dioxo-1,4,7,8-tetrahydro-1,8-naphthyridine-3-carboxylic acid	>300, IR, NMR	1162
1-Cyclopropyl-6-fluoro-4-oxo-7-piperidino-1,4-dihydro-1,8-naphthyridine-3-carboxylic acid	260–261, IR, NMR	1162
1-Cyclopropyl-6-fluoro-4-oxo-7-vinyl-1,4-dihydro-1,8-naphthyridine-3-carboxylic acid	solid, IR, MS, NMR	1158
1,3-Diacetyl-4-hydroxy-1,8-naphthyridin-2(1*H*)-one	170–171, NMR	1225
2,4-Diamino-5-bromo-7-methyl-1,8-naphthyridine-3-carbonitrile (?)	>250	189
2,7-Diamino-1,8-naphthyridine-3-carboxamide	203–206, IR, MS, NMR	123, 531
3,7-Diamino-1,8-naphthyridin-2(1*H*)-one	>320	721
6,7-Diamino-1,8-naphthyridin-2(1*H*)-one	>320	721
5,6-Diamino-7-oxo-6-phenylazo-7-thioxo-1,2,7,8-tetrahydro-1,8-naphthyridine-3-carbonitrile	300, IR	538

TABLE A.4. (*Continued*)

1,8-Naphthyridine	Melting Point (°C) etc.	Reference(s)
4,5-Diamino-7-oxo-2-thioxo-1,2,7,8-tetrahydro-1,8-naphthyridine-3-carbonitrile	180, IR, NMR	538
3,4-Diamino-1-phenyl-1,8-naphthyridin-2(1*H*)-one	crude	892
2,5-Diazido-7-methyl-1,8-naphthyridine	154–156, IR	701
2,7-Diazido-3-methyl-1,8-naphthyridine	~170	686, 743
2,7-Diazido-4-methyl-1,8-naphthyridine	200–205, anal	34, 761
2,7-Diazido-1,8-naphthyridine	190–200, IR	686, 764
2,7-Diazido-3-phenyl-1,8-naphthyridine	230–233, IR	972
2,7-Diazido-4-phenyl-1,8-naphthyridine	200–202 or 203, IR	70, 74, 718
6-Diazonio-5-oxo-5,8-dihydro-1,8-naphthyridine-2-carboxylate	187–188	726
1,3-Dibenzoyl-4-hydroxy-1,8-naphthyridin-2(1*H*)-one	268–270, IR, NMR	912
2,6-Dibromo-4-hydroxy-1,8-naphthyridin-2(1*H*)-one	95, IR	693
3,6-Dibromo-1,8-naphthyridin-2-amine	MS	546
2,3-Dibromo-1,8-naphthyridine	MS	272, 303
2,4-Dibromo-1,8-naphthyridine	134–135, MS	303, 386, 628
2,6-Dibromo-1,8-naphthyridine	—	390
2,7-Dibromo-1,8-naphthyridine	MS	303
3,4-Dibromo-1,8-naphthyridine	MS	303, 346
3,6-Dibromo-1,8-naphthyridine	>300	173, 985
3,6-Dibromo-1,8-naphthyridine 1-oxide	273–275, NMR	317
2,4-Dibromo-3-nitro-1,8-naphthyridine	233–234	1273
Dibutyl 1,8-naphthyridine-2,7-dicarboxylate	liq, IR, MS, NMR, UV	92, 536
2,4-Dichloro-5,7-dimethyl-1,8-naphthyridine	148	1301
3,4-Dichloro-2,7-dimethyl-1,8-naphthyridine	168–170	1085
2,4-Dichloro-3,5-dimethyl-7-phenyl-1,8-naphthyridine	168	1301
2,4-Dichloro-5,7-dimethyl-3-phenyl-1,8-naphthyridine	156	1301
6,7-Dichloro-1-ethyl-4-oxo-1,4-dihydro-1,8-naphthyridine-3-carboxylic acid	262–267 or 270–273, IR	364, 387
2,5-Dichloro-7-methyl-1,8-naphthyridine	178–180, NMR	701, 1186
2,7-Dichloro-3-methyl-1,8-naphthyridine	250	686
2,7-Dichloro-4-methyl-1,8-naphthyridine	193–194, IR, UV	34, 74, 88, 689, 759
2,7-Dichloro-3-methyl-6-phenyl-1,8-naphthyridine	210–212, NMR	1226
5,7-Dichloro-1,8-naphthyridin-2-amine	245–247, NMR	533
2,3-Dichloro-1,8-naphthyridine	—	272
2,4-Dichloro-1,8-naphthyridine	125–126	33, 386
2,5-Dichloro-1,8-naphthyridine	131–132, NMR	1186
2,7-Dichloro-1,8-naphthyridine	258 or 259, NMR	389, 690, 1222, 1257
3,4-Dichloro-1,8-naphthyridine	—	346
2,4-Dichloro-1,8-naphthyridine-3-carbonitrile	212	31
2,7-Dichloro-1,8-naphthyridine-4-carboxylic acid	287	690
3,7-Dichloro-1,8-naphthyridin-2(1*H*)-one	310–315, IR, NMR	715
5,7-Dichloro-1,8-naphthyridin-2(1*H*)-one	>320, NMR	533
2,4-Dichloro-3-nitro-1,8-naphthyridine	188–189	1273

TABLE A.4. (*Continued*)

1,8-Naphthyridine	Melting Point (°C) etc.	Reference(s)
2,7-Dichloro-3-nitro-1,8-naphthyridine	201–203, NMR	1273, 1401
2,7-Dichloro-3-phenyl-1,8-naphthyridine	218–220	972
2,7-Dichloro-4-phenyl-1,8-naphthyridine	158	70, 73
2,4-Dichloro-3,5,7-trimethyl-1,8-naphthyridine	132	1301
2,4-Diethoxy-5,7-dimethyl-3-phenyl-1,8-naphthyridine	196	1301
2,7-Diethoxy-4-methyl-1,8-naphthyridine	72–73	74
2,5-Diethoxy-7-phenyl-1,8-naphthyridine	176–178, NMR	263
2,7-Diethoxy-*N*-phenyl-1,8-naphthyridine	87–88	74
6-(2-Diethylaminoethyl)-5,7-dimethyl-1,8-naphthyridin-2(1*H*)-one	HCl: 226–228	338
7-Diethylamino-1-ethyl-6-nitro-4-oxo-1,4-dihydro-1,8-naphthyridine-3-carboxylic acid	150	364
Diethyl 8-ethyl-5-oxo-5,8-dihydro-1,8-naphthyridine-2,6-dicarboxylate	181–183, NMR	1135
5,7-Diethyl-1-methyl-4-phenyl-1,8-naphthyridin-2(1*H*)-one	68–69, NMR	1196
5,8-Diethyl-1,8-naphthyridin-2-amine	187–189	338
5,7-Diethyl-1,8-naphthyridin-2(1*H*)-one	159–161	338
1,7-Diethyl-4-oxo-1,4-dihydro-1,8-naphthyridine-3-carboxylic acid	174–176	154
1,7-Diethyl-4-phenyl-1,8-naphthyridin-2(1*H*)-one	85–86	1196
2,7-Diethynyl-1,8-naphthyridine	140, IR, MS, NMR	466, 505
2,7-Difluoro-1,8-naphthyridine	215–217, IR, NMR, pol, th	821, 918, 1313
2,7-Dihydrazino-4-methyl-1,8-naphthyridine	2HCl: anal; 2PhCH=: 263	34
2,7-Dihydrazino-1,8-naphthyridine	—	702
2,7-Dihydrazino-4-phenyl-1,8-naphthyridine	2HCl: 253–254; 2(*p*-O$_2$NC$_6$H$_4$CH=): 327–328	70, 74
2,7-Diisovaleramido-1,8-naphthyridine-3-carbonitrile	NMR	531
2,4-Dimethoxy-5,7-dimethyl-1,8-naphthyridine	174	1301
2,4-Dimethoxy-5,7-dimethyl-3-phenyl-1,8-naphthyridine	179	1301
2,7-Dimethoxy-4-methyl-1,8-naphthyridine	102–103	74
5,7-Dimethoxy-1-methyl-1,8-naphthyridin-2(1*H*)-one	137–138	64
2,4-Dimethoxy-1,8-naphthyridine	145–146; 8-MeI: 212	31
2,7-Dimethoxy-1,8-naphthyridine	73, NMR	1222
2,7-Dimethoxy-3-nitro-6-phenyl-1,8-naphthyridine	210–212	1226
2,7-Dimethoxy-6-phenyl-1,8-naphthyridin-3-amine	226–228, NMR	1226
2,7-Dimethoxy-3-phenyl-1,8-naphthyridine	113–115, NMR	1226
2,7-Dimethoxy-4-phenyl-1,8-naphthyridine	156	74
2,4-Dimethoxy-3,5,7-trimethyl-1,8-naphthyridine	108	1301
1-(4-Dimethylaminobutyl)-5,7-dimethyl-1,8-naphthyridin-2(1*H*)-one	2HBr: 222–225	338
1-(2-Dimethylaminoethyl)-5,7-diethyl-1,8-naphthyridin-2(1*H*)-one	59–60	338
1-(2-Dimethylaminoethyl)-5,7-dimethyl-1,8-naphthyridin-2(1*H*)-one	HCl: 231–233	338

TABLE A.4. (*Continued*)

1,8-Naphthyridine	Melting Point (°C) etc.	Reference(s)
1-(2-Dimethylaminoethyl)-7-ethyl-1,8-naphthyridin-2(1*H*)-one	HCl: 146–149	338
1-(2-Dimethylaminoethyl)-7-isobutyl-1,8-naphthyridin-2(1*h*)-one	HCl: 189–191	338
1-(Dimethylaminoethyl)-7-isopropyl-1,8-naphthyridin-2(1*H*)-one	HCl: 209–211	338
1-(2-Dimethylaminoethyl)-7-methoxy-5-methyl-1,8-naphthyridin-2(1*H*)-one	98–101	338
1-(2-Dimethylaminoethyl)-5-methyl-1,8-naphthyridin-2(1*H*)-one	HCl: 229–231	338
1-(2-Dimethylaminoethyl)-6-methyl-1,8-naphthyridin-2(1*H*)-one	HCl: 207–210	338
1-(2-Dimethylaminoethyl)-7-methyl-1,8-naphthyridin-2(1*H*)-one	HCl: 201–204	338
1-(2-Dimethylaminoethyl)-1,8-naphthyridin-2(1*H*)-one	HCl: 202–203	338
1-(2-Dimethylaminoethyl)-5,6,7-trimethyl-1,8-naphthyridin-2(1*H*)-one	HCl: 180–185	338
4-Dimethylamino-7-methyl-2-phenyl-1,8-naphthyridine	147–149, NMR	221
1-(2-Dimethylamino-2-methylpropyl)-6,7-dimethyl-1,8-naphthyridin-2(1*H*)-one	109–111	338
2-Dimethylamino-1,8-naphthyridine-3-carbonitrile	172–174	278
2-Dimethylamino-1,8-naphthyridine-3-carboxamide	235–237	278
7-Dimethylamino-2-phenyl-1,8-naphthyridin-4(1*H*)-one	257–260, NMR	263
1-(3-Dimethylaminopropyl)-6,7-dimethyl-1,8-naphthyridin-2(1*H*)-one	HCl: 232–234	338
5,7-Dimethyl-1-(2-methylaminiethyl)-1,8-naphthyridin-2(1*H*)-one	HCl: 106–109	338
5,7-Dimethyl-1,8-naphthyridin-2-amine	215 to 220, IR, pK_a, UV	67, 69, 88, 98, 689, 759, 763, 980, 1043
2,3-Dimethyl-1,8-naphthyridine	135 to 142, IR, NMR, pK_a, UV	278, 983, 1013
2,4-Dimethyl-1,8-naphthyridine	84–85 or 85–86, NMR; HCl: 240; pic: 204–206	67, 88, 177, 269, 964, 1297
2,5-Dimethyl-1,8-naphthyridine	85–87, NMR	269, 964
2,6-Dimethyl-1,8-naphthyridine	84–86 or 161–163, NMR; 8-EtI: 145; 8-MeI: 211	269, 667, 964
2,7-Dimethyl-1,8-naphthyridine	193–194 or 194–195, IR, NMR, pK_a, UV; EtI: 209–210; MeI: 234–235	185, 269, 287, 348, 473, 642, 667, 678, 714, 961, 964, 1133, 1306
3,5-Dimethyl-1,8-naphthyridine	86–87, NMR	269, 964
3,6-Dimethyl-1,8-naphthyridine	191–192 or 192–193, IR, NMR; pic: 210–211	269, 939, 964

TABLE A.4. (*Continued*)

1,8-Naphthyridine	Melting Point (°C) etc.	Reference(s)
4,5-Dimethyl-1,8-naphthyridine	152–154, NMR	269, 880, 964
Dimethyl-1,8-naphthyridine-2,7-dicarboxylate	215–217, IR, MS, NMR	473
1,4-Dimethyl-1,8-naphthyridine-2,7(1H,8H)-dione	288–290	1392
1,7-Dimethyl-1,8-naphthyridin-4(1H)-one	140–142, NMR	1380
2,3-Dimethyl-1,8-naphthyridin-4(1H)-one	120, IR	1073
2,7-Dimethyl-1,8-naphthyridin-4(1H)-one	320–323 or >320	1076, 1085, 1133
3,7-Dimethyl-1,8-naphthyridin-4(1H)-one	280	1076
4,7-Dimethyl-1,8-naphthyridin-2(1H)-one	232–233, NMR	1333
5,7-Dimethyl-1,8-naphthyridin-2(1H)-one	251 or 253–254, NMR, UV; H_2SO_4: 225–230	67, 69, 88, 759, 763, 938
5,7-Dimethyl-1,8-naphthyridin-4(1H)-one	198–199, NMR	1098
5,7-Dimethyl-1,8-naphthyridin-2(1H)-one 8-oxide	250, NMR	1008
1,7-Dimethyl-4-oxo-1,4-dihydro-1,8-naphthyridine-3-carbaldehyde	233–235, IR, NMR; PhNHN=: 260–263, NMR	1394
1,7-Dimethyl-4-oxo-1,4-dihydro-1,8-naphthyridine-3-carbohydrazide	243–244	1116
5,7-Dimethyl-2-oxo-1,2-dihydro-1,8-naphthyridine-3-carbonitrile	299–300 or 302–305, IR	276, 1099
5,7-Dimethyl-2-oxo-1,2-dihydro-1,8-naphthyridine-3-carboxamide	325–330	276
2,7-Dimethyl-4-oxo-1,4-dihydro-1,8-naphthyridine-3-carboxylic acid	288	249
2,7-Dimethyl-3-phenoxy-1,8-naphthyridin-4(1H)-one	>300, IR, NMR, UV	909
5,7-Dimethyl-3-phenyl-1,8-naphthyridin-2-amine	214–215	276
5,7-Dimethyl-2-phenyl-1,8-naphthyridine-3-carbonitrile	170–172	276
1,7-Dimethyl-4-phenyl-1,8-naphthyridin-2(1H)-one	117–118, NMR	1196
5,7-Dimethyl-3-phenylsulfonyl-1,8-naphthyridin-2-amine	271–273	276
2,7-Dimethyl-3-propyl-1,8-naphthyridin-4(1H)-one	255–257	976
2-Dimethylsulfimido-1,8-naphthyridine	152–154, NMR	692
5,7-Dimethyl-1-(2-trimethylammonioethyl)-1,8-naphthyridin-2(1H)-one iodide	219–220	338
3,6-Dinitro-1,8-naphthyridin-2-amine	>350, IR, NMR	860, 1180, 1273
3,6-Dinitro-1,8-naphthyridin-4-amine	315–317, IR, MS, NMR	692, 1273
3,6-Dinitro-1,8-naphthyridine	309–310, IR, MS, NMR	692, 1273
3,6-Dinitro-1,8-naphthyridine-2,4-diamine	>350, NMR	1180, 1273
3,6-Dinitro-1,8-naphthyridine-2,5-diamine	crude, NMR	1180
3,6-Dinitro-1,8-naphthyridine-2,5(1H,8H)-dione	>320, IR; NH_4: >320, IR	758, 1273
3,6-Dinitro-1,8-naphthyridine-2,7(1H,8H)-dione	>320, IR	721, 1273
3,6-Dinitro-1,8-naphthyridin-2(1H)-one	302–310, NMR	682, 860, 1128, 1273
3,6-Dinitro-4-phenyl-1,8-naphthyridine-2,7(1H,8H)-dione	305–310, IR	1273
2,7-Dioxo-1,2,7,8-tetrahydro-1,8-naphthyridine-4-carboxylic acid	>360	690

TABLE A.4. (*Continued*)

1,8-Naphthyridine	Melting Point (°C) etc.	Reference(s)
4,7-Dioxo-1,4,7,8-tetrahydro-1,8-naphthyridine-2-carboxylic acid	>320	758
4,7-Dioxo-1,4,7,8-tetrahydro-1,8-naphthyridine-3-carboxylic acid	>300 or >320, NMR	758, 1029
2,7-Diphenylazo-1,8-naphthyridine	286–287, NMR	494, 1222
2,6-Diphenyl-1,8-naphthyridine	207–208, IR	1112
2,7-Diphenyl-1,8-naphthyridine	207–208, IR	409, 1112
2,7-Diphenyl-1,8-naphthyridine-3-carbonitrile	>340	1400
2,7-Dipiperidino-1,8-naphthyridin-3-amine	103–105, NMR	1401
2,7-Divaleramido-1,8-naphthyridine-3-carbonitrile	211, IR, MS, NMR	123, 531
1-(1-Ethoxycarbonylethyl)-7-methyl-1,8-naphthyridin-4(1*H*)-one	80–82	1380
3-(2-Ethoxycarbonylethyl)-7-methyl-1,8-naphthyridin-4(1*H*)-one	192, IR, NMR, UV	1185
2-(2-Ethoxycarbonylethyl)-1,8-naphthyridine	55–56 or 61–63, IR, NMR	121, 145
3-Ethoxycarbonylmethyl-2-methyl-1,8-naphthyridine	101–102, NMR	121
1-Ethoxycarbonylmethyl-7-methyl-1,8-naphthyridin-4(1*H*)-one	151–152	1380
3-Ethoxycarbonylmethyl-7-methyl-1,8-naphthyridin-4(1*H*)-one	212 or 228–230, IR, NMR, UV	305, 1185
4-Ethoxycarbonylmethyl-1,8-naphthyridine-2,7(1*H*,8*H*)-dione	205, IR	681
1-Ethoxycarbonylmethyl-3-phenyl-1,8-naphthyridin-2(1*H*)-one	126, IR, MS, NMR	740
6-Ethoxycarbonyl-5-oxo-5,8-dihydro-1,8-naphthyridine-2-carboxylic acid	>279, NMR	1186
1-(1-Ethoxycarbonylpropyl)-7-methyl-1,8-naphthyridin-4(1*H*)-one	250/1.5	1380
1-(3-Ethoxycarbonylpropyl)-7-methyl-1,8-naphthyridin-4(1*H*)-one	53–56, 200/4	1380
2-(3-Ethoxycarbonylpropyl)-1,8-naphthyridine	45–47, IR, NMR	145
4-Ethoxy-5,7-dimethyl-1,8-naphthyridin-2-amine	228–230, IR, NMR	1071
2-Ethoxy-5,7-dimethyl-1,8-naphthyridine	105 or 108, NMR, UV	96, 763
4-Ethoxy-3,7-dimethyl-1,8-naphthyridine	75–77, NMR, UV	1076
2-Ethoxy-5,7-dimethyl-1,8-naphthyridine 1,8-dioxide	105	769
2-Ethoxy-5,7-dimethyl-1,8-naphthyridine 8-oxide	134–135, NMR	1008
2-Ethoxy-6,7-dimethyl-4-phenyl-1,8-naphthyridine-3-carbonitrile	221–222, IR, MS, NMR	896
2-Ethoxy-3,6-dinitro-1,8-naphthyridin-4-amine	239–240, NMR	1180, 1273
7-Ethoxy-3,6-dinitro-1,8-naphthyridin-4-amine	243–244, NMR	1180, 1273
2-Ethoxy-3,6-dinitro-1,8-naphthyridine	160–161, NMR	1128, 1273
7-*d*-2-Ethoxy-3,7-dinitro-1,8-naphthyridine	—	1128
2-Ethoxy-7-ethyl-4-phenyl-1,8-naphthyridine-3-carbonitrile	192–193, IR, MS, NMR	896
5-Ethoxy-1-ethyl-7-phenyl-1,8-naphthyridin-2(1*H*)-one	149–150, NMR	263
7-Ethoxy-1-ethyl-2-phenyl-1,8-naphthyridin-4(1*H*)-one	158–159, NMR	263

TABLE A.4. (*Continued*)

1,8-Naphthyridine	Melting Point (°C) etc.	Reference(s)
6-Ethoxy-4-hydroxy-1,8-naphthyridin-2(1*H*)-one	325	114
2-Ethoxy-6-methyl-4,7-diphenyl-1,8-naphthyridine-3-carbonitrile	159–160, IR, MS, NMR	896
7-Ethoxy-3-methyl-1,8-naphthyridin-2-amine	170–175	743
7-Ethoxy-4-methyl-1,8-naphthyridin-2-amine	171–173	761
7-Ethoxy-5-methyl-1,8-naphthyridin-2-amine	202–204	761
7-Ethoxy-6-methyl-1,8-naphthyridin-2-amine	155–157	743
4-Ethoxy-7-methyl-1,8-naphthyridine	102–103, NMR	1380
7-Ethoxy-3-methyl-1,8-naphthyridin-2(1*H*)-one	163–165	743
7-Ethoxy-4-methyl-1,8-naphthyridin-2(1*H*)-one	190–192	761
7-Ethoxy-5-methyl-1,8-naphthyridin-2(1*H*)-one	173–175	761
7-Ethoxy-6-methyl-1,8-naphthyridin-2(1*H*)-one	184–185	743
4-Ethoxy-5-methyl-7-phenyl-1,8-naphthyridin-2-amine	227–228, IR	1071
2-Ethoxy-7-methyl-4-phenyl-1,8-naphthyridine-3-carbonitrile	244–246, IR, MS, NMR	896
7-Ethoxy-1,8-naphthyridine-2-amine	177–179	764
2-Ethoxy-1,8-naphthyridine	—	427
3-*d*-2-Ethoxy-1,8-naphthyridine	99–101	838
7-Ethoxy-1,8-naphthyridin-2(1*H*)-one	189–190	764
7-Ethoxy-1,8-naphthyridin-4(1*H*)-one	—	304
2-Ethoxy-3-nitro-1,8-naphthyridin-2-amine	230–231, IR, MS, NMR	492, 1273, 1320
2-Ethoxy-3-nitro-1,8-naphthyridine	130–131, MS, NMR	492, 1273
7-Ethoxy-4-oxo-1,4-dihydro-1,8-naphthyridine-3-carboxylic acid	225–230	102
2-Ethoxy-7-oxo-4-phenyl-7,8-dihydro-1,8-naphthyridine-3-carbonitrile	283–285, IR, MS, NMR	896
2-Ethoxy-7-oxo-4-phenyl-7,8-dihydro-1,8-naphthyridine-3,6-dicarbonitrile	>300, IR, MS, NMR	896
7-Ethoxy-3-phenyl-1,8-naphthyridin-2-amine	196–198	972
7-Ethoxy-4-phenyl-1,8-naphthyridin-2-amine	196–197	718
7-Ethoxy-5-phenyl-1,8-naphthyridin-2-amine	204–205	718
7-Ethoxy-6-phenyl-1,8-naphthyridin-2-amine	124–126	972
7-Ethoxy-2-phenyl-1,8-naphthyridin-4(1*H*)-one	223–225, NMR	221
7-Ethoxy-3-phenyl-1,8-naphthyridin-2(1*H*)-one	180–182	972
7-Ethoxy-4-phenyl-1,8-naphthyridin-2(1*H*)-one	191–193	718
7-Ethoxy-5-phenyl-1,8-naphthyridin-2(1*H*)-one	181–183	718
7-Ethoxy-6-phenyl-1,8-naphthyridin-2(1*H*)-one	228–230	972
2-Ethoxy-4-phenyl-7-styryl-1,8-naphthyridine-3-carbonitrile	233–235, IR, MS, NMR	896
Ethyl 7-acetamido-1-(2-chloroethyl)-4-oxo-1,4-dihydro-1,8-naphthyridine-3-carboxylate	111–113, NMR	1029
Ethyl 7-acetamido-2-chloro-1,8-naphthyridine-3-carboxylate	>300, NMR	482, 600
Ethyl 4-acetamido-1-ethyl-2-oxo-1,2-dihydro-1,8-naphthyridine-3-carboxylate	148–150	612
Ethyl 7-acetamido-1-(2-hydroxyethyl)-4-oxo-1,4-dihydro-1,8-naphthyridine-3-carboxylate	263–265	1029

TABLE A.4. (*Continued*)

1,8-Naphthyridine	Melting Point (°C) etc.	Reference(s)
Ethyl 7-acetamido-2-oxo-1,2-dihydro-1,8-naphthyridine-3-carboxylate	>350, NMR	482
Ethyl 7-acetamido-4-oxo-1,4-dihydro-1,8-naphthyridine-3-carboxylate	>320	720
Ethyl 7-acetoxymethyl-1-ethyl-4-oxo-1,4-dihydro-1,8-naphthyridine-3-carboxylate	132–133, NMR	1248
Ethyl 7-acetoxymethyl-1-(2-fluoroethyl)-4-oxo-1,4-dihydro-1,8-naphthyridine-3-carboxylate	162–163	1248
Ethyl 1-acetyl-4-hydroxy-2-oxo-1,2-dihydro-1,8-naphthyridine-3-carboxylate	157–159, NMR	1225
Ethyl 6-(2-acetylvinyl)-1-ethyl-7-methyl-4-oxo-1,4-dihydro-1,8-naphthyridine-3-carboxylate	146, IR, MS, NMR	905
Ethyl 1-allyl-4-amino-2-oxo-1,2-dihydro-1,8-naphthyridine	161–164	612
Ethyl 1-allyl-4-chloro-7-methyl-2-oxo-1,2-dihydro-1,8-naphthyridine-3-carboxylate	80–82	612
Ethyl 1-allyl-4-hydroxy-7-methyl-2-oxo-1,2-dihydro-1,8-naphthyridine-3-carboxylate	113–115 or 122–125	612, 1127
Ethyl 6-amino-7-chloro-1-(2-chloroethyl)-4-oxo-1,4-dihydro-1,8-naphthyridine-carboxylate	>250	1029
Ethyl 6-amino-1-(2-chloroethyl)-7-methylamino-4-oxo-1,4-dihydro-1,8-naphthyridine-3-carboxylate	crude, 156–160	1029
Ethyl 6-amino-7-chloro-1-ethyl-4-oxo-1,4-dihydro-1,8-naphthyridine-3-carboxylate	289–291, NMR	364
Ethyl 7-amino-6-chloro-1-ethyl-4-oxo-1,4-dihydro-1,8-naphthyridine-3-carboxylate	206–208	364
Ethyl 2-amino-6-cyano-5,8-dimethyl-7-oxo-7,8-dihydro-1,8,-naphthyridine-3-carboxylate	262–264, IR, MS, NMR	618
Ethyl 2-amino-6-cyano-7-ethoxy-5-phenyl-1,8-naphthyridine-3-carboxylate	293–295, IR, MS, NMR	896
Ethyl 4-amino-1-cyclohexyl-2-oxo-1,2-dihydro-1,8-naphthyridine-3-carboxylate	172–175	612
Ethyl 2-amino-5,7-dimethyl-1,8-naphthyridine-3-carboxylate	268–270	276
Ethyl 2-amino-5,7-dimethyl-4-oxo-1,4-dihydro-1,8-naphthyridine-3-carboxylate	366	1301
Ethyl 4-amino-5,7-dimethyl-2-oxo-1,2-dihydro-1,8-naphthyridine-3-carboxylate	288–290	612
Ethyl 7-amino-2-ethoxycarbonylmethyl-1,8-naphthyridine-3-carboxylate	180–183, IR, MS, NMR	482
Ethyl 4-amino-1-ethoxycarbonylmethyl-2-oxo-1,2-dihydro-1,8-naphthyridine-3-carboxylate	164–166, IR	725
Ethyl 6-amino-7-ethoxy-1-ethyl-4-oxo-1,4-dihydro-1,8-naphthyridine-3-carboxylate	249–253, IR	1211

TABLE A.4. (*Continued*)

1,8-Naphthyridine	Melting Point (°C) etc.	Reference(s)
Ethyl 4-amino-1-ethyl-7-ethylamino-5-methyl-2-oxo-1,2-dihydro-1,8-naphthyridine-3-carboxylate	195–197	612
Ethyl 6-amino-1-ethyl-7-methoxy-4-oxo-1,4-dihydro-1,8-naphthyridine-3-carboxylate	250–252, IR, NMR	387
Ethyl 4-amino-1-ethyl-7-methyl-2-oxo-1,2-dihydro-1,8-naphthyridine-3-carboxylate	196–198	612
Ethyl 7-amino-1-ethyl-6-nitro-4-oxo-1,4-dihydro-1,8-naphthyridine-3-carboxylate	262–264	364
Ethyl 4-amino-1-ethyl-2-oxo-1,2-dihydro-1,8-naphthyridine-3-carboxylate	210–212	612
Ethyl 4-amino-1-(2-hydroxyethyl)-7-methyl-2-oxo-1,2-dihydro-1,8-naphthyridine-3-carboxylate	168–170	612
Ethyl 4-amino-1-isobutyl-2-oxo-1,2-dihydro-1,8-naphthyridine-3-carboxylate	157–160	612
Ethyl 7-amino-2-methyl-1,8-naphthyridine-3-carboxylate	240–243, IR, NMR	482
Ethyl 4-amino-1-methyl-2-oxo-1,2-dihydro-1,8-naphthyridine-4-carboxylate	203–206	612
Ethyl 7-amino-3-methyl-2-oxo-1,2-dihydro-1,8-naphthyridine-4-carboxylate	320, IR	686, 687
Ethyl 7-amino-3-methyl-4-oxo-1,4-dihydro-1,8-naphthyridine-2-carboxylate	crude	115
Ethyl 2-amino-1,8-naphthyridine-3-carboxylate	187–189, IR, UV	358, 1030, 1054
4-Ethylamino-3-nitro-1-phenyl-1,8-naphthyridin-2(1*H*)-one	189–193	1191
Ethyl 4-amino-2-oxo-1,2-dihydro-1,8-naphthyridine-3-carboxylate	264–267 or 269–271, IR	612, 725
Ethyl 7-amino-2-oxo-1,2-dihydro-1,8-naphthyridine-3-carboxylate	>320 or >350, NMR	482, 702
Ethyl 7-amino-2-oxo-1,2-dihydro-1,8-naphthyridine-4-carboxylate	320, IR	687, 690
Ethyl 7-amino-4-oxo-1,4-dihydro-1,8-naphthyridine-3-carboxylate	292–295	720
Ethyl 4-amino-2-oxo-1-propyl-1,2-dihydro-1,8-naphthyridine-3-carboxylate	165–168	612
Ethyl 6-benzoyl-5-dimethylamino-7-methylthio-4-oxo-1,4-dihydro-1,8-naphthyridine-3-carboxylate	158–161, NMR	841
Ethyl 2-(N'-benzoylhydrazino)-1,8-naphthyridine-3-carboxylate	220, IR, MS, NMR	728
Ethyl 1-benzoyl-4-hydroxy-2-oxo-1,2-dihydro-1,8-naphthyridine-3-carboxylate	188–190, IR, NMR	912
Ethyl 1-benzyl-7-chloro-4-oxo-,1,4-dihydro-1,8-naphthyridine-3-carboxylate	128–130, NMR	1284
Ethyl 1-benzyl-4-methyl-2-oxo-1,2-dihydro-1,8-naphthyridine-3-carboxylate	125–127, NMR	500
Ethyl 1-benzyl-7-methyl-4-oxo-1,4-dihydro-1,8-naphthyridine-3-carboxylate	140–141, NMR	305

TABLE A.4. (*Continued*)

1,8-Naphthyridine	Melting Point (°C) etc.	Reference(s)
Ethyl 6-benzyloxy-7-(2-dimethylaminovinyl)-1-ethyl-4-oxo-1,4-dihydro-1,8-naphthyridine-3-carboxylate	158	1060
Ethyl 6-benzyloxy-1-ethyl-7-methyl-4-oxo-1,4-dihydro-1,8-naphthyridine-3-carboxylate	192	1060
Ethyl 6-bromo-1-ethyl-7-methyl-4-oxo-1,4-dihydro-1,8-naphthyridine-3-carboxylate	168, IR, MS, NMR	905
Ethyl-7-bromomethyl-4-oxo-1,4-dihydro-1,8-naphthyridine-3-carboxylate	213–215, NMR	1064
Ethyl 7-(2-bromovinyl)-6-ethoxy-1-ethyl-4-oxo-1,4-dihydro-1,8-naphthyridine-3-carboxylate	142–144	1060
Ethyl 7-(2-bromovinyl)-1-ethyl-6-hydroxy-4-oxo-1,4-dihydro-1,8-naphthyridine-3-carboxylate	235–238	1060
Ethyl 1-*tert*-butyl-7-chloro-6-fluoro-5-methyl-4-oxo-1,4-dihydro-1,8-naphthyridine-3-carboxylate	178, NMR	454
Ethyl 1-*tert*-butyl-7-chloro-6-fluoro-4-oxo-1,4-dihydro-1,8-naphthyridine-3-carboxylate	158–160	493
Ethyl 7-(2-carboxyvinyl)-6-ethoxy-1-ethyl-4-oxo-1,4-dihydro-1,8-naphthyridine-3-carboxylate	285–286, NMR	1060
Ethyl 7-(2-carboxyvinyl)-6-ethoxy-4-oxo-1,4-dihydro-1,8-naphthyridine-3-carboxylate	>300	1060
Ethyl 7-chloro-1-(2-chloroethyl)-6-nitro-4-oxo-1,4-dihydro-1,8-naphthyridine-3-carboxylate	130–132	1029
Ethyl 7-chloro-1-(2-chloroethyl)-4-oxo-1,4-dihydro-1,8-naphthyridine-3-carboxylate	148–149, IR, NMR	1049
Ethyl 7-chloro-6-cyano-1-ethyl-4-oxo-1,4-dihydro-1,8-naphthyridine-3-carboxylate	243 or 252–256, IR	387, 1211
Ethyl 7-chloro-1-cyclopropyl-5-ethyl-6-fluoro-4-oxo-1,4-dihydro-1,8-naphthyridine-3-carboxylate	158, NMR	452
Ethyl 7-chloro-1-cyclopropyl-6-fluoro-5-formyl-4-oxo-1,4-dihydro-1,8-naphthyridine-3-carboxylate	158–163, NMR	1172
Ethyl 7-chloro-1-cyclopropyl-6-fluoro-5-methyl-4-oxo-1,4-dihydro-1,8-naphthyridine-3-carboxylate	solid, IR, MS, NMR	1159
Ethyl 7-chloro-1-cyclopropyl-6-fluoro-4-oxo-1,4-dihydro-1,8-naphthyridine-3-carboxylate	174–176, IR, NMR	1159, 1172
Ethyl 7-chloro-1-cyclopropyl-6-fluoro-4-oxo-5-phenyl-1,4-dihydro-1,8-naphthyridine-3-carboxylate	253, NMR	454
Ethyl 7-chloro-1-cyclopropyl-6-fluoro-4-oxo-5-trimethylsilyl-1,4-dihydro-1,8-naphthyridine-3-carboxylate	185–189, NMR	1172
Ethyl 6-chloro-7-diethylamino-1-ethyl-4-oxo-1,4-dihydro-1,8-naphthyridine-3-carboxylate	109–110	364
Ethyl 4-chloro-7-diethylamino-6-nitro-1,8-naphthyridine-3-carboxylate	74–75	364

TABLE A.4. (*Continued*)

1,8-Naphthyridine	Melting Point (°C) etc.	Reference(s)
Ethyl 4-chloro-1,7-dimethyl-2-oxo-1,2-dihydro-1,8-naphthyridine-3-carboxylate	180–184	612
Ethyl 7-chloro-1-ethyl-6-fluoro-2-methyl-4-oxo-1,4-dihydro-1,8-naphthyridine-3-carboxylate	anal, MS, NMR	
Ethyl 7-chloro-1-ethyl-6-fluoro-4-oxo-1,4-dihydro-1,8-naphthyridine-3-carboxylate	191–194	418
Ethyl 7-chloro-1-ethyl-6-formyl-4-oxo-1,4-dihydro-1,8-naphthyridine-3-carboxylate	172–173, IR	1211
Ethyl 6-chloro-1-ethyl-7-methoxy-4-oxo-1,4-dihydro-1,8-naphthyridine-3-carboxylate	156–157, IR, NMR	387
Ethyl 1-(2-chloroethyl)-7-methylamino-6-*n*-tro-4-oxo-1,4-dihydro-1,8-naphthyridine-3-carboxylate	239–242	1029
Ethyl 4-chloro-1-ethyl-7-methyl-2-oxo-1,2-dihydro-1,8-naphthyridine-3-carboxylate	143–144	612
Ethyl 6-chloro-1-ethyl-7-morpholino-4-oxo-1,4-dihydro-1,8-naphthyridine-3-carboxylate	194–195	364
Ethyl 1-(2-chloroethyl)-6-nitro-4,7-dioxo-1,4,7,8-tetrahydro-1,8-naphthyridine-3-carboxylate	275–285, NMR	1029
Ethyl 4-chloro-1-ethyl-2-oxo-1,2-dihydro-1,8-naphthyridine-3-carboxylate	134–136	612
Ethyl 7-chloro-1-ethyl-4-oxo-1,4-dihydro-1,8-naphthyridine-3-carboxylate	164–165, NMR	1284
Ethyl 6-chloro-1-ethyl-4-oxo-7-piperidino-1,4-dihydro-1,8-naphthyridine-3-carboxylate	169–171	364
Ethyl 7-chloro-6-fluoro-4-oxo-1-phenyl-1,4-dihydro-1,8-naphthyridine-3-carboxylate	219–220	613
Ethyl 7-chloro-6-fluoro-4-oxo-1-vinyl-1,4-dihydro-1,8-naphthyridine-3-carboxylate	150–151	418
Ethyl 7-chloro-1-(2-hydroxyethyl)-4-oxo-1,4-dihydro-1,8-naphthyridine-3-carboxylate	197–198	1284
Ethyl 7-chloromethyl-1-ethyl-4-oxo-1,4-dihydro-1,8-naphthyridine-3-carboxylate	135–137, NMR	1248
Ethyl 7-chloromethyl-1-(2-fluoroethyl)-4-oxo-1,4-dihydro-1,8-naphthyridine-3-carboxylate	139–140	1248
Ethyl 4-chloro-1-methyl-2-oxo-1,2-dihydro-1,8-naphthyridine-3-carboxylate	132–135	612
Ethyl 4-chloro-7-methyl-2-oxo-1,2-dihydro-1,8-naphthyridine-3-carboxylate	160–163	612
Ethyl 2-chloro-1,8-naphthyridine-3-carboxylate	125, IR, MS, NMR	728
Ethyl 7-chloro-6-nitro-4-oxo-1,4-dihydro-1,8-naphthyridine-3-carboxylate	195–197	364
Ethyl 7-chloro-4-oxo-1,4-dihydro-1,8-naphthyridine-3-carboxylate	287–288, IR, NMR	1284
Ethyl 6-cyano-2-(2-dimethylaminovinyl)-7-ethoxy-5-phenyl-1,8-naphthyridine-3-carboxylate	250–251, IR, MS, NMR	900
Ethyl 6-cyano-7-ethoxy-2,5-diphenyl-1,8-naphthyridine-3-carboxylate	243–245, IR, MS, NMR	896

TABLE A.4. (*Continued*)

1,8-Naphthyridine	Melting Point (°C) etc.	Reference(s)
Ethyl 6-cyano-7-ethoxy-1-ethyl-4-oxo-1,4-dihydro-1,8-naphthyridine-3-carboxylate	175–180, IR	1211
Ethyl 6-cyano-7-ethoxy-2-methyl-5-phenyl-1,8-naphthyridine-3-carboxylate	224–225, IR, MS, NMR	896, 900
Ethyl 6-cyano-7-ethoxy-2-oxo-5-phenyl-1,2-dihydro-1,8-naphthyridine-3-carboxylate	290–291, IR, MS, NMR	896
Ethyl 6-cyano-1-ethyl-7-methoxy-4-oxo-1,4-dihydro-1,8-naphthyridine-3-carboxylate	221–223, IR, NMR	387
Ethyl 6-(2-cyanovinyl)-1-ethyl-7-methyl-4-oxo-1,4-dihydro-1,8-naphthyridine-3-carboxylate	214, IR, MS, NMR	905
Ethyl 7-cyclohexylamino-1-ethyl-6-nitro-4-oxo-1,4-dihydro-1,8-naphthyridine-3-carboxylate	215–216	364
Ethyl 1-cyclopropyl-6-fluoro-4,7-dioxo-1,4,7,8-tetrahydro-1,8-naphthyridine-3-carboxylate	257–259, IR, NMR	1162
Ethyl 1-cyclopropyl-6-fluoro-7-(3-hydroxycyclopent-1-enyl)-4-oxo-1,4-dihydro-1,8-naphthyridine-3-carboxylate	solid, IR, MS, NMR	1158
Ethyl 1-cyclopropyl-6-fluoro-7-(1-methoxyvinyl)-4-oxo-1,4-dihydro-1,8-naphthyridine-3-carboxylate	151–152, IR, MS, NMR	1158
Ethyl 1-cyclopropyl-6-fluoro-4-oxo-7-(3-oxocyclopent-1-enyl)-1,4-dihydro-1,8-naphthyridine-3-carboxylate	183–184, IR, MS, NMR	1158
Ethyl 1-cyclopropyl-6-fluoro-4-oxo-7-piperidino-1,4-dihydro-1,8-naphthyridine-3-carboxylate	158–161, IR, NMR	1162
Ethyl 1-cyclopropyl-6-fluoro-4-oxo-7-vinyl-1,4-dihydro-1,8-naphthyridine-3-carboxylate	solid, IR, MS, NMR	1158
Ethyl 6-diazonio-1-ethyl-7-methoxy-4-oxo-1,4-dihydro-1,8-naphthyridine-3-carboxylate tetrafluoroborate	148–149, IR	387
Ethyl 6,7-dichloro-1-ethyl-4-oxo-1,4-dihydro-1,8-naphthyridine-3-carboxylate	225–227, NMR	364
Ethyl 4-diethylamino-1-ethyl-7-methyl-2-oxo-1,2-dihydro-1,8-naphthyridine-3-carboxylate	82–85	612
Ethyl 7-diethylamino-1-ethyl-6-nitro-4-oxo-1,4-dihydro-1,8-naphthyridine-3-carboxylate	126–128	364
Ethyl 7-diethylamino-6-nitro-4-oxo-1,4-dihydro-1,8-naphthyridine-3-carboxylate	254–256	364
Ethyl 7-dimethylamino-1-ethyl-6-nitro-4-oxo-1,4-dihydro-1,8-naphthyridine-3-carboxylate	186–188	364
Ethyl 7-(2-dimethylaminovinyl)-1-ethyl-6-fluoro-4-oxo-1,4-dihydro-1,8-naphthyridine-3-carboxylate	175–176, IR, NMR	1163
1-Ethyl-3,7-dimethyl-1,8-naphthyridin-4(1*H*)-one	112–113, NMR, UV	1076
1-Ethyl-5,7-dimethyl-1,8-naphthyridin-2(1*H*)-one	90–92	338
3-Ethyl-2,7-dimethyl-1,8-naphthyridin-4(1*H*)-one	268–270, IR, NMR	976
Ethyl 2,7-dimethyl-4-oxo-1,4-dihydro-1,8-naphthyridine-3-carboxylate	223, NMR	249

TABLE A.4. (*Continued*)

1,8-Naphthyridine	Melting Point (°C) etc.	Reference(s)
Ethyl 5,7-dimethyl-4-oxo-1,4-dihydro-1,8-naphthyridine-3-carboxylate	232–234	1076
1-Ethyl-2,7-dimethyl-4-oxo-1,4-dihydro-1,8-naphthyridine-3-carboxylic acid	201–203	249
1-Ethyl-5,7-dimethyl-4-phenyl-1,8-naphthyridine-2(1*H*)-thione	148–149, NMR	1196
1-Ethyl-5,7-dimethyl-4-phenyl-1,8-naphthyridin-2(1*H*)-one	113–114, NMR	1196
7-Ethyl-1,5-dimethyl-4-phenyl-1,8-naphthyridin-2(1*H*)-one	88–89, NMR	1196
1-Ethyl-3,6-dinitro-1,8-naphthyridine-4,7(1*H*,8*H*)-dione	282–284, NMR	364
1-Ethyl-3,6-dinitro-1,8-naphthyridin-2(1*H*)-one	168–170, IR, MS, NMR	692, 1273
Ethyl 2,7-dioxo-1,2,7,8-tetrahydro-1,8-naphthyridine-4-carboxylate	264–265	690
Ethyl 2-ethoxycarbonylmethyl-1,8-naphthyridine-3-carboxylate	79–80, NMR	1419
Ethyl 6-ethoxy-1-ethyl-7-formyl-4-oxo-1,4-dihydro-1,8-naphthyridine-3-carboxylate	169–170	1060
Ethyl 7-ethoxy-1-ethyl-6-formyl-4-oxo-1,4-dihydro-1,8-naphthyridine-3-carboxylate	199–202, IR, NMR	1211
Ethyl 7-ethoxy-1-ethyl-6-hydroxymethyl-4-oxo-1,4-dihydro-1,8-naphthyridine-3-carboxylate	200–202, NMR	1211
Ethyl 4-ethoxy-1-ethyl-7-methyl-2-oxo-1,2-dihydro-1,8-naphthyridine-3-carboxylate	77–79	612
Ethyl 6-ethoxy-1-ethyl-7-methyl-4-oxo-1,4-dihydro-1,8-naphthyridine-3-carboxylate	163–164	1060
Ethyl 7-ethoxy-1-ethyl-6-nitro-4-oxo-1,4-dihydro-1,8-naphthyridine-3-carboxylate	184–185	1211
Ethyl 1-ethoxy-6-formyl-4,7-dioxo-1,4,7,8-tetrahydro-1,8-naphthyridine-3-carboxylate	232–235	1211
Ethyl 6-ethoxy-7-methyl-4-oxo-1,4-dihydro-1,8-naphthyridine-3-carboxylate	279–282	1060
1-Ethyl-7-ethylcarbamoyl-4-oxo-1,4-dihydro-1,8-naphthyridine-3-carboxylic acid	276–278	1012
Ethyl 1-ethyl-6,7-dimethyl-4-oxo-1,4-dihydro-1,8-naphthyridine-3-carboxylate	181–183, IR, MS, NMR	905
Ethyl 1-ethyl-4-ethylamino-7-methyl-2-oxo-1,2-dihydro-1,8-naphthyridine-3-carboxylate	132–136	612
Ethyl 1-ethyl-7-ethylsulfonyl-6-morpholino-4-oxo-1,4-dihydro-1,8-naphthyridine-3-carboxylate	169–172, IR, NMR	1256
Ethyl 1-ethyl-7-ethylsulfonyl-4-oxo-6-piperidino-1,4-dihydro-1,8-naphthyridine-3-carboxylate	95–98, IR, NMR	1256
Ethyl 1-ethyl-7-ethylthio-6-fluoro-4-oxo-1,4-dihydro-1,8-naphthyridine-3-carboxylate	135–137	1120
Ethyl 1-ethyl-6-fluoro-7-formyl-4-oxo-1,4-dihydro-1,8-naphthyridine-3-carboxylate	crude, 170–175, NMR	1163

TABLE A.4. (*Continued*)

1,8-Naphthyridine	Melting Point (°C) etc.	Reference(s)
Ethyl 1-ethyl-6-fluoro-7-methyl-4-oxo-1,4-dihydro-1,8-naphthyridine-3-carboxylate	146–147, IR, NMR	1091
Ethyl 1-ethyl-7-fluoromethyl-4-oxo-1,4-dihydro-1,8-naphthyridine-3-carboxylate	137–139, NMR	1148
Ethyl 1-ethyl-6-fluoro-7-morpholino-4-oxo-1,4-dihydro-1,8-naphthyridine-3-carboxylate	169–171, IR, NMR	1256
Ethyl 1-ethyl-6-fluoro-4-oxo-7-piperidino-1,4-dihydro-1,8-naphthyridine-3-carboxylate	175–176, IR, NMR	1256
Ethyl 1-ethyl-6-fluoro-4-oxo-7-vinyl-1,4-dihydro-1,8-naphthyridine-3-carboxylate	solid, IR, MS, NMR	1158
Ethyl 1-ethyl-7-formyl-6-hydroxy-4-oxo-1,4-dihydro-1,8-naphthyridine-3-carboxylate	243–245	1060
Ethyl 1-ethyl-7-formyl-4-oxo-1,4-dihydro-1,8-naphthyridine-3-carboxylate	158–159; HON=: 222–224; PhN=: 169–170; sc: 273–275; tsc: 264–266; PhNHN=: 283–285	1066
1-Ethyl-7-(*N*-ethylhydrazino)-4-oxo-1,4-dihydro-1,8-naphthyridine-3-carboxylate	223–225	1129
Ethyl 1-ethyl-4-hydroxy-7-methyl-2-oxo-1,2-dihydro-1,8-naphthyridine-3-carboxylate	147–151	612
Ethyl 1-ethyl-6-hydroxy-7-methyl-4-oxo-1,4-dihydro-1,8-naphthyridine-3-carboxylate	184–185, NMR	1060
Ethyl 1-ethyl-7-hydroxymethyl-4-oxo-1,4-dihydro-1,8-naphthyridine-3-carboxylate	171–172	1248
Ethyl 1-ethyl-4-hydroxy-2-oxo-1,2-dihydro-1,8-naphthyridine-3-carboxylate	122–125 or 124–125	612, 1127
Ethyl 1-ethyl-6-(2-methoxycarbonylvinyl)-7-methyl-4-oxo-1,4-dihydro-1,8-naphthyridine-3-carboxylate	140–142, IR, MS, NMR	905
Ethyl 1-ethyl-7-methoxy-6-nitro-4-oxo-1,4-dihydro-1,8-naphthyridine-3-carboxylate	224–227, IR, MS	387
Ethyl 1-ethyl-7-methoxy-4-oxo-1,4-dihydro-1,8-naphthyridine-3-carboxylate	142–143, IR	387
Ethyl 1-ethyl-7-methyliminomethyl-4-oxo-1,4-dihydro-1,8-naphthyridine-3-carboxylate	177–178	1066
Ethyl 1-ethyl-7-methyl-4-oxo-1,4-dihydro-1,8-naphthyridine-3-carboxylate	118 to 122, biol	154, 1276, 1380
Ethyl 1-ethyl-7-methyl-4-oxo-6-phenyl-1,4-dihydro-1,8-naphthyridine-3-carboxylate	144–145, IR, MS, NMR	905
Ethyl 1-ethyl-7-methyl-4-thioxo-1,4-dihydro-1,8-naphthyridine-3-carboxylate	121–122	1380
Ethyl 1-ethyl-7-morpholino-6-nitro-4-oxo-1,4-dihydro-1,8-naphthyridine-3-carboxylate	193–195	364
Ethyl 1-ethyl-6-nitro-4,7-dioxo-1,4,7,8-tetrahydro-1,8-naphthyridine-3-carboxylate	239–241, NMR	364
Ethyl 8-ethyl-6-nitro-5-oxo-5,8-dihydro-1,8-naphthyridine-2-carboxylate	205–207, IR	364

TABLE A.4. (*Continued*)

1,8-Naphthyridine	Melting Point (°C) etc.	Reference(s)
Ethyl 1-ethyl-6-nitro-4-oxo-7-piperidino-1,4-dihydro-1,8-naphthyridine-3-carboxylate	173–175	364
1-Ethyl-7-(*N*-ethyl-*N*-nitrosoamino)-4-oxo-1,4-dihydro-1,8-naphthyridine-3-carboxylic acid	242–243	1129
Ethyl 7-ethyl-4-oxo-1,4-dihydro-1,8-naphthyridine-3-carboxylate	249–250	1076
Ethyl 7-ethylsulfonyl-6-fluoro-1-(2-fluoroethyl)-4-oxo-1,4-dihydro-1,8-naphthyridine-3-carboxylate	176–178	418
Ethyl 7-ethylthio-6-fluoro-1-(2-fluoroethyl)-4-oxo-1,4-dihydro-1,8-naphthyridine-3-carboxylate	174–175	418
Ethyl 7-ethylthio-6-fluoro-4-oxo-1,4-dihydro-1,8-naphthyridine-3-carboxylate	280–284	1120
1-Ethyl-6-fluoro-4,7-dioxo-1,4,7,8-tetrahydro-1,8-naphthyridine-3-carboxylic acid	>300, IR, NMR	387, 1091
Ethyl 1-(2-fluoroethyl)-7-fluoromethyl-4-oxo-1,4-dihydro-1,8-naphthyridine-3-carboxylate	136–138, NMR	1248
Ethyl 1-(2-fluoroethyl)-7-hydroxymethyl-4-oxo-1,4-dihydro-1,8-naphthyridine-3-carboxylate	153–154, NMR	1248
1-Ethyl-6-fluoro-7-formyl-4-oxo-1,4-dihydro-1,8-naphthyridine-3-carboxylic acid	crude, NMR; HON=: 258–262, NMR	395
Ethyl 6-fluoro-7-methyl-4-oxo-1,4-dihydro-1,8-naphthyridine-3-carboxylate	293–296, NMR	1091
1-Ethyl-6-fluoro-7-methyl-4-oxo-1,4-dihydro-1,8-naphthyridine-3-carboxylic acid	248–252 or 252–254, IR, NMR	395, 1091
1-Ethyl-7-fluoromethyl-4-oxo-1,4-dihydro-1,8-naphthyridine-3-carboxylic acid	224–227, NMR	1248
1-Ethyl-6-fluoro-7-morpholino-4-oxo-1,4-dihydro-1,8-naphthyridine-3-carboxylic acid	261–263	387
1-Ethyl-6-fluoro-4-oxo-7-piperidino-1,4-dihydro-1,8-naphthyridine-3-carboxylic acid	211–213	387
1-Ethyl-7-formyl-4-oxo-1,4-dihydro-1,8-naphthyridine-3-carboxylic acid	266–268; HON=: 246–248; MeN=: 223–225; PhN=: >300; tsc: 276–278; PhNHN=: 285–287	1066
3-Ethyl-4-hydroxy-5,7-dimethyl-1,8-naphthyridin-2(1*H*)-one	276	1301
Ethyl 4-hydroxy-1,7-dimethyl-2-oxo-1,2-dihydro-1,8-naphthyridine-3-carboxylate	143–145 or 145–151	612, 1127
Ethyl 4-hydroxy-5,7-dimethyl-2-oxo-1,2-dihydro-1,8-naphthyridine-3-carboxylate	330	1301
1-Ethyl-4-hydroxy-3,7-diphenyl-1,8-naphthyridin-2(1*H*)-one	216–218, IR	1112
1-Ethyl-3-hydroxy-7-methyl-1,8-naphthyridin-4(1*H*)-one	128–130, IR, MS, NMR	1010

TABLE A.4. (*Continued*)

1,8-Naphthyridine	Melting Point (°C) etc.	Reference(s)
1-Ethyl-4-hydroxy-7-methyl-1,8-naphthyridin-2(1*H*)-one	205–206	344
Ethyl 4-hydroxy-1-methyl-2-oxo-1,2-dihydro-1,8-naphthyridine-3-carboxylate	158–160 or 163–166, IR, NMR	612, 1127
Ethyl 4-hydroxy-7-methyl-2-oxo-1,2-dihydro-1,8-naphthyridine-3-carboxylate	190–193	612
1-Ethyl-6-hydroxy-7-methyl-4-oxo-1,4-dihydro-1,8-naphthyridine-3-carboxylic acid	325–332	219
1-Ethyl-7-hydroxymethyl-4-oxo-1,4-dihydro-1,8-naphthyridine-3-carboxylic acid	246–247 or 253–256	151, 154, 192, 219, 295, 1277
Ethyl 4-hydroxy-5-methyl-2-oxo-7-phenyl-1,2-dihydro-1,8-naphthyridine-3-carboxylate	236	1301
Ethyl 1-hydroxy-2-oxo-1,2-dihydro-1,8-naphthyridine-3-carboxylate	168–170, IR	1112
Ethyl 4-hydroxy-2-oxo-1-phenyl-1,2-dihydro-1,8-naphthyridine-3-carboxylate	247–252, IR, NMR	453, 497
1-Ethyl-4-hydroxy-7-phenyl-1,8-naphthyridin-2(1*H*)-one	260–261	1112
3-Ethyl-4-hydroxy-1-phenyl-1,8-naphthyridin-2(1*H*)-one	314–315	607
1-Ethyl-3-iodo-7-methyl-1,8-naphthyridin-4(1*H*)-one	204–205, IR, NMR	905, 1213
1-Ethyl-7-isopropylcarbamoyl-4-oxo-1,4-dihydro-1,8-naphthyridine-3-carboxylic acid	293–295	1012
1-Ethyl-7-(*N'*-isopropylidene-*N*-methylhydrazino)-4-oxo-1,4-dihydro-1,8-naphthyridine-3-carboxylic acid	227–230	1129
1-Ethyl-3-(2-methoxycarbonylvinyl)-7-methyl-1,8-naphthyridin-4(1*H*)-one	202, IR, NMR	905
1-Ethyl-7-methylcarbamoyl-4-oxo-1,4-dihydro-1,8-naphthyridine-3-carboxylic acid	303–305	1012
Ethyl 3-methyl-2,7-dioxo-1,3,7,8-tetrahydro-1,8-naphthyridine-4-carboxylate	244	687
1-Ethyl-7-(*N*-methylhydrazino)-4-oxo-1,4-dihydro-1,8-naphthyridin-3-carboxylic acid	268–271	1129
1-Ethyl-7-[*N*-methyl-*N'*-(1-methylpropylidene)hydrazine]-4-oxo-1,4-dihydro-1,8-naphthyridine-3-carboxylic acid	169–172	1129
2-Ethyl-3-methyl-1,8-naphthyridine	93–94, NMR	1013
3-Ethyl-2-methyl-1,8-naphthyridine	93–94, NMR	1013
Ethyl 2-methyl-1,8-naphthyridine-3-carboxylate	85–86 or 90–91, IR, NMR	658, 1054, 1112, 1440
1-Ethyl-4-methyl-1,8-naphthyridine-2,7-(1*H*,8*H*)-dione	228–230	1392
1-Ethyl-7-methyl-1,8-naphthyridin-4-(1*H*)-one	—	344, 518, 1267, 1393
2-Ethyl-7-methyl-1,8-naphthyridin-4-(1*H*)-one	214–215	1076
3-Ethyl-7-methyl-1,8-naphthyridin-4(1*H*)-one	215–216	1076
7-Ethyl-2-methyl-1,8-naphthyridin-4(1*H*)-one	242–243	1076

TABLE A.4. (*Continued*)

1,8-Naphthyridine	Melting Point (°C) etc.	Reference(s)
1-Ethyl-7-methyl-6-nitro-4-oxo-1,4-dihydro-1,8-naphthyridine-3-carboxylic acid	248–250, IR, NMR	395
1-Ethyl-7-(*N*-methyl-*N*-nitrosoamino)-4-oxo-1,4-dihydro-1,8-naphthyridine-3-carboxylic acid	251–252	1129
1-Ethyl-7-methyl-4-oxo-1,4-dihydro-1,8-naphthyridine-3-carboguanide	HCl: >260, IR, NMR	635
1-Ethyl-7-methyl-4-oxo-1,4-dihydro-1,8-naphthyridine-3-carbohydrazide	192–193, NMR	1116
1-Ethyl-7-methyl-4-oxo-1,4-dihydro-1,8-naphthyridine-3-carbonitrile	221–223 or 225–226, IR, NMR	1213, 1405
1-Ethyl-7-methyl-4-oxo-1,4-dihydro-1,8-naphthyridine-3-carboxanilide	214–215	212
Ethyl 2-methyl-4-oxo-1,4-dihydro-1,8-naphthyridine-3-carboxylate	—	162
Ethyl 4-methyl-2-oxo-1,2-dihydro-1,8-naphthyridine-3-carboxylate	226–229, NMR	500
Ethyl 7-methyl-4-oxo-1,4-dihydro-1,8-naphthyridine-3-carboxylate	259 to 277, IR, MS, NMR	179, 286, 350, 1076, 1186, 1235
1-Ethyl-7-methyl-4-oxo-1,4-dihydro-1,8-naphthyridine-3-carboxylic acid (nalidixic acid)	225 to 230, boil, IR, NMR, pK_a, UV, xl st	151, 154, 171, 179, 191, 192, 227, 228, 294, 295, 311, 342, 350, 373, 528, 799, 883, 1012, 1213, 1359
1-Ethyl 7-methyl-4-oxo-1,4-dihydro-1,8-naphthyridine-3-sulfonamide	228, IR, NMR, UV	1045
Ethyl 2-methyl-7-phenyl-1,8-naphthyridine-3-carboxylate	130–131, IR	1112
1-Ethyl-7-methyl-2-phenyl-1,8-naphthyridin-4(1*H*)-one	125–126, NMR	221
1-Ethyl-7-methyl-3-phenyl-1,8-naphthyridin-4(1*H*)-one	155, IR, NMR	905
1-Ethyl-7-methyl-4-phenyl-1,8-naphthyridin-2(1*H*)-one	109, NMR	1196
1-Ethyl-7-methyl-3-styryl-1,8-naphthyridin-4(1*H*)-one	145, NMR	905
Ethyl 7-methyl-4-thioxo-1,4-dihydro-1,8-naphthyridine-3-carboxylate	208, NMR	1380
1-Ethyl-7-methyl-4-thioxo-1,4-dihydro-1,8-naphthyridine-3-carboxylic acid	249–251, NMR	1380
1-Ethyl-7-methyl-3-vinyl-1,8-naphthyridin-4(1*H*)-one	130–132, IR, NMR	905
1-Ethyl-7-morpholino-4-oxo-1,4-dihydro-1,8-naphthyridine-3-carboxylic acid	288–290	1284
7-Ethyl-1,8-naphthyridin-2-amine	169–172	338
2-Ethyl-1,8-naphthyridine	78–79, IR, NMR	145
4-Ethyl-1,8-naphthyridine	liquid NMR	1441

TABLE A.4. (*Continued*)

1,8-Naphthyridine	Melting Point (°C) etc.	Reference(s)
1-Ethyl-1,8-naphthyridine-2,7(1H,8H)-dione	202–204	1392
7-Ethyl-1,8-naphthyridin-2(1H)-one	133–134	338
1-Ethyl-6-nitro-4,7-dioxo-1,4,7,8-tetrahydro-1,8-naphthyridine-3-carboxylic acid	275–277, NMR	364
1-Ethyl-3-nitro-1,8-naphthyridin-2(1H)-one	166–168, or 285–286, IR, MS, NMR	692, 1273
1-Ethyl-3-nitro-1,8-naphthyridin-4(1H)-one	207–209 or 213–215, IR, MS, NMR	364, 692, 1273
8-Ethyl-6-nitro-5-oxo-5,8-dihydro-1,8-naphthyridine-2-carboxylic acid	274–276, NMR	364
1-Ethyl-7-(*N*-nitroso-*N*-propylamino)-4-oxo-1,4-dihydro-1,8-naphthyridine-3-carboxylic acid	223–224	1129
Ethyl 2-oxo-1,2-dihydro-1,8-naphthyridine-3-carboxylate	200–202 or 205–207, IR, NMR	658, 674, 1112, 1440
Ethyl 4-oxo-1,4-dihydro-1,8-naphthyridine-3-carboxylate	—	350
Ethyl 7-oxo-7,8-dihydro-1,8-naphthyridine-2-carboxylate	207–208, IR, MS, NMR	473
1-Ethyl-2-oxo-1,2-dihydro-1,8-naphthyridin-3-carboxylic acid	191–192, IR	1112
8-Ethyl-5-oxo-5,8-dihydro-1,8-naphthyridine-2,6-dicarboxylic acid	—	151
1-Ethyl-2-oxo-7-phenyl-1,2-dihydro-1,8-naphthyridine-3-carbonitrile	274–276, IR	1112
Ethyl 4-oxo-2-phenyl-1,4-dihydro-1,8-naphthyridine-3-carboxylate	—	162
1-Ethyl-2-oxo-7-phenyl-1,2-dihydro-1,8-naphthyridine-3-carboxylic acid	248–249, IR	1112
1-Ethyl-4-oxo-7-piperidino-1,4-dihydro-1,8-naphthyridine-3-carboxylic acid	260–262	1284
1-Ethyl-4-oxo-7-propylcarbamoyl-1,4-dihydro-1,8-naphthyridin-3-carboxylic acid	143–145	1012
Ethyl 4-oxo-2-propyl-1,4-dihydro-1,8-naphthyridine-3-carboxylate	—	162
1-Ethyl-4-oxo-7-(*N*-propylhydrazino)-1,4-dihydro-1,8-naphthyridine-3-carboxylic acid	200–201	1129
3-Ethyl-2-phenyl-1,8-naphthyridine	125–126, NMR	1013, 1030
7-Ethyl-5-phenyl-1,8-naphthyridine-3-carboxamide	>275, IR, NMR	1326
Ethyl 2-phenyl-1,8-naphthyridine-3-carboxylate	104, IR, MS, NMR	658, 674, 859, 1440
1-Ethyl-3-phenyl-1,8-naphthyridin-2(1H)-one	95–96, IR	1112
N-(1-Ethylpropyl)-7-methyl-1-(2-morpholinoethyl)-4-oxo-1,4-dihydro-1,8-naphthyridine-3-carboxamide	140–142	305
N-(1-Ethylpropyl)-7-methyl-4-oxo-1,4-dihydro-1,8-naphthyridine-3-carboxamide	213–215, NMR	305
Ethyl 5,6,7-trichloro-1-ethyl-4-oxo-1,4-dihydro-1,8-naphthyridine-3-carboxylate	165–167, NMR	364
3-Ethyl-2,5,7-trimethyl-1,8-naphthyridin-4(1H)-one	236–239	976

Appendix

TABLE A.4. (*Continued*)

1,8-Naphthyridine	Melting Point (°C) etc.	Reference(s)
6-Fluoro-4,7-dioxo-1-vinyl-1,4,7,8-tetrahydro-1,8-naphthyridine-3-carboxylic acid	256–259, IR, NMR	387, 418
1-(2-Fluoroethyl)-7-fluoromethyl-4-oxo-1,4-dihydro-1,8-naphthyridine-3-carboxylic acid	195–197	1248
2-Fluoro-5-methyl-1,8-naphthyridine	—	893
7-Fluoro-2-phenyl-1,8-naphthyridin-4(1*H*)-one	>300, NMR	221
Hexachloro-1,8-naphthyridine	271–273	918
Hexafluoro-1,8-naphthyridine	165–166, IR, NMR, th	918, 1313
3-Hexyl-4-hydroxy-7-methyl-1,8-naphthyridin-2(1*H*)-one	240 or 242	810, 1301
1-Hydrazinocarbonylmethyl-3-phenyl-1,8-naphthyridin-2(1*H*)-one	148, IR, MS, NMR; MeCH: 285, IR, MS, NMR	740
2-Hydrazino-5,7-dimethyl-1,8-naphthyridine	242–243; PhCH=: 244–245; PhMeC=: 245–246	763
4-Hydrazino-5,7-dimethyl-1,8-naphthyridine	NMR	954
2-Hydrazino-3,6-dinitro-1,8-naphthyridine	crude, 138–145	692
4-Hydrazino-7-methyl-1,8-naphthyridine	208–210, NMR	153, 954
2-Hydrazino-7-methyl-5-phenyl-1,8-naphthyridine	PhCH=: 253–254	771
4-Hydrazino-7-methyl-2-phenyl-1,8-naphthyridine	NMR; PhCH=:—	221, 1308
2-Hydrazino-1,8-naphthyridine	182–183, IR, UV; PhCH=: 218–220, IR, UV	1100
4-Hydrazino-1,8-naphthyridin-2(1*H*)-one	—	386
2-Hydrazino-3-phenyl-1,8-naphthyridine	174, IR, MS, NMR; PhCH=: 208; PhMeC=: 123, IR, MS, NMR	850
3-(4-Hydroxybutyl)-1-methyl-1,8-naphthyridin-2(1*H*)-one	solid, NMR	1298
1-Hydroxy-5,7-dimethyl-1,8-naphthyridin-2(1*H*)-one	250	769
3-Hydroxy-1,7-dimethyl-1,8-naphthyridin-4(1*H*)-one	237–241, IR, NMR	1394
4-Hydroxy-1,3-dimethyl-1,8-naphthyridin-2(1*H*)-one	238–240, IR, MS, NMR, UV	565
4-Hydroxy-3,7-dimethyl-1,8-naphthyridin-2(1*H*)-one	280, IR, NMR	1267
4-Hydroxy-5,7-dimethyl-1,8-naphthyridin-2(1*H*)-one	370	1301
4-Hydroxy-1,7-dimethyl-3-nitro-1,8-naphthyridin-2(1*H*)-one	159–161, IR, NMR	1127
4-Hydroxy-3,5-dimethyl-7-phenyl-1,8-naphthyridin-2(1*H*)-one	231	1301
4-Hydroxy-5,7-dimethyl-3-phenyl-1,8-naphthyridin-2(1*H*)-one	280	1301
4-Hydroxy-6,7-dimethyl-3-phenyl-1,8-naphthyridin-2(1*H*)-one	290	1301
4-Hydroxy-1,3-diphenyl-1,8-naphthyridin-2(1*H*)-one	248–249	
4-Hydroxy-3,7-diphenyl-1,8-naphthyridin-2(1*H*)-one	326–327, IR	1112

TABLE A.4. (*Continued*)

1,8-Naphthyridine	Melting Point (°C) etc.	Reference(s)
1-(2-Hydroxyethyl)-4-oxo-7-piperidino-1,4-dihydro-1,8-naphthyridine-3-carboxylic acid	>300	1284
4-Hydroxy-3-(3-hydroxybutyl)-1-phenyl-1,8-naphthyridin-2(1*H*)-one	234–235	607
4-Hydroxy-3-iodo-1,8-naphthyridin-2(1*H*)-one	220–222, IR	693
4-Hydroxy-6-methoxy-1-phenyl-1,8-naphthyridin-2(1*H*)-one	285–287	607
4-Hydroxy-3-(3-methylbut-2-enyl)-1-phenyl-1,8-naphthyridin-2(1*H*)-one	Na: 250–260	607
4-Hydroxy-5-methyl-2,7-diphenyl-1,8-naphthyridin-2(1*H*)-one	266	1301
7-Hydroxymethyl-1,8-naphthyridin-2-amine	240, IR, NMR	1445
2-Hydroxymethyl-1,8-naphthyridine	99–100, IR, NMR	473
4-Hydroxy-1-methyl-1,8-naphthyridin-2(1*H*)-one	308–310, NMR	1127
4-Hydroxy-6-methyl-1,8-naphthyridin-2(1*H*)-one	300	114
4-Hydroxy-7-methyl-1,8-naphthyridin-2(1*H*)-one	260, IR, NMR	1269
4-Hydroxy-1-methyl-3-nitro-1,8-naphthyridin-2(1*H*)-one	156–158	1127
4-Hydroxy-*N*-methyl-2-oxo-1-phenyl-1,2-dihydro-1,8-naphthyridine-3-carboxamide	>300	453
4-Hydroxy-7-methyl-3-pentyl-1,8-naphthyridin-2(1*H*)-one	260 or 261	810, 1301
4-Hydroxy-3-methyl-1-phenyl-1,8-naphthyridin-2(1*H*)-one	>320	607
4-Hydroxy-3-methyl-7-phenyl-1,8-naphthyridin-2(1*H*)-one	288–290, IR	1112
4-Hydroxy-5-methyl-7-phenyl-1,8-naphthyridin-2(1*H*)-one	350	1301
4-Hydroxy-7-methyl-3-phenyl-1,8-naphthyridin-2(1*H*)-one	311 to 317, IR	810, 1269, 1301
4-Hydroxy-1,8-naphthyridin-2(1*H*)-one	292–294, IR, MS, NMR	33, 693, 1253
4-Hydroxy-3-nitro-1,8-naphthyridin-2(1*H*)-one	288–289, NMR	312, 682, 1273
4-Hydroxy-3-nitro-1-phenyl-1,8-naphthyridin-2(1*H*)-one	296–298, IR, NMR	892, 1191, 1273
4-Hydroxy-3-nitro-7-phenyl-1,8-naphthyridin-2(1*H*)-one	245–246, IR	1112, 1273
4-Hydroxy-3-nitroso-1-phenyl-1,8-naphthyridin-2(1*H*)-one	205–210, IR, MS, NMR	892
4-Hydroxy-2-oxo-1,2-dihydro-1,8-naphthyridine-3-carbaldehyde	222; HON=: 210	1281
4-Hydroxy-2-oxo-1,2-dihydro-1,8-naphthyridine-3-carbonitrile	solid, NMR	912
4-Hydroxy-2-oxo-1,2-dihydro-1,8-naphthyridine-3-carboxamide	315	31
1-Hydroxy-2-oxo-1,2-dihydro-1,8-naphthyridine-3-carboxylic acid	>315, IR	1112
4-Hydroxy-2-oxo-1-phenyl-1,2-dihydro-1,8-naphthyridine-3-carboxanilide	>300	453
4-Hydroxy-3-pentyl-1-phenyl-1,8-naphthyridin-2(1*H*)-one	226–228	607

TABLE A.4. (Continued)

1,8-Naphthyridine	Melting Point (°C) etc.	Reference(s)
4-Hydroxy-1-phenyl-1,8-naphthyridin-2(1H)-one	300 or 309–310, IR, NMR, UV	453, 603, 607
4-Hydroxy-3-phenyl-1,8-naphthyridin-2(1H)-one	306–308, IR, NMR	796
4-Hydroxy-7-phenyl-1,8-naphthyridin-2(1H)-one	360, IR	1112
3-(2-Hydroxypropyl)-1-methyl-1,8-naphthyridin-2(1H)-one	solid, NMR	1298
3-(2-Hydroxypropyl)-1,8-naphthyridin-2(1H)-one	crude, NMR	1298
4-Hydroxy-3,5,7-trimethyl-1,8-naphthyridin-2(1H)-one	325	1301
4-Hydroxy-3,6,7-trimethyl-1,8-naphthyridin-2(1H)-one	318	1301
3-Iodo-7-methyl-1,8-naphthyridin-4(1H)-one	>260, IR, NMR	905
1-Isobutyl-7-methyl-1,8-naphthyridin-4(1H)-one	84–86	1380
7-Isobutyl-1,8-naphthyridin-2-amine	125–127	338
2-Isobutyl-1,8-naphthyridine	56–57, IR, NMR	145
7-Isobutyl-1,8-naphthyridin-2(1H)-one	135–137	338
2-Isobutyramido-5-isobutyroxy-7-phenyl-1,8-naphthyridine	crude, NMR	472
2-Isobutyramido-5-methoxy-7-phenyl-1,8-naphthyridine	194–195, IR, MS, NMR	472
7-Isobutyramido-2-phenyl-1,8-naphthyridin-4(1H)-one	266–267, IR, MS, NMR	472
2-Isocyanato-5,7-dimethyl-1,8-naphthyridine	256–258, IR, MS, NMR	1190
1-Isopentyl-3-methyl-1,8-naphthyridin-2(1H)-one	solid (?), NMR	
4-Isopropylamino-3-nitro-1-phenyl-1,8-naphthyridin-2(1H)-one	259–261	1191
7-(N'-Isopropylidene-N-methylhydrazino)-1-propyl-4-oxo-1,4-dihydro-1,8-naphthyridine-3-carboxylic acid	178–180	1129
1-Isopropyl-7-methyl-4-phenyl-1,8-naphthyridin-2(1H)-one	137–139, NMR	1196
7-Isopropyl-1,8-naphthyridin-2-amine	158–160	338
2-Isopropyl-1,8-naphthyridine	65–66, NMR	1018
7-Isopropyl-1,8-naphthyridin-2(1H)-one	131–133	338
2-Methoxy-4,5-bismethylamino-3,6-dinitro-1,8-naphthyridine	245–247, IR, MS, NMR	860
2-(4-Methoxycarbonylbutyl)-1,8-naphthyridine	anal, NMR	264
2-Methoxy-5,7-dimethyl-1,8-naphthyridine	65; pic: 188–189	67, 88
4-Methoxy-2,7-dimethyl-1,8-naphthyridine	130 or 148–149, IR, NMR	185, 1445
2-Methoxy-3,6-dinitro-1,8-naphthyridine	172–173, IR, MS, NMR	860, 1273
2-Methoxy-4,6-diphenyl-1,8-naphthyridine-3-carbonitrile	227–229, IR, MS, NMR	896
2-Methoxy-4,7-diphenyl-1,8-naphthyridine-3-carbonitrile	235–237, IR, MS, NMR	896
2-Methoxy-4-methylamino-3-nitro-1,8-naphthyridine	240–242, IR, NMR	682, 785, 1273
4-Methoxy-7-methyl-1,8-naphthyridine	111–113, NMR	1380
7-Methoxy-1-methyl-1,8-naphthyridin-2(1H)-one	114–116, NMR	1380
7-Methoxy-5-methyl-1,8-naphthyridin-2(1H)-one	214–216	338
2-Methoxy-7-methyl-5-phenyl-1,8-naphthyridine	124–125	771
4-Methoxy-7-methyl-2-phenyl-1,8-naphthyridine	101–102, NMR	221

TABLE A.4. (*Continued*)

1,8-Naphthyridine	Melting Point (°C) etc.	Reference(s)
2-Methoxy-6-methyl-4-phenyl-1,8-naphthyridine-3-carbonitrile	239–241, IR, MS, NMR	896
5-Methoxy-1-methyl-7-phenyl-1,8-naphthyridin-2(1H)-one	192–194, NMR	221
2-Methoxy-7-morpholino-3-nitro-6-phenyl-1,8-naphthyridine	190–192, NMR	1304
2-Methoxy-7-morpholino-6-phenyl-1,8-naphthyridine	128–130, NMR	1304
7-Methoxy-1,8-naphthyridin-2-amine	150–151, IR, NMR	1257
2-Methoxy-1,8-naphthyridine	53–55 or 61–62, IR, NMR, UV	225, 1100
4-Methoxy-1,8-naphthyridine	liq, anal, NMR; pic: 183–185	225
2-Methoxy-1,8-naphthyridine-3-carbonitrile	193–195	278
4-Methoxy-1,8-naphthyridine-2,7-dicarbaldehyde	210, IR, NMR	1445
7-Methoxy-1,8-naphthyridin-2(1H)-one	171–172, IR, NMR	1257
2-Methoxy-3-nitro-1,8-naphthyridine	166–167, IR, NMR	682, 1273
2-Methoxy-3-nitro-6-phenyl-7-piperidino-1,8-naphthyridine	148–150, NMR	1304
2-Methoxy-7-phenethyl-4-phenyl-1,8-naphthyridine-3-carbonitrile	214–216, IR, MS, NMR	896
5-Methoxy-7-phenyl-1,8-naphthyridin-2-amine	247–248, IR, MS, NMR	472
7-Methoxy-5-phenyl-1,8-naphthyridin-2-amine	203–205, NMR	1226
7-Methoxy-6-phenyl-1,8-naphthyridin-2-amine	187–189, NMR	1226
7-Methoxy-2-phenyl-1,8-naphthyridin-4(1H)-one	244–246, NMR	221
7-Methoxy-5-phenyl-1,8-naphthyridin-2(1H)-one	220–222, NMR	1226
7-Methoxy-6-phenyl-1,8-naphthyridin-2(1H)-one	294–296, NMR	1226
2-Methoxy-6-phenyl-7-piperidino-1,8-naphthyridine	168–170, NMR	1304
2-Methoxy-7-piperidino-5-piperidinomethyl-1,8-naphthyridine	140–141	79
Methyl 7-acetamido-4-oxo-1,4-dihydro-1,8-naphthyridine-2-carboxylate	260–261	758
Methyl 1-acetyl-4-hydroxy-2-oxo-1,2-dihydro-1,8-naphthyridine-3-carboxylate	173–175, NMR	1225
Methyl 2-amino-5,7-dimethyl-1,8-naphthyridine-3-carboxylate	230–232, NMR	276
4-Methylamino-3,6-dinitro-1,8-naphthyridin-2-amine	>280, IR, MS, NMR	860
2-Methylamino-3,6-dinitro-1,8-naphthyridine	234–236, IR, MS, NMR	860, 1273
2-Methylamino-3,6-dinitro-1,8-naphthyridin-4(1H)-one	>280, IR, MS, NMR	860
4-Methylamino-3,6-dinitro-1,8-naphthyridin-2(1H)-one	>290, IR, MS, NMR	860
2-Methylamino-1,8-naphthyridine-3-carbonitrile	238–239	278
2-Methylamino-1,8-naphthyridine-3-carboxamide	257–258	278
Methyl 2-amino-1,8-naphthyridine-3-carboxylate	191–192	278
4-Methylamino-3-nitro-1,8-naphthyridin-2-amine	266–268, IR, NMR	682, 785, 1273
2-Methylamino-3-nitro-1,8-naphthyridine	199–200, IR, MS, NMR	682, 1273
4-Methylamino-3-nitro-1,8-naphthyridine	223–225, IR, NMR	682, 785, 1273
2-d-4-Methylamino-3-nitro-1,8-naphthyridine	—	682

TABLE A.4. (Continued)

1,8-Naphthyridine	Melting Point (°C) etc.	Reference(s)
2-Methylamino-3-nitro-1,8-naphthyridin-4(1H)-one	320–322, IR, NMR	682, 1273
4-Methylamino-3-nitro-1,8-naphthyridin-2(1H)-one	>290, IR, NMR	682, 785, 1273
4-Methylamino-3-nitro-1-phenyl-1,8-naphthyridin-2(1H)-one	>300, IR, NMR	1191
Methyl 2-amino-7-phenyl-1,8-naphthyridine-3-carboxylate	242–244	1400
Methyl 1-benzoyl-4-hydroxy-2-oxo-1,2-dihydro-1,8-naphthyridine-3-carboxylate	209–212, IR, NMR	912
Methyl 7-(2-dimethylaminovinyl)-1-ethyl-4-oxo-1,4-dihydro-1,8-naphthyridine-3-carboxylate	182, NMR	1407
3-Methyl-2,7-dioxo-1,2,7,8-tetrahydro-1,8-naphthyridine-4-carboxylic acid	340	687
7-Methyl-2,3-diphenyl-1,8-naphthyridin-2(1H)-one	290–292	1082
Methyl 1-ethyl-7-methyl-4-oxo-1,4-dihydro-1,8-naphthyridine-3-carboxylate	154	1407
7-(N-Methylhydrazino)-4-oxo-1,4-dihydro-1,8-naphthyridine-3-carboxylic acid	>290	1129
7-(N-Methylhydrazino)-4-oxo-1-propyl-1,4-dihydro-1,8-naphthyridine-3-carboxylic acid	232–233	1129
Methyl 4-hydroxy-2-oxo-1,2-dihydro-1,8-naphthyridine-3-carboxylate	236	33
Methyl 2-methoxycarbonylmethyl-1,8-naphthyridine-3-carboxylate	146–148, IR, NMR	1041
1-Methyl-7-(N-methylhydrazino)-4-oxo-1,4-dihydro-1,8-naphthyridine-3-carboxylic acid	>290	1129
Methyl 3-methyl-4-oxo-1,4-dihydro-1,8-naphthyridine-3-carboxylate	270–272	1249
2-Methyl-5-methylthio-1,8-naphthyridine	135–138, NMR	263
5-Methyl-1,8-naphthyridin-2-amine	198 to 213, MS	153, 680, 980, 1227
6-Methyl-1,8-naphthyridin-2-amine	197–198; pic: 254–256	300, 680
7-Methyl-1,8-naphthyridin-2-amine	217–218, IR, MS, NMR	153, 1071, 1227
7-Methyl-1,8-naphthyridin-4-amine	211–212	153
2-Methyl-1,8-naphthyridine	97 to 115, IR, NMR, UV; 1-MeI: 165–167	153, 177, 287, 302, 348, 667, 873, 880, 961, 964, 1013, 1100, 1112, 1441
3-Methyl-1,8-naphthyridine	117–118 or 119–129, NMR, pK_a; MeI:—	302, 873, 880, 961, 964, 1306, 1441
4-Methyl-1,8-naphthyridine	56–57, ~135/0.1, NMR, pK_a; pic: ~204; $HClO_4$: 180–181	66, 67, 153, 177, 302, 873, 880, 961, 964, 1297, 1306, 1441
2-Methyl-1,8-naphthyridine-3-carbohydrazide	202; PhMeC=: 241	738, 871
2-Methyl-1,8-naphthyridine-3-carbonylazide	120, IR, NMR	738
2-Methyl-1,8-naphthyridine-3-carbonyl chloride	crude	738

TABLE A.4. (Continued)

1,8-Naphthyridine	Melting Point (°C) etc.	Reference(s)
2-Methyl-1,8-naphthyridine-3-carboxanilide	215 or 280	706, 727, 1258, 1439
Methyl 1,8-naphthyridine-2-carboxylate	146–150, IR, MS, NMR	473
2-Methyl-1,8-naphthyridine-3-carboxylic acid	185, IR	738
1-Methyl-1,8-naphthyridine-2,7(1H,8H)-dione	277–279	1380
3-Methyl-1,8-naphthyridine-4,7(1H,8H)-dione	320 or 350–355, IR	687, 1071
4-Methyl-1,8-naphthyridine-2,7(1H,8H)-dione	>320 to >350, IR, UV	34, 88, 681, 687, 689, 690, 759
7-Methyl-1,8-naphthyridine-2,5(1H,8H)-dione	>320, NMR	701, 1186
7-Methyl-1,8-naphthyridine-2(1H)-thione	solid, IR, NMR	520
1-Methyl-1,8-naphthyridin-2(1H)-one	94–96, IR, NMR, UV	1100, 1214
3-Methyl-1,8-naphthyridin-2(1H)-one	231–232 or 234–236, IR, NMR	477, 865
5-Methyl-1,8-naphthyridin-2(1H)-one	245–247	680, 1005
6-Methyl-1,8-naphthyridin-2(1H)-one	245–246 or 254–255, IR, NMR	477, 680
7-Methyl-1,8-naphthyridin-2(1H)-one	176–177 or 180–182, IR, NMR	153, 477
7-Methyl-1,8-naphthyridin-4(1H)-one	231 to 240, IR, NMR	153, 726, 1076, 1098, 186, 1235, 1383
4-Methyl-5-nitro-1,8-naphthyridine-2,7(1H,8H)-dione	—	69
7-Methyl-4-oxo-1,4-dihydro-1,8-naphthyridine-3-carbaldehyde	>300, IR, NMR	1394
7-Methyl4-oxo-1,4-dihydro-1,3-naphthyridine-3-carboguanide	HCl: >260, IR, NMR	635
7-Methyl-4-oxo-1,4-dihydro-1,8-naphthyridine-3-carbonitrile	>275 or >300, IR, NMR, UV	292, 1076, 1407
Methyl 2-oxo-1,2-dihdyro-1,8-naphthyridine-3-carboxylate	219–220	278
7-Methyl-4-oxo-1,4-dihydro-1,8-naphthyridine-3-carboxylic acid	274 to 286, IR	102, 153, 179
Methyl 2-oxo-7-phenyl-1,2-dihydro-1,8-naphthyridine-3-carboxylate	220–222	1400
7-Methyl-4-oxo-1,propyl-1,4-dihydro-1,8-naphthyridine-3-carboxylic acid	209–210	154
5-Methyl-7-pentyl-1,8-naphthyridin-2(1H)-one	171–173, NMR	938
7-Methyl-5-pentyl-1,8-naphthyridin-2(1H)-one	158–161, NMR	938
5-Methyl-7-phenoxy-1,8-naphthyridin-2-amine	216–217	1052
2-Methyl-5-phenoxy-7-phenyl-1,8-naphthyridine	154–155, NMR	221
2-Methyl-7-phenoxy-4-phenyl-1,8-naphthyridine	156–157	1052
2-Methyl-5-(N'-phenylhydrazino)-1,8-naphthyridine	161–162, NMR	1097
5-Methyl-7-phenyl-1,8-naphthyridin-2-amine	247–248 or 253–255, NMR, UV	980, 1071
7-Methyl-2-phenyl-1,8-naphthyridin-4-amine	268–270, NMR	221
7-Methyl-5-phenyl-1,8-naphthyridin-2-amine	247–248; pic: 284	771, 1052
2-Methyl-3-phenyl-1,8-naphthyridine	128–129, IR	1112, 1440
2-Methyl-7-phenyl-1,8-naphthyridine	112–113 or 113–114, IR, NMR	221, 1112, 1390

Appendix 405

TABLE A.4. (*Continued*)

1,8-Naphthyridine	Melting Point (°C) etc.	Reference(s)
3-Methyl-2-phenyl-1,8-naphthyridine	113–114, NMR	1013, 1030
3-Methyl-4-phenyl-1,8-naphthyridine	190–191	879
7-Methyl-5-phenyl-1,8-naphthyridine-3-carboxamide	>275, IR, NMR	1326
1-Methyl-7-phenyl-1,8-naphthyridine-2,5(1*H*,8*H*)-dione	303–305, NMR	221
7-Methyl-2-phenyl-1,8-naphthyridine-4(1*H*)-thione	204–205, NMR	221
1-Methyl-3-phenyl-1,8-naphthyridin-2(1*H*)-one	solid, NMR	1298
1-Methyl-4-phenyl-1,8-naphthyridin-2(1*H*)-one	133–135, NMR	1196
5-Methyl-7-phenyl-1,8-naphthyridin-2(1*H*)-one	solid, IR, NMR; HCl: xl st	505
7-Methyl-2-phenyl-1,8-naphthyridin-4(1*H*)-one	266–267 or 273–275, IR, NMR	221, 1082, 1390
7-Methyl-3-phenyl-1,8-naphthyridin-4(1*H*)-one	>260 or 315–317, IR, NMR	905, 1076
7-Methyl-5-phenyl-1,8-naphthyridin-2(1*H*)-one	250 or 252–253	771, 1052
2-Methyl-4-phenyl-7-piperidino-1,8-naphthyridine	131–132; pic: 220–221	1052
7-Methyl-4-phenyl-1-propyl-1,8-naphthyridin-2(1*H*)-one	92–93, NMR	1196
2-Methyl-3-phenylsulfonyl-1,8-naphthyridine	171–173	278
5-Methyl-7-piperidino-1,8-naphthyridin-2-amine	221–222	1052
7-Methyl-2-piperidino-1,8-naphthyridin-4(1*H*)-one	210–212	1076
7-Methyl-1-propyl-1,8-naphthyridin-4(1*H*)-one	72–74	1380
7-Methyl-2-propyl-1,8-naphthyridin-4(1*H*)-one	215–217 or 220–222	976, 1085
1-Methylthiomethyl-1,8-naphthyridin-2(1*H*)-one	—	412
2-Methylthio-1,8-naphthyridine	—	427
7-Methyl-4-thioxo-1,4-dihydro-1,8-naphthyridine-3-carboxylic acid	275–280	1380
2-Morpholino-1,8-naphthyridine	—	427
7-Morpholino-3-nitro-6-phenyl-1,8-naphthyridin-2(1*H*)-one	288–290, NMR	1226
7-Morpholino-5-phenyl-1,8-naphthyridin-2-amine	280–283	1381
7-Morpholino-6-phenyl-1,8-naphthyridin-2-amine	174–176, NMR	1226
7-Morpholino-2-phenyl-1,8-naphthyridin-2(1*H*)-one	258–261, NMR	263
7-Morpholino-6-phenyl-1,8-naphthyridin-2(1*H*)-one	220–222, NMR	1226
1,8-Naphthyridin-2-amine	135 to 144, IR, NMR, UV; pic: 267–269; HClO$_4$: 177–178, IR, NMR, UV	173, 174, 275, 278, 300, 425, 427, 643, 647, 680, 830, 1100, 1218
1,8-Naphthyridin-3-amine	141–142 or 142–144, IR, MS, NMR	346, 647, 830
1,8-Naphthyridin-4-amine	185–187, IR, NMR, UV	628, 830
1,8-Naphthyridine	95–97 or 98–99, ESR, IR, MS, N_D, NMR, NQR, pK_a, pol, Raman, th, xl st, UV; pic: 207–208; MeI: 180–181; HClO$_4$: 214–215, IR; also others	30, 42–45, 173, 177, 250, 287, 302, 365, 386, 488, 616, 665, 671, 676, 806, 807, 809, 813, 829, 832, 840, 873, 880, 906,

TABLE A.4. (*Continued*)

1,8-Naphthyridine	Melting Point (°C) etc.	Reference(s)
		922, 923, 957, 961, 975, 991, 998, 1002, 1016, 1061, 1079, 1083, 1093, 1100, 1101, 1124, 1126, 1137, 1145, 1173, 1174, 1176, 1181, 1189, 1192, 1214, 1228, 1214, 1247, 1297, 1303, 1306, 1312, 1313, 1319, 1329, 1355, 1426, 1441
2,7-d_2-1,8-Naphthyridine	MS	302
1,8-Naphthyridine-2-carbaldehyde	143–145, IR, NMR, UV; dnp: 310–312	724, 953
1,8-Naphthyridine-4-carbaldehyde	IR, UV; H_2O: 129–130	953
1,8-Naphthyridine-2-carboxylic acid	175, IR; $HCl.H_2O$: 229–230, NMR	874
1,8-Naphthyridine-2,7-diamine	222–223 or 270–273, NMR	494, 1438, 1445
1,8-Naphthyridine-3,4-diamine	—	346
1,8-Naphthyridine-2,7-dicarbaldehyde	224 to 227, IR, NMR, UV; dnp: >360	642, 724, 1133
1,8-Naphthyridine-2,7-dicarbonyl chloride	196–198, IR, MS, NMR	1133
1,8-Naphthyridine-2,7-dicarboxylic acid	242, IR, MS, NMR	1133
1,8-Naphthyridine-2,5(1H,8H)-dione	355–357 or >360, NMR	758, 1076, 1186
1,8-Naphthyridine-2,7(1H,8H)-dione	320 or 321–323, IR, NMR	389, 687, 690, 702, 1222
1,8-Naphthyridine 1-oxide	98 or 132–133, IR, NMR	225, 1011, 1212
1,8-Naphthyridine-2(1H)-thione	253–254, st	427, 591, 862
1,8-Naphthyridine-4(1H)-thione	st	591
1,8-Naphthyridin-2(1H)-one	192 to 203, IR, MS, NMR, UV	427, 473, 477, 643, 647, 680, 865, 1005, 1008, 1044, 1100, 1253
7-d-1,8-Naphthyridin-2(1H)-one	—	1128
1,8-Naphthyridin-4(1H)-one	239 or 241–242, IR, MS, NMR, st, UV; $HCl.H_2O$: 243–245	16, 591, 628, 1186, 1253, 1441

TABLE A.4. (*Continued*)

1,8-Naphthyridine	Melting Point (°C) etc.	Reference(s)
2-Nitroamino-1,8-naphthyridine	210–211, IR, MS, NMR	860
6-Nitro-4,7-dioxo-1,4,7,8-tetrahydro-1,8-naphthyridine-2-carboxylic acid	>320	758
6-Nitro-4,7-dioxo-1,4,7,8-tetrahydro-1,8-naphthyridine-3-carboxylic acid	>320, IR	758
3-Nitro-2,7-dipiperidino-1,8-naphthyridine	131–133, NMR	1273, 1401
3-Nitro-1,8-naphthyridin-2-amine	276–277, NMR	1320
3-Nitro-1,8-naphthyridin-4-amine	>350	346, 1273, 1320
6-Nitro-1,8-naphthyridin-2-amine	337–339	300, 1273
2-Nitro-1,8-naphthyridine	250–251, MS, NMR	692, 1273
3-Nitro-1,8-naphthyridine	247–248, IR, MS, NMR	346, 647, 682, 1273, 1320
2-*d*-3-Nitro-1,8-naphthyridine	247–248	1273, 1320
3-Nitro-1,8-naphthyridine-2,4-diamine	>350, IR, MS, NMR	492, 1273, 1320
3-Nitro-1,8-naphthyridine-2,5(1*H*,8*H*)-dione	>320, NMR	758, 1273
3-Nitro-1,8-naphthyridine-2,7(1*H*,8*H*)-dione	>320	721, 1273
3-Nitro-1,8-naphthyridin-2(1*H*)-one	>265 or 324–328, IR, NMR	647, 860, 1112, 1273
3-Nitro-1,8-naphthyridin-4(1*H*)-one	>320	346, 682, 1273
6-Nitro-5-oxo-5,8-dihydro-1,8-naphthyridine-2-carboxylic acid	314–315	726, 1273
3-Nitro-7-phenyl-1,8-naphthyridin-2(1*H*)-one	274–278, IR	1112, 1273
3-Nitro-7-piperidino-1,8-naphthyridin-2(1*H*)-one	276–277, NMR	1401
6-Nitro-7-piperidino-1,8-naphthyridin-2(1*H*)-one	268–270, NMR	1401
3-Nitro-7-piperidino-5-piperidinomethyl-1,8-naphthyridin-2(1*H*)-one	233	79
2-Oxo-1,2-dihydro-1,8-naphthyridine-3-carbohydrazide	>300, IR, MS, NMR; PhCH=: >300, IR, MS	278, 632
2-Oxo-1,2-dihydro-1,8-naphthyridine-3-carbonitrile	>300, IR, MS, NMR	278, 367, 384
2-Oxo-1,2-dihydro-1,8-naphthyridine-3-carboxamide	>300	278
2-Oxo-1,2-dihydro-1,8-naphthyridine-3-carboxanilide	>300	728
2-Oxo-1,2-dihydro-1,8-naphthyridine-3-carboxylic acid	358–360, IR	1100, 1112
4-Oxo-1,4-dihydro-1,8-naphthyridine-3-carboxylic acid	304	16
7-Oxo-7,8-dihydro-1,8-naphthyridine-2,4-dicarbaldehyde	222–223, IR, NMR	1084
4-Oxo-1,4-dihydro-1,8-naphthyridine-2,3-dicarboxylic acid	304	16
2-Oxo-4-phenyl-1,2-dihydro-1,8-naphthyridine-3-carbonitrile	328–330, IR	1400
2-Oxo-7-phenyl-1,2-dihydro-1,8-naphthyridine-3-carbonitrile	278–280, IR	1112
2-Oxo-7-phenyl-1,2-dihydro-1,8-naphthyridine-3-carboxamide	304–305 or 307–308, IR	1112, 1400

TABLE A.4. (*Continued*)

1,8-Naphthyridine	Melting Point (°C) etc.	Reference(s)
2-Oxo-7-phenyl-1,2-dihydro-1,8-naphthyridine-3-carboxylic acid	339, IR	1112
2-Pentyl-1,8-naphthyridine	45–47, NMR	121
1-Pentyl-1,8-naphthyridin-2(1*H*)-one	solid, NMR	1298
2-Phenethyl-1,8-naphthyridine	83–84, IR, NMR	145
2-Phenethyl-1,8-naphthyridine-4-carboxylic acid	145	71
2-Phenoxy-1,8-naphthyridine	—	427
7-Phenoxy-2-phenyl-1,8-naphthyridin-4(1*H*)-one	245–247, NMR	221
3-Phenyl-1,8-naphthyridin-2-amine	252 to 264, NMR	278, 658, 674, 680, 1112
4-Phenyl-1,8-naphthyridin-2-amine	207–209	680
5-Phenyl-1,8-naphthyridin-2-amine	248–250	680
6-Phenyl-1,8-naphthyridin-2-amine	H_3PO_4: 264–266	300
7-Phenyl-1,8-naphthyridin-2-amine	225 or 229–230, IR, NMR, UV	1071, 1100
2-Phenyl-1,8-naphthyridine	97 to 118, IR, NMR, UV	68, 278, 306, 409, 659, 707, 1013, 1030, 1054, 1100
3-Phenyl-1,8-naphthyridine	126–127, IR	1112, 1440
2-Phenyl-1,8-naphthyridine-3-carbohydrazide	206, IR, MS, NMR; PhCH=:—	450, 859
2-Phenyl-1,8-naphthyridine-3-carbonitrile	224–225, IR	278, 1100
2-Phenyl-1,8-naphthyridine-3-carboxamide	233–234	278, 1100
2-Phenyl-1,8-naphthyridine-3-carboxanilide	279 or 280, IR, MS, NMR	727, 859, 1258
2-Phenyl-1,8-naphthyridine-3-carboxylic acid	259–260, IR	762
2-Phenyl-1,8-naphthyridine-4-carboxylic acid	145	68
3-Phenyl-1,8-naphthyridine-2,7(1*H*,8*H*)-dione	293–295	972
3-Phenyl-1,8-naphthyridine-4,7(1*H*,8*H*)-dione	>360	1076
4-Phenyl-1,8-naphthyridine-2,7(1*H*,8*H*)-dione	271 or 272–273	70, 73, 771
7-Phenyl-1,8-naphthyridin-2,5(1*H*,8*H*)-dione	>300, NMR	221
2-Phenyl-1,8-naphthyridin-4(1*H*)-one	225–227, NMR	221
3-Phenyl-1,8-naphthyridin-2(1*H*)-one	242–244 or 252–254, IR, NMR	117, 680, 1112
4-Phenyl-1,8-naphthyridin-2(1*H*)-one	150 to 260, NMR; HCl: 211–216; pic: 183	7, 18, 680, 1196
5-Phenyl-1,8-naphthyridin-2(1*H*)-one	278–280	680
7-Phenyl-1,8-naphthyridin-2(1*H*)-one	244–245, IR, UV	1100
2-Phenyl-7-piperidino-1,8-naphthyridin-4(1*H*)-one	247–249, NMR	263
2-Phenyl-3-propyl-1,8-naphthyridine	95–96, NMR	1013
1-Phenyl-3-(prop-2-ynyl)-4-(prop-2-ynyloxy)-1,8-naphthyridin-2(1*H*)-one	173–175	607
2-Phenyl-7-styryl-1,8-naphthyridine	196–198, IR	1112
3-Phenylsulfonyl-1,8-naphthyridin-2-amine	235–237	278
3-Phenylsulfonyl-1,8-naphthyridin-2(1*H*)-one	294–296	278
2-Phenylthio-1,8-naphthyridine	—	427
2-Piperidino-1,8-naphthyridine	—	427
7-Piperidino-5-piperidinomethyl-1,8-naphthyridin-2-amine	215–217	79

TABLE A.4. (*Continued*)

1,8-Naphthyridine	Melting Point (°C) etc.	Reference(s)
7-Piperidino-5-piperidinomethyl-1,8-naphthyridin-2(1*H*)-one	178–180	79
7-Propoxy-1-propyl-1,8-naphthyridin-2(1*H*)-one	46–47, NMR	1380
2-Propyl-1,8-naphthyridine	80–81, IR, NMR	145
2-Styryl-1,8-naphthyridine	133	874
2-Styryl-1,8-naphthyridine-4-carboxylic acid	186	71
2,3,5,7-Tetramethyl-1,8-naphthyridine	144–146	276
2,3,5,7-Tetramethyl-1,8-naphthyridin-4(1*H*)-one	261–262	1085
2,3,6,7-Tetraphenyl-1,8-naphthyridine	301–302, IR, NMR	284
4,5,7-Triamino-6-benzylidene-2-thioxo-2,6-dihydro-1,8-naphthyridine-3-carbonitrile (?)	220–222, IR, NMR	1075
4,5,7-Triamino-6-phenylazo-2-thioxo-1,2-dihydro-1,8-naphthyridine-3-carbonitrile	>300, IR	538
4,5,7-Triamino-2-thioxo-1,2-dihydro-1,8-naphthyridine-3-carbonitrile	260, IR, NMR	538
2,3,4-Tribromo-1,8-naphthyridine	MS	303, 386
5,6,7-Trichloro-1-ethyl-4-oxo-1,2-dihydro-1,8-naphthyridine-3-carboxylic acid	252–254	364
2,3,7-Trichloro-1,8-naphthyridine	258, IR, NMR	715
5,6,7-Trimethyl-1,8-naphthyridin-2-amine	liq	338
2,4,5-Trimethyl-1,8-naphthyridine	160–163; pic: 134–135	269, 864
2,4,6-Trimethyl-1,8-naphthyridine	146–147, NMR	269
2,4,7-Trimethyl-1,8-naphthyridine	98–99, NMR	269, 964
2,3,7-Trimethyl-1,8-naphthyridin-4(1*H*)-one	300	976, 1085
2,5,7-Trimethyl-1,8-naphthyridin-4(1*H*)-one	285–290	1085
5,6,7-Trimethyl-1,8-naphthyridin-2(1*H*)-one	crude	338
1,5,7-Trimethyl-4-phenyl-1,8-naphthyridin-2(1*H*)-one	126–127, NMR	345
3-Trimethylsilylmethyl-1,8-naphthyridine	93–95, NMR	1441

TABLE A.5. ALPHABETICAL LIST OF SIMPLE 2,6-NAPHTHYRIDINES

2,6-Naphthyridine	Melting Point (°C) etc.	Reference(s)
3-Acetamido-1-bromo-4-methyl-2,6-naphthyridine	147–148	1007
3-Acetamido-1-bromo-2,6-naphythyridine	243, NMR, UV	186, 1007
3-Acetamido-2,6-naphythyridine	245–246, NMR	186
4-Allyl-1-morpholino-2,6-naphythyridin-3-amine	149–150, IR, NMR	133
4-Allyl-1-piperidino-,6-naphythyridin-3-amine	118–119, IR, NMR	133
7-Amino-5-bromo-3-methyl-2,6-naphythyridin-4(6*H*)-one	—	441
3-Bromo-1-methoxy-4-methyl-2,6-naphthyridine	145–146, UV	1007
3-Bromo-1-methoxy-2,6-naphthyridine	129–130, NMR	1007
1-Bromo-4-methyl-2,6-naphthyridin-3-amine	197–198, NMR, UV	1007
3-Bromo-4-methyl-2,6-naphythyridin-1(2*H*)-one	296, UV	1007
1-Bromo-2,6-naphthyridin-3-amine	199–200, NMR, UV	186, 1007
1-Bromo-2,6-naphthyridine	94–95, NMR	439
5-*d*-1-Bromo-2,6-naphythyridine	—	439
3-Bromo-2,6-naphythyridin-1(2*H*)-one	296–298	1007

TABLE A.5. (*Continued*)

2,6-Naphthyridine	Melting Point (°C) etc.	Reference(s)
4-(But-2-enyl)-1-morpholino-2,6-naphythyridin-3-amine	liq, IR, NMR	133
4-(But-2-enyl)-1-phenyl-2,6-naphythyridin-3-amine	117–118, IR, NMR	133
1-Butylamino-2,6-naphythyridin-3-amine	liq, IR, NMR	133
1-*tert*-Butylamino-2,6-naphthyridin-3-amine	liq, IR, NMR	133
1-Chloro-2,6-naphthyridine	92–93, NMR	439, 1115
5-*d*-1-Chloro-2,6-naphythyridine	—	439
3-Chloro-2,6-naphthyridine	149–150, NMR, UV	1007
2,6-Dibenzyl-3,7-diphenyl-2,6-naphythyridine-1,5(2*H*,6*H*)-dione	297–298 fl sp, NMR, UV	914
1,3-Dibromo-4-methyl-2,6-naphthyridine	150–151, NMR, UV	1007
1,4-Dibromo-2,6-naphthyridin-3-amine	212–213, NMR, UV	1007
1,3-Dibromo-2,6-naphthyridine	132–133, NMR, UV	186, 1007
1,5-Dibutyl-2,6-naphthyridin-3-amine	liq, IR, NMR	133
1,3-Dichloro-2,6-naphthyridine	116	677
2,6-Diethyl-3,7-diphenyl-2,6-naphthyridine-1,5(2*H*,6*H*)-dione	286–287, fl sp, NMR, UV	914
1,3-Dihydrazino-4-methyl-2,6-naphthyridine	197–198	1007
1,3-Dihydrazino-2,6-naphthyridine	214–216 or >300	186, 677, 1007
2,6-Dimethyl-3,7-diphenyl-2,6-naphthyridine-1,5(2*H*,6*H*)-dione	>300, fl sp, NMR, xl st, UV	788, 914
5,7-Dimethyl-3-phenyl-2,6-naphthyridin-1(2*H*)-one	253–255, IR, NMR	1055
1-Ethoxy-2,6-naphthyridin-3-amine	159–160, NMR, UV	186, 710, 1007
3-Ethoxy-2,6-naphthyridin-1-amine	134–135, NMR	710
Ethyl 1-anilino-2,6-naphthyridine-4-carboxylate	181–182, IR, MS, NMR	442
Hexaphenyl-2,6-naphthyridine	287–288, NMR	1132
3-Hexyl-2,6-naphthyridine	79/3, NMR	621
3-Hexyl-2,6-naphthyridine 2-oxide	84–85, NMR	621
3-Hydroxy-2,6-naphthyridin-1(2*H*)-one	229–230	677
1-Isopropoxy-2,6-naphthyridin-3-amine	137–138, NMR	710
3-Isopropoxy-2,6-naphthyridin-1-amine	177–178, NMR	710
1-Isopropylamino-2,6-naphthyridin-3-amine	liq, IR, NMR	133
1-Methoxy-4-methyl-2,6-naphthyridin-3-amine	195–197, UV	1007
1-Methoxy-2,6-naphthyridin-3-amine	140, NMR	710
3-Methoxy-2,6-naphthyridin-1-amine	122, NMR	710
1-Methylamino-2,6-naphthyridin-3-amine	167–168, IR, NMR	133
4-(3-Methylbut-2-enyl)-1-morpholino-2,6-naphthyridin-3-amine	liq, IR, NMR	133
4-(3-Methylbut-2-enyl)-1-phenyl-2,6-naphthyridin-3-amine (?)	153–154, IR, NMR	133
4-Methyl-2,6-naphthyridin-3-amine	197–198, NMR, UV	1007
4-Methyl-2,6-naphthyridine	79 or 94–95, IR, NMR, UV; pic: 186–187	226, 241, 1007
1-Morpholino-2,6-naphthyridin-3-amine	191–192, IR, NMR	133
1-Morpholino-4-(prop-1-enyl)-2,6-naphthyridin-3-amine	liq, IR, NMR	133
2,6-Naphthyridin-1-amine	243–244, NMR	1115
2,6-Naphthyridin-3-amine	153–154, NMR, UV	186, 1007

TABLE A.5. (Continued)

2,6-Naphthyridine	Melting Point (°C) etc.	Reference(s)
2,6-Naphthyridine	114–115 or 118–119, ESR, IR, NMR, pK_a, pol, th, UV, xl st; pic: 206 or 222–224; di-pic: 207–209	186, 193, 621, 669, 676, 677, 798, 813, 840, 878, 995, 1007, 1079, 1083, 1115, 1126, 1173, 1174, 1176, 1181, 1192, 1241, 1250, 1312, 1319, 1329, 1426
2,6-Naphthyridine-1,5-diamine	>300, NMR; pic: >300	439
2,6-Naphthyridine 2-oxide	198–201	621
2,6-Naphthyridin-1(2H)-one	235–237 or 248–251, IR, NMR	1055, 1115
1,3,4,5,8-Pentaphenyl-2,6-naphthyridine	262, NMR	1132
3-Phenyl-2,6-naphthyridine	76–80, NMR	147, 621
3-Phenyl-2,6-naphthyridine 2-oxide	196–199, NMR	621
1-Phenyl-4-(prop-1-enyl)-2,6-naphthyridin-3-amine	102–103, IR, NMR	133
1-Piperidino-2,6-naphthyridin-3-amine	123–124, IR, NMR	133
1-Piperidino-4-(prop-1-enyl)-2,6-naphthyridin-3-amine	89–90, IR, NMR	133
1,4,5,8-Tetraphenyl-2,6-naphthyridine	285–286, NMR	1132
1,3,4-Tribromo-2,6-naphthyridine	160–162, NMR, UV	186, 1007

TABLE A.6. ALPHABETICAL LIST OF SIMPLE 2,7-NAPHTHYRIDINES

2,7-Naphthyridine	Melting Point (°C) etc.	Reference(s)
2-Allyl-5-(1-hydroxyethyl)-2,7-naphthyridine-1,3(2H,7H)-dione	205–206, NMR	206
3-Amino-2-benzyl-1-oxo-1,2-dihydro-2,7-naphthyridine-4-carbonitrile	247, IR, NMR	1220
1-Amino-2,7-dimethyl-3,8-dioxo-6-phenyl-2,3,7,8-tetrahydro-2,7-naphthyridine-4-carbonitrile	>380	699, 1183
1-Amino-6-hydroxy-8-imino-2,7-diphenyl-4-phenylazo-7,8-dihydro-2,7-naphthyridin-3(2H)-one	278, IR	802
8-Benzylamino-3,6-diphenyl-2,7-naphthyridin-1(2H)-one	240–241, MS, NMR, UV	1239
2-Benzyl-8-benzylamino-3,6-diphenyl-2,7-naphthyridin-1(2H)-imine	184–185, MS, NMR, UV	1239
2-Benzyl-5-(1-hydroxyethyl)-2,7-naphthyridine-1,3(2H,7H)-dione	NMR	206

TABLE A.6. (*Continued*)

2,7-Naphthyridine	Melting Point (°C) etc.	Reference(s)
7-Benzyl-4-(3-methoxycarbonylpropyl)-2,7-naphthyridine-1,3(2*H*,7*H*)-dione	228–231, IR, NMR	410
7-Benzyl-8-oxo-1-phenacyl-3-phenyl-7,8-dihydro-2,7-naphthyridine-4-carbonitrile	277–278, IR, NMR	987
1-Bromo-2,7-naphthyridine	128–129, NMR	838
4-Bromo-2,7-naphthyridine	anal, NMR	1374
8-Butylamino-3,6-diphenyl-2,7-naphthyridin-1(2*H*)-one	199–200, MS, UV	1239
7-Butyl-8-imino-3,6-diphenyl-2,7-naphthyridin-1(2*H*)-one	264–265, MS, NMR, UV	1239
1-Chloro-3-methyl-2,7-naphthyridine	105–106, IR, UV	1251, 1279, 1431
1-Chloro-2,7-naphthyridine	117–118, IR, NMR, UV	838, 1251, 1279
4-Chloro-1,3,5,6,8-pentafluoro-2,7-naphthyridine	41–43, IR	918
1-Chloro-3-phenyl-2,7-naphthyridine	128–130	189
1,8-Diamino-3-(*N*-butyl-*N*-ethylamino)-6-ethoxy-2,7-naphthyridine-4-carbonitrile	148, NMR	668
1,8-Diamino-3-diethylamino-6-ethoxy-2,7-naphthyridine-4-carbonitrile	145, NMR	668
1,8-Diamino-3-diethylamino-6-methoxy-2,7-naphthyridine-4-carbonitrile	167, NMR	668
1,8-Diamino-3,6-dimethoxy-2,7-naphthyridine-4-carbonitrile	258–260, NMR	1074
1,8-Diamino-6-ethoxy-3-morpholino-2,7-naphthyridine-4-carbonitrile	157, NMR	668
1,8-Diamino-6-ethoxy-3-piperidino-2,7-naphthyridine-4-carbonitrile	143, NMR	668
1,8-Diamino-6-methoxy-3-morpholino-2,7-naphthyridine-4-carbonitrile	243, NMR	668
1,8-Diamino-6-methoxy-3-piperidino-2,7-naphthyridine-4-carbonitrile	129, NMR	668
4,5-Dibromo-2,7-naphthyridine	anal, NMR	1374
4,5-Dichloro-1,3,6,8-tetrafluoro-2,7-naphthyridine	102–104, IR, NMR	918
3,6-Dihydroxy-2,7-naphthyridine-1,8(2*H*,7*H*)-dione	—	1309
1,6-Dimethyl-3,8-dioxo-7-phenyl-2,3,7,8-tetrahydro-2,7-naphthyridine-4-carbonitrile	>350, IR, MS, NMR, UV	742
3,6-Dimethyl-2,7-naphthyridine-1,8-diamine	220, NMR	736
2,3-Dimethyl-2,7-naphthyridin-1(2*H*)-one	138	1431
3,7-Dimethyl-2,7-naphthyridin-1(7*H*)-one	HCl: MS, NMR, UV	360, 1315, 1431
6,8-Dimethyl-3-phenyl-2,7-naphthyridin-1(2*H*)-one	252–253, IR, NMR	1055
3,6-Diphenyl-1,8-dipiperidino-2,7-naphthyridine	195–196, NMR, UV	1239
3,6-Diphenyl-2,7-naphthyridine	162–163, IR, MS, NMR	1065
3,6-Diphenyl-2,7-naphthyridine-1,8(2*H*,7*H*)-dione	296–297 or 307–308, UV	190, 1239
3,6-Diphenyl-8-piperidino-2,7-naphthyridin-1(2*H*)-one	237–238, MS, NMR	1239
Ethyl 1,6-dimethyl-3,8-dioxo-7-phenyl-2,3,7,8-tetrahydro-2,7-naphthyridine-4-carboxylate	284–286	742
Ethyl 1-oxo-1,2-dihydro-2,7-naphthyridine-3-carboxylate	229–230 or 232–234, IR, NMR	253, 986, 989
7-Ethyl-3-phenyl-2,7-naphthyridin-1(2*H*)-one	—	598
1-Ethyl-3,6,8-trimethyl-2,7-naphthyridine	36–37, MS, NMR	585, 1168

TABLE A.6. (*Continued*)

2,7-Naphthyridine	Melting Point (°C) etc.	Reference(s)
3-Ethyl-1,6,8-trimethyl-2,7-naphthyridine	MS, NMR	1168
Hexachloro-2,7-naphthyridine	145–147, IR	918
Hexafluoro-2,7-naphthyridine	61–63, IR, NMR	918
Hexaphenyl-2,7-naphthyridine	321–322, NMR	1132
3-Hexyl-2,7-naphthyridine	90/3, NMR	621
3-Hexyl-2,7-naphthyridine 2-oxide	74–75, NMR	621
1-Hydrazino-3-methyl-2,7-naphthyridine	208–211	1279
1-Hydrazino-2,7-naphthyridine	crude	1279
1-Hydrazino-3-phenyl-2,7-naphthyridine	crude, 214–216	188
6-Hydroxy-1,8-dioxo-2,7-diphenyl-6-phenylazo-1,2,7,8-tetrahydro-2,7-naphthyridine-4-carbonitrile	—	456
6-Hydroxy-1,8-dioxo-2,7-diphenyl-1,2,7,8-tetrahydro-2,7-naphthyridine-4-carbonitrile	—	456
5-(1-Hydroxyethyl)-2,7-naphthyridine-1,3(2H,7H)-dione	>300, NMR	206
2-(2-Hydroxyethyl)-3-phenyl-2,7-naphthyridin-1(2H)-one	227–229, IR, NMR	902
4-Hydroxy-2,7-naphthyridin-1(2H)-one	>240; pic: 195	13
1-Isopropyl-3,6,8-trimethyl-2,7-naphthyridine	liq, MS, NMR	585, 1168
3-Isopropyl-1,6,8-trimethyl-2,7-naphthyridine	MS, NMR	1168
1-Methoxy-3-methyl-2,7-naphthyridine	93–94, IR, NMR, UV	1148, 1431
2-Methyl-3,6-diphenyl-2,7-naphthyridinediium diperchlorate	308–309, UV	1065
Methyl 5-ethyl-1-oxo-1,2-dihydro-2,7-naphthyridine-3-carboxylate	200–201, IR, NMR, UV	172
Methyl 1-methyl-2,7-naphthyridine-4-carboxylate (neozeylanicine)	IR, MS, NMR, UV	502, 597
Methyl 5-methyl-2,7-naphthyridine-4-carboxylate	—	502
Methyl 8-methyl-6-oxo-6,7-dihydro-2,7-naphthyridine-4-carboxylate	>300, NMR	206
3-Methyl-2,7-naphthyridine	36–38, IR; pic: 220–221	1251, 1279, 1297, 1361, 1431
5-Methyl-2,7-naphthyridine-4-carbaldehyde	181–182, IR, MS, NMR	595
3-Methyl-2,7-naphthyridin-1(2H)-one	256–258 or 271–272, IR, UV	1148, 1279
2-Methyl-1-oxo-1,2-dihydro-2,7-naphthyridine-4-carbaldehyde	254–256	413
2-Methyl-8-oxo-7,8-dihydro-2,7-naphthyridin-2-ium-6-carboxylate	265, fl sp, UV; HCl: 270	253
7-Methyl-3-phenyl-2,7-naphthyridin-1(2H)-one	—	598
8-Morpholino-2,3,6-triphenyl-2,7-naphthyridin-1(2H)-one	280–285, NMR, UV	1239
2,7-Naphthyridin-1-amine	227–238, NMR	1374
2,7-Naphthyridine	92–94 or 96, ESR, IR, MS, NMR, pK_a, th, xl st; pic: 240; 2MeI: pK_a	621, 630, 676, 776, 813, 840, 1007, 1079, 1083, 1126, 1173, 1174, 1176, 1181,

TABLE A.6. (*Continued*)

2,7-Naphthyridine	Melting Point (°C) etc.	Reference(s)
		1192, 1251, 1279, 1297, 1299, 1312, 1313, 1319, 1329, 1374, 1426
2,7-Naphthyridine-4-carbaldehyde	215–216, IR, MS, NMR	595, 825
2,7-Naphthyridine-3-carboxylic acid	—	676
2,7-Naphthyridine-4-carboxylic acid	289	630
2,7-Naphthyridine 2,7-dioxide	>300, NMR	1212
2,7-Naphthyridine 2-oxide	213–215 or 228–229, NMR	621, 1212
2,7-Naphthyridin-1(2*H*)-one	255 to 265, IR, NMR, UV	339, 1055, 1279
8-Nitromethyl-3,6-diphenyl-2,7-naphthyridin-1(2*H*)-one	288–289, UV	1239
1-Oxo-1,2-dihydro-2,7-naphthyridine-3-carboxylic acid	330	253
8-Oxo-1-phenacyl-3-phenyl-7-propyl-7,8-dihydro-2,7-naphthyridine-4-carbonitrile	258–260, NMR	987
1,3,4,5,8-Pentafluoro-2,7-naphthyridine	221–222, NMR	1132
4-Phenyl-2,7-naphthyridin-1-amine	NMR	1435
3-Phenyl-2,7-naphthyridine	126–128, IR, NMR, UV; pic: 210–211	188, 621
3-Phenyl-2,7-naphthyridine 2-oxide	203–205, NMR	621
3-Phenyl-2,7-naphthyridin-1(2*H*)-one	237–238 or 239–340, IR	15, 188
4-Phenyl-2,7-naphthyridin-1(2*H*)-one	211–213, NMR	1435
1,3,6,8-Tetrabutyl-2,7-naphthyridine	140/0.1, MS, NMR	1178
1,3,6,8-Tetrachloro-2,7-naphthyridine	157–160	32, 1374
2,3,6,8-Tetraethyl-2,7-naphthyridine	80/0.1, MS, NMR	1178
1,3,6,8-Tetrafluoro-2,7-naphthyridine	58–59, IR, NMR	821, 918
1,3,6,8-Tetraisopropyl-2,7-naphthyridine	liq, MS, NMR	1178
1,3,6,8-Tetramethyl-2,7-naphthyridine	62, IR, MS, NMR	388, 585, 1178
4,5-d_2-1,3,6,8-Tetramethyl-2,7-naphthyridine	MS, NMR	388
1,4,5,6-Tetraphenyl-2,7-naphthyridine	solid, MS, NMR	1132
1,3,6,8-Tetrapropyl-2,7-naphthyridine	110/0.06, MS, NMR	1178
1,3,6-Triethyl-8-methyl-2,7-naphthyridine	MS, NMR	1168
1,3,8-Triethyl-6-methyl-2,7-naphthyridine	MS, NMR	1168
1,3,6-Triisopropyl-8-methyl-2,7-naphthyridine	MS, NMR	1168
1,3,8-Triisopropyl-6-methyl-2,7-naphthyridine	MS, NMR	1168
1,3,6-Trimethyl-8-propyl-2,7-naphthyridine	50, MS, NMR	585

Index

Acylamino-1,5-naphthyridines, from amino-1,5-naphthyridines, 59
Acylamino-1,6-naphthyridines, from amino-1,6-naphthyridines, 132
Acylamino-1,8-naphthyridines, hydrolysis, 240
Acylamino-2,6-naphthyridines, from amino-2,6-naphthyridines, 272
C-Acyl-1,5-naphthyridines, see 1,5-Naphthyridine ketones
C-Acyl-1,6-naphthyridines, see 1,6-Naphthyridine ketones
Acyloxy-1,5-naphthyridines, 46
Acyloxy-1,6-naphthyridines, 120
  rearrangement, 121
Acyloxy-1,7-naphthyridines, 170
  aminolysis, 171
  deacyloxylation 170
  preparation, 170
  reactions, 171
Alkoxy-1,5-naphthyridines, 48
  aminolysis, 48
  from halogeno-1,5-naphthyridines, 31
  from 1,5-naphthyridinones, 46
  to 1,5-naphthyridinones, 45, 49
  rearrangement, 48, 49
Alkoxy-1,6-naphthyridines, 120
  from halogeno-1,6-naphthyridines, 108
  from 1,6-naphthyridinones, 118
  to 1,6-naphthyridinones, 117
  preparation, 121
  reactions, 121
Alkoxy-1,7-naphthyridines, 170
  from halogeno-1,7-naphthyridines, 163
  preparation, 170
  reactions, 170
Alkoxy-1,8-naphthyridines, 227
  from azido-1,8-naphthyridines, 227
  from halogeno-1,8-naphthyridines, 214
  hydrolysis, 222
  from 1,8-naphthyridinones, 224
  preparation, 227
  reactions, 228

Alkoxy-2,6-naphthyridines, 270
  from halogeno-2,6-naphthyridines, 268
  preparation, 270
  reactions, 270
  reduction, 270
Alkoxy-2,7-naphthyridines, 290
  from halogeno-2,7-naphthyridines, 286
  from 2,7-naphthyridinones, 290
Alkyl-1,5-naphthyridines, 13, 18
  acylation, 22
  by C-alkylation, 20
  amination, 22
  halogenation, 22
  from halogeno-1,5-naphthyridines, 38
  by interconversion, 21
  oxidation, 23
  preparation, 20
  by quaternization, 21
  reactions, 22
  by rearrangement, 21
Alkyl-1,6-naphthyridines, 97
  by C-alkylation, 94, 98
  by N-alkylation, 98
  alkylidenation, 99
  cyclocondensations, 100
  dealkylation, 100
  by Grignard addition, 99
  from halogeno-1,6-naphthyridines, 99
  oxidation, 101
  preparation, 97
  reactions, 99
  reduction, 102
Alkyl-1,7-naphthyridines, 157
  by C-alkylation, 158
  by N-alkylation, 158
  N-dealkylation, 158
  from methanesulfonyloxy-1,7-naphthyridines, 158
  preparation, 157
  reactions, 158

416   Index

Alkyl-1,8-naphthyridines, 203
　acylation, 206
　by alkylation, 203
　cyclizations, 207
　halogenation, 205
　from halogeno-1,8-naphthyridines, 203
　from 1,8-naphthyridine N-oxides, 204
　oxidation, 206
　preparation, 203
　reactions, 205
Alkyl-2,6-naphthyridines, 265
Alkyl-2,7-naphthyrinines, 283
　preparation, 283
Alkylsulfinyl-1,8-naphthyridines, 232
　from alkylthio-1,8-naphthyridines, 232
Alkylsulfonyl-1,5-naphthyridines, 54
　from alkylthio-1,5-naphthyridines, 54
Alkylsulfonyl-1,8-naphthyridines, 232
　aminolysis, 233
　from alkylthio-1,8-naphthyridines, 232
Alkylthio-1,5-naphthyridines, 54
　from halogeno-1,5-naphthyridines, 40
　from 1,5-naphthyridinethiones, 53
　oxidation, 54
　preparation, 54
　reactions, 54
Alkylthio-1,6-naphthyridines, 125
Alkylthio-1,8-naphthyridines, 232
　from 1,8-naphthyridinethiones, 231
　oxidation, 232
　preparation, 232
　reactions, 232
Amino-1,5-naphthyridines, 55, 57
　acylation, 59
　by amination, 58
　bioactivities, 57
　complexation, 60
　cyclizations, 58
　to dimethylsulfimido-1,5-naphthyridines, 56
　from halogeno-1,5-naphthyridines, 31
　ionization, 57
　from 1,5-naphthyridinones, 49
　to 1,5-naphthyridinones, 44
　from nitro-1,5-naphthyridines, 57
　preparation, 58
　reactions, 58
Amino-1,6-naphthyridines, 129
　acylation, 132
　alkylation, 131
　by amination, 95, 130
　carbamoylation, 133
　by Curtius or Hofmann reactions, 130
　cyclizations, 133
　diazotization, 133
　　from halogeno-1,6-naphthyridines, 110
　　to halogeno-1,6-naphthyridines, 108
　　hydrolysis, 116
　　ionization, 129
　　from nitro-1,6-naphthyridines, 128
　　preparation, 130
　　reactions, 131
　　reviews, 130
　　spectra, 129
　　by transamination, 131
Amino-1,7-naphthyridines, 176
　by amination, 176
　cyclizations, 177
　deamination, 177
　from halogeno-1,7-naphthyridines, 163
　hydrolysis, 168
　from nitro-1,7-naphthyridines, 175
　preparation, 176
　reactions, 177
　transamination, 177
Amino-1,8-naphthyridines, 238
　from acylamino-1,8-naphthyridines, 240
　acylation, 241
　alkylation, 243
　alkylidenation, 243
　from alkylsulfonyl-1,8-naphthyridines, 233
　by amination, 239
　azo coupling, 241
　to azo-1,8-naphthyridines, 244
　complex formation, 244
　by a Curtius, reaction, 238
　cyclizations, 245
　from halogeno-1,8-naphthyridines, 215
　to halogeno-1,8-naphthyridines, 212
　to 1,8-naphthyridinecarbonitriles, 244
　from nitro-1,8-naphthyridines, 237
　to nitro-1,8-naphthyridines, 236
　from nitroso-1,8-naphthyridines, 238
　preparation, 238
　properties, 238
　reactions, 241
　by transamination, 240
　to ureido-1,8-naphthyridines, 244
Amino-2,6-naphthyridines, 271
　acylation, 272
　by amination, 271
　dehydrazination, 272
　from halogeno-2,6-naphthyridines, 268
　preparation, 271
　reactipns, 272
Amino-2,7-naphthyridines, 290
　by amination, 285, 291
　dehydrazination, 283
　from halogeno-2,7-naphthyridines, 288

Index

preparation, 290
reactions, 291
Apalcillin, 45
Arenesulfonyloxy-1,7-naphthyridines, see
    Acyloxy-1,7-naphthyridines
Aryl-1,5-naphthyridines, 13, 19
    preparation, 20
    reactions, 22
Aryl-1,6-naphthyridines, 97
    preparation, 97
    reactions, 99
Aryl-1,7-naphthyridines, see Alkyl-1,7-naphthyridines
Aryl-1,8-naphthyridines, see Alkyl-1,8-naphthyridines
Aryloxy-1,6-naphthyridines, see
    Alkoxy-1,6-naphthyridines
Aryloxy-1,7-naphthyridines, see
    Alkoxy-1,7-naphthyridines
Arylthio-1,7-naphthyridines, 155
Austrodimerine, 289
Azido-1,8-naphthyridines, from
    halogeno-1,8-naphthyridines, 216

Bis (1,5-naphthyridinyl) sulfides, 54
    preparation, 54
Bromonaphthyridines, see
    Halogenonaphthyridines

Chloronaphthyridines, see
    Halogenonaphthyridines
Copyrine, xi

Dihydrojasminine, 289
Dimethylsulfimido-1,5-naphthyridines, from
    amino-1,5-naphthyridines, 56
    to nitro-1,5-naphthyridines, 56
Dimethylsulfimido-1,8-naphthyridines, from
    amino-1,8-naphthyridines, 236
    to nitro-1,8-naphthyridines, 236

Enoxacin, 209

Fluoronaphthyridines, see
    Halogenonaphthyridines

Halogeno-1,5-naphthyridines, 25
    alcoholysis, 31
    alkanelysis, 38
    alkanethiolysis, 38
    aminolysis, 34
    bioactivities, 25
    dehalogenation, 39
    by halogenation, 25
    hydrolysis, 40

from 1,5-naphthyridine N-oxides, 29
from 1,5-naphthyridinones, 26, 31
phenolysis, 31
positional activation, 25
preparation, 25
reactions, 31
transhalogenation, 31
Halogeno-1,6-naphthyridines, 103
    alcoholysis, 108
    aminolysis, 110
    from amino-1,6-naphthyridines, 108
    cyclocondensations, 114
    dehalogenation, 111
    by halogenation, 96, 103
    hydrolysis, 113
    MS patterns, 103
    from naphthyridine N-oxides, 107
    from naphthyridinones, 104
    phenolysis, 110
    preparation, 103
    reactions, 108
    thiolysis, 113
Halogeno-1,7-naphthyridines, 158
    alcoholysis, 163
    aminolysis, 163
    dehalogenation, 165
    by halogenation, 161
    from 1,7-naphthyridine N-oxides, 163
    from 1,7-naphthyridinones, 162
    positional activation, 161
    preparation, 161
    reactions, 163
    ring fission, 165
Halogeno-1,8-naphthyridines, 209
    alcoholysis, 215
    alkanelysis, 203
    from alkyl-1,8-naphthyridines, 205
    aminolysis, 215
    azidolysis, 216
    bioactivities, 209
    cyanolysis, 221
    cyclizations, 217
    dehalogenation, 217
    by halogenation, 210
    heteroarenolysis, 219
    hydrolysis, 214
    from 1,8-naphthyridinamines, 212
    from 1,8-naphthyridinecarboxylic acids, 213
    from 1,8-naphthyridine N-oxides, 213
    from 1,8-naphthyridinones, 210
    phosphinolysis, 218
    preparation, 209
    reactions, 214
    thiolysis, 218

Halogeno-2,6-naphthyridines, 267
　alcoholysis, 269
　aminolysis, 268
　dehalogenation, 266, 269
　hydrolysis, 269
　from 2,6-naphthyridinamines, 267
　from 2,6-naphthyridinones, 266
　preparation, 267
　reactions, 268
Halogeno-2,7-naphthyridines, 285
　alcoholysis, 287
　aminolysis, 288
　dehalogenation, 283, 289
　by halogenation, 285, 286
　from 2,7-naphthyridinones, 286
　positional activation, 285
　preparation, 286
　reactions, 287
　transhalogenation, 287
Hydrazinonaphthyridines, see
　Aminonaphthyridines
Hydro-1,5-naphthyridines, preparation, 14
Hydro-1,6-naphthyridines, preparation, 92
Hydro-1,8-naphthyridines, preparation, 197
Hydroxy-1,5-naphthyridines (extranuclear), 43
　from 1,5-naphthyridinecarboxylic esters, 45
Hydroxy-1,6-naphthyridines (extranuclear), 115
　halogenolysis, 107
　oxidation, 120
Hydroxy-1,8-naphthyridines (extranuclear),
　from 1,8-naphthyridinecarboxylic
　　acids, 224

Iodonaphthyridines, see Halogenonaphthyridines
Isonaphthyridine, xi

Jasminidine, 289
　　Jasminine, 283, 289

Lophocladine A, 289

Medorinone, 115
Methyl 5-methyl-2,7-naphthyridine, natural
　occurrence, 292
4-Methyl-2,6-naphthyridine, natural occurrence,
　265

Nalidixic acid, 221
1,5 Naphthyridine, 13
　C-Alkylation, 16
　N-Alkylation, 16
　amination, 15
　complexes, 14
　crystal structure, 14

　cycloaddition, 17
　deuteration, 18
　electron density, 15
　ESR, 15
　halogenation, 18, 26
　ionization, 15
　IR/Raman, 15
　MS, 15
　NMR, 15
　N-oxidation, 18
　phosphorescence, 15
　polarography, 15
　preparation, 13
　properties, 14
　reactions, 16
　reduction, 14
　Reissert reactions, 19
　resonance energy, 16
　trimethylsilylation, 19
1,6-Naphthyridine, 91
　C-alkylation, 94
　N-alkylation, 94
　amination, 95
　cycloaddition, 95
　deuteration, 95
　electron density, 93
　halogenation, 96
　ionization, 93
　IR/Raman, 93
　MS, 93
　NMR, 93
　N-oxidation, 96
　polarography, 93
　preparation, 91
　properties, 93
　reactions, 94
　reduction, 92
　Reissert reactions, 96
　UV, 93
1,7-Naphthyridine, 153
　amination, 155
　C-arylation, 155
　cycloaddition, 156
　electron density, 154
　ESR, 154
　halogenation, 156
　ionization, 155
　IR, 153
　NMR, 155
　N-oxidation, 156
　polarography, 155
　preparation, 153
　properties, 154
　quaternization, 157

reactions, 155
Reissert reactions, 157
1,8-Naphthyridine, 197
  amination, 200
  aromaticity, 198
  complexes, 197, 200
  cycloadditions, 201
  halogenation, 201
  MS, 199
  NMR, 199
  NQR, 199
  N-oxidation, 201
  preparation, 197
  properties, 198
  reactions, 200
  reduction, 197
  Reissert reactions, 202
  tautomerism, 199
  UV, 199
  X-ray analysis, 199
2,6-Naphthyridine, 265
  aromaticity, 266
  ESR, 266
  ionization, 266
  NMR, 266
  polarography, 266
  preparation, 266
  UV, 266
  X-ray analysis, 266
2,7-Naphthyridine, 283
  amination, 285
  aromaticity, 283
  electron density, 283
  ESR, 283
  halogenation, 285
  ionization, 284
  NMR, 284
  N-oxidation, 284
  polarography, 284
  preparation, 283
  properties, 283
  quaternization, 285
  reactions, 284
  reduction, 284
  X-ray analysis, 284
1,5-Naphthyridinecarbaldehydes, 65
  hydration, 65
  oxidation, 61
  to traditional derivatives, 66
1,6-Naphthyridinecarbaldehydes, 141
  preparation, 141
  reactions, 141
1,7-Naphthyridinecarbaldehydes, 182
  oxidation, 180

1,8-Naphthyridinecarbaldehydes, 257
  to anils, 259
  cyclizations, 260
  by C-formylation, 258
  to hydrazones, 260
  from 1,3-naphthyridinecarbonitriles, 257
  to 1,8-naphthyridine ketones, 258
  oxidation, 249
  by oxidation, 206
  to oximes, 259
  preparation, 258
  reactions, 259
  reduction, 260
  from trimethylsilyl-1,8-naphthyridines, 258
2,7-Naphthyridinecarbaldehydes, 293
  from alkyl-2,7-naphthyridines, 283
  oxidation, 283
  preparation, 293
  reactions, 293
1,8-Naphthyridinecarbohydrazides, to
  amino-1,8-naphthyridines, 240
1,5-Naphthyridinecarbonitriles, 65
1,6-Naphthyridinecarbonitriles, 141
  aminolysis, 140
  by cyanation, 141
  hydrolysis, 136, 140
  preparation, 141
  thiolysis, 140
1,7-Naphthyridinecarbonitriles, 182
  hydrolysis, 182
  reactions, 182
  by Reissert reactions, 117
1,8-Naphthyridinecarbonitriles, 257
  from halogeno-1,8-naphthyridines, 217
  hydrolysis, 251, 254
  to 1,8-naphthyridinecarbaldehydes, 257
  preparation, 257
  reactions, 257
2,7-Naphthyridinecarbonitriles, 293
1,7-Naphthyridinecarbonyl halides, 182
  to 1,7-naphthyridinecarboxamides, 181
  from 1,7-naphthyridinecarboxylic acids, 181
1,8-Naphthyridinecarbonyl halides, 251
  to 1,8-naphthyridinecarboxamides, 251
  from 1,8-naphthyridinecarboxylic acids, 250
  from 1,8-naphthyridinecarboxylic esters, 250
1,6-Naphthyridinecarbothioamides, from 1,6-naphthyridinecarbonitriles, 140
1,5-Naphthyridinecarboxamides, 65
  from 1,5-naphthyridinecarboxylic acids, 64

1,6-Naphthyridinecarboxamides, 139
    N-alkylation, 140
    intramolecular cyclization, 140
    from 1,6-naphthyridinecarbonitriles, 140
    from 1,6-naphthyridinecarboxylic acids, 138
    from 1,6-naphthyridinecarboxylic esters, 140
    preparation, 140
    reactions, 140
    transamination (of amide group), 140
1,7-Naphthyridinecarboxamides, 182
    hydrolysis, 180
    from 1,7-naphthyridinecarbonitriles, 182
    from 1,7-naphthyridinecarboxylic acids, 181
    preparation, 182
    from Reissert reactions, 157
1,8-Naphthyridinecarboxamides, 254
    N-alkylidenation, 255
    N-carbamoylation, 256
    cyclizations, 256
    from 1,8-naphthyridinecarbonitriles, 254
    from 1,8-naphthyridinecarbonyl halides, 251
    from 1,8-naphthyridinecarboxylic esters, 253
    preparation, 254
    reactions, 255
    from trichloromethyl-1,8-naphthyridines, 255
2,6-Naphthyridinecarboxamides, 272
    from 2,6-naphthyridinecarboxylic acids, 272
2,7-Naphthyridinecarboxamides, 293
1,6-Naphthyridinecarboxamidines, from
    1,6-naphthyridinecarbonitriles, 140
1,5-Naphthyridihecarboxylic acids, 60
    from alkyl-1,5-naphtnyridines, 23
    decarboxylation, 63
    esterification, 63
    from 1,5-naphthyridinecarbaldehydes, 61
    to 1,5-naphthyridinecarboxamides, 64
    from 1,5-naphthyridinecarboxylic esters, 61
    preparation, 61
    reactions, 63
1,6-Naphthyridinecarboxylic acids, 135
    from alkyl-1,6-naphthyridines, 100
    decarboxylation, 137
    esterification, 138
    from 1,6-naphthyridinecarbonitriles, 136
    to 1,6-naphthyridinecarboxamides, 135
    from 1,6-naphthyridinecarboxylic esters, 135
    preparation, 135
    reactions, 133
        1,7-Naphthyridinecarboxylic acids, 179
    decarboxylation, 180
    from 1,7-naphthyridinecarbaldehydes, 180
    to 1,7-naphthyridinecarbonyl halides, 181
    from 1,7-naphthyridinecarboxamides, 180
    to 1,7-naphthyridinecarboxamides, 181

from 1,7-naphthyridinecarboxylic esters, 129, 130
    preparation, 180
    reactions, 180
1,8-Naphthyridinecarboxylic acids, 247
    from alkyl-1,8-naphthyridines, 206
    complex formation, 251
    decarboxylation, 249
    degradation, 251
    esterification, 250
    from 1,8-naphthyridinecarbaldehydes, 249
    from 1,8-naphthyridinecarbonitriles, 248
    to 1,8-naphthyridinecarbonyl halides, 250
    to 1,8-naphthyridinecarboxamides, 251
    from 1,8-naphthyridinecarboxylie esters, 247
    to 1,8-naphthyridinones, 223
    preparation, 247
    reactions, 249
2,6-Naphthyridinecarboxylic acids, 272
    to 2,6-naphthyridinecarboxamides, 272
    from 2,6-naphthyridinecarboxylic asters, 272
2,7-Naphthyridinecarboxylic acids, 291
    decarboxylation, 283, 292
    from 2,7-naphthyridinecarbaldehydes, 283
    from 2,7-naphthyridinecarboxylie esters, 291
    preparation, 291
    reactions, 293
1,5-Naphthyridinecarboxylic esters, 64
    hydrolysis, 61
    from 1,5 naphthyridinecarboxylic acids, 63
    preparation, 64, 65
    reactions, 65
    reduction, 45
    by Reissert reactions, 19
1,6-Naphthyridinecarboxylic esters, 139
    cyclizations, 139
    hydrolysis, 135
    to 1,6-naphthyridinecarboxamides, 139
    preparation, 139
    reactions, 139
    by Reissert reactions, 96
1,7-Naphthyridinecarboxylic esters, 182
    hydrcgenolysis, 180
    hydrolysis, 179
    1,8-Naphthyridinecarboxylic esters, 252
    from halogeno-1,8-naphthyridines, 218
    hydrolysis, 247
    hydrolysis and decarboxylation, 252
    from 1,8-naphthyridinecarbonyl halides, 250
    to 1,8-naphthyridinecarboxamides, 253
    from 1,8-naphthyridinecarboxylic acids, 250
    preparation, 252
    reactions, 252

Index

2,6-Naphthyridinecarboxylic esters, 272
  hydrolysis, 272
2,7-Naphthyridinecarboxylic esters, 292
  hydrolysis, 291
  natural occurrence, 292
  preparation, 292
  reactions, 292
1,5-Naphthyridine ketones, 66
  from alkyl-1,5-naphthyridines, 22
  reactions, 66
1,6-Naphthyridine ketones, 141
  deacylation, 141
    1,8-Naphthyridine ketones, 259. *See also*
    1,8-Naphthyrininecarbaldehydes
  alkylidenation, 260
  from alkyl-1,8-naphthyridines, 248
  from 1,8-naphthyridinecarbaldehydes, 260
2,7-Naphthyridine ketones, 293
1,5-Naphthyridine *N*-oxides, 42
  complexation, 51
  deoxygenation, 51
  to halogeno-1,5-naphthyridines, 29
  preparation, 18, 50
  reactions, 51
1,6-Naphthyridine *N*-oxides, 121
  cyanation, 123
  deoxygenation, 123
  Meissenheimer reaction, 107
  by oxidation, 91, 122
  preparation, 96, 121
  quaternization, 122
  reactions, 122
  rearrangement, 117
1,7-Naphthyridine *N*-oxides, 171
  deoxygenation, 171
  to halogeno-1,7-naphthyridines, 163
  by oxidation, 156
  preparation, 171
  reactions, 171
1,8-Naphthyridine *N*-oxides, 228
  to alkyl-1,8-naphthyridines, 204
  complex formation, 228
  deoxygenation, 228
  to halogeno-1,8-naphthyridines, 213
  by oxidation, 201, 228
  preparation, 228
  reactions, 228
2,6-Naphthyridine *N*-oxides, 270
  primary synthesis, 270
2,7-Naphthyridine *N*-oxides, 290
  by oxidation, 284
1,5-Naphthyridines, from aliphatics, 1
  from benzo[*b*]-1,5-naphthyridines, 10
  from 1,3-dioxolanes, 11
  by primary syntheses, 1
  from pyridines, 2
  from pyridines with synthon(s), 5
  reviews, 13
  from 1,2,4-triazines, 11
1,6-Naphthyridines, from aliphatics, 67
  from benzo[*de*] [1,6]naphthyridines, 89
  from benzo-1,4-thiazines, 88
  from cyclopenta[*b*]pyridines, 85
  by primary syntheses, 67
  from pyrano[4,3-*b*]pyridines, 86
  from pyrans, 85
  from pyridines, 69
  from pyridines with synthon(s), 75, 83
  from pyrido[1,2-*c*]pyrimidines, 86
  from pyrimidines, 87
  from pyrrolo[3,4-*b*]pyridines, 87
  reviews, 91
  from thiazolo[2,3-*f*] [1,6]naphthyridines, 88
  from 1,3,5-triazines, 88
  from 1,2,3-triazoles, 88
1,7-Naphthyridines, from aliphatics, 143
  from cyclopenta[*b*]pyridihes, 151
  from diazabicycloundecanes, 150
  from a Maillard reaction, 143
  by primary syntheses, 143
  from pyrano[3,4-*b*]pyridines, 151
  from pyrazines, 151
  from pyridines, 143
  from pyridines with synthon(s), 147
  from pyrrolo[3,4-*b*]pyridines, 151
  reviews, 143,153
1,8-Naphthyridines, from aliphatics, 183
  bioactivities, 183
  from isoxazolo[4,3-*c*] [1,8]naphthyridijes, 192
  from oxazoles, 192
  primary synthesis, 183
  from pyrans, 192
  from pyrazolo[2,4,5-*ij*] [1,8]naphthyridines, 195
  from pyridines, 184
  from pyridines with synthon(s), 187
  from pyrido[2,3-*b*] [1,8]naphthyridines, 193
  from pyrido[2,3-*d*] [1,3]oxazines, 193
  from pyrido[1,2-*a*]pyrimidines, 193
  from pyrido[2,3-*d*]pyrimidines, 194
  from pyrrolo[2,3-*b*] pyridines, 195
  from tetrazolo[2,4,5-*ij*] [1,8]naphthyridines, 195
  from 1,2,4-triazines, 195

2,6-Naphthyridines, 261
  from diazacyclodeca-3,8-diynes, 264
  primary syntheses, 261
  from pyrano[4,3-c]pyridines, 265
  from pyridines, 261
  from pyridines with synthon(s), 264
  from pyrroles, 265
  reviews, 265
  from 1,2,4-triazines, 265
2,7-Naphthyridines, 275
  from isoquino[2,1-b] [2,7]naphthyridines, 280
  from nonheterocyclics, 275
  from oxazoles, 280
  primary syntheses, 275
  from pyrano[3,4-c]pyridines, 281
  from pyrano[4,3-b]pyridines, 282
  from pyrans, 281
  from pyrazines, 282
  from pyridines, 277
  from pyridines with synthon(s), 278
  from pyrrolo[3,4-c]pyridines, 282
1,5-Naphthyridine sulfones, see
  Alkylsulfonyl-1,5-naphthyridines
1,8-Naphthyridine sulfones, see
  Alkylsulfonyl-1,8-naphthyridines
1,8-Naphthyridinesulfonic acids, 233
  by sulfonation, 233
1,8-Naphthyridinesulfoxides, see
  alkylsulfinyl-1, 8-naphthyridines
1,5-Naphthyridinethiones, 53
  S-alkylation, 53
  from halogeno-1,5-naphthyridines, 41
  preparation, 53
  reactions, 53
1,6-Naphthyridinethiones, 125
  S-alkylation, 125
  cyclocondensations, 126
  from halogeno-1,6-naphthyridines, 113
  preparation, 125
  reactions, 125
1,8-Naphthyridinethiones, 231
  S-alkylation, 231
  complex formation, 232
  cyclocondensations, 232
  from halogeno-1,8-naphthyridines, 218
  preparation, 231
  reactions, 231
  tautomerism, 231
1,5-Naphthyridinols, see 1,5-Naphthyridinones,
Naphthyridinomycin, 43
1,5-Naphthyridinones, 43
  from alkoxy-1,5-naphthyridines, 45
  N-alkylation, 46
  O-alkylation, 46

  aminolysis, 47
  arylation, 46
  complex formation, 44
  deoxygenation, 48
  halogenolysis, 26
  from halogeno-1,5-naphthyridines, 40
  MS, 44
  from 1,8-naphthyridinamines, 44
  from 1,5-naphthyridinecarboxylic esters, 45
  by oxylation, 44
  preparation, 44
  reactions, 46
1,6-Naphthyridinones, 115
  from alkoxy-1,6-naphthyridines, 117
  to alkoxy-1,6-naphthyridines, 118
  alkylation, 118, 119
  from alkyl-1,6-naphthyridinium salts, 117
  aminolysis, 119
  from amino-1,6-naphthyridines, 116
  cyclization, 120
  deoxygenation, 117
  halogenolysis, 104
  from halogeno-1,6-naphthyridines, 113
  from 1,6-naphthyridine N-oxides, 117
  properties, 115
1,7-Naphthyridinones, 167
  N-alkylation, 168
  bioactivities, 167
  cyclizations, 169
  halogenolysis, 162
  from 1,7-naphthyridinamines, 168
  by oxidation, 168
  preparation, 167
  reactions, 168
  tautomerism, 167
1,8-Naphthyridinones, 221
  from alkoxy-1,8-naphthyridines, 222
  N-alkylation, 225
  O-alkylation, 224
  aminolysis, 225
  complex formation, 227
  cyclizations, 226
  halogenolysis, 210
  from halogeno-1,8-naphthyridines 214
  from 1,8 naphthyridinecarboxylic acids, 223
  by oxidation, 222
  preparation, 221
  reactions, 224
  thiation, 226
  from trichloromethyl-1,8-naphthyridines, 223
2,6-Naphthyridinones, 270
  halogenolysis, 267
  from halogeno-2,6-naphthyridines, 268
  preparation, 270

2,7-Naphthyridinones, 289
  alkylation, 290
  halogenolysis, 286
  natural occurrence, 289
  preparation, 289
  reactions, 289
1,5-Naphthyridinones (nontautomeric), 49
  preparation, 49
  reactions, 50
1,6-Naphthyridinones (nontautomeric), from 1,6-naphthyridinones, 118
1,7-Naphthyridinones (nontautomeric), N-dealkylation, 169
  preparation, 168
1,8-Naphthyridinones (nontautomeric), from 1,8-naphthyridinones, 225
2,7-Naphthyridinones (nontautomeric), from 2,7-naphthyridinones, 290
Neozoeylanicine, 292
Nitro-1,5 naphthyridines, 55
  from dimethylsulfimido-1,5-naphthyridines, 56
  halogenolysis, 56
  MS, 55
  by nitration, 55
  preparation, 55
  reactions, 56
  reduction, 56
Nitro-1,6-naphthyridines, 127
  from dimethylsulfimido-1,6-naphthyridines, 127
  by nitration, 127
  preparation, 127
  reactions, 128
  rearrangement, 129
  reduction, 128
  review, 127
Nitro-1,7-naphthyridines, 175
  by nitration, 175
  preparation, 175
  reactions, 175
  reduction, 176

Nitro-1,8-naphthyridines, 235
  from dimethylsulfimido-1,8-naphthyridines, 236
  by nitration, 235
  from nitroso-1,8-naphthyridines, 236
  preparation, 235
  reactions, 237
  reduction, 237
  review, 235
Nitro -2,6-naphthyridines, 271
Nitro-2,7-naphthyridines, 290
Nitroso-1,8-naphthyridines, 238
  by nitrosation, 238
  oxidation, 236, 238
  reduction, 238

Oxy-1,5-naphthyridines, 43
Oxy-1,6-naphthyridines, 114
Oxy-1,7-naphthyridines, 167
Oxy-1,8-naphthyridines, 221
  bioactivities, 221
Oxy-2,6-naphthyridines, 269
Oxy-2,7-naphthyridines, 289
  natural products, 289

Phosphino-1,8-naphthyridines, from halogeno-1,8-naphthyridines, 218

Thio-1,5-naphthyridines, 53
Thio-1,6-naphthyridines, 125
Thio-1,7-naphthyridines, 173
Thio-1,8-naphthyridines, 231
Thio-2,6-naphthyridines, 270
Thio-2,7-naphthyridines, 290
Tocufloxacin, 209
Trovafloxacin, 209

Ureido-1,6-naphthyridines, from amino-1,6-naphthyridines, 133